CALCULUS: Basic concepts and applications

CALCULUS

Basic concepts and applications

R. A. ROSENBAUM & G. PHILIP JOHNSON
Wesleyan University   *Oakland University*

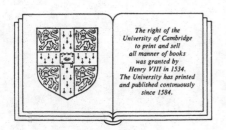

The right of the
University of Cambridge
to print and sell
all manner of books
was granted by
Henry VIII in 1534.
The University has printed
and published continuously
since 1584.

CAMBRIDGE UNIVERSITY PRESS

Cambridge London New York New Rochelle Sydney Melbourne

CAMBRIDGE UNIVERSITY PRESS
Cambridge, New York, Melbourne, Madrid, Cape Town, Singapore, São Paulo, Delhi

Cambridge University Press
The Edinburgh Building, Cambridge CB2 8RU, UK

Published in the United States of America by Cambridge University Press, New York

www.cambridge.org
Information on this title: www.cambridge.org/9780521250122

First published 1984
This digitally printed version 2008

A catalogue record for this publication is available from the British Library

Library of Congress Cataloguing in Publication data
Rosenbaum, R. A. (Robert A.)
Calculus: basic concepts and applications.
1. Calculus. I. Johnson, G. P. II. Title.
QA303.R69   1984   515   83–14257

ISBN 978-0-521-25012-2 hardback
ISBN 978-0-521-09590-7 paperback

Contents

Preface   *page* xi

How this book is organized, and how it can be used   xiii

Some study hints   xv

0   Prerequisites   1
0.1   Fundamental operations; parentheses   1
0.2   Zero and negatives   3
0.3   Fractions and rational numbers   5
0.4   Integral exponents   8
0.5   Radicals, fractional exponents, and real numbers   9
0.6   Notation for implication   11
0.7   Equalities   11
0.8   Inequalities   13
0.9   Linear equations   16
0.10 Quadratic equations   18
0.11 Higher-degree equations   20
0.12 Progressions   24
0.13 Logarithms   27
0.14 Keeping track of units   29
0.15 Mensuration formulas   30

1   Functional relationships   33
1.1   Introduction   33
1.2   An example   33
1.3   Variation of one quantity with another;
      graphical interpolation   34

Contents

1.4	More on graphing, interpolation, and extrapolation	37
1.5	Linear interpolation	41
1.6	Relations expressed by formulas	44
★ 1.7	Formulas (*continued*)	53
1.8	Relationships between science and mathematics	55
1.9	Functions	56
1.10	Further discussion of functions: notation and natural domains	59
1.11	Inverse functions	62
★ 1.12	Absolute values	65
1.13	Summary	65

2	**Rate of change**	72
2.1	Average speed and average velocity	72
2.2	Instantaneous velocity and limits	74
2.3	Theorems on limits	81
★ 2.4	Proofs of some results on limits	84
2.5	Average slope in an interval and slope at a point	87
2.6	Tangent to a curve	91
2.7	The derivative	93
★ 2.8	Guessing limits with a calculator	101
2.9	Review	103

3	**Applications of the derivative**	110
3.1	The Mean-Value Theorem	110
3.2	Increasing and decreasing functions	112
3.3	Approximate increments	114
3.4	Applications to economics: marginal cost and unit cost	119
3.5	Maxima and minima: the basic idea	123
3.6	How do we know whether we have a maximum or a minimum?	125
3.7	Further questions about maxima and minima	129
3.8	Applied maxima and minima	130
3.9	Maxima and minima in some problems in economics	136

3.10 Approximate solution of equations: the Newton–Raphson method and the bisection method   139

3.11 Review   144

4 Further differentiation   148

4.1 Repeated differentiation and derived curves   148

4.2 Points of inflection and third test for maxima and minima   152

4.3 Extreme rates   156

4.4 Derivative of a function of a function: the Chain Rule   157

4.5 Continuity   161

★ 4.6 Proof that differentiability implies continuity and proof of the Chain Rule   164

4.7 Notation   165

4.8 Related rates   168

4.9 Functions in implicit form and implicit differentiation   171

4.10 Derivatives of fractional powers   173

4.11 Implicit differentiation applied to related rates   174

4.12 Differentials   176

4.13 Formulas for derivatives of products and quotients   178

4.14 Marginal cost, marginal revenue, and optimal production levels   182

★ 4.15 Maxima and minima using implicit differentiation   185

4.16 Summary   186

5 Antidifferentiation and integration   194

5.1 The reverse of differentiation   194

5.2 The antiderivatives of a given function differ by at most a constant   196

5.3 Formulas for antiderivatives   197

5.4 Repeated antidifferentiation: projectiles thrown vertically   202

5.5 The limit of a sum   205

5.6 Further limits of sums   209

5.7 The Fundamental Theorem   213

5.8 Applications of the Fundamental Theorem   217

Contents

5.9   Use of the Chain Rule in integration
       (antidifferentiation)   222
5.10 The indefinite integral   224
5.11 Summary   225

6 Exponential functions   231
6.1   Introduction to exponential functions   231
6.2   The rate of change: preliminary remarks   235
6.3   Compound interest   237
6.4   Continuous compounding   240
6.5   The derivative of the exponential function   243
6.6   Relative errors and relative rates   246
6.7   Antiderivatives of the exponential   249
6.8 e^u: derivative and antiderivative   253
6.9   Summary   255

7 Logarithmic functions   257
7.1   Introduction   257
7.2   Inverse functions and the inverse of the exponential   257
7.3   Laws of logarithms   259
7.4   The derivative of the log function   264
7.5   Antiderivatives of $1/x$ 266
7.6   Derivatives of b^x and $\log_b x$ 268
7.7   Log–log and semilog graphs   269
7.8   Summary   274

8 Differential equations   282
8.1   Introduction   282
8.2   An approximate solution of a differential equation   284
8.3   Variables separable   286
8.4   Comparison of approximate and exact solutions   287
8.5   Population changes   289
8.6   The logistic equation   290
8.7   The method of partial fractions   291

8.8	The logistic equation (*continued*)	292
8.9	Linear differential equations with constant coefficients	294
8.10	Linear differential equations with constant coefficients (*continued*)	299
★ 8.11	Approximating the solutions of a pair of simultaneous differential equations	302

9 Further integration 308

9.1	Introduction	308
9.2	Review of the use of the Chain Rule in integration (antidifferentiation)	308
9.3	Force of attraction	310
9.4	Loads	312
9.5	Moment of a force	314
9.6	Consumers' and producers' surpluses	315
9.7	Horizontal rectangular strips and circular strips	317
9.8	The idea of an average	320
9.9	Average velocity	321
9.10	The average of a function defined on an interval	322
★ 9.11	Further averages	325
9.12	Summary	329
9.13	Quadrature	334
9.14	More on quadrature: the trapezoidal rule and its adjustment	336

10 Trigonometric functions 343

10.1	Introduction	343
10.2	Angle measure	346
10.3	The sine and cosine functions	349
10.4	The tangent function, and application of the basic functions to triangles	354
10.5	Differentiation of the trigonometric functions	358
10.6	Antidifferentiation and integration of trigonometric functions	363
10.7	Inverse trigonometric functions	365
10.8	Further integration involving trigonometric functions	371

Contents

⋆ 10.9 Other periodic functions   375

10.10 A return to differential equations   380

10.11 Summary   386

Answers to selected problems   391

Appendix: *Tables*   411

  A Compound interest: $(1 + r)^n$ 411

  B_1 Values of e^x and e^{-x} 412

  B_2 Natural logarithms ($\ln x$)   413

  C Logarithms, base 10   415

  D Trigonometric functions   417

Index   421

Preface

Mathematics, with its origin in problems of land measurement and the keeping of accounts, has grown in complexity and power as the needs of society have required ever more sophisticated reasoning and techniques. Over a period of 2000 years mathematics developed, in some periods slowly, in others rapidly, until in the seventeenth century there was a dramatic advance – the invention of calculus – to match and facilitate equally dramatic achievements in science and, somewhat later, in technology.

Indeed, calculus proved essential for the handling of difficult problems in astronomy, physics, and engineering, as well as in other branches of mathematics itself, such as the determination of tangent lines to curves and the computation of volumes bounded by closed surfaces. In the eighteenth and nineteenth centuries, the demands of the physical sciences and technology stimulated rapid and far-reaching developments of the branch of mathematics called *analysis*, growing out of calculus; and, reciprocally, the mathematical developments contributed to the further growth of those sciences.

Comparable progress was not made in the applications of mathematics, including calculus, to the social and life sciences, largely because problems in these areas proved difficult to formulate in mathematical terms. With what precision – or even meaning – can one assign a number to the degree of a person's conviction on a controversial subject, such as the investment of more money to reduce the size of classes in public schools?

In the last 50 years, however, much of the mathematics that has proved so useful in the physical sciences has been applied successfully in the social and life sciences. The examples and problems presented in this book recognize this fact. We shall often find that the major difficulty is the proper *formulation* or *modeling* of the problem: Starting with a complicated and perhaps vaguely defined situation, how do we sharpen the definition of the problem, removing inessential features and adding data as required, so as to make it feasible to apply mathematical methods? Providing practice in the formulation of problems, as well as in their solution, is one of our goals in this book.

Thus far, we have mentioned mathematics solely in the context of its applications to various fields. But mathematics has also been created and studied for its own sake, as a system of thought with great appeal because of its logical and aesthetic qualities. In this sense, mathematics is an art, in addition to being a tool of the sciences. In its dual roles, mathematics holds a central position in our cultural heritage, and an appreciation of mathematics should contribute significantly to our intellectual development, in the same way that an appreciation of literature, music, and philosophy, for example, contribute to that development. We hope that study of this book will prove rewarding through an increased appreciation of the power and the beauty of mathematics.

How this book is organized, and how it can be used

The spirit of the development of calculus in this book is *intuitive*, with "real-world" problems and concrete examples to provide motivation and to clarify concepts. We have chosen data to minimize arithmetic and algebraic complexities while you are absorbing new ideas, and we use plausibility arguments rather than formal proofs to justify most of the conclusions. However, we have tried to make careful statements, so that you will never have to unlearn anything later.

Chapter 0 provides a review of algebra, graphing, and related topics for those who need it. The core material of Chapters 1 through 7 is appropriate for a one-semester, three-hour course for an average class. Students with a strong background in mathematics may work the starred problems and study the starred sections, which go more deeply than the core material into some of the topics. (There are some proofs in starred sections.) Omission of the starred sections, however, does not interrupt the basic development of the subject.

Students with a more limited background can skip some sections as well:
Sections 3.4, 3.9, and 4.14 involving applications to economics;
Section 3.10 on approximate solution of equations; and
Sections 4.3, 4.8, 4.11, and 6.6 on extreme rates of change, related rates, and relative rates.

These can all be omitted without giving you trouble with subsequent material.

Chapter 8, Chapter 9, and the core of Chapter 10 can be studied, in any order, after Chapter 7 has been completed. Moreover, at the end of Chapter 7 we have made suggestions for independent projects that you may find of interest. On the assumption that you have developed some mathematical maturity in working through the first seven chapters, we have somewhat condensed the exposition in the later chapters. In those chapters, too, you will find considerable emphasis on numerical methods.

In addition to problem sets at the end of sections, there are exercises embedded in the expository material itself. Be sure to do them, for they are designed to help you understand the material that follows. Other exercises

Significance of ★

Significance of ○

xiii

are marked with a small open circle ○; you should do all these exercises and save your solutions, because results later in the book depend upon them.

At the end of each chapter, in addition to review problems, there is a set of questions constituting a "sample test," which should help you to check whether you have mastered the material. Answers to selected problems, and to virtually all the "sample test" questions, appear at the end of the book.

For problems marked with the letter **c** a modern calculator will be useful. This is not to say that you should avoid using a calculator on other problems, but rather that some of the numerical experimentation suggested by the **c** problems can be extremely time-consuming if done by "hand."

The use of calculators in the study of calculus has both advantages and drawbacks. Among the advantages are these: (1) many concepts can be well illustrated and simply explained through calculation, and (2) the practical applications of calculus often call for delicate numerical work. Among the drawbacks, we note that the intricacies of calculators can become a study in itself and a distraction from learning basic concepts.

This book seeks a middle ground. Its primary thrust is calculus, and the material is presented so that you can acquire all the essential content without the use of a calculator. At the same time, there is additional content – and opportunity for additional insights – for those who choose to use calculators.

Even the simplest calculator, with nothing more than a square-root key, will come in handy in saving you time and in providing you with significant insights. More useful is a calculator with the usual features of scientific-engineering models: logarithmic, exponential, and trigonometric functions, and floating point representation. Better yet is a programmable calculator, preferably with branching ability. Best of all is a computer – and a little knowledge of how to use it.

Some study hints

Your secondary-school course in geometry has given you some experience with the careful statement, the attention to detail, and the concern for logical argument that are typical of mathematics. You may or may not have approached algebra in a similar spirit. In calculus we *must* be precise in language, alert to the niceties of seemingly "minor" points, and prepared to follow a rigorous argument, because the material involves subtleties that must be appreciated if you are to learn to handle novel situations (as contrasted with merely solving routine problems) and to realize the aesthetic satisfactions and intellectual rewards that can come from a critical study of calculus and related topics.

In order to learn as much as possible from the text, you should form the habit of reading it slowly, while seated at a desk, with paper and pencil at hand. It is probably worthwhile to "skim" the assigned reading once, to obtain a general idea of the subject matter. Then study the material, reading each sentence slowly, and doing your best to fill in any details that we have left to you. Next, review the material, analyzing its relation to what you have previously learned and trying to put the main results into your own words. Finally, do problems and exercises.

Throughout your study you should maintain a critical attitude. Constantly ask questions: Why is it done this way? Could it not have been accomplished more easily as follows? Is this hypothesis really needed? Doesn't the following example contradict the statement in the text? Is this a significant or a trivial point, and what relation does it have to the entire structure that I'm trying to understand? And so forth. By cultivating an active involvement in the course you will greatly increase the satisfaction obtained from it.

Here is still another suggestion: When you finish working a problem, spend a few minutes thinking through what you have done. As soon as you have come out with a neat $r = 5$; $h = 10$, there is a temptation to think That's done; what's next in my assignment? But there is great value in reviewing a completed problem in terms like these:

Some study hints

What was asked for in this problem? Have I answered the question(s)? Does my answer sound right – does it make sense?

How do this problem and its answer compare with other problems I have solved and with situations I know apart from my math course?

Are there any general conclusions that I can draw from my work on this problem?

Now that I have solved the problem, do I see some easier way that it could have been done?

In short, what have I *learned* from this problem?

Such an analytical, reflective approach will pay big dividends in understanding and enjoyment.

Prerequisites

0

This chapter begins with a rapid review of elementary arithmetic and algebra, emphasizing only those techniques essential to an understanding of the calculus. No attempt is made to provide a complete logical development of the subject.

0.1 Fundamental operations; parentheses

We begin with a brief statement of familiar properties of the numbers of arithmetic.

It makes no difference in what order we add numbers: $3+4=7$ and $4+3=7$, and, in general, for a and b any numbers,

$$a + b = b + a. \tag{1}$$

Likewise, the way in which numbers are grouped for addition does not affect the result: $3+(4+5)=3+9=12$, and $(3+4)+5=7+5=12$. In general,

$$a+(b+c)=(a+b)+c. \tag{2}$$

Multiplication of natural (i.e., counting) numbers may be thought of as repeated addition. Instead of $4+4+4$, we write $3\cdot4$, and for $3+3+3+3$, we write $4\cdot3$. But both are equal to 12, and, in general,

$$a\cdot b=b\cdot a. \tag{3}$$

As with addition, the way in which numbers are grouped for multiplication does not matter: $3\cdot(4\cdot5)=3\cdot20=60$, and $(3\cdot4)\cdot5=12\cdot5=60$. In general,

$$a\cdot(b\cdot c)=(a\cdot b)\cdot c. \tag{4}$$

Adding two numbers and multiplying the result by a third number gives the same result as multiplying each of the first two by the third and then adding. For example, $3\cdot(4+5)=3\cdot9=27$, and $3\cdot4+3\cdot5=12+15=27$. In general,

$$a\cdot(b+c)=(a\cdot b)+(a\cdot c). \tag{5}$$

Experience with sets of objects makes properties (1) through (5) intuitively clear for the natural numbers. As the number system is extended to include fractions, negative numbers, and so forth, definitions are so chosen that these properties hold for them also. Remember that the letters in algebra stand for numbers; hence, these properties are basic to all the manipulations of algebra.

Parentheses, and other grouping symbols such as brackets, [ ], and braces, { }, are essentially punctuation marks. Indicated operations inside parentheses are to be thought of as performed first. In equations (2) and (4) they are used to indicate special ways of looking at the expression. If we care only about the result, not how it is obtained, equation (2) says that we can omit the parentheses and write $a + b + c$ without ambiguity. Similarly, in (4), $a \cdot b \cdot c$ represents the same number whichever way of associating factors is chosen. In (5), the situation is different. Writing $a \cdot b + c$ gives no indication, without some further agreement, whether this means $a \cdot (b + c)$ or $(a \cdot b) + c$. But these are different: $3 \cdot (4 + 5) = 3 \cdot 9 = 27$, and $(3 \cdot 4) + 5 = 12 + 5 = 17$. The universal convention is to choose the second. That is:

Unless there is notation to the contrary, multiplications (and divisions) are performed before additions (and subtractions).

The convention permits removal of parentheses on the right side of (5): $a \cdot (b + c) = a \cdot b + a \cdot c$.

So far we have used the dot to indicate multiplication. When there is no ambiguity, we can omit the dot. Obviously, 24 and $2 \cdot 4$ have different meanings, but $2 \cdot x$ can be written as $2x$ and $a \cdot b$ as ab. Equation (4) can be written

$$(ab)c = a(bc), \qquad (4')$$

and (5) as

$$a(b + c) = ab + ac. \qquad (5')$$

Property (5) can be used to "expand" $(a + b)(p + q)$ as follows:

$$(a + b)(p + q) = a(p + q) + b(p + q) = ap + aq + bp + bq. \qquad (6)$$

If the two factors are alike, we use the shorter notation $(a + b)^2$ for $(a + b)(a + b)$, and likewise $(a + b)^3$ for $(a + b)(a + b)(a + b)$, and so forth. Then, as a special case of (6), we have

$$(a + b)^2 = a^2 + 2ab + b^2.$$

The difference $a - b$ is defined as the number d such that $a = b + d$, and the quotient $a \div b$, also written a/b, as the number q such that $a = bq$. Properties such as the following can be understood intuitively for the natural numbers by dealing with sets of objects and can be proved formally on the basis of these definitions and properties (1)–(5):

$$a-(b+c)=a-b-c,$$
$$a-(b-c)=a-b+c,$$
$$a(b-c)=ab-ac,$$
$$a\div(bc)=(a\div b)\div c.$$

Calculators are designed to make it easy and natural to follow conventional arithmetic and algebraic usage. However, there are variations in the way different calculators work, and the beginner on any calculator must study its characteristics carefully. With practice, one soon learns to observe the standard conventions just as instinctively as in hand calculation.

PROBLEMS

1. Evaluate each of the following expressions:
   (a) $5+3\cdot5$ (b) $2+\dfrac{12}{3}$ (c) $\dfrac{3+7}{1+4}$ (d) $3\cdot96+3\cdot4$ (e) $10-2\cdot5$
   (f) $(10-2)\cdot5$ (g) $(6\div2)\div3$ (h) $6\div(2\div3)$ (i) $12-(4+2)$
   (j) $(9+4)-(7-3)$ (k) $2\cdot[7+3-(2\cdot4)]$ (l) $(4+6\cdot4)\div7$
2. Simplify each of the following expressions:
   (a) $(a+p)(p+q)-p^2-pq$ (b) $(c+2d)^2-4cd$
   (c) $(a+b)^2-2(a^2+ab)$ (d) $x^2-xy+y^2-x(x-y)$
○ 3. Expand each of the following expressions, and keep your results for future reference:
   (a) $(x+h)^3$ (b) $(x+h)^4$ (c) $(x+h)^5$
4. Evaluate the expressions in each of the following pairs.
   (a) $5\cdot10-8,$ $5(10-8)$
   (b) $17-6+5,$ $17-(6+5)$
   (c) $8+4\div2,$ $(8+4)\div2$
   (d) $2+3^2,$ $(2+3)^2$
   (e) $x\cdot x+y,$ $x(x+y)$
   (f) $sr-rr+s,$ $s[r-r(r+s)]$

0.2 Zero and negatives

The number 0 is defined by the property $a+0=a$ for all numbers a. We have, from this definition and (5),

$$a\cdot0=0 \quad \text{for all } a. \tag{7}$$

An important consequence follows:

$$\text{If } ab=0, \quad \text{then } a=0 \quad \text{or} \quad b=0. \tag{8}$$

Division by 0 is impossible. For suppose that $2\div0=q$. Then, by the definition of division, $q\cdot0=2$. But, by (7), there is no such number q. The

3

same argument applies for $a \div 0$, a being any number other than 0. Now suppose $0 \div 0 = q$. Again, by the definition of division, $0 \cdot q = 0$. Here q could be anything: $\frac{1}{2}$, 0, 1, 100, and 9999 all satisfy this condition; there is no way of picking out one number as the "answer."

Dividing 0 by any number different from 0 gives no trouble: Suppose that $0 \div a = q$, with $a \neq 0$. Then $aq = 0$ and $q = 0$, by (8).

The negative of a number a is the number x such that $a + x = 0$, and it is denoted by $-a$. The negatives of the positive numbers are the negative numbers, and the negatives of the negative numbers are the positive numbers. Positive numbers are greater than 0, negative numbers less than 0. Note that 0 is neither positive nor negative and that the negative of 0 is 0. The natural numbers, their negatives, and 0 compose the set of integers. The following familiar results follow from this definition and the properties (1)–(5) in Section 0.1.

For all p, q, $p + (-q) = p - q$. (Note that the minus sign plays two roles – as a label for the negative of a number and as the symbol for subtraction. This equation means that $p - q$ can be thought of either as indicating the subtraction of q from p or as indicating the addition of p and $-q$.) Similarly, for all p, q, r,

$$p - (-q) = p + q,$$

$$-(-p) = p,$$

$$-p = (-1)p,$$

$$(-p) \cdot q = -(pq),$$

$$(-p)(-q) = pq,$$

$$\frac{p}{-q} = \frac{-p}{q} = -\left(\frac{p}{q}\right),$$

$$\frac{-p}{-q} = \frac{p}{q},$$

$$r(p - q) = rp - rq.$$

PROBLEMS

1. Evaluate, if possible, for $x = 0$, $x = 1$, and $x = 3$:
   (a) $\dfrac{x^2 - 9}{x + 3}$ (b) $\dfrac{x^2 + 9}{x - 3}$ (c) $\dfrac{x^2 - 9}{x - 3}$ (d) $\dfrac{x}{x^2 - 4x + 3}$

2. Evaluate each of the following expressions:
   (a) $5 + (-8)$ (b) $5 - (-8)$ (c) $4 \cdot 106 - 4 \cdot 6$
   (d) $12 - \dfrac{10}{-2}$ (e) $12 + \dfrac{-10}{2}$ (f) $12 - \dfrac{-10}{-2}$
   (g) $\dfrac{12 - 10}{-2}$ (h) $\dfrac{12 + (-10)}{2}$ (i) $\dfrac{12 - (-10)}{-2}$

(j) $\dfrac{16-10}{4-1}$ (k) $\dfrac{-16-10}{1-14}$ (l) $\dfrac{1-3\cdot5}{5-(-2)}$

(m) $\dfrac{56-8(7+5)}{2(1-6)}$

3. Simplify each of the following expressions:

  (a) $(a+b)(p+q)-(a+b)p$ (b) $x(u+v)-2xu$

  (c) $(x-y)^2+2xy$ (d) $(c-d)-(a-d)-(c-a)$

  (e) $ab-[cd-(ef-ab)]$ (f) $\dfrac{ac-bc}{a-(a+c)}$

4. Simplify each of the following expressions:

  (a) $(x-y)(x+y)$ (b) $(x-y)(x^2+xy+y^2)$

  (c) $(x-y)(x^3+x^2y+xy^2+y^3)$

5. Expand each of the following expressions:

  (a) $(x-h)^2$ (b) $(x-h)^3$ (c) $(x-h)^4$ (d) $(x-h)^5$

0.3   Fractions and rational numbers

A rational number is one that can be expressed as the quotient of two integers; that is, as a fraction with numerator and denominator integers. Every integer satisfies this definition because it can be expressed (in many ways) as such a quotient; for example, $3 = 3 \div 1 = 6 \div 2 = 15 \div 5$, and so on. The way we read a common fraction like two-thirds indicates that we are thinking of it as $2 \cdot (\frac{1}{3})$. But $3(2)(\frac{1}{3}) = 2(3)(\frac{1}{3}) = 2 \cdot 1 = 2$; hence, it satisfies the definition of the quotient $2 \div 3$. We can choose whichever interpretation of $\frac{2}{3}$ suits us.

Common sense assures us that $2 \cdot (\frac{1}{3}) = 4 \cdot (\frac{1}{6})$ – twice as many parts, each half as big. That is, $\frac{2}{3} = \frac{2\cdot2}{3\cdot2} = \frac{4}{6}$. In general, for $k \neq 0$,

$$\frac{a}{b} = \frac{ka}{kb}. \qquad (9)$$

(In this and the formulas that follow, assume that the denominators of the given fractions are not 0.) We can use this property to reduce a fraction to "lower terms" – $\frac{10}{15} = \frac{2\cdot5}{3\cdot5} = \frac{2}{3}$ – or to change to "higher terms" – $\frac{3}{25} = \frac{3\cdot4}{25\cdot4} = \frac{12}{100} = 0.12$.

If a decimal terminates, it can be written as a fraction whose denominator is a power of 10 (e.g., $0.12 = \frac{12}{100}$), and hence it is rational. The converse is *not* true; for example, $\frac{1}{3} = 0.333\ldots$, continued indefinitely.

Multiplying simple fractions like $\frac{1}{2} \cdot \frac{1}{3} = \frac{1}{6}$, and $\frac{5}{2} \cdot \frac{7}{3} = (5 \cdot \frac{1}{2})(7 \cdot \frac{1}{3}) = (5 \cdot 7)(\frac{1}{2} \cdot \frac{1}{3}) = \frac{35}{6}$, leads to the general rule

$$\frac{a}{b} \cdot \frac{c}{d} = \frac{ac}{bd}. \qquad (10)$$

The rule for dividing fractions can be obtained by applying (9) to

$$\frac{a/b}{c/d} = \frac{(a/b)\cdot(d/c)}{(c/d)\cdot(d/c)} = \frac{ad/bc}{1} = \frac{ad}{bc}.$$

In other words:

To divide by a fraction, invert the divisor and multiply.

To see how to add fractions, we look again at cases in which it is easy to see what the answer must be: $\frac{1}{3} + \frac{2}{3} = \frac{3}{3} = 1$; $\frac{4}{5} + \frac{3}{5} = (4+3) \cdot \frac{1}{5} = \frac{7}{5}$. Clearly, if the fractions have the same denominator, the numerator of the sum is the sum of the numerators, and the denominator is that common denominator. But by the use of (9) we can always express each of the fractions so that they do have a common denominator: For example, $\frac{1}{2} + \frac{4}{3} = \frac{1 \cdot 3}{2 \cdot 3} + \frac{4 \cdot 2}{3 \cdot 2} = \frac{3}{6} + \frac{8}{6} = \frac{11}{6}$. In general,

$$\frac{a}{b} + \frac{c}{d} = \frac{a \cdot d}{b \cdot d} + \frac{b \cdot c}{b \cdot d} = \frac{ad + bc}{bd}. \tag{11}$$

Whereas (11) always gives the correct result, that result can sometimes be obtained more easily: For example, $\frac{5}{6} + \frac{3}{4} = \frac{10}{12} + \frac{9}{12} = \frac{19}{12}$, whereas (11) would have us say $\frac{20}{24} + \frac{18}{24} = \frac{38}{24}$, which reduces to $\frac{19}{12}$, the same result, of course. The work will be simplest if we use as the denominator the least number that contains both denominators as factors; to see what this is in a less obvious case, we write the denominators in factored form.

Example 1

$$\frac{7}{60} + \frac{5}{72} = \frac{7}{2^2 \cdot 3 \cdot 5} + \frac{5}{2^3 \cdot 3^2}.$$

The least common denominator is $2^3 \cdot 3^2 \cdot 5$. We have then

$$\frac{7 \cdot 2 \cdot 3}{2^2 \cdot 3 \cdot 5 \cdot 2 \cdot 3} + \frac{5 \cdot 5}{2^3 \cdot 3^2 \cdot 5} = \frac{42}{360} + \frac{25}{360} = \frac{67}{360},$$

which is in lowest terms. Rule (11) would give $\frac{7}{60} + \frac{5}{72} = \frac{804}{4320}$, which reduces to $\frac{67}{360}$.

Obviously the arithmetic needed involves larger numbers than the first method. We could replace (11) by the following rule, but it is awkward to write it as a formula:

To add fractions (with minimum labor), change each fraction to one whose denominator is the least common denominator for all the fractions; then add the numerators and set that result over the common denominator. (11′)

Example 2

$$\frac{1}{3+h} - \frac{1}{3} = \frac{1 \cdot 3}{(3+h) \cdot 3} - \frac{1 \cdot (3+h)}{3(3+h)} = \frac{3 - (3+h)}{3(3+h)}$$

$$= \frac{3 - 3 - h}{3(3+h)} = \frac{-h}{3(3+h)}.$$

Example 3

0.3
Fractions and
rational numbers

$$\frac{1}{2x(x+h)} - \frac{1}{2x^2}.$$

The least common denominator is $2x^2(x+h)$.

$$\frac{1}{2x(x+h)} - \frac{1}{2x^2} = \frac{1 \cdot x}{2x(x+h) \cdot x} - \frac{1 \cdot (x+h)}{2x^2(x+h)} = \frac{x - x - h}{2x^2(x+h)}$$

$$= \frac{-h}{2x^2(x+h)}.$$

In the preceding examples there is no point in multiplying out in the denominators (except possibly in the last step for some purposes, especially in numerical examples); on the other hand, it *is* necessary to multiply out in the numerators in order to combine like terms.

In the rational numbers we have a set closed under the operations of addition, subtraction, multiplication, and division (i.e., combining any two numbers of the set by any of these operations gives again a member of the set), with the single exception of division by 0. Although we have by no means done so, it can be shown that properties (1) through (5) hold for the rational numbers.

PROBLEMS

1. Reduce each of the following fractions to a simpler form, if possible:

(a) $\dfrac{6}{15}$ (b) $\dfrac{100}{24}$ (c) $\dfrac{504}{108} = \dfrac{2^3 \cdot 3^2 \cdot 7}{2^2 \cdot 3^3}$ (d) $\dfrac{90}{675}$ (e) $\dfrac{12}{3/4}$

(f) $\dfrac{12/3}{4}$ (g) $\dfrac{36h}{6(6+h)}$ (h) $\dfrac{2x^2}{10x(x+2)}$ (i) $\dfrac{3y}{(y+3)/y^2}$

(j) $\dfrac{3y/(y+3)}{y^2}$

2. Perform the indicated additions and subtractions:

(a) $\dfrac{5}{6} + \dfrac{2}{7}$ (b) $\dfrac{11}{30} - \dfrac{2}{21}$ (c) $\dfrac{5b}{a^2 + ab} + \dfrac{5}{a+b}$

(d) $\dfrac{y}{x+y} - \dfrac{z}{x+z}$ (e) $\dfrac{3x}{5} + \dfrac{7}{2x}$ (f) $a + \dfrac{b}{c}$

○ 3. Simplify each of the following expressions:

(a) $\left(\dfrac{1}{4+h} - \dfrac{1}{4} \right) \Big/ h$ (b) $\left(\dfrac{3}{(x+h)^2} - \dfrac{3}{x^2} \right) \Big/ h$

(c) $\left(\dfrac{1}{(x+h)^3} - \dfrac{1}{x^3} \right) \Big/ h$

0.4 Integral exponents

We have assumed familiarity with the definition

$$x^m = x \cdot x \cdot \ldots \cdot x \quad (m \text{ factors}), \quad m \text{ a positive integer.}$$

This definition leads immediately to the following results, where m and n are positive integers:

$$(xy)^m = x^m y^m \tag{12}$$

$$x^m x^n = x^{m+n} \tag{13}$$

$$(x^m)^n = x^{mn} \tag{14}$$

$$\frac{x^m}{x^n} = x^{m-n}, \quad \text{if } m > n \quad (\text{read } m \text{ greater than } n), \quad \text{and } x \neq 0. \tag{15}$$

We define x^m for zero and negative m in such a way that (12)–(15) hold. If we ignore the restriction $m > n$ in (15), we have, for example, $x^4/x^7 = x^{4-7} = x^{-3}$, $x \neq 0$. On the other hand, $x^4/x^7 = 1/x^{7-4} = 1/x^3$, $x \neq 0$. In general, we define

$$x^{-m} = \frac{1}{x^m}, \quad x \neq 0, \quad m \text{ a positive integer.} \tag{16}$$

Similarly, if $m = n$, $x^m/x^m = x^{m-m} = x^0$, $x \neq 0$. But $x^m/x^m = 1$, $x \neq 0$, and $0^m/0^m = 0/0$ is a meaningless symbol. Hence, we define

$$x^0 = 1, \quad x \neq 0. \tag{17}$$

Note that 0^0 is undefined.

PROBLEMS

Simplify the following expressions, writing each of them without negative exponents.

1. $x + x^{-1}$

2. $(x + x^{-1})^2$

3. $\dfrac{x^2 - 1}{2x^{-1}}$

4. $(7x)^0 + \dfrac{7}{x^0}$

5. $\dfrac{3x}{(2x)^2 - 1} - \dfrac{1}{2x - 1}$

6. $\dfrac{x^{-2}}{2^{-1}x - 1}$

7. $\dfrac{x + y}{x^{-1} + y^{-1}}$

8. $\dfrac{a^{-2} - b^{-2}}{a - b}$

9. $(x^{-1} + y^{-1})^{-2}$

10. $\left(1 + \dfrac{1}{x}\right)^0$

11. $\dfrac{a^{-2}}{2b^{-2}}$

12. $k(r + k)^{-2} - kr^{-2}$

13. $\dfrac{(a^3 \cdot a^2)^4}{(ab)^7}$

14. $(rs)^{2x} r^{-x} s^{-3x}$

15. $(w^{-2} + w^{-3})^2$

16. $(w^{-2} + w^{-3})^{-2}$

0.5 Radicals, fractional exponents, and real numbers

Recall now the familiar notation using radicals to denote roots of numbers. By definition,

$$\left(\sqrt[q]{x}\right)^q = x, \quad \text{where } q \text{ is a positive integer.}$$

For example, $(\sqrt{5})^2 = 5$, $(\sqrt{9})^2 = 9$. Now, both $3^2 = 9$ and $(-3)^2 = 9$. It is agreed that $\sqrt{9} = 3$, the positive root only; then $-3 = -\sqrt{9}$. (There is a discrepancy here, which may be confusing, between the way we read $\sqrt{x}$ and the precise definition of the symbol; strictly speaking, we should say "the positive square root of x" instead of simply "the square root of x," as we usually do.) Likewise, $\sqrt{5}$ stands for the positive root only. For q any even number and $x > 0$, we have a similar situation; for example, $\sqrt[4]{16} = 2$, not -2. For q even and $x < 0$, there is no real root; nevertheless, we shall later find meanings for such expressions. For q odd, there is exactly one real root, and so we need no such convention; for example, $\sqrt[3]{8} = 2$, $\sqrt[3]{-8} = -2$.

It is easily shown that

$$\sqrt[q]{ab} = \sqrt[q]{a} \cdot \sqrt[q]{b} \quad \text{and} \quad \sqrt[q]{\frac{a}{b}} = \frac{\sqrt[q]{a}}{\sqrt[q]{b}}. \tag{18}$$

This gives a means of simplifying radical expressions, or changing them to more convenient forms (e.g., without radicals in denominators). Examples:

$$\sqrt[3]{54} = \sqrt[3]{27} \cdot \sqrt[3]{2} = 3\sqrt[3]{2}.$$

$$\sqrt{\frac{3}{2}} = \sqrt{\frac{6}{4}} = \frac{\sqrt{6}}{\sqrt{4}} = \frac{\sqrt{6}}{2}.$$

$$\sqrt{5a^2 + 10ab + 5b^2} = \sqrt{5(a^2 + 2ab + b^2)}$$
$$= \sqrt{5} \cdot \sqrt{(a+b)^2} = \sqrt{5}(a+b), \quad \text{if } a + b \geq 0.$$

$$\sqrt[4]{16a^4 + 16b^4} = \sqrt[4]{16} \cdot \sqrt[4]{a^4 + b^4} = 2\sqrt[4]{a^4 + b^4}.$$

$$\frac{\sqrt{x+y}}{\sqrt{x-y}} = \frac{\sqrt{x+y}}{\sqrt{x-y}} \cdot \frac{\sqrt{x-y}}{\sqrt{x-y}} = \frac{\sqrt{x^2 - y^2}}{x - y}, \quad \text{if } x - y > 0.$$

We return now to exponents. If we apply (14), disregarding the restriction that m be an integer, we have $(x^{1/q})^q = x^{q/q} = x$, and so we define

$$x^{1/q} = \sqrt[q]{x}, \quad q \text{ a positive integer.} \tag{19}$$

This means that everything we have said about radical expressions can be stated in terms of fractional exponents. For example, (18) becomes

$$(ab)^{1/q} = a^{1/q}b^{1/q} \quad \text{and} \quad \left(\frac{a}{b}\right)^{1/q} = \frac{a^{1/q}}{b^{1/q}}.$$

Now, using (14) and neglecting the restriction that m be an integer, we have $x^{p/q} = (x^{1/q})^p$, p and q integers, $q > 0$. For example, $8^{2/3} = (8^{1/3})^2 = 2^2 = 4$. If $x > 0$, then also $x^{p/q} = (x^p)^{1/q}$; if $x < 0$ and if p/q is not in lowest terms, this may lead to error. For example, $(-8)^{2/6} = (-8)^{1/3} = -2$, but $[(-8)^2]^{1/6} = (64)^{1/6} = 2$, not -2. We therefore take as the definition of $x^{p/q}$,

$$x^{p/q} = (x^{1/q})^p, \quad p \text{ and } q \text{ integers without a common factor, } q > 0. \quad (20)$$

With definitions (16), (17), (19), and (20) it can be shown that properties (12) through (15) hold for m and n any rational numbers. The only restrictions that are retained are that $x \neq 0$ in (15) and that 0^0 remains undefined.

In the preceding discussion we mentioned some symbols (e.g., $\sqrt{3}, \sqrt{5}, \sqrt[3]{2}$) that have no meaning in the rational number system. That is, it can be proved that there is no rational number whose square is 3, and so forth. The set of real numbers can be defined as the set of all decimal representations, terminating and nonterminating. The rationals compose the subset with decimal representation, either terminating or periodic from some point on. (This is not hard to show.) All other real numbers are irrational. For example, the following numbers are all rational:

$$\tfrac{51}{4} = 12.75,$$

$$\tfrac{5}{3} = 1.66\ldots \quad \text{(the dots mean "continued indefinitely"),}$$

$$\tfrac{1}{22} = 0.0454\dot{5}\ldots \quad \text{(the dots above 45 indicate the period),}$$

$$\tfrac{22}{7},$$

$$3.1416.$$

It can be shown that $\sqrt{2}, \sqrt[3]{9}$, and the number π are irrational.

The real numbers can be put into one-to-one correspondence with the points of a line, once a zero point, a unit point, and a positive direction have been chosen, so that numbers on the number line increase in the positive direction and decrease in the opposite (negative) direction.

Even in the set of real numbers, we have no number whose square is a negative number. Later we shall make one more extension of our number system which will remedy that lack.

PROBLEMS

1. Simplify:

(a) $\sqrt[4]{48}$ (b) $\sqrt[3]{\dfrac{4}{9}}$ (c) $\sqrt{a^3 - 2a^2b + ab^2}$

(d) $\sqrt[3]{8a^3 - 8b^3}$ (e) $\dfrac{5}{\sqrt{5(a+b)^3}}$

2. Simplify each of the following expressions, writing them without negative exponents.

(a) $(16x)^{3/4}$ (b) $\left(\dfrac{16}{x}\right)^{-3/4}$ (c) $(x^2+25)^{-1/2}$ (d) $\dfrac{a^2+a^3}{a^{1/2}}$

(e) $\dfrac{x^2}{(9-x^2)^{1/2}}+(9-x^2)^{1/2}$ (f) $\left[\dfrac{(a^2+b^2)^2}{(a^2-b^2)^2}-1\right]^{1/2}$

3. (a) Is $(ab)^2$ always equal to a^2b^2? If not, is it ever equal to …?
   (b) Is $(p^2)^{1/2}$ always equal to p? If not, is it ever equal to …?
   (c) Is $(a+b)^2$ always equal to a^2+b^2? If not, is it ever equal to …?
   (d) Is $(p^2+q^2)^{1/2}$ always equal to $p+q$? If not, is it ever equal to …?
   (e) Is $(p^2-q^2)^{1/2}$ always equal to $p-q$? If not, is it ever equal to …?
   (f) Is $(xy)^{1/3}$ always equal to $x^{1/3}\cdot y^{1/3}$? If not, is it ever equal to …?
   (g) Is $(p+q)^{-1}$ always equal to $p^{-1}+q^{-1}$? If not, is it ever equal to …?
   (h) Is $(x^{-1}+y^{-1})^{-1}$ always equal to $x+y$? If not, is it ever equal to …?

4. (a) Simplify $Q=\dfrac{1}{2}\left[\dfrac{2x}{x^2-25}-\dfrac{2x}{x^2+25}\right]$.

   (b) Simplify $P=\sqrt{\dfrac{x^2-25}{x^2+25}}\cdot Q$,   where Q is as in part (a).

5. (a) Simplify $R=\dfrac{1}{2}\left(\dfrac{2y}{9+y^2}+\dfrac{2y}{9-y^2}\right)$.

   (b) Simplify $S=\sqrt{\dfrac{9+y^2}{9-y^2}}\cdot R$,   where R is as in part (a).

0.6   Notation for implication

We say that the hypothesis $x=7$ *implies* the conclusion $x+3=10$, or, in abbreviated notation, $x=7\Rightarrow x+3=10$. In general, if P and Q are statements, $P\Rightarrow Q$ is read "P implies Q," or "if P, then Q."

Likewise, $x+3=10\Rightarrow x=7$; that is, the implication goes both ways. We can combine the two statements in the form $x=7\Leftrightarrow x+3=10$. In general, $P\Leftrightarrow Q$ is read "P implies Q and conversely," or "P implies and is implied by Q," or "P if and only if Q," or "P and Q are logically equivalent."

We have also $x=7\Rightarrow x^2=49$. However, if $x^2=49$, then $x=7$ or $x=-7$; so we *cannot* say $x^2=49\Rightarrow x=7$. In symbols, $x^2=49 \nRightarrow x=7$.

0.7   Equalities

So far, the equalities we have used have been statements that hold for all numbers of the set under consideration. We shall have occasion now to deal with "conditional equalities," that is, to solve equations, or, in the language

11

of "new math," to "find solution sets of open sentences." In this we rely heavily on the following properties of equality:

$$r = s \Leftrightarrow r + a = s + a, \quad \text{or, more generally,}$$

$$\text{if } a = b, \quad \text{then } r = s \Leftrightarrow r + a = s + b, \tag{21}$$

that is, equals added to (or subtracted from) equals give equal results; and,

$$\text{if } k \neq 0, \quad r = s \Leftrightarrow kr = ks, \tag{22}$$

that is, multiplying (or dividing) both sides of an equality by the same number gives a valid equality.

Property (8) in Section 0.2, which we repeat in the notation for implication, will also be useful:

$$pq = 0 \Leftrightarrow p = 0 \quad \text{or} \quad q = 0, \tag{23}$$

i.e., a product is 0 if and only if one of its factors is 0.

Example 1

$$3x - 2 = 10 \Leftrightarrow 3x = 12 \quad \text{by (21)}$$

$$\Leftrightarrow \quad x = 4 \quad \text{by (22).}$$

Example 2

$$ax + b = c \Leftrightarrow ax = c - b \quad \text{by (21)}$$

$$\Leftrightarrow x = \frac{c - b}{a} \quad \text{by (22), provided } a \neq 0.$$

Example 3

$$\frac{x}{2} = \frac{3}{5} \Leftrightarrow x = \frac{6}{5} \quad \text{by (22).}$$

Example 4

$$\frac{4}{7} = \frac{3}{(x - 1)} \Leftrightarrow 4(x - 1) = 21 \quad \text{by (22)}$$

$$\Leftrightarrow \quad 4x - 4 = 21$$

$$\Leftrightarrow \quad 4x = 25 \quad \text{by (21)}$$

$$\Leftrightarrow \quad x = \tfrac{25}{4} \quad \text{by (22).}$$

Example 5

$$x - \frac{27}{x^2} = 0 \Leftrightarrow x^3 - 27 = 0 \quad \text{(multiplying each side by } x^2\text{)} \quad \text{by (22)}$$

$$\Leftrightarrow \quad x^3 = 27$$

$$\Leftrightarrow \quad x = 3.$$

Example 6

$$x^2 - 5x + 6 = 0 \Leftrightarrow (x-2)(x-3) = 0$$

$$\Leftrightarrow x - 2 = 0 \quad \text{or} \quad x - 3 = 0 \quad \text{by (23)}$$

$$\Leftrightarrow \qquad x = 2 \quad \text{or} \qquad x = 3.$$

PROBLEMS

Find all real numbers satisfying each of the following equations:

1. $\dfrac{1}{2}x + 3 = 8$ 2. $8 = \dfrac{2}{3}x - \dfrac{1}{2}$ 3. $\dfrac{5}{3} = \dfrac{y-1}{2}$

4. $\dfrac{x}{2} = \dfrac{3x}{x+2}$ 5. $\dfrac{x+2}{2x-3} = \dfrac{5}{3}$ 6. $\dfrac{7}{2x+1} = \dfrac{10}{3x}$

7. $x^2 - 3x - 4 = 0$ 8. $u^2 - 3u = 10$ 9. $\dfrac{x-2}{5} = \dfrac{4}{x-3}$

10. $8y + 90 = 2y^2$ 11. $v^2 + \dfrac{125}{v} = 0$ 12. $x^3 = 16x$

13. $\dfrac{4}{3x} + \dfrac{5}{3} = 2x$ 14. $\dfrac{1}{2} + \dfrac{1}{y-2} - \dfrac{y}{2y-4} = 0$

0.8   Inequalities

In equation (15) in Section 0.4 we used the symbol $m > n$ for "m is greater than n." Clearly this is equivalent to "n is less than m," written $n < m$. For example, we can write $2 > -3$ or $-3 < 2$. (Remember positions on the number line.) For the statement "x is less than or equal to 7," we write $x \le 7$, or, equivalently, $7 \ge x$ for "7 is greater than or equal to x"; for "u is positive and less than 10," $0 < u < 10$; for "v is greater than -5 and less than or equal to -2," $-5 < v \le -2$.

There are properties of inequalities analogous to those for equalities:

$$r < s \Leftrightarrow r + a < s + a, \quad \text{or, more generally,}$$
$$\text{if } a \le b, \quad \text{then } r < s \Leftrightarrow r + a < s + b. \tag{24}$$

Also,

$$r < s \quad \text{and} \quad k > 0 \Rightarrow kr < ks, \tag{25a}$$
$$r < s \quad \text{and} \quad k < 0 \Rightarrow kr > ks. \tag{25b}$$

In words: If both sides of an inequality are multiplied by the same positive number, the sense of the inequality is preserved; if multiplied by a negative number, the sense is reversed. We have

$$rs > 0 \Leftrightarrow r > 0 \quad \text{and} \quad s > 0 \quad \text{or} \quad r < 0 \quad \text{and} \quad s < 0, \tag{26a}$$
$$rs < 0 \Leftrightarrow r > 0 \quad \text{and} \quad s < 0 \quad \text{or} \quad r < 0 \quad \text{and} \quad s > 0, \tag{26b}$$

that is, the product of two numbers is positive if and only if both numbers

13

are positive or both are negative; the product of two numbers is negative if and only if one of the numbers is positive and the other is negative.

Example 1

$$3x + 2 < 14 \Leftrightarrow 3x < 12 \quad \text{by (24)}$$
$$\Leftrightarrow \quad x < 4 \quad \text{by (25a).}$$

Example 2

$$2 - 3x \geq 14 \Leftrightarrow -3x \geq 12 \quad \text{by (24)}$$
$$\Leftrightarrow \quad x \leq -4 \quad \text{by (25b).}$$

Example 3

$$x^2 + 6 > 5x \Leftrightarrow \quad x^2 - 5x + 6 > 0 \quad \text{by (24)}$$
$$\Leftrightarrow (x - 2)(x - 3) > 0 \quad \text{(factoring)}$$
$$\Leftrightarrow x - 2 > 0 \quad \text{and} \quad x - 3 > 0$$
$$\text{or} \quad x - 2 < 0 \quad \text{and} \quad x - 3 < 0 \quad \text{by (26a)}$$
$$\Leftrightarrow x > 2 \quad \text{and} \quad x > 3$$
$$\text{or} \quad x < 2 \quad \text{and} \quad x < 3 \quad \text{by (24).}$$

Now, if $x > 2$ and $x > 3$, it must be that $x > 3$. Likewise, if $x < 2$ and $x < 3$, it must be that $x < 2$. Hence, our conclusion is that

$$x^2 + 6 > 5x \Leftrightarrow x > 3 \quad \text{or} \quad x < 2.$$

Example 4

$$x^2 < 2x + 8 \Leftrightarrow \quad x^2 - 2x - 8 < 0 \quad \text{by (24)}$$
$$\Leftrightarrow (x + 2)(x - 4) < 0 \quad \text{(factoring)}$$
$$\Leftrightarrow x + 2 > 0 \quad \text{and} \quad x - 4 < 0$$
$$\text{or} \quad x + 2 < 0 \quad \text{and} \quad x - 4 > 0 \quad \text{by (26b)}$$
$$\Leftrightarrow x > -2 \quad \text{and} \quad x < 4$$
$$\text{or} \quad x < -2 \quad \text{and} \quad x > 4 \quad \text{by (24).}$$

Now, there is no number x such that $x < -2$ and $x > 4$. Hence, the only possibility is that $x > -2$ and $x < 4$, or $-2 < x < 4$. Thus, our conclusion is that

$$x^2 < 2x + 8 \Leftrightarrow -2 < x < 4.$$

⋆ Example 5

$$\frac{x-3}{x-2} > \frac{2}{3}$$

We must multiply both sides by $3(x-2)$ in order to "clear of fractions." But we must distinguish the case in which this multiplier is positive from that in which it is negative.

Case 1. $x > 2$. Then

$$\frac{x-3}{x-2} > \frac{2}{3} \Leftrightarrow 3(x-2)\cdot\frac{x-3}{x-2} > 3(x-2)\frac{2}{3}$$

$$\Leftrightarrow 3(x-3) > 2(x-2)$$

$$\Leftrightarrow 3x - 9 > 2x - 4$$

$$\Leftrightarrow x > 5.$$

If $x > 2$ and $x > 5$, then $x > 5$.

Case 2. $x < 2$. Then

$$\frac{x-3}{x-2} > \frac{2}{3} \Leftrightarrow 3(x-2)\frac{x-3}{x-2} < 3(x-2)\frac{2}{3}$$

$$\Leftrightarrow 3x - 9 < 2x - 4$$

$$\Leftrightarrow x < 5.$$

If $x < 2$ and $x < 5$, then $x < 2$. Hence, the given inequality holds for $x > 5$ or $x < 2$.

PROBLEMS

Find all real numbers satisfying each of the following inequalities:

1. $x^2 > 2x + 3$ 2. $3x + 10 > x^2$ 3. $-2x^2 - 14x > 12$

4. $x^2 > \frac{1}{2}x$ 5. $\frac{2x+3}{-5} > \frac{x+6}{2}$ 6. $7x - 12x - x^2 \geq 0$

⋆ 7. $\frac{5}{2x-6} < \frac{2}{3}$

⋆ 8. $(x-1)(x-2)(x-3) > 0$

9. $x^2 > 9$ 10. $-6x + 3 > 9$ 11. $x^2 + 7 < 11$

⋆ 12. $\frac{3}{x-2} < \frac{2x+1}{x+6}$

15

0.9 Linear equations

A. Linear equations in one unknown

A linear equation in one unknown, x, is one that, by the use of (21) and (22), can be written in the form $ax + b = 0$, with a and b constants (i.e., free of x), and $a \neq 0$. Then $x = -(b/a)$ is the one and only solution (see Examples 1 and 2, **0.7**).

If we set $y = ax + b$ and draw the graph, we shall understand the reason for the term "linear." Remember that in the ordinary graphical system there is a one-to-one correspondence between the points of the plane and the *ordered* pairs of real numbers; for example, $(1,2)$ and $(2,1)$ represent different points. (By convention, the x-value appears first.)

Example 1   (See Example 1, **0.7**.)

$$3x - 2 = 10 \Leftrightarrow 3x - 12 = 0.$$

Let $y = 3x - 12$. We see in Table 0-1 that for an increase of one unit in x there is an increase of three units in y, and if we look at the formula we see that this is true for *any* increase of one unit in x. This means that the graph is a straight line, of "slope" 3, as in Figure 0-1. The line crosses the x axis at $(4,0)$; that is, $x = 4$ is the solution of $3x - 12 = 0$.

In general, the graph of $y = ax + b$ is a straight line of "slope" a, crossing the x axis where $x = -(b/a)$. The graph of an equation consists of those points and only those points whose coordinates satisfy the equation.

B. Simultaneous linear equations in two unknowns

A linear equation in two unknowns can be written in the form $ax + by = c$, in which a, b, and c are constants, with at least one of a, b not 0. A simultaneous solution of two such equations is a pair of values, one for x and one for y, that satisfies both equations. Such a solution, if there is one, can be found by using (21) and (22) so as to eliminate one variable.

Example 2

$$\begin{cases} 2x - 3y = 7 \\ 3x + 4y = 2 \end{cases} \Leftrightarrow \begin{cases} 8x - 12y = 28 \\ 9x + 12y = 6 \end{cases} \text{ by (22)}$$

$$\Rightarrow 17x = 34 \quad \text{by (21)}$$

$$\Leftrightarrow \quad x = 2 \quad \text{by (22).}$$

Now,

$$\begin{cases} x = 2 \\ 3x + 4y = 2 \end{cases} \Rightarrow 6 + 4y = 2 \Leftrightarrow y = -1.$$

Table 0-1

x	y
-1	-15
0	-12
1	-9
2	-6
3	-3
4	0
5	3
6	6

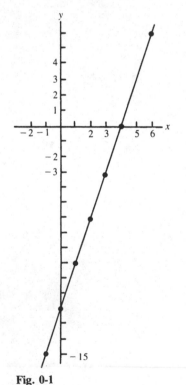

Fig. 0-1

The solution is $x = 2$, $y = -1$. (Check by substitution in the original equations.)

Example 3

$$\begin{cases} 2x - 3y = 7 \\ 4x - 6y = -9 \end{cases} \Leftrightarrow \begin{cases} 4x - 6y = 14 \\ 4x - 6y = -9 \end{cases} \Rightarrow 0 = 23.$$

No pair of values for x and y can make $0 = 23$. There is no solution.

A graphical interpretation of these examples may be illuminating (Figure 0-2). We know that each equation has a straight-line graph; so we plot only a few points.

$$2x - 3y = 7 \qquad\qquad 3x + 4y = 2 \qquad\qquad 4x - 6y = -9$$

$$\Leftrightarrow y = \tfrac{2}{3}x - \tfrac{7}{3} \qquad \Leftrightarrow y = -\tfrac{3}{4}x + \tfrac{1}{2} \qquad \Leftrightarrow y = \tfrac{2}{3}x + \tfrac{3}{2}$$

x	y
-1	-3
0	$\frac{-7}{3}$
$\frac{7}{2}$	0

x	y
-1	$\frac{5}{4}$
0	$\frac{1}{2}$
$\frac{2}{3}$	0

x	y
-1	$\frac{5}{6}$
0	$\frac{3}{2}$
$\frac{-9}{4}$	0

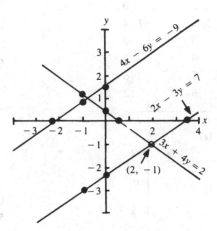

Fig. 0-2

The solution $x = 2$, $y = -1$ of the equations in Example 2 corresponds to the point of intersection of their graphs. The equations of Example 3, which have no solution, have graphs that are parallel lines; that is, they have no point of intersection.

PROBLEMS

Find all real numbers satisfying each of the following equations:

1. $\dfrac{1}{x-2} = \dfrac{2}{3}$ 2. $\dfrac{1}{5} - \dfrac{1}{x} = \dfrac{2}{x}$

Find both algebraically and graphically all solutions of the following pairs of simultaneous equations:

3. $\begin{cases} x + 2y = 5 \\ 2x - 3y = 3 \end{cases}$ 4. $\begin{cases} 3x + 4y = 8 \\ x + 6y = 12 \end{cases}$

5. $\begin{cases} 4x - 3y = 12 \\ \tfrac{4}{3}x - y = 4 \end{cases}$ 6. $\begin{cases} 4x - 3y = 12 \\ \tfrac{4}{3}x - y = 5 \end{cases}$

Solve algebraically the following pairs of simultaneous equations:

7. $\begin{cases} 5x + 2y = 3 \\ 3x - 6y = 9 \end{cases}$ 8. $\begin{cases} y - 3x = -2 \\ 2x - 5y = 7 \end{cases}$ 9. $\begin{cases} 9x = 13 + y \\ x + y = 5 \end{cases}$

★ 10. $\begin{cases} x^2 + 4\sqrt{y} = 8 \\ 3x^2 - \sqrt{y} = 11 \end{cases}$

0
Prerequisites

Table 0-2

x	y
-1	12
0	6
2	0
5/2	$-1/4$
3	0
4	2
5	6
6	12

(a)

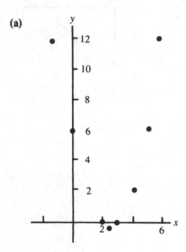

(b)

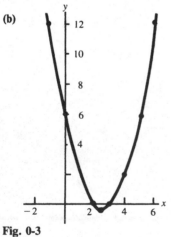

Fig. 0-3

18

0.10 Quadratic equations

A quadratic (second-degree) equation in one unknown is one that can be put into the form

$$ax^2 + bx + c = 0, \quad \text{where } a, b, \text{ and } c \text{ are constants, and } a \neq 0. \quad (27)$$

If the left side can be factored, the equation can be solved by (23). (See Example 6 in Section 0.7 and Problems 7–10 following Section 0.7.)

Every quadratic equation can be solved by "completing the square," a method that was known to the ancient Babylonians.

Example 1

$$2x^2 + 2x - 1 = 0$$

$$\Leftrightarrow x^2 + x - \tfrac{1}{2} = 0 \qquad \text{by (22)}$$

$$\Leftrightarrow x^2 + x \qquad = \tfrac{1}{2} \qquad \text{by (21)}$$

$$\Leftrightarrow x^2 + x + \tfrac{1}{4} = \tfrac{3}{4} \qquad \text{by (21)}$$

(Adding the square of half the coefficient of x makes the left side a perfect square; hence the name.)

$$\Leftrightarrow \left(x + \tfrac{1}{2}\right)^2 = \tfrac{3}{4}$$

$$\Leftrightarrow x + \tfrac{1}{2} = \sqrt{\tfrac{3}{4}} \quad \text{or} \quad x + \tfrac{1}{2} = -\sqrt{\tfrac{3}{4}} \quad \text{(taking square roots of both sides)}$$

$$\Leftrightarrow x = \frac{-1 + \sqrt{3}}{2} \quad \text{or} \quad x = \frac{-1 - \sqrt{3}}{2} \quad \text{by (21).}$$

Applying this method to (27) yields the familiar formula

$$x = \frac{-b \pm \sqrt{b^2 - 4ac}}{2a},$$

which can then be used to find the roots without going through the steps of completing the square. Factoring, when it works, remains the easiest method.

If the expression $b^2 - 4ac < 0$, the roots involve square roots of negative numbers, which do not belong to the set of "real" numbers. We extend the number system once more. The set of complex numbers is the set $m + ni$, m and n real, where $i = \sqrt{-1}$, with addition and multiplication defined as if these were any algebraic expressions. Then it can be shown that properties (1)–(5) hold and that the set is closed under each of the fundamental operations, with the exception of division by 0. The subset for which $n = 0$ is the set of real numbers; all others are, unfortunately, called imaginary. (Because all numbers are abstract mental concepts, the term *imaginary* is not more appropriate for these than for any other numbers.) The set of complex numbers is sufficient for the purposes of ordinary algebra and analysis.

Every quadratic equation has two roots, which may be real or imaginary. We turn again to graphical methods.

0.10 Quadratic equations

Example 2

We use the quadratic x^2-5x+6 of Example 3 in Section 0.8. If we write $y=x^2-5x+6$ and make a table of values, we obtain Table 0-2. A plot of the points corresponding to the pairs of values in the table is shown in Figure 0-3(a), and a smooth curve through these points is shown in Figure 0-3(b). We see that $y=x^2-5x+6$ is 0 for $x=2$ and $x=3$ (as found in Example 6 of Section 0.7), is negative for $2<x<3$, and is positive for $x>3$ or $x<2$, as we have already found in Example 3, **0.8**.

Table 0-3

x	y
-2	3
-1	-1
0	-1
1	3
$-1/2$	$-3/2$
$-3/2$	$1/2$
$1/2$	$1/2$

Example 3

Let $y=2x^2+2x-1$, the quadratic in Example 1. (The last three values of Table 0-3 were added after plotting the others because it was hard to see how to draw a smooth curve with only the first four.) Here we can read off only approximately the values of x where the curve crosses the x-axis: Say $x=-1.4$ and $x=0.4$. In Example 1, we found algebraically the roots to be the irrational numbers $(-1+\sqrt{3})/2$ and $(-1-\sqrt{3})/2$. Now, $(-1+\sqrt{3})/2 \approx 0.366$, and $(-1-\sqrt{3})/2 \approx -1.366$ ($\approx$ means approximately equal). These values are rational approximations to the irrational values, but closer than can be read from a graph (Figure 0-4). The graph of $y=ax^2+bx+c$, where a, b, and c are real numbers, and $a>0$, is always a curve with a shape of this one, called a *parabola*, symmetric with respect to an axis parallel to the y axis, extending indefinitely upward more and more steeply. Remember that the graph consists of those points with *real* coordinates satisfying the equation. If the curve does not cut or touch the x axis, the equation $ax^2+bx+c=0$ has no real roots; hence, its two roots are imaginary.

Fig. 0-4

PROBLEMS

Solve algebraically; draw and interpret appropriate graphs:

1. $2x^2-7x+3=0$
2. $x^2-5x+\frac{25}{4}=0$
3. $x^2-5x+8=0$
4. $x^2-4x-5=0$
5. $x^2-4x=0$
6. $x^2-4x+1=0$
7. $x^2-4x+4=0$
8. $x^2-4x+5=0$
9. $x^2-7x+2=4$
10. $3x^2+12=-12x$
11. $6x+56=2x^2$
12. $\frac{4}{3}x^2-4=\frac{8}{6}$
13. $8x^2-24x+8=0$

★ 14. By "completing the square," show that

$$ax^2+bx+c=0 \Leftrightarrow x=\frac{-b\pm\sqrt{b^2-4ac}}{2a}, \quad \text{if } a\neq 0.$$

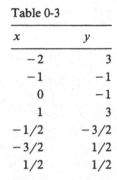

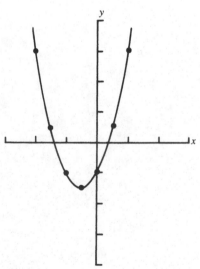

0.11 Higher-degree equations

A *polynomial* is an expression of the form

$$a_n x^n + a_{n-1} x^{n-1} + \cdots + a_1 x + a_0,$$

in which the a's are constants, $a_n \neq 0$, and n is a positive integer. We use a's with subscripts for the coefficients instead of different letters because it is easier to remember which goes with which power of x. In this notation, the linear polynomial will be $a_1 x + a_0$; the quadratic will be $a_2 x^2 + a_1 x + a_0$. The degree of the polynomial is n. An algebraic (or polynomial) equation of degree n is such an expression set equal to 0.

There are formulas analogous to the quadratic formula for the roots of the general cubic (third-degree) and the quartic (fourth-degree) equations (discovered by the sixteenth-century Italian mathematicians Cardan, Tartaglia, and Ferrari), but they are extremely awkward to use and will not be dealt with here. There is no such formula for the general algebraic equation of degree 5 or more. This was proved in 1824 by the Norwegian mathematician Abel, who was 22 years old at the time. In 1829 the French mathematician Galois characterized those equations that can be solved algebraically; shortly thereafter he was killed in a duel, before his 21st birthday.

In relatively few cases can we find the exact values of the roots of equations of degree greater than 2; however, we can find *approximate* values of the real roots by graphical and numerical methods. We shall state a few theorems that are helpful. We shall deal only with polynomials with real coefficients.

A. Every equation of degree n has n roots (which may be real or imaginary). (This was proved by the German mathematician Gauss in 1799, when he was 22 years old.)

B. Imaginary roots occur in pairs.
From A and B follows C:

C. Every equation of odd degree has at least one real root. Another very useful result is more easily expressed if we use the notation $P(x)$ (read "P of x") for the general polynomial given earlier and $P(k)$ for its value when k is substituted for x. For example,

$$\text{if } P(x) = 6x^3 - 29x^2 - x + 84,$$
$$P(2) = 48 - 116 - 2 + 84 = 14.$$

Dividing $P(x)$ by $(x - k)$ means finding a quotient $Q(x)$, which is a polynomial of degree $n - 1$, and a constant remainder R such that

$$P(x) = (x - k)Q(x) + R.$$

Then

$$P(k) = (k - k)Q(k) + R = 0 \cdot Q(k) + R = 0 + R = R.$$

That is:

D. The remainder in the division of $P(x)$ by $(x - k)$ is the value of $P(x)$ when $x = k$; if $R = 0$, k is a root of $P(x) = 0$, and $x - k$ is a factor of $P(x)$.

Example 1
Let $P(x) = 6x^3 - 29x^2 - x + 84$. Divide by $(x - 2)$:

$$
\begin{array}{r}
6x^2 - 17x \quad - 35 \qquad \text{(quotient)} \\
\overline{6x^3 - 29x^2 - x + 84} \;\big|\; x - 2 \quad \text{(divisor)} \\
6x^3 - 12x^2 \\
\overline{\quad -17x^2 - x} \\
-17x^2 + 34x \\
\overline{\qquad -35x + 84} \\
-35x + 70 \\
\overline{\qquad\qquad 14}
\end{array}
$$

The remainder, 14, is the value we found earlier when we substituted 2 for x in the polynomial.

There is much unnecessary writing in the foregoing division. In the first place, we need not write any x's if we write the coefficients in order of descending powers of x (being sure to put in the 0 coefficient of any missing term). The first terms in the second, fourth, and sixth lines are necessarily the same as the terms above them, so they need not be rewritten; likewise, the second terms in the third and fifth lines are copied from the first line, where we can read them just as well. The terms of the quotient are the first terms in the first, third, and fifth lines. We can remember that the first term in the divisor is x, without writing it. We are left with this skeleton:

$$
\begin{array}{r}
6 \;-29 \;\;-1 \;+84 \;\;)-2 \\
-12 \\
\overline{-17} \\
+34 \\
\overline{-35} \\
+70 \\
\overline{14}
\end{array}
$$

This can be written on three lines:

$$
\begin{array}{r}
6 \;-29 \;\;-1 \;+84 \;\;)-2 \\
-12 \;+34 \;+70 \\
\overline{-17 \;-35 \;+14}
\end{array}
$$

The -12 is obtained by multiplying 6 by -2, the -17 is obtained by subtracting this result from -29, and successive terms in the third line are obtained by multiplying by -2 and subtracting from successive numbers in the first line. We can get exactly the same results by multiplying each time by *positive* 2 and *adding*. Finally, then, we write positive 2 in the position of the divisor, bring down the first coefficient, 6, to the bottom line, multiply by 2 and add to the next term, -29, in the first line, multiply the result, -17, by 2 and add, and so on, getting

$$
\begin{array}{rrrr|r}
6 & -29 & -1 & +84 & \underline{2} \\
  & +12 & -34 & -70 & \\
\hline
6 & -17 & -35 & +14 &
\end{array}
$$

The last term in the bottom line, 14, is the remainder, and the first three terms are the coefficients of the quotient, $6x^2 - 17x - 35$.

This process is called *synthetic division* or *synthetic substitution*. This method is usually quicker than direct substitution, and it has other advantages as well. Suppose that we wish to solve

$$6x^3 - 29x^2 - x + 84 = 0.$$

Let

$$y = 6x^3 - 29x^2 - x + 84$$

and make Table 0-4. We have incidentally found one solution: $x = 4$. Because y is negative for $x = -2$ and positive for $x = -1$, we expect that there is a root between -2 and -1. Likewise, we expect that there is a root between 2 and 3. If we plot the pairs of numbers in our table, we obtain the dots in Figure 0-5(a).

Let us look at the synthetic division that gives y for $x = 4$:

$$
\begin{array}{rrrr|r}
6 & -29 & -1 & +84 & \underline{4} \\
  & 24 & -20 & -84 & \\
\hline
6 & -5 & -21 & +0 &
\end{array}
$$

We see that 4 is a root of $P(x) = 0$, and $x - 4$ is a factor of $P(x)$; that is,

$$P(x) = (x - 4)(6x^2 - 5x - 21),$$

so that the other two roots of the cubic are the roots of the quadratic equation

$$6x^2 - 5x - 21 = 0.$$

This quadratic expression happens to factor:

$$6x^2 - 5x - 21 = (2x + 3)(3x - 7),$$

so $x = -\frac{3}{2}$ and $x = \frac{7}{3}$ are roots. Thus, we are able to find all three roots exactly, so the graph in Figure 0-5(b) was not really needed.

Synthetic substitution is helpful in deciding "how far we have to go" in testing for the roots of a polynomial equation. The method can be il-

Table 0-4

x	y
-2	-78
-1	50
0	84
1	60
2	14
3	-18
4	0
5	104

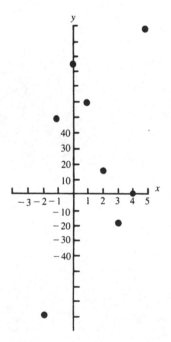

Fig. 0-5a

lustrated by the example we used earlier: $6x^3 - 29x^2 - x + 84$. If we apply synthetic substitution to find the value of this expression for $x = 5$, we have the following:

$$
\begin{array}{rrrr|}
6 & -29 & -1 & +84 \quad)5 \\
 & +30 & +5 & +20 \\
\hline
6 & +1 & +4 & +104 \\
\end{array}
$$

Because all the numbers in the last row of this schema are nonnegative, we realize that substituting a number larger than 5 would *increase* each of the numbers in the last row, and therefore we could not end with 0, which is what we need for a root.

Similarly, let us substitute -2 for x:

$$
\begin{array}{rrrr}
6 & -29 & -1 & +84 \quad)-2 \\
 & -12 & +82 & -162 \\
\hline
6 & -41 & +81 & -78 \\
\end{array}
$$

Because the numbers in the last row *alternate* in sign, we realize that substituting a number smaller than -2 (such as -3 or -4) would result in alternation with greater amplitudes, so again we could not end with 0. We state the general result as follows:

> In testing for the roots of an equation, we have gone far enough in the positive direction when all the numbers in the last row in synthetic substitution are nonnegative, and far enough in the negative direction when the numbers in the last row alternate in sign.

PROBLEMS

Use a graph to find approximate values of the roots of each of the following equations. When possible, find exact values of the roots as well.

1. $6x^3 - 31x^2 - 10x + 75 = 0$ 2. $6x^3 - 32x^2 - 5x + 75 = 0$
3. $6x^3 - 31x^2 - 10x + 74 = 0$ 4. $6x^3 - 11x^2 - x - 2 = 0$

Find all roots of each of the following equations by synthetic substitution:

5. $x^3 + 2x^2 - 11x - 12 = 0$ 6. $x^3 + 4x^2 - 11x - 30 = 0$
7. $2x^3 + 8x^2 - 38x + 28 = 0$ 8. $x^3 + 3x^2 - 31x + 12 = 0$
9. $2x^3 + 15x^2 + 17x - 6 = 0$ 10. $x^3 - 14x - 8 = 0$

★ 11. $x^4 + 2x^3 - 8x^2 - 26x - 12 = 0$
★ 12. Note that the polynomial $6x^3 - 29x^2 - x + 84$ from the example worked out in this section can be written as $\{[(6x - 29)x - 1]x + 84\}$, or more simply, without danger of misinterpretation, as

$$(((6x - 29)x - 1)x + 84).$$

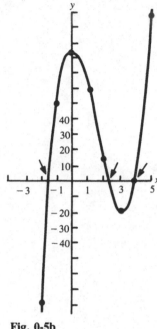

Fig. 0-5b

23

If $x = 2$, the expression $6x - 29$, in the innermost parentheses, equals -17; the next is -35; and the outermost is 14. These are exactly the numbers (except for the 6) of the third line of our first synthetic division, which is repeated here:

$$
\begin{array}{r}
6 \;-29 \;-1 \;+84 \quad)2 \\
12 \;-34 \;-70 \\
\hline
6 \;-17 \;-35 \;+14
\end{array}
$$

Explain why these numbers are the same, and why synthetic division works in general.

C★ 13. Synthetic division, or the equivalent "nested-parentheses" process of Problem 12, lends itself well to evaluation of polynomials on calculators or computers. With a little practice you will not have to rewrite $P(x)$ each time you wish to calculate $P(k)$ for a new k. Use a calculator to evaluate the following:

(a) $4x^4 - 13x^3 + 2x^2 - 19x + 12$ at $x = 0, -2, 2, 3$

(b) $27x^7 - 0.1x^5 + 34x^4 + 17x^3 + 2x + \pi$ at $x = -1, -7, 100$

0.12 Progressions

A. An *arithmetic progression* is a sequence of numbers such that the difference, d, between any term and its predecessor is the same; that is,

$$
a, a + d, a + 2d, \ldots, l - d, l,
$$

where the nth term, l, is given by $l = a + (n-1)d$. A formula for the sum, A_n, of the first n terms of such a progression can be obtained as follows:

$$
A_n = a + (a+d) + (a+2d) + \cdots + (l-2d) + (l-d) + l.
$$

Also

$$
A_n = l + (l-d) + (l-2d) + \cdots + (a+2d) + (a+d) + a.
$$

Adding,

$$
2A_n = (a+l) + (a+l) + (a+l) + \cdots + (a+l) + (a+l) + (a+l)
$$
$$
= n(a+l);
$$

so

$$
A_n = \frac{n}{2}(a+l). \tag{28}
$$

Example 1

Find the sum of the first 20 odd integers. The 20th odd integer is $1 + 19 \cdot 2 = 39$, so

$$
A_{20} = \tfrac{20}{2}(1 + 39) = 10 \cdot 40 = 400.
$$

Example 2

Find the sum of the first 10 terms of an arithmetic progression of which the first two terms are $\frac{1}{3}$ and $\frac{1}{2}$. Here,

$$a = \tfrac{1}{3}, \quad d = \tfrac{1}{2} - \tfrac{1}{3} = \tfrac{1}{6}, \quad l = \tfrac{1}{3} + 9 \cdot \tfrac{1}{6} = \tfrac{1}{3} + \tfrac{3}{2} = \tfrac{11}{6}.$$

Hence,

$$A_{10} = \tfrac{10}{2}\left(\tfrac{1}{3} + \tfrac{11}{6}\right) = 5 \cdot \tfrac{13}{6} = \tfrac{65}{6}.$$

  B. A *geometric progression* is a sequence of numbers such that the ratio, r, of any term to its predecessor is the same; that is,

$$a, ar, ar^2, \ldots, l = ar^{n-1}.$$

A formula for the sum, G_n, of the first n terms of such a progression can be obtained as follows:

$$G_n = a + ar + ar^2 + \cdots + ar^{n-2} + ar^{n-1},$$

$$rG_n = ar + ar^2 + ar^3 + \cdots + ar^{n-1} + ar^n.$$

Subtracting gives

$$G_n - rG_n = a - ar^n.$$

Thus, $(1 - r)G_n = a(1 - r^n)$, and

$$G_n = \frac{a(1 - r^n)}{1 - r}. \tag{29}$$

Example 3

Find the fifth term and the sum of the first five terms of the geometric progression whose first two terms are 3 and 6. Here $r = 2$. Hence, the fifth term is $3 \cdot 2^4 = 48$, and the sum is

$$G_5 = \frac{3(1 - 2^5)}{1 - 2} = 3 \cdot 31 = 93.$$

Example 4

Rework Example 3 if the first two terms are 6 and -3. Here $r = -\frac{1}{2}$. The fifth term is $6(-\frac{1}{2})^4 = 6 \cdot \frac{1}{16} = \frac{3}{8}$.

$$G_5 = \frac{6\left[1 - \left(-\tfrac{1}{2}\right)^5\right]}{1 - \left(-\tfrac{1}{2}\right)} = \frac{6\left(1 + \tfrac{1}{32}\right)}{\tfrac{3}{2}} = 6 \cdot \frac{33}{32} \cdot \frac{2}{3} = \frac{33}{8}.$$

  If r is numerically greater than 1 (i.e., if $r > 1$ or $r < -1$), the terms increase in numerical value as n increases, as in Example 3. If r is numerically less than 1 (i.e., if $-1 < r < 1$), the terms decrease in numerical value, as in Example 4, and the expression r^n can be made as close to 0 (but never equal to 0) as we choose. The formula for G_n then approaches $a/(1 - r)$ as n increases without limit. We then write

$$G_\infty = \frac{a}{1 - r} \quad \text{(read "sum to infinity")}.$$

Example 5

For the progression $1, \frac{1}{3}, \frac{1}{9}, \frac{1}{27}, \ldots$,

$$G_\infty = \frac{1}{1 - \frac{1}{3}} = \frac{1}{\frac{2}{3}} = \frac{3}{2}.$$

Example 6

$0.333\ldots$ means $0.3 + 0.03 + 0.003 + \cdots$, a geometric progression with $r = 0.1$. So

$$G_\infty = \frac{0.3}{1 - 0.1} = \frac{0.3}{0.9} = \frac{1}{3}.$$

This means that as we take more 3's in the decimal we get closer and closer to the rational number $\frac{1}{3}$ (but never reach it with a finite number of terms). Similar treatment of any periodic decimal yields a rational number in fractional form.

The expression "increasing geometrically" means increasing like the terms in a geometric progression. Geometric progressions are basic to handling problems in investments at compound interest and in many scientific and practical problems.

PROBLEMS

1. The first two terms in a progression are 2 and 4. Find the sixth term and the sum of the first six terms (a) if the progression is arithmetic and (b) if the progression is geometric; also find G_∞, if it exists.
2. Rework Problem 1 if the first two terms are 4 and 2.
3. Rework Problem 1 if the first two terms are 2 and -4.
4. Rework Problem 1 if the first two terms are 4 and -2.
5. Rework Problem 1 if the first two terms are 1 and 3.
6. Rework Problem 1 if the first two terms are 3 and 1.
7. Find the fraction represented by the repeating decimal $0.18\dot{1}\dot{8}\ldots$.
8. A method basically equivalent to that described in Example 6 for finding the fraction represented by the repeating decimal $0.18\dot{1}\dot{8}\ldots$ is to set

$$N = 0.18\dot{1}\dot{8}\ldots.$$

Then $\qquad\qquad 100N = 18.18\dot{1}\dot{8}\ldots.$

So $\qquad 100N - N = 18.18\dot{1}\dot{8}\ldots - 0.18\dot{1}\dot{8}\ldots = 18.$

Complete the process by solving this equation for N. Compare with your answer to Problem 7.
9. Use the method outlined in Problem 8 to find the fraction represented by the following repeating decimals:
    (a) $0.36\dot{3}\dot{6}\ldots$ (b) $0.7\dot{7}\ldots$ (c) $0.189\dot{1}8\dot{9}\ldots$ (d) $0.74\dot{7}\dot{4}\ldots$

0.13   Logarithms

If $M = b^x$, $b > 0$, then x is called the logarithm (log for short) of M to the base b, written $\log_b M = x$. For example,

$$\text{because} \quad 8 = 2^3, \quad \log_2 8 = 3;$$
$$\text{because} \quad \sqrt{10} = 10^{1/2}, \quad \log_{10}\sqrt{10} = \tfrac{1}{2}.$$

Because logs are really exponents, their rules of operation are just restatements of those of exponents: Let $M = b^x$, $N = b^y$; then $\log_b M = x$, $\log_b N = y$; $MN = b^x \cdot b^y = b^{x+y}$; that is,

$$\log_b MN = x + y = \log_b M + \log_b N. \tag{30}$$

Similarly,

$$\log_b \frac{M}{N} = \log_b M - \log_b N, \tag{31}$$

and

$$\log_b M^n = n \log_b M. \tag{32}$$

Logs to base 10, called common logs, are helpful in calculations because our numerals are also based on 10. In the remainder of this section we shall use only logs to base 10 and shall not write the base; that is, by $\log M$ we shall mean $\log_{10} M$. We get the common logs of integral powers of 10 by translating from the exponential form:

$$
\begin{aligned}
0.01 &= 10^{-2} & \log 0.01 &= -2 \\
0.1 &= 10^{-1} & \log 0.1 &= -1 \\
1 &= 10^0 & \log 1 &= 0 \\
10 &= 10^1 & \log 10 &= 1 \\
100 &= 10^2 & \log 100 &= 2 \\
1000 &= 10^3 & \log 1000 &= 3 \\
& \text{etc.}
\end{aligned}
$$

Every number can be written as a number between 1 and 10 multiplied by an integral power of 10 (the standard scientific notation); for example,

$$
\begin{aligned}
0.0124 &= 1.24 \times 0.01 & = 1.24 \times 10^{-2}, \\
12.4 &= 1.24 \times 10 & = 1.24 \times 10^1, \\
12{,}400 &= 1.24 \times 10{,}000 & = 1.24 \times 10^4.
\end{aligned}
$$

Then, by (30),

$$
\begin{aligned}
\log 0.0124 &= \log 1.24 - 2, \\
\log 12.4 &= \log 1.24 + 1, \\
\log 12{,}400 &= \log 1.24 + 4.
\end{aligned}
$$

In every case the log of a number is the log of a number between 1 and 10 plus an integer (called the characteristic) that is exactly the exponent of 10 in the scientific notation. Tables give approximate values of the logs of numbers between 1 and 10. The tabular entries are numbers between 0 and 1 (a decimal point is understood before each entry). To find $\log 1.24$ from the table at the back of this book or in any four-place common log table, find 12 under N, then in this row find the entry under 4 at the top. There you find the entry 0934. This means that

$$\log 1.24 = 0.0934, \text{ approximately.}$$

Then

$$\log 12{,}400 = 4.0934 \quad \text{and} \quad \log 12.4 = 1.0934.$$

Similarly,

$$\log 0.0124 = 0.0934 - 2,$$

which is usually more convenient to use in the form $\log 0.0124 = 8.0934 - 10$.

The decimal part of the logarithm is called the *mantissa*, and it is desirable to keep the positive mantissa in evidence, rather than to carry out the indicated subtraction.

Example 1
Find $N = 758 \times 0.416$ if each factor is a three-significant-figure approximation.

$$\log 758 = 2.8797$$
$$+ \log 0.416 = 9.6191 - 10$$
$$\overline{\log N = 12.4988 - 10 = 2.4988.}$$

Find the entry in the table closest to the mantissa 0.4988. It is 4983, which is the log of 3.15. Hence, to three figures,

$$N = 3.15 \times 10^2 = 315.$$

Example 2
Find $(0.814)^5$.

$$\log 0.814 = 9.9106 - 10;$$
$$5 \log 0.814 = 49.5530 - 50 = 9.5530 - 10 = \log 0.357$$

Hence, $(0.814)^5 = 0.357$.

Example 3
Find $\sqrt[3]{100}$ to three significant figures.

$$\log 100 = 2.0000;$$
$$\tfrac{1}{3} \log 100 = 0.6667.$$

Hence, $\sqrt[3]{100} = 4.64$.

All computations with log tables are approximate, because the entries are themselves approximate. We should, therefore, really have used " ≈ " rather than " = " throughout these computations. Remember also that products, quotients, and powers should be given to no more significant figures than the least accurate of the given numbers. If the numbers involved in a computation are given to four figures, then with four-place log tables we can interpolate to get a fourth digit in the result. Interpolation will be explained in Chapter 1.

PROBLEMS

Compute, using the table of logs at the back of this book, assuming that all numbers are given to three significant figures:

1. 0.0791×8.17 2. $\dfrac{6.77}{25.4}$ 3. $\sqrt[4]{39.6}$ 4. $(1.06)^{12}$
5. Give the values of the following without use of tables:
    (a) $\log_2 32$ (b) $\log_7 \frac{1}{49}$ (c) $\log_3 81$ (d) $\log_4 2$
    (e) $\log_a a^4$ (f) $\log_{ab} a^3 + \log_{ab} b^3$ (g) $\log_4 64 \cdot \log_{64} 4$
★   (h) $\log_q p \cdot \log_p q$
6. Solve each of the following equations for x without use of tables:
    (a) $\log_{10} x = -3$ (b) $\log_5 x = 2$ (c) $\log_x \frac{1}{27} = -3$ (d) $\log_2 x = 0$
    (e) $\log_a 1 = x$ (f) $\log_x 8 = -3$ (g) $\log_a a^2 = x$
7. Simplify each of the following expressions:
    (a) $\log_{10}(x^2 - 4) + 2\log_{10} x - [\log_{10}(x - 2) + \log_{10} x^2]$
    (b) $3\log_p 2 + \log_p 9 - \log_p 144$
    (c) $\dfrac{\log_6 12 - \log_6 3}{\log_6 8}$
    (d) $\log_a(x^2 + 2x + 1) + 3\log_x x^2 - \log_a(x + 1)$

0.14   Keeping track of units

Operations in arithmetic and algebra are performed on numbers or on letters that stand for numbers. If we drive 50 miles per hour for 3 hours, the number of miles we travel is
$$50 \cdot 3 = 150;$$
or if we drive v miles per hour for t hours, the number of miles we travel is
$$s = v \cdot t.$$
Clearly, 50, 3, and 150 are numbers; also, v, t, and s represent numbers.

It is often helpful in solving a problem to keep track of the units associated with the numbers. We can do this by writing a schematic "equation" in the units and then putting in the appropriate numbers. For example, if we write miles per hour as miles/hours, we have
$$\frac{\text{miles}}{\text{hours}} \times \text{hours} = \text{miles}$$

29

to indicate the correct units. Then $50 \times 3 = 150$, the number of miles traveled.

Similarly, if we travel 150 miles at a uniform speed of 50 miles per hour, then to find the time for the trip, we write first the "equation" in the units:

$$\frac{\text{miles}}{\text{miles/hours}} = \text{hours}; \text{ then } \frac{150}{50} = 3,$$

the number of hours.

Here is still another example. Suppose that in a certain period, heat from the sun provides 75 calories per square inch on a field. What is the total number of calories per acre? The schematic "equation" in the units involved is

$$\frac{\text{calories}}{\text{in.}^2} \cdot \frac{\text{in.}^2}{\text{ft}^2} \cdot \frac{\text{ft}^2}{\text{acre}} = \frac{\text{calories}}{\text{acre}}.$$

Then the result is

$$75 \cdot 144 \cdot 43,560 = 470,448,000,$$

the number of calories per acre.

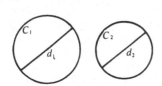

Fig. 0-6

PROBLEMS

1. (a) If the temperature in a room increases from 61° to 73° between 2 and 5 p.m., at what average rate does the temperature change during this period? (b) If the average rate of increase of the temperature in a room is 2° per hour, by how much does the temperature change in 4 hr? (c) If the average rate of increase of the temperature in a room is 3° per hour, how long does it take the temperature to increase by 24°?

2. If there are 640 acres in a square mile, and 5280 ft in a mile, how many square feet in an acre? How many square yards in an acre? If there are $5\frac{1}{2}$ yards in a rod, how many square rods per acre?

3. If a cubic foot of water weighs 62.4 lb, how many cubic yards of water weigh 1 ton (2000 lb)?

4. If a deck of cards is approximately $\frac{1}{2}$ in. thick, approximately how many miles high would a stack of a billion cards be? (Assume, for the approximation, that there are 50 cards in a deck, 10 in. in a foot, and 5000 ft in a mile.)

(a)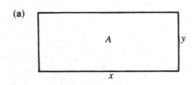

(b)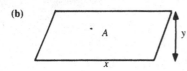

(c)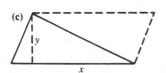

Fig. 0-7

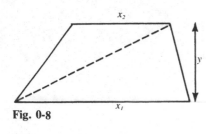

Fig. 0-8

0.15   Mensuration formulas

Because all circles are similar to each other (have the same shape), the ratio of the circumference of a circle to its diameter is the same for all circles; that is, $C_1/d_1 = C_2/d_2$ in Figure 0-6. This constant ratio is designated by π, and its value is approximately 3.14159. Note that because π is defined as the

30

ratio of two lengths, it is a "pure number" – it has no units associated with it. Thus, $C/d = \pi$, or $C = \pi d$, or

$$C = 2\pi r,$$

where the radius, r, is half the diameter. In this equation, the right side represents a length, and the left side does too.

The area (A units of length, squared) of a rectangle is given by

$$A = xy,$$

where x units of length and y units of length are the dimensions of the rectangle (Figure 0-7a). The same formula applies to a parallelogram, with x representing the base and y the altitude of the figure (Figure 0-7b). For a triangle, the formula is

$$A = \tfrac{1}{2}xy,$$

where x represents the base and y the altitude. This can be seen from the fact that two congruent triangles can be put together to form a parallelogram (Figure 0-7c). In each of these three cases, the right side of the equation is the product of two lengths (units of length, squared), and the left side has the same units.

A trapezoid with bases x_1 and x_2 and altitude y can be divided into two triangles (Figure 0-8), one having area $\tfrac{1}{2}x_1 y$ and the other having area $\tfrac{1}{2}x_2 y$. Hence, the area, A, of the trapezoid is given by

$$A = \tfrac{1}{2}(x_1 + x_2)y.$$

The areas of other polygons can be found by dividing the figures into triangles (Figure 0-9).

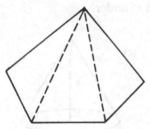

Fig. 0-9

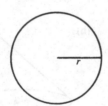

Fig. 0-10

The area of a circle (Figure 0-10) is given by

$$A = \pi r^2.$$

Note that the units of each side of this equation are (length)^2.

A prism (Figure 0-11a) and a cylinder (Figure 0-11b) have the same formula for volume (V units of length, cubed):

$$V = (\text{area of base}) \times \text{height}.$$

It makes no difference how many sides there are in the polygons forming the lower and upper bases of the prism, and the cylinder can be thought of as a prism whose bases are polygons with an infinite number of sides. Note that the units of both sides of the equation for the volume are (length)^3.

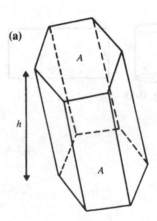

(a)

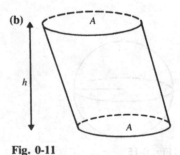

(b)

Fig. 0-11

31

0
Prerequisites

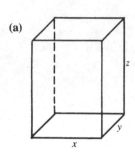

(a)

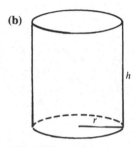

(b)

Fig. 0-12

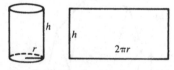

Fig. 0-13

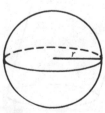

Fig. 0-15

Important special cases of this volume formula are illustrated in Figure 0-12, where part (a) shows a rectangular box, whose volume is given by

$$V_{\text{box}} = (\text{area of base}) \cdot z = xyz;$$

Figure 0-12(b) shows a right circular cylinder (bases are circles, and generators are perpendicular to the bases). The volume is given by $V = (\text{area of base}) \cdot h$, or

$$V_{\text{cylinder}} = \pi r^2 h.$$

The formula for the surface area of the rectangular box is clearly

$$\text{surface area of rectangular box} = 2xy + 2xz + 2yz.$$

The formula for the lateral surface area of the right circular cylinder (Figure 0-13) can be derived by visualizing the cylinder as formed from paper, cutting it along a generator, and unrolling the paper to form a rectangle of dimensions $2\pi r$ and h. Hence,

$$\text{lateral surface area of cylinder} = 2\pi rh.$$

The total surface area (including the top and bottom of the cylinder) is given by

$$\text{total surface area of cylinder} = 2\pi rh + 2\pi r^2.$$

Note that the units in both these area formulas are $(\text{length})^2$.

A pyramid (Figure 0-14a) and a cone (Figure 0-14b) have the same formula for volume:

$$V = \tfrac{1}{3}(\text{area of base}) \times \text{height},$$

$$= \tfrac{1}{3} \text{ volume of corresponding prism or cylinder}.$$

(a) 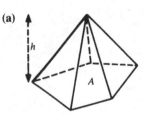 (b) 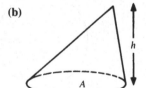 (c)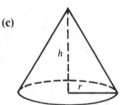

Fig. 0-14

For the special case of a right circular cone (Figure 0-14c),

$$V_{\text{cone}} = \tfrac{1}{3}\pi r^2 h.$$

Again, the units in these volume formulas are $(\text{length})^3$.

Finally, for a sphere (Figure 0-15), the surface area is

$$S_{\text{sphere}} = 4\pi r^2,$$

and the volume is

$$V_{\text{sphere}} = \tfrac{4}{3}\pi r^3.$$

32

Functional relationships

1

1.1 Introduction

Mathematics has traditionally been divided into three branches: algebra, geometry, and analysis. The first two of these are somewhat familiar to you from earlier courses; analysis may be briefly described as the branch of mathematics dealing with *limiting processes*. The term *calculus* is applied to the beginning portions of analysis.

You will recognize immediately that the three branches overlap. Your experience with graphs, for example, involved a certain blending of geometrical and algebraic ideas. The definition of the circumference of a circle, in terms of the perimeters of polygons with ever-increasing numbers of sides (Figure 1-1), exemplifies the necessary application of notions from analysis to topics in geometry. Similarly, the algebraic-appearing expression $S = 1 + \frac{1}{2} + \frac{1}{4} + \frac{1}{8} + \frac{1}{16} + \cdots$ (where "$\cdots$" means "continue indefinitely") depends for its very *meaning* on the concept of *limit*, as does the equation $\frac{1}{3} = 0.3333\ldots.$

This course will deal, to some extent, with all three branches, but principal emphasis will be placed on calculus.

Fig. 1-1

Problem 1

What do you think is an appropriate value of S? Here is an argument justifying the answer: What is the sum of the first two terms on the right? The first three terms? The first four terms? In each case, what is the difference between 2 and the sum? Can you continue the argument?

1.2 An example

We can obtain some indication of the subject matter and the type of problems to be studied by considering the following example. A colony of a certain bacterial strain is observed in a petri dish over a number of hours. In particular, the size of the colony – measured as the area (in square

33

1
Functional relationships

Table 1-1

t	A
0	4.4
1	6.8
2	10.2
3	14.4
4	19.2
5	24.2
6	28.6
7	32.2
8	34.8
9	36.7
10	38.0

millimeters) that it occupies on the surface of the nutrient medium – is recorded at the end of each hour after the start of the experiment. At the beginning, the size of the colony is 4.4 mm^2, and for various values of t (hours after the start), the area A (mm^2) is shown in Table 1-1.

Problem 1

Before reading further, answer these questions: (a) What interesting or useful information can be obtained from this table? (b) What further information about the size of the colony would be significant?

There are several basic questions that are suggested by Table 1-1.

I. *How* does A vary with t? That is, what values of A correspond to values of t that do not appear in the table? What happens to A after $t = 10$?

II. *How fast* does A change with t? That is, what is the average rate of increase of A in various time intervals? What is the rate of increase of A at various instants?

III. Is there a *maximum* value of A? Or, if A does not attain a maximum value, is there a value that A does not exceed? (Such a value is called an *upper bound* of A.) Is there a *minimum*? A *lower bound*?

These questions have both mathematical and scientific importance. It is often necessary to *interpolate between* known values to determine how a quantity behaves. In this instance it is particularly desirable to be able also to *extrapolate beyond* tabulated values to predict how the bacterial colony will behave in the future. In this connection, the rate of increase of A is significant: In the first hour the area increases by 2.4 mm^2; in the fifth hour it increases by 5.0 mm^2; in the tenth hour it increases by 1.3 mm^2. In other words, after a period during which the rate of change of A grows to be quite large, we see that the rate of change "tapers off," leading us to predict that A probably will never exceed 40 or so. (In some situations of population growth, this tapering off in rate of increase might result from exhaustion of the food supply; in this case, it is probably a consequence of the inhibiting effect of the accumulation of metabolites, the chemical products of the growth activity.)

We shall devote the remainder of this chapter to the foregoing question I and its mathematical ramifications. The remaining questions will be treated in subsequent chapters.

1.3 Variation of one quantity with another; graphical interpolation

Graphing is an appealing means of displaying the relationship between two quantities, because we can literally see what is going on. As in Figure 1-2, we choose an appropriate scale on each of two perpendicular lines and proceed to plot the pairs of numbers in Table 1-1, finally assuming that the variation of A with t is represented by a smooth curve through these points.

Some practical hints on sketching: Turn the paper, if necessary, to keep your hand on the concave side of the curve; sketch relatively short arcs at first, with light pencil strokes; then "fair in" the entire curve to obtain a smooth curve. For increased accuracy, draw as large a graph as practicable.

We can now interpolate from the graph, obtaining such results as the following: If $t = 1.7$, A is approximately 9.0; if $t = 4.7$, A is approximately 22.8; if $A = 29.5$, t is approximately 6.3. We can also extrapolate beyond the points in the table (see the dotted portion of the graph) to estimate that if $t = 10.3$, A is approximately 38.3; if $t = 11.0$, A is approximately 38.8.

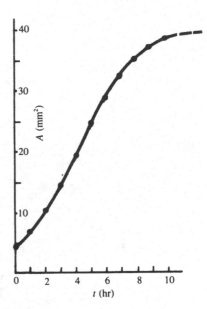

Fig. 1-2

Problem 1

What is your feeling about the accuracy, in general, of such extrapolation as compared with interpolation? Can you give reasons for that feeling?

The assumption "that the variation of A with t is represented by a smooth curve" through the points obtained from the table has far-reaching theoretical and practical importance. After all, the only *explicit* information we have is contained in the pairs of numbers in Table 1-1. Anything more comes from knowledge of or guesses about the behavior of A. Is there a value of A corresponding to every t between 0 and 10? Suppose that we select any real number, R, between the smallest and largest values of A in the table (4.4 and 38.0). Is there some time at which the value of A equals R? We are answering yes to both these questions when we draw a smooth curve.

Problem 2

Suppose that a table gives the number of undergraduates enrolled in your school at various times. Would it make sense to draw a smooth curve through the points plotted from the table? Discuss.

Problem 3

For data like those of Table 1-1, some people tend to join pairs of adjacent points by straight-line segments rather than by a smooth curve. What would this imply about the variation?

You should realize that there is no *single* smooth curve through a finite set of points – indeed, there are always infinitely many such curves – so we cannot expect a unique result. We may make serious mistakes if we have insufficient data. For example, plotting y, the number of counts per minute of protons and electrons, against x, the number of thousands of kilometers from the earth at which the radiation is measured, gives rise to Figure 1-3(a), which might be extended to the smooth graph shown in Figure 1-3(b). Further data, however, show the situation actually to be as in Figure

35

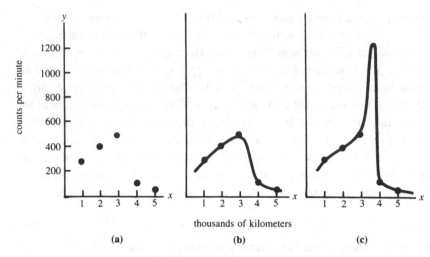

thousands of kilometers

(a)   (b)   (c)

Fig. 1-3

1-3(c). The existence of a peak between 3000 and 4000 km from the earth (the Van Allen belt) is unsuspected from the five points plotted in Figure 1-3(a).

Table 1-2

t	y
0	48
1	120
2	160
3	168
4	144
5	88
6	0

Table 1-3

t	v
0	60
1	38.2
2	16.4
3	−5.5
4	−27.3
5	−49.1
6	−70.9

PROBLEMS

○ 4. A toy rocket at the edge of a flat roof 48 ft above the ground is propelled vertically upward by a compressed spring that gives it an initial velocity of 60 mph. The rocket will reach a height of y ft in t seconds, as shown in Table 1-2 and will hit the ground (provided it just misses the roof on the way down) when $t = 6$. Choose a scale on each of two perpendicular lines, plot the points of Table 1-2, sketch a smooth curve through them, and estimate

(a) y, corresponding to $t = 1.4$;

(b) y, corresponding to $t = 3.3$;

(c) t, corresponding to $y = 144$ (just one answer?);

(d) the number t for which the rocket is highest, and what the maximum height is.

Explain why a smooth graph of the data in Table 1-2 is justified.

Table 1-3 shows the velocity (v mph) of the rocket at time t (sec). Negative values mean that the rocket is moving downward.

(e) Plot the points corresponding to this table and sketch a smooth curve through them.

(f) From your graph, estimate the number t corresponding to $v = 0$.

(g) Relate your answers to (d) and (f), and note the relationship of zero velocity to maximum height.

○ 5. The temperature ($T°$ Celsius) of a pan of water varies with the time (t min) after it has been put on the stove, as in Table 1-4.

    (a) Plot a graph to show how T varies with t.

    (b) What T corresponds to $t = 8$? To $t = 11$?

    (c) What t's correspond to $T = 100$? (Try to represent all of them.)

    (d) If the table had stopped with the pair $(15, 94)$, what value of T would you have extrapolated for $t = 20$?

    (e) What physical principles are involved in the variation exhibited in Table 1-4? How did you apply them in drawing the graph in the interval $t = 15$ to $t = 20$?

★ 6. Try to represent the variation in Problem 5 by means of one or more formulas.

Table 1-4

t	T
0	49
5	64
10	79
15	94
20	100
25	100
30	100

1.4 More on graphing, interpolation, and extrapolation

In some instances there may be no significance to values intermediate to those listed in a table. Consider, for example, the number (N) of rooms at various prices (p dollars) in a city hotel. The values in a table might be plotted to give the points shown in Figure 1-4, but there would be no significance to a smooth curve drawn through these points. In such cases we sometimes join the points by straight-line segments to carry the eye from point to point. Indeed, it is likely that the information about numbers of rooms at various prices might be pictured in a different way, such as in a bar graph (Figure 1-4′).

It is obvious that drawing a figure and reading distances from it can give only approximate results. But beyond that, if we are dealing with physical measurements, there are inaccuracies inherent in the data, because any measuring instrument can indicate only that a value lies within a certain interval. For this reason, plotting the tabulated values is unlikely to give points all lying neatly on a smooth curve (as in Figure 1-2), but more likely clustering near such a curve. What curve to draw, then, is a matter of judgment, and of knowledge of the quantity being studied. In particular, if the points seem to be reasonably close to a straight line, and especially if we have a theoretical reason for expecting one, we probably will draw a straight line with some points on each side of it, not necessarily through any.

Data based on counting or arithmetic processes (as for Figure 1-4) are not subject to the same kind of experimental error, but of course mistakes can be made. In any case, if one pair of values deviates markedly from the pattern suggested by the others, we can suspect a gross error. For example, if in Table 1-1 the pair $t = 5$, $A = 27.2$ had occurred, we probably would have discarded it. However, the pair might be correct, indicating some unexpected phenomenon. Here the insight (and luck!) of the investigator plays a role. Remember Figure 1-3.

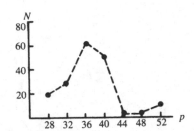

Fig. 1-4

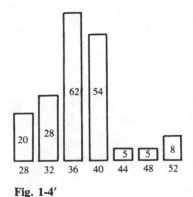

Fig. 1-4′

37

1
Functional relationships

Table 1-5

t	N
0	2.5
5	4.0
10	6.5
15	10.4

A principal concern of human beings in general, and of business people in particular, is to attempt to predict the future. Hence, extrapolation from sets of data is of great interest, but often unreliable, on account of unforeseen factors that keep things from going on as they did in the past. This is illustrated by the following problem.

Problem 1

The number of overseas telephone calls originating in the United States (N million per year) varied with the time (t years after 1950) as in Table 1-5. Plot these data and extrapolate from the graph to get N for 1970 and for 1975. Also interpolate from the graph to find when N equaled 8.

The results of your interpolation are doubtless in good agreement with the historical situation, but the predictions are far from the mark. From the graph, you probably read N as about 16.6 in 1970 and about 26.6 in 1975. But in 1970, N was actually 21, and by 1972, N was already 45 (*New York Times*, June 9, 1974, p. 6E).

Problem 2

Can you suggest an explanation for this large discrepancy between prediction and actuality?

The following amusing example may be instructive.

Rate of industrialization. Buckminster Fuller, in *World* magazine of July 3, 1973, made the observations shown in Table 1-6. On the basis of the table, he predicted in 1947 that China (4) would industrialize in 25 years – which, he says, happened. When he wrote the magazine article, he predicted that India, Africa, and South America (5) would be industrialized by 1985 (i.e., in $12\frac{1}{2}$ years). His argument was that each geographical area builds on the know-how of predecessors, so as to lead to ever shorter time intervals for industrialization.

Roderick L. Hall, in a letter to *World* magazine, published in the issue of August 28, 1973, with tongue somewhat in cheek, made the observations shown in Figure 1-5. Observing that these points lie nearly on a straight line, he extrapolated to conclude that "by waiting until the year 2000, nations will be developing overnight – literally." In other words, he suggested that the *date of beginning of industrialization* is of prime importance, whereas Fuller put his main focus on the number of areas that have already been industrialized. Clearly, there are many other factors that also play roles.

Note, by the way, the plot of Fuller's data in Figure 1-6. This "die-away" curve never reaches the horizontal axis, although, if there were enough new geographical areas to "develop," the time interval for the later ones would become very short, if not literally "overnight." In this case, there is no

Table 1-6

A: Geographical area	I: Interval from beginning of significant industrialization to achievement of "full" industrialization
(1) Europe	200 years
(2) United States	100 years
(3) Russia	50 years

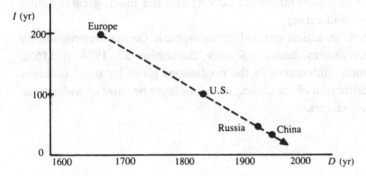

Fig. 1-5

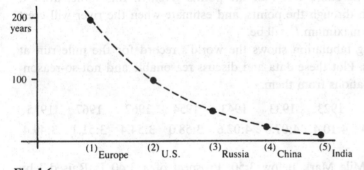

Fig. 1-6

Table 1-7

A: Geographical area	D: Date of beginning of significant industry	I
(1) Europe	1670	200 (i.e., full industrialization by 1870)
(2) U.S.	1835	100 (i.e., full industrialization by 1935)
(3) Russia	1917	50 (i.e., full industrialization by 1967)
(4) China	1947	25 (i.e., full industrialization by 1972)

1

Functional relationships

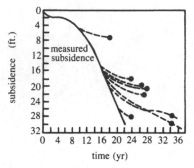

Fig. 1-7 Predictions made for the subsidence at Long Beach, California, associated with oil field pumping in the Wilmington oil fields.

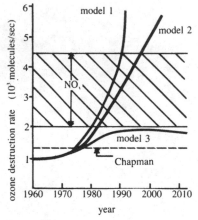

Fig. 1-8

Table 1-8

t	L
0	0.5
1	2.5
2	4.3
3	5.9
4	7.3
5	8.5
6	9.5
7	10.3

significance to values of A other than $1, 2, 3, 4, \ldots$; there is no area "between" Europe and the United States.

We have here a highly questionable use of mathematics. Building a table of pairs of numbers in a meaningful way is the first step in applying mathematics to a particular problem. One may ask if either Fuller's serious approach or Hall's facetious treatment does this.

Here are two dramatic examples of the difficulties associated with important problems of prediction. Figure 1-7, from an article entitled "Subsidence of Venice: Predictive Difficulties" (*Science*, September 27, 1974, p. 1185), shows the large differences between various predictions of subsidence at Long Beach, Calif. (several dotted curves) and the much greater actual subsidence there (solid curve).

Figure 1-8, from an article entitled "Stratospheric Ozone Destruction by Man-Made Chlorofluormethanes" (*Science*, September 27, 1974, p. 1165), shows the enormous differences in the predictions given by three different models of the destruction of the protective ozone layer because of widespread use of aerosol propellants.

PROBLEMS

○ 3. During the spring runoff, the levels (L ft) of the Connecticut River above a certain stage at various times (t hr) after midnight Sunday were recorded, as in Table 1-8. Plot the points given in this table, draw a smooth graph through the points, and estimate when the river will crest and what the maximum L will be.

4. The following tabulation shows the world's record for the mile run at various dates. Plot these data and discuss reasonable and not-so-reasonable extrapolations from them.

Date	1913	1923	1933	1943	1954	1962	1967	1975
Time	4:14.4	4:10.4	4:07.6	4:02.6	3:58.0	3:54.4	3:51.1	3:49.4

(See "With Mile Mark Below 3:50, Prospect of a 3:40 Is Raised," by James O. Dunaway, on p. 6 of the Sports Section of the *New York Times*, August 17, 1975. See also "The First 4-Minute Mile, 25 Years Later," on p. 1 of the Sports Section of the same paper, May 6, 1979.)

5. The pressure (p lb/in.^{2}) in a pump cylinder was read from a gauge at hourly intervals, giving Table 1-9. One of the values for p seems to be in error. Find it by drawing a graph, and correct the error.

6. The rate of flow (v cm/sec) in a blood vessel was found to vary with the distance (x mm) from the center of the vessel, as in Table 1-10. Plot the points corresponding to this table, draw the most reasonable smooth graph through these points, and use the graph to estimate the value of x at which $v = 2.2$.

* 7. Find a formula for v in terms of x corresponding to the graph you drew in Problem 6.

○ 8. A sausage manufacturer has facilities for producing a limited quantity of specialty smoked sausage. There is a fixed daily cost of the operation of $128, whether or not any sausage is made. Beyond that, the costs rise with the output, mainly because of raw material, energy, and labor costs. By carefully calculating and averaging costs over a number of days of operations at a given output level, the manufacturer determines that the total daily operating expense E (in dollars) corresponding to an output x (in pounds) is as given in Table 1-11. Draw a graph of these data and a smooth curve indicating the value of E for each x between 0 and 100. Extend this curve in order to estimate E corresponding to $x = 120$ and $x = 140$. Would you believe, on the basis either of Table 1-11 or of the nature of the manufacturing process, that there is an upper bound to the value of E?

* 9. Assume that the numbers of overseas calls (N million per year) varied with the time (t years after 1950) as in Table 1-12. Obtain a reasonable formula for this variation.

1.5 Linear interpolation

A straight line is characterized by the fact that it rises (or falls) at the same rate everywhere; in other words, corresponding changes in the two quantities represented are proportional. This is shown geometrically in Figure 1-9. If P, T, and R are three points on a straight line, if PQ is parallel to the x axis, and if ST and QR are parallel to the y axis, then triangles PST and PQR are similar. Hence, corresponding sides are proportional, so $PS/PQ = ST/QR$.

Suppose that P and R are obtained from tabulated values as indicated

$$3\left[1\left\{\begin{matrix}2\\3\\5\end{matrix}\right.\quad\begin{matrix}4\\?\\6\end{matrix}\right\}ST\right]2$$

and that we want to find y for $x = 3$. We have $PS = 3 - 2 = 1$, $PQ = 5 - 2 = 3$, $QR = 6 - 4 = 2$. Then $\frac{1}{3} = ST/2$, whence $3 \cdot ST = 2$, and $ST = \frac{2}{3}$. Hence, $VT = VS + ST = 4 + \frac{2}{3} = \frac{14}{3}$. Clearly the computation for this interpolation could have been done without drawing the line at all.

For a nonlinear relation, in an interval where the graph is nearly straight, *linear interpolation* (i.e., interpolating as if the graph were really straight) may give a satisfactory approximation. For example, in Table 1-1, to find A

Table 1-9		Table 1-10	
t	p	x	v
0	60.1	0	4.01
1	29.8	0.2	3.61
2	19.9	0.4	3.18
3	15.1	0.6	2.82
4	12.9	0.8	2.40
5	10.1	1.0	2.01
6	8.6	1.2	1.58
7	7.5	1.4	1.19
8	6.7	1.6	0.81

Table 1-11		Table 1-12	
x	E	t	N
0	128	0	4
20	180	5	6
40	225	10	9
60	260	15	13.5
80	290	20	20.25
100	310		

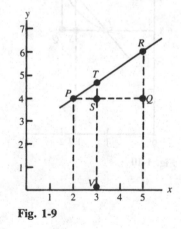

Fig. 1-9

41

for $t = 1.7$, we proceed as follows: The relevant portion of Table 1-1 reads

t	A
1	6.8
2	10.2

and we write

$$
\begin{array}{c|c}
t & A \\
\hline
1 \left[0.7 \left\{ \begin{array}{c} 1 \\ 1.7 \\ 2 \end{array} \right. \right. & \left. \left. \begin{array}{c} 6.8 \\ A \\ 10.2 \end{array} \right\} A - 6.8 \right] 3.4
\end{array}
$$

We argue that

$$
\frac{0.7}{1} = \frac{A - 6.8}{3.4}.
$$

This leads to $A = 9.18$, which we round to 9.2, because the tabulated values are themselves measurements rounded to the nearest tenth.

Problem 1
Do the algebra and arithmetic that result in this value of A.

Problem 2
In this case, the interpolated value, 9.2, is greater than 9.0, the result obtained by graphical interpolation in Section 1.3. Explain by reference to the graph (Figure 1-10) why this should be expected.

Similarly, but working the other way, we estimate the value of t corresponding to $A = 29.5$ as follows: The relevant portion of Table 1-1 reads

t	A
6	28.6
7	32.2

and we write

$$
\begin{array}{c|c}
t & A \\
\hline
1 \left[t - 6 \left\{ \begin{array}{c} 6 \\ t \\ 7 \end{array} \right. \right. & \left. \left. \begin{array}{c} 28.6 \\ 29.5 \\ 32.2 \end{array} \right\} 0.9 \right] 3.6
\end{array}
$$

so

$$
\frac{t - 6}{1} = \frac{0.9}{3.6} = \frac{1}{4}.
$$

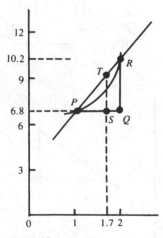

Fig. 1-10

Hence, $t = 6.25$, which we round to 6.3.

Problem 3
Once again, do the algebra and the arithmetic.

Problem 4
Figure 1-11 shows a portion of Figure 1-2, the graph corresponding to Table 1-1. Is the result we have found by linear interpolation ($t = 6.3$) larger or smaller than the value for t we would find from the smooth graph?

It is difficult to know how accurate the results of linear interpolation are. One needs to keep in mind that the approximation to the true value will be good only if the graph is "nearly straight" in the interval used. Surely this is more likely to be true in small intervals than in large. The method is particularly useful in tables of roots, logarithms, trigonometric functions, and the like, where values are given at small intervals, and where computation of the "true" value is not feasible.

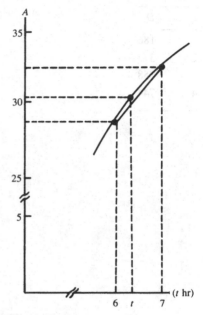

Fig. 1-11

PROBLEMS

5. Use linear interpolation with Table 1-1, **1.2** to estimate the value of
   (a) A, corresponding to $t = 8.3$,
   (b) t, corresponding to $A = 18$.

○ 6. Use linear interpolation with Table 1-2, **1.3**, to estimate the value of
   (a) y, corresponding to $t = 1.4$,
   (b) y, corresponding to $t = 3.3$ (keep your plus and minus signs straight!),
   (c) t, corresponding to $y = 144$ (just one answer?).
   Compare with your graphical results in Problem 4, **1.3**. Which do you think are more reliable?

7. Use linear interpolation with these data (taken from a table of cube roots) to estimate

		x	$\sqrt[3]{x}$
(a)	$\sqrt[3]{10.27}$,	10.2	2.1687
(b)	$\sqrt[3]{10.32}$,	10.3	2.1758
(c)	x, if $\sqrt[3]{x} = 2.17$.	10.4	2.1828

8. Use linear interpolation with these data (taken from a table of common logarithms) to estimate

		x	$\log x$
(a)	$\log 25.59$,	25.5	1.40654
(b)	$\log 25.64$,	25.6	1.40824
(c)	x, if $\log x = 1.407$.	25.7	1.40993

43

Table 1-13

t	D
1	186
2	372
3	558
4	744

9. The distance (*D* thousand miles) of a light signal from its source after *t* sec is given in Table 1-13. What value of *D* corresponds to *t* = 2.8? Do you have any hesitancy about using linear interpolation in this case?

10. Make a sketch similar to Figure 1-10 of the "top" portion of Figure 1-2 with *P* the point for *t* = 9, *R* the point for *t* = 10, and *T* the point on *PR* produced for which *t* = 10.3. With this sketch as a guide, use linear extrapolation to obtain an estimate of the value of *A* corresponding to *t* = 10.3. Compare this estimate with the estimate from graphical extrapolation in Section 1.3.

11. As in Problem 10, use linear extrapolation with Table 1-8, **1.4**, to obtain an estimate of the value of *L* corresponding to *t* = 9.7. Compare this estimate with your answer to Problem 3, **1.4**.

1.6 Relations expressed by formulas

A formula that expresses the variation of one quantity with another gives theoretically complete information about that variation. If the formula is not too complicated, we can dispense with graphs, methods of approximate interpolation, and so forth, and work only with the formula. On the other hand, if we want to make the relationship "graphic," we can get as extensive a table as we want by use of the formula. It may be difficult to find a formula to fit given data, especially in the social and life sciences, but it is usually worthwhile to try.

Example 1
Variation expressed by a polynomial formula

In the case of the rocket in Problem 4, **1.3**, physical principles (which we shall discuss later) lead to the equation

$$y = -16t^2 + 88t + 48. \tag{1}$$

Problem 1
Check that this formula does indeed yield Table 1-2, **1.3**.

Now interpolation is straightforward:

$$\text{at } t = 1.4, y = -16(1.4)^2 + 88(1.4) + 48 = 139.84.$$

Problem 2
Compare this value with those obtained in Problem 4(a), **1.3**, and Problem 6(a), **1.5**, and explain any differences.

Equation (1) suggests the possibility of considering all positive and negative numbers, and zero, as values of *t*. But in our problem, *t* was

44

restricted: The rocket began its flight at $t = 0$, and it hit the ground at $t = 6$. Thus, the complete formula is

$$y = -16t^2 + 88t + 48, \quad 0 \le t \le 6. \tag{2}$$

This can be put, "y is given by the equation $y = -16t^2 + 88t + 48$, with t taking on all values between 0 and 6, both 0 and 6 included."

The right side of equation (1) is called a *quadratic polynomial* in t. The general quadratic (i.e., second-degree) polynomial in x can be written

$$y = ax^2 + bx + c, \quad \text{with } a \ne 0.$$

The general linear (first-degree) polynomial is

$$ax + b, \quad \text{with } a \ne 0;$$

the general cubic (third-degree) polynomial is

$$ax^3 + bx^2 + cx + d, \quad \text{with } a \ne 0, \quad \text{etc.}$$

The general polynomial of the nth degree is

$$a_n x^n + a_{n-1} x^{n-1} + \cdots + a_1 x + a_0,$$

where $a_n, a_{n-1}, \cdots, a_1, a_0$ are constants, and $a_n \ne 0$.

Variation that can be expressed through a polynomial formula can be thoroughly analyzed, as we shall see later.

Example 2
Variation given by different formulas in different regions of the independent variable

In Problem 5, **1.3**, the graph of Table 1-4 perhaps looks like Figure 1-12 ("perhaps," because the table gives no information about the crucial period between $t = 15$ and $t = 20$). If the graph is correct, it is clear that its horizontal section can be represented by $T = 100$. For the sloping portion of the graph, we note that T increases by 15 for each increase of 5 in t. Because we are assuming a uniform rate of increase for T, this implies that T increases by 3 for each increase of 1 in t. Putting this result together with the observation that $T = 49$ at $t = 0$, we conclude that $T = 3t + 49$ for the sloping portion of the graph.

To find the value of t corresponding to the point where the sloping and horizontal segments join, we set $3t + 49 = 100$, giving $t = 17$. Thus, we have the following formula for the graph:

$$T = \begin{cases} 3t + 49, & 0 \le t \le 17, \\ 100, & 17 < t \le 30. \end{cases} \tag{3}$$

It makes no essential difference whether the value 17 is assigned to t with the first equation or the second, because both expressions yield the value 100 for $t = 17$.

In either case, we use two formulas to describe how T varies with t. It might be possible to find a single formula to do the job, but it probably

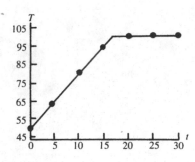

Fig. 1-12

would be more cumbersome and might not exhibit the variation as clearly. It is not uncommon to use more than one equation to express variation.

Example 3

Exponential variation

An important type of variation is illustrated by the growth of overseas telephone calls with time, as set out in Problem 1 and its accompanying Table 1-5, **1.4**. We observe that at $t = 0$, $N = 2.5$; at the end of one 5-year period, $N = 4.0 = (2.5)(1.6)$; at the end of two 5-year periods, $N = 6.5 =$ (approx.)$(4.0)(1.6) = (2.5)(1.6)^2$; and at the end of three 5-year periods, $N = 10.4 = 6.5(1.6) = (4.0)(1.6)^2 = (2.5)(1.6)^3$. If we let x equal the number of 5-year periods elapsed since 1950, we then have

$$N = (2.5)(1.6)^x, \quad \text{for } x = 1, 2, 3.$$

Indeed, because $a^0 = 1$ for all $a \neq 0$ (if this seems unfamiliar, see Chapter 0), the formula gives, for $t = 0$, $N = (2.5)(1.6)^0 = 2.5$, which, of course, is also in agreement with the table.

If the variation continued in this fashion, we would have by the end of the twentieth century ($t = 50$, or $x = 10$), $N = (2.5)(1.6)^{10}$, which is approximately 275. As was remarked after Problem 1, **1.4**, N grows much more rapidly than this prediction (because of the price reduction permitted by the introduction of communications satellites in 1965), and it is estimated that by 1985 (long before the end of the twentieth century) N will be approximately 2000; that is, about 2 billion overseas telephone calls will originate in the United States in that year. But even the more modest prediction of our formula gives, for the end of the twenty-first century ($t = 150$, or $x = 30$), $N = (2.5)(1.6)^{30}$, which is approximately 3,500,000. This means 3.5 trillion overseas telephone calls that year!

Such extraordinary growth is typical of *exponential functions* – those given by formulas like

$$y = c \cdot b^x.$$

Whether or not the formula $N = (2.5)(1.6)^x$ is a good predictor, it has a clear *mathematical meaning* for all nonnegative integral (whole-number) values of x. We can indicate this symbolically by writing

$$N = (2.5)(1.6)^x, \quad x = 0, 1, 2, 3, \ldots.$$

The formula is also meaningful for many other values of x. For example, for $x = \frac{2}{5}$,

$$N = (2.5)(1.6)^{2/5} = (2.5)\sqrt[5]{(1.6)^2}.$$

(By use of a calculator, or by logarithms, we find this number to be approximately 3.02, which, incidentally, is probably a good estimate of the number of millions of overseas telephone calls in 1952 – compare this with your answer to Problem 1, **1.4**.)

An interpretation similar to that just given for $x = \frac{2}{3}$ is possible for any rational x, but if x is an irrational number, like $\sqrt{3}$ or π, the definition of $(1.6)^x$ requires the concept of limit, which comes later.

Example 4

Straight lines and linear equations

The relationship between a temperature on the Celsius scale (C degrees) and the same temperature on the Fahrenheit scale (F degrees) is shown in Table 1-14. If we plot these data, the points appear to lie on a straight line (Figure 1-13). Assuming that they do, what is the formula expressing F in terms of C?

We shall solve the problem in two ways. First, we follow the method used in finding the equation corresponding to the sloping line in the example of the heated pan of water (Example 2 in this section): We observe that, for every increase of 10 in C there is an increase of 18 in F, or, because the rate of increase is uniform, an increase of $\frac{18}{10} = \frac{9}{5}$ in F for each increase of 1 in C. Moreover, at C = 0, F = 32. Putting these statements together gives $F = \frac{9}{5}C + 32$.

The second method consists in noting that, for straight-line variation (a uniform rate of change), the formula must be of the type $F = aC + b$, and our problem is to determine the constants a and b. This can be done by substituting any two pairs of values of C and F from the table into this equation, say

$$50 = 10a + b, \text{ and}$$

$$86 = 30a + b.$$

Problem 3

Solve this pair of equations to obtain $a = \frac{9}{5}$, $b = 32$, thus obtaining the same result as by the first method. Choose two other pairs from the table and verify that they lead to the same values for a and b.

Example 5

The power law

If the relationship between two quantities, x and y, is such that *y is proportional to x*, then an equation expressing the relationship is $y = kx$, for some constant k. If the value of y corresponding to some value of x is known, we can determine k. Another phrase for this variation is "*y varies as x.*" Note that "varies as" implies a specific kind of variation, whereas the phrase "*y varies with x*" means simply that there is some relationship (unspecified) between y and x.

Similarly, to say that *y is proportional to x^2, or y varies as x^2*, is equivalent to $y = cx^2$, for some constant c. Likewise for any positive power of x. The power of x does not have to be a whole number. For example, one of

Table 1-14

C	F
−20	−4
−10	14
0	32
10	50
20	68
30	86
100	212

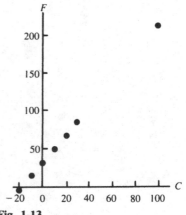

Fig. 1-13

47

Kepler's laws states that the periods of revolution (T years) of our planets are proportional to the $\frac{3}{2}$ power of the semimajor axes (x miles) of their orbits (i.e., $T = kx^{3/2}$).

If y varies as the inverse (or reciprocal) of x, the formula is $y = k(1/x) = k/x$, and we say that y is inversely proportional to x. If $y = k/x^3$, y is inversely proportional to x^3. Formulas of the type $y = k/x^2$ (the "inverse square law") describe variations that occur in many different contexts (e.g., in Newton's law of universal gravitation). By way of contrast with inverse proportionality, and for the sake of emphasis, the formula $y = cx^2$ is sometimes phrased "y is *directly* proportional to x^2," or "y varies *directly* as x^2."

All the foregoing can be subsumed under the one formula

$$y = kx^n$$

for some constants k and n, in which n can be any real number, positive or negative (or even zero, but that doesn't give very interesting variation!). This general formula, called the *power law*, appears in a host of applications.

For example, suppose that we know that the force of attraction (F dynes) between any two particles with unlike electric charges *varies inversely as the square* (second power) of the distance (x cm) between them. This means that $F = k/x^2 = k \cdot x^{-2}$, for some k. If we know that, for two specific particles, the force of attraction between them is 150 dynes when the particles are 2 cm apart, we can write $150 = k/4$, so $k = 600$, and the formula in this case is $F = 600/x^2$. Thus, if these particles are placed 10 cm apart, $F = 600/100 = 6$ – the force of attraction is 6 dynes.

It should be noted that a formula is not always helpful in determining values of one of two related quantities. In the problem of Section 1.5, for example, we had delightfully compact equations: $y = \sqrt[3]{x}$, $z = \log x$. These equations are useful, but it is not simple to determine from them what values of y or z correspond to certain values of x. Hence, we often have to resort to tables or other computational devices.

PROBLEMS

4. Use formula (2), **1.6**, to determine the value (or values) of
   (a) y corresponding to $t = 2.8$,
   (b) y corresponding to $t = 3.3$,
   (c) t corresponding to $y = 144$,
   (d) t corresponding to $y = 0$,
   (e) t corresponding to $y = 169$.
   Compare your answers to (b), (c), and (e) with your graphical results in Problem 4, **1.3**. Also compare your answers to (b) and (c) with the results of linear interpolation in Problem 6, **1.5**.

5. (a) Would any of your answers in Problem 4 be changed if equation (1), 1.6, were used, with no restriction on values of t?

   (b) Suggest a modification of the rocket flight experiment (Problem 4, 1.3) to make sense of a value of t less than 0 or greater than 6.

6. Find an equation of the straight line

   (a) passing through $(-1, 4)$ and $(3, 12)$,

   (b) passing through $(5, 1)$ and $(13, -5)$,

   (c) passing through $(-6, 7)$ and $(0, 7)$,

   (d) passing through $(0, 0)$ and $(1, c)$ for some constant c,

   (e) passing through $(-2, -3)$ and rising uniformly 2 units for every increase of 1 unit horizontally,

   (f) passing through $(1, -1)$ and falling uniformly 3 units for every increase of 4 units horizontally,

   (g) passing through $(-2, 0)$ and $(0, 2)$,

   (h) passing through $(0, 0)$ and rising uniformly $\frac{2}{3}$ unit for every increase of 1 unit horizontally,

   (i) passing through $(1, 0)$ and rising uniformly a units for every increase of b units horizontally, for some constants a and b with $b \neq 0$.

○ 7. To help fix in mind the distinction between the power law and exponential functions, fill in Table 1-15. Use the entries of your table to sketch smooth graphs of $y = x^2$ and $y = 2^x$ on the same set of axes.

8. Ralph Wreckless, European sportsman, drives so that his speedometer always reads 10 kilometers per hour above the posted speed limit. He does this on the basis of two assumptions: (i) His speedometer reads high by 10% of the speedometer reading (e.g., when the speedometer reads 70, he is actually traveling at 63 kph). (ii) The police will not mind if he travels a bit over the limit. Complete Table 1-16 using assumption (i).

   (a) Write a formula for E (kph), the excess of actual speed over the posted limit, in terms of P, the posted limit, for all P from 0 to 200 inclusive.

   (b) Draw a graph showing the relation of E to P.

   (c) For what values of P is Wreckless law-abiding?

Table 1-15

x	x^2	2^x
0		
1		
2		
2.5		
3		
4		
5		
6		
10		
-1		
-2		
-3		

Table 1-16

Posted speed limit	Speedometer reading	Actual speed	Excess of actual speed over posted limit
40			
60			
80			
100			
P			

Table 1-17

T	n
42	7
48	33
53	50
60	82
67	105
72	128
79	158

(d) For what value of P from 0 to 200 inclusive is E a maximum?

(e) For what value of P from 0 to 200 inclusive is E a minimum?

9. Crickets chirp more often in warm weather than in cold. At each of various temperatures (T degrees Fahrenheit) the numbers of chirps are counted for five separate 1-min periods and averaged to get the values n in Table 1-17.

(a) Plot the points corresponding to this table and draw the graph suggested by these points, with the understanding that there may be slight inaccuracies in the counts.

(b) Find a formula that fits your graph.

(c) For what values of T do you think this formula may be valid?

(d) Solve the formula for T in terms of n. You now have a cricket thermometer.

○ 10. If a sum of P dollars is invested at 10% interest, it will increase to an amount $P + 0.10P = 1.1P$ after 1 year.

(a) On the assumption that the whole amount is left to draw interest, find the amount, A, after 2 years, after 3 years, and after 4 years.

(b) Find a formula for the amount A after n years, that is, for the amount accrued by investing P dollars for n years at 10% interest compounded annually.

(c) Draw a graph of the variation of A with n, on the assumption that if you withdraw your money at any time during a year, you will receive no interest for the period since the start of that year.

11. (a) Use Table 1-14 to derive a formula expressing any temperature (C) on the Celsius scale in terms of the same temperature (F) on the Fahrenheit scale.

(b) Verify your answer to part (a) by solving for C in the formula $F = \frac{9}{5}C + 32$.

★ 12. Because the form of the graph in Problem 3, **1.4**, is similar to that of Problem 4, **1.3**, for which the formula is given in Example 1 of this section, assume that the formula for L in terms of t is of the same type; that is,

$$L = at^2 + bt + c.$$

Use three pairs of values from Table 1-8 to determine a, b, and c. Verify that this formula gives agreement with the other pairs of the table.

13. The cost ($\$C$) per hour of operating a motorboat varies as the cube (third power) of the speed (v knots), and $C = 40$ when $v = 12$. Find the formula.

14. At constant temperature, the volume (V cc) of gas in a cylinder is inversely proportional to the pressure applied (p dynes/cm^2). If $V = 50$ when $p = 20$, find the formula for V in terms of p.

15. The distance traveled (s m) by a stone dropped from some height above the ground varies as the square of the time (t sec) since release. If

$s = 1000$ when $t = 10$, find a formula for s in terms of t. Make a table of values for $t = 0$, 2, 4, 6, 8, and 10, and sketch a reasonable graph showing the variation of s with t for all t between 0 and 10 inclusive. Interpolate in your table to approximate the value of s for $t = 4.5$. By reference to your graph, explain whether you think the result of this interpolation is less than or greater than the "true" value of s. (Of course, in this case, you could use your formula to obtain the value of s corresponding to $t = 4.5$.)

16. Temperatures close to a heat lamp are high, and they fall off with increasing distance from the lamp: The heat intensity (H calories per second) at a point varies inversely as the square of the distance (x cm) of the point from the lamp. If for a certain heat lamp $H = 45$ at $x = 4$, find a formula for H in terms of x. At what distance from this lamp does $H = 20$?

17. For an observer h ft above the surface of the sea, the distance to the horizon (d miles) varies as the square root of h, and $d = 24.6$ for $h = 400$. Find a formula for d in terms of h.

★ 18. Use geometry to demonstrate the validity of the first sentence in Problem 17. (The statement is only *approximately* valid.)

19. The demand for a certain style of tennis shoe (D pairs) varies inversely as the square root of the price ($\$\, p$ per pair), and $D = 6000$ at $p = 16$. Find a formula for D in terms of p. What, then, is the demand at $p = 9$? At $p = 25$? For what price would the demand be 10,000?

20. The demand for a certain type of lawn chair (D chairs) varies inversely as the $\frac{3}{2}$ power of the price ($\$\, p$ per chair), and $D = 6000$ at $p = 16$. Find a formula for D in terms of p. What, then, is the demand at $p = 9$? At $p = 25$? For what price would the demand be 48,000?

★ 21. (a) Suppose that a quantity z varies with two others x and y in such fashion that for any fixed x, z is proportional to y^n, and for any fixed y, z is proportional to x^m. Show that $z = kx^m y^n$, for some constant k.

(b) The strength of a beam with rectangular cross section (Figure 1-14) and fixed length is proportional to the product of its width (x) and the square of its depth (y). How many times as strong is a beam 2×10 in. of a certain length if placed as in (i) rather than as in (ii)? How many times as strong is the beam of (i) as a beam 4×4 in. of the same length?

22. *Average decreases in body functions with age.* If various body functions are calibrated as 100% at age 35, then on the average they decrease t years later in accordance with Table 1-18. On the same axes, plot graphs of the relationships of v, p, and c to t, and find formulas for each of these three body functions in terms of t.

23. In the United States, the amount of fertilizer used (x kg per hectare of arable land) and the average yield for all cereals (y tons per hectare)

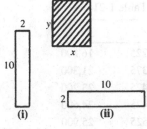

Fig. 1-14

Table 1-18

Function	t			
	0	5	25	45
Velocity of nerve signal transmission, v	100	98.9	94.5	90
Pumping efficiency of heart, p	100	95.6	78	60
Maximum breathing capacity, c	100	93.2	66	39

Table 1-19

	Year				
	1950	1963	1966	1969	1974
x	22	50	60	68	86
y	1.63	2.70	2.96	3.38	3.87

Table 1-20

	SAT average						
	Below 350	350–399	400–449	450–549	550–599	600–649	650+
Mean parental income	16,800	21,300	23,200	25,900	27,400	28,900	31,100

Table 1-21

x	y
225	16,800
375	21,300
425	23,200
475	24,600
525	25,900
575	27,400
625	28,900
725	31,100

were as shown in Table 1-19 ("The Natural History of Crop Yields," *American Scientist*, 68:390, 1980). Plot a graph of the relationship between x and y, draw the straight line that seems a reasonable fit to these points, and find the formula for y in terms of x for the straight line.

24. In 1978 the average scores on SAT exams and the mean incomes of the parents of the students who took the exams were as shown in Table 1-20. Let us take the midpoints of the ranges of the SAT scores as the independent variable, x, choosing the average of 100 and 350 as the smallest value, and the average of 650 and 800 as the largest value. Let us denote by y the mean parental income, so that the data are as shown in Table 1-21. Plot a graph of the relationship between x and y, draw the straight line that seems a reasonable fit to these points, and find the formula for y in terms of x for this straight line.

C 25. Newton's law of universal gravitation states that the force of attraction between any two particles is proportional to the product of their masses and inversely proportional to the square of the distance between them.

The law also applies to spheres (such as the earth and the moon) if we use the distance between their centers. The mass of the earth is 5.97×10^{24} kg, the mass of the moon is 7.35×10^{22} kg, and the average distance between their centers is 384,403 km. The force of attraction between these two bodies, when at average distance apart, is 1.98×10^{17} km-kg/sec^2. Use this information to find the constant of proportionality (often called the "gravitational constant," G) in Newton's law.

★ 1.7 Formulas (*continued*)

A formula for our initial example – the variation with time of the area covered by a colony of bacteria – is more difficult to obtain than those for the examples of the preceding section.

We look at the kinds of formulas we have already met. Formulas of the type $A = at + b$ correspond to straight lines, but the graph in Figure 1-2 is by no means straight, so this does not look hopeful. Exponential formulas, like $A = c \cdot b^t$, grow more and more rapidly (or else decrease more and more slowly), unlike the curve in Figure 1-2. Quadratic formulas, like $A = at^2 + bt + c$, correspond to graphs (called *parabolas*) of the general shape indicated in Figure 1-15, opening upward if a is positive, downward if a is negative.

Again, neither of these fits Figure 1-2, but to the left of about $t = 4$, Figure 1-2 looks roughly like part of Figure 1-15(i), and to the right, like part of Figure 1-15(ii). What we can do, then, is to approximate Figure 1-2 by joining parts of two parabolas in much the way we joined two line segments in Figure 1-12.

Because our curve changes from "opening upward" to "opening downward" when t is about 4, we choose to join the parabolas at that point. It takes three equations to determine the three coefficients in a quadratic, so we choose the pairs (0, 4.4), (2, 10.2), and (4, 19.2) from Table 1-1 to determine the first, and the pairs (4, 19.2), (6, 28.6), and (8, 34.8) to determine the second. Substituting in turn the first three pairs in

$$A = at^2 + bt + c$$

gives three equations in the unknowns a, b, and c:

$$4.4 = a \cdot 0 + b \cdot 0 + c$$
$$10.2 = a \cdot 4 + b \cdot 2 + c$$
$$19.2 = a \cdot 16 + b \cdot 4 + c.$$

Problem 1
Solve these equations to obtain $a = 0.4$, $b = 2.1$, and $c = 4.4$, so that the first polynomial is $0.4t^2 + 2.1t + 4.4$.

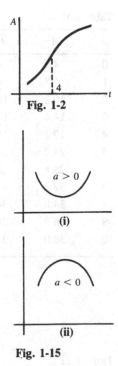

Fig. 1-2

(i)

(ii)

Fig. 1-15

Problem 2

In the same way, using the three pairs $(4, 19.2)$, $(6, 28.6)$, and $(8, 34.8)$, determine the second polynomial to be $-0.4t^2 + 8.7t - 9.2$.

It is not to be expected that combination of the two polynomials will give us A exactly, except for the five values of t for which we forced agreement. To respect this distinction, let us call the new quantity $\bar{A}$. Then our formula reads

$$\bar{A} = \begin{cases} 0.4t^2 + 2.1t + 4.4, & 0 \leq t \leq 4, \\ -0.4t^2 + 8.7t - 9.2, & 4 < t \leq 10. \end{cases}$$

Table 1-1′, an enlargement of Table 1-1, indicates the success of our effort. In only one case – that for $t = 10$ – does $\bar{A}$ differ from A by more than 0.1, and then the difference is only 0.2, but one can guess that the curve for $\bar{A}$ is "topping out." We shall later learn a method that will show that $\bar{A}$ reaches a maximum value at $t = 10.875$, and thereafter the values of $\bar{A}$ decrease. Thus, our formula is not much good for extrapolation; but, for interpolation, it does rather well.

You are likely to conclude, on the basis of the foregoing discussion, that "curve fitting" is more art than science. There is, however, extensive literature on the subject, making the process less a matter of trial than the example in this section suggests, and providing a means of considerable importance in making practical problems amenable to mathematical treatment.

Table 1-1′

t	A	$\bar{A}$
0	4.4	4.4
1	6.8	6.9
2	10.2	10.2
3	14.4	14.3
4	19.2	19.2
5	24.2	24.3
6	28.6	28.6
7	32.2	32.1
8	34.8	34.8
9	36.7	36.7
10	38.0	37.8

PROBLEMS

3. Calculate the values of $\bar{A}$ corresponding to $t = 1.7$ and $t = 4.7$. How do your results compare with the values of A we obtained by graphical interpolation in Section 1.3?
4. Draw a graph showing the relationship of $\bar{A}$ to t.
5. Information additional to that provided by a table of data can simplify the curve-fitting process. The sausage manufacturer in Problem 8, **1.4**, estimates a cost of \$2.80 per pound of sausage, at least at low output levels, in addition to the fixed cost of \$128. A first approximation to the daily operating expense would then be given by the formula

$$\tilde{E} = 128 + 2.8x.$$

Table 1-11

x	E
0	128
20	180
40	225
60	260
80	290
100	310

(a) Calculate the values of $\tilde{E}$ and of the errors, $E - \tilde{E}$, for each of the values of x in Table 1-11, which is repeated here.

(b) Observe that $E - \tilde{E}$ is given approximately by an expression of the form kx^2. Find k. (Keep your signs straight!)

(c) For the value of k found in part (b), calculate the values of $\bar{E} = \tilde{E} + kx^2$ for each of the values of x in Table 1-11 and also for $x = 120, 140,$ and 160.

(d) Draw a smooth curve showing the relation of $\overline{E}$ to x for $0 \le x \le 160$.

(e) Discuss the appropriateness of $\overline{E}$ as a substitute for E. Note especially what starts to happen to $\overline{E}$ at $t = 140$. Is this realistic?

1.8 Relationships between science and mathematics

Equation (2), **1.6**, gives only an approximation to the actual physical situation of the toy rocket. In the derivation of equation (2), the effects of air friction were ignored; moreover, the coefficient of t^2 (-16) is related to the acceleration due to the earth's gravitational attraction, and this varies from place to place on the earth's surface and with the distance of an object from the earth's center. We choose to "idealize" the experiment, describing a situation that is admittedly not entirely realistic. To incorporate some of the many minor factors into our mathematical formulation would make the mathematics dreadfully complicated, and our simple approximation is sufficiently precise for many practical purposes.

Likewise, careful observation of the heating of a pan of water (Example 2, **1.6**), involving the recording of temperatures at many more times, especially between $t = 15$ and $t = 20$, might lead to Figure 1-12' as a modification of Figure 1-12, because of a change in pressure above the surface of the water, and so forth. But the graph of Figure 1-12 and the associated equations (3) give reasonably good approximations.

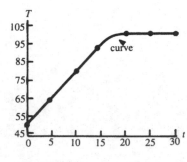

Fig. 1-12'

The situation was described succinctly by Warren Weaver in the 1958 annual report of the Rockefeller Foundation:

> Speaking roughly, one may say that the seventeenth, eighteenth, and nineteenth centuries formed the period in which physical science learned how to analyze two-variable problems. During that three hundred years, science developed the experimental and analytical techniques for handling problems in which one quantity – say, a gas pressure – depends primarily upon a second quantity – say, the volume of the gas. The essential character of these problems rests in the fact that, at least under a significant range of circumstances, the first quantity depends wholly upon the second quantity, and not upon a large number of other factors. Or in any event, and to be somewhat more precise, the behavior of the first quantity can be described with a useful degree of accuracy by taking into account only its dependence upon the second quantity, and by neglecting the minor influence of other factors.

Much of the mathematics we shall study in this course had its origin in situations in the physical, biological, and social sciences that could adequately be treated as "two-variable problems." Often the mathematics progressed to a life of its own, independent of the scientific problems for which it was developed. Some of the most attractive parts of our subject are those that stemmed from general intellectual curiosity and then were found to be useful in solving scientific problems. The constant interplay between the practical needs of science and technology and the aesthetic satisfactions of free inquiry gives mathematics a unique appeal.

1.9 Functions

Scientific "two-variable problems" usually begin, as in the preceding sections, with some statement of *how* one variable changes with another (see question I, **1.2**). The mathematical concept, *function*, is very useful in the study of this topic. We shall complete this introduction to the question of exhibiting variation by a discussion of this and related ideas.

In everyday language we make statements like these: "The area covered by the bacterial colony is a *function* of the time." "The size of the wheat crop is a *function* of the amount of rainfall." "The demand for a product is a *function* of its selling price." There may be an intuitive feeling that *causation* is involved in the last two statements, but surely not in the first. Rather, what we have in mind in this case might be put as follows: "For each number designating a time (in hours) after the beginning of the experiment there *corresponds just one* number designating the area (in square millimeters) covered by the bacterial colony."

In each of the examples in this chapter showing variation there is a set of ordered pairs of numbers (obtainable from a table or a formula). Each pair gives *the* value (only one) of a quantity (the *dependent variable*) corresponding to a chosen value of another quantity (the *independent variable*). By convention, in each ordered pair the value of the independent variable is written first. Such a correspondence is called a *function*. The set of values of the independent variable is called the *domain* of the function; the set of values of the dependent variable is its *range*. The notion of correspondence *from* the members of one set *to* those of another set is the central feature of the mathematical definition of function.

A **function** is a correspondence *from* the members of one set (the domain) *to* those of a second set (the range) in which with each member of the domain is paired just one member of the range.

For the toy rocket, the correspondence given by Table 1-2, repeated here, specifies a function (call it f) in which the domain, D_f, is $\{0,1,2,3,4,5,6\}$ and the range, R_f, is $\{48,120,160,168,144,88,0\}$, or, what is the same thing, $\{0,48,88,120,144,160,168\}$. We say that y is a "function of" t.

For the heating water in the pan, Table 1-4, repeated here, specifies a function (call it w) with a domain $D_w = \{0,5,10,15,20,25,30\}$ and range $R_w = \{49,64,79,94,100\}$. We say that T is a "function of" t.

Table 1-2

t	y
0	48
1	120
2	160
3	168
4	144
5	88
6	0

Table 1-4

t	T
0	49
5	64
10	79
15	94
20	100
25	100
30	100

Note that distinct members of the domain (e.g., 20, 25, and 30 in D_w) may have the same correspondent in the range (e.g., 100). Note also that 100 need be listed only once as a member of R_w.

Interpolation and extrapolation are means of *extending* a function. Adding one or more pairs to the table that defines a function $\mathscr{F}$ defines a new function that includes all the pairs of $\mathscr{F}$ and at least one new pair. Earlier we extended Tables 1-2 and 1-4 by means of formulas. Here are those examples, set out in our new language:

As we saw in Example 1, **1.6**, the values in Table 1-2 (the pairs of the function f) satisfy

$$y = -16t^2 + 88t + 48, \quad 0 \le t \le 6, \tag{2}$$

which defines another function (call it F). The domain of F is the set of *all* real numbers 0 to 6, both 0 and 6 included. It can be shown, by methods that we shall learn later, that the maximum value of y is 169. (See Problem 4e, **1.6**, for a hint of this.) Hence, the range of F is given by $0 \le y \le 169$. We have, then, a function, F, with a domain that includes that of f ($D_F \supset D_f$). Moreover, for the members of D_f, both F and f determine the same pairings. In such a case we say that F is an *extension* of f and that f is the *restriction* of F to D_f. The graph of F is seen as the curve in Figure 1-16, whereas the graph of f consists of only the seven dots.

Similarly, for the heating pan of water, we saw in Example 2, **1.6**, that Table 1-4 (the pairs of the function w) satisfies the formulas

$$T = \begin{cases} 3t + 49, & 0 \le t \le 17, \\ 100, & 17 < t \le 30. \end{cases} \tag{3}$$

These define another function (call it W). The domain, D_W, is $0 \le t \le 30$, the set of all real numbers between 0 and 30, both end points included. The range, R_W, is $49 \le T \le 100$, the set of all real numbers between 49 and 100, both end points included. The function W is an *extension* of w, and w is the

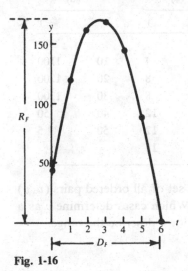

Fig. 1-16

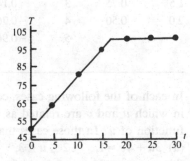

Fig. 1-12

(a)

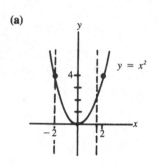

(b)

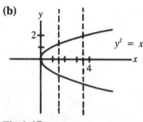

Fig.1-17

restriction of W to D_w. The graph of W consists of the two line segments in Figure 1-12, whereas the graph of w is the set of seven dots.

If the relationship between two quantities is shown on a graph, it is easy to decide whether a function is defined. For example, the correspondence *from* $-2 \le x \le 2$ *to* $0 \le y \le 4$ shown in Figure 1-17(a) *does* define a function: Any vertical line (like the dotted lines of the sketch) meets the graph at no more than one point – "with each member of the domain is paired *just one* member of the range." On the other hand, the correspondence *from* $0 \le x \le 4$ *to* $-2 \le y \le 2$ shown in Figure 1-17(b) *does not* define a function: Some vertical lines (like the dotted lines of the sketch) meet the graph at *more than* one point.

PROBLEMS

1. If a function G is an extension of a function g, then D_G contains D_g as a proper part – all the members of D_g are found in D_G, and D_G has at least one member not in D_g. What can you say about the relation between R_G and R_g – are all members of R_g necessarily in R_G? Are there necessarily any other members of R_G? Give examples.

2. A table of ordered pairs of real numbers (x, y), in which no x value corresponds to two or more different y values, determines y as a function of x. If we plot these pairs as points and then draw a curve through the points, we are drawing a graph of an extension of the function. It is unlikely that two people will draw exactly the same graph. Plot points for each of the parts of Table 1-22. Sketch what seems to you an appropriate curve through these points, and compare your results with those of a classmate.

Table 1-22

(a)		(b)		(c)		(d)	
x	y	x	y	x	y	x	y
0	0	0	0				
0.50	0.50	1	0.84	0.6	4	10	1100
1.0	0.25	2	0.91	1.2	8	20	1200
1.5	0.75	3	0.14	1.8	8	30	1300
2.0	0.50	4	-0.76	2.4	12	40	50
		5	-0.96	3.0	12	50	5
				3.6	16		

3. In each of the following cases, consider the set of all ordered pairs (u, v) in which u and v are related as indicated. Which cases determine v as a function of u? In those cases that are functions, describe the range.
   (a) $3u + 4v = 1$, $\quad -2 \le u \le 3$,
   (b) $u^2 + v^2 = 25$, $\quad -5 \le u \le 5$,

(c) $uv = 1$, $\quad 1 < u \le 10$,

(d) $uv = -1$, $\quad -2 \le u \le -0.01$ $\quad$ or $\quad$ $0.01 < u < 2$,

(e) $uv - v = 3$, $\quad 0 \le u \le 2$,

(f) $u^3 = v$, $\quad -1 < u \le 1$,

(g) $v = u^2$, $\quad 0 \le u < 5$,

(h) $v = u^n$, $\quad 0 \le u \le 2$ $\quad$ (consider separately $n = 1, 2, 3, \ldots$).

4. (a)–(h) For each of the parts of Problem 3, sketch two graphs of the points (u, v) determined by the given equation and inequalities, the first with the u axis horizontal and the v axis vertical, the second with the v axis horizontal and the u axis vertical.

★ 5. (a) If $Q = 15 - (7 - 2x)^2$, what is the maximum value of Q, and for what x is the maximum Q attained? Is there a minimum value of Q? If so, find it; if not, explain.

(b) If $P = 20 + (x + 3)^2$, what is the minimum value of P, and for what x is the minimum P attained? Is there a maximum value of P? If so, find it; if not, explain.

(c) Use the ideas of parts (a) and (b) and the process of "completing the square" to find the value of x corresponding to an extreme (maximum or minimum) value of R, if R equals

(i) $x^2 - 2x - 3$, $\qquad$ (ii) $-4x^2 + 4x - 5$,

(iii) $48 + 88x - 16x^2$, $\qquad$ (iv) $3x^2 - 19.2x + 30.72$.

(d) For each case in part (c) in which $R = 0$ for certain values of x, what is the relationship of the value of x corresponding to an extreme value of R and the values of x corresponding to $R = 0$? Show the reasonableness of this result in terms of the graphs of the equations for R.

(e) Using the methods suggested in the foregoing numerical examples, discuss the extreme value (or values) of $Q = ax^2 + bx + c$. (Hint: It will be convenient to consider separately the cases $a > 0$, $a = 0$, and $a < 0$.)

1.10 Further discussion of functions: notation and natural domains

The correspondence defining a function is often given by one or more equations. If f is the function defined by the equation $y = x^2$, this might be written

$$f : y = x^2.$$

Another notation, perhaps more precise, is

$$f : x \to x^2,$$

which could be read "the function f defined by the correspondence from a number to its square." In both cases, of course, the domain and range of f should be specified, if they are not obvious from the context. It would be natural to specify the domain here as all real numbers, in which case the range would be the set of all nonnegative real numbers.

Another notation, particularly useful for the calculus, is

$$f(x) = x^2,$$

which is read "f of x equals x^2." Actually, $f(x)$ is the member of R_f corresponding to x as a member of D_f. It is common to speak of the "function, $f(x)$," and although this is not really correct – f is the function, and $f(x)$ is a member of its range – no confusion results. The notation $f(2)$, for example, means "the value of the function f for $x = 2$."

Here is some practice with the notation for this function: $f(3) = 3^2 = 9$; $f(0) = 0$; $f(-\frac{1}{2}) = \frac{1}{4}$; $f(a) = a^2$; $f(3+h) = (3+h)^2 = 9 + 6h + h^2$; $f(a-h) = (a-h)^2 = a^2 - 2ah + h^2$.

In some cases, the expression or expressions used to define a function automatically restrict the possible domain. Suppose, for example, that $f(x) = \sqrt{25 - x^2}$. If we consider only real numbers (as we shall always do unless otherwise stated), then x must belong to the interval $-5 \le x \le 5$. Thus, $-5 \le x \le 5$ is the largest possible domain for the function f; we call this the *natural domain of the function*. If no domain is specified, the natural domain will be understood.

Sometimes the natural domain of a function is the set of all real numbers. For instance, in the problem of the toy rocket, where $y = -16t^2 + 88t + 48$, there is no bound for t so far as *mathematics* is concerned – t can assume any real value. The restriction, $0 \le t \le 6$, comes from the physical application.

Functions can be built as combinations of other functions. For example, if $f(x) = x^2 - 1$ and $g(x) = 1/x$, then

$$f(x) + g(x) = x^2 - 1 + \frac{1}{x}, \quad = r(x), \text{ say,}$$

$$f(x) \cdot g(x) = (x^2 - 1) \cdot \frac{1}{x} = \frac{x^2 - 1}{x}, \quad = s(x), \text{ say,}$$

$$2f(x) - g(x) = 2(x^2 - 1) - \frac{1}{x}, \quad = t(x), \text{ say,}$$

$$\frac{f(x)}{g(x)} = \frac{x^2 - 1}{1/x} = (x^2 - 1)x, \quad = u(x), \text{ say, etc.}$$

Another important combination is represented by the symbol $f(g(x))$, which is read "f of $g(x)$" and is the result of replacing the x in $f(x)$ by $g(x)$. Thus, for the f and g of this example,

$$f(g(x)) = [g(x)]^2 - 1 = \left(\frac{1}{x}\right)^2 - 1 = \frac{1}{x^2} - 1 = \frac{1 - x^2}{x^2}.$$

For the f and g of this example,

$$\text{"}g\text{ of }f(x)\text{"} = g(f(x)) = \frac{1}{f(x)} = \frac{1}{x^2 - 1}.$$

Note that the natural domain of this $f(g(x))$ is the set of all real numbers except 0, and the natural domain of this $g(f(x))$ is the set of all real numbers except ± 1.

There are some convenient notations we shall often use: The set of all real numbers from a to b, a and b included, is represented by $[a, b]$. Other similar notations are as follows:

$[a, b)$ is the set $a \leq x < b$;

$(a, b]$ is the set $a < x \leq b$;

(a, b) is the set $a < x < b$; note that (a, b) is also used for number pair, but which interpretation is meant should be clear from the context;

$[a, \infty)$ is the set $a \leq x$ (i.e., the set of all real numbers greater than or equal to a);

$(-\infty, b]$ is the set $x \leq b$ (i.e., the set of all real numbers less than or equal to b);

$(-\infty, \infty)$ is the set of all real numbers.

PROBLEMS

1. If $f(x) = x^3 + 2x^2 - 4x + 1$, compute $f(0)$, $f(1)$, $f(-1)$, and $f(a)$ for any number a; compute $f(1 + h)$ for any number h.

2. If $g(x) = \frac{3}{2}x^2$, compute $g(5) - g(4)$, $[g(4.5) - g(4)]/0.5$, $[g(4.1) - g(4)]/0.1$, $[g(4.01) - g(4)]/0.01$, and $[g(4 + h) - g(4)]/h$ for any number $h \neq 0$.

3. If $p(x) = \sqrt{16 + x^2}$,
   (a) find the values of $p(0)$, $p(3)$, and $p(4) - p(3)$;
   (b) use a calculator to find approximate values of $[p(3.1) - p(3)]/0.1$ and $[p(3.01) - p(3)]/0.01$;
   (c) write a reasonably simple expression for $[p(3 + h) - p(3)]/h$ for any number $h \neq 0$.

4. If $F(x) = \sqrt{x}$ and $G(x) = x^2$, what are the values of $G(F(9))$ and $F(G(x))$? Remember that our convention is that $\sqrt{x}$ stands for the nonnegative square root of x. What is the value of $F(G(-9))$? Is $G(F(-9))$ a real number? Explain.

5. If $f(x) = 1 - x$, $g(x) = 1/x$, and $p(x) = x/(x - 1)$, find reasonably simple expressions for
   (a) $f(f(x))$ (b) $g(g(x))$ (c) $p(p(x))$
   (d) $f(p(x))$ (e) $g(f(x))$ (f) $p(g(x))$

61

6. Find the natural domain for each of the following functions:

(a) $F(x) = \dfrac{1}{\sqrt{25 - x^2}}$ (b) $G(x) = x - \dfrac{1}{x}$ (c) $H(x) = \dfrac{\sqrt{10 + x^2}}{x - 3}$

(d) $f(x) = \sqrt{\dfrac{9 - x^2}{16 - x^2}}$

7. (a)–(b) Determine the ranges for the functions in Problem 6 (a) and (b).

C 8. If $f(x) = 2x^3 - 15x^2 + 27x - 10$, compute the values of $f(x)$ for integral (whole-number) values of x from 0 to 6, and use your results to sketch the graph of $y = f(x)$.

C 9. Rework Problem 8 for $f(x) = x^3 + 3x^2 - x - 3$, for integral x from -4 to 2.

Table 1-23(a)		Table 1-23(b)	
C	F	F	C
−10	14	14	−10
−5	23	23	−5
0	32	32	0
5	41	41	5
10	50	50	10
15	59	59	15
20	68	68	20

1.11   Inverse functions

Which of two related quantities we wish to consider the "independent variable" sometimes depends on the use we have in mind for the relationship. The expression $y = f(x)$ indicates a correspondence *from* the set of numbers x *to* the set of numbers y. We can also think of the reverse correspondence *from* the set of numbers y *to* the set of numbers x. The reverse correspondence is called the *inverse* of the original correspondence. As we shall see, this inverse may or may not be a function.

For example, if we want to determine the Fahrenheit readings corresponding to various Celsius readings for the same temperatures, we can use Table 1-23(a), plot C horizontally and F vertically to get the solid line in Figure 1-18, and use the formula $F = \frac{9}{5}C + 32$. On the other hand, if we want the Celsius readings corresponding to Fahrenheit readings, we can use Table 1-23(b), plot F horizontally and C vertically to get the dotted line in Figure 1-18, and use the formula $C = \frac{5}{9}F - \frac{160}{9}$.

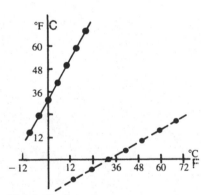

Fig. 1-18

From the point of view of mathematics, we have two functions here: The line segment of Figure 1-19 is the graph of a function, f, say, determined by the equation $y = \frac{9}{5}x + 32$, with D_f given by $-10 \le x \le 20$ and R_f given by $14 \le y \le 68$; whereas the line segment of Figure 1-20 is the graph of a function, g, say, determined by the equation $y = \frac{5}{9}x - \frac{160}{9}$, with D_g given by $14 \le x \le 68$ and R_g given by $-10 \le y \le 20$.

Each of the functions f and g is the inverse of the other. If we "reflect" the graph of f in the line $y = x$, we obtain the graph of g, as in Figure 1-21. Likewise, the reflection of the graph of g in the line $y = x$ is the graph of f – the graphs are "mirror images" of each other, with the line $y = x$ as the "mirror." $D_f = R_g$ and $R_f = D_g$.

The algebraic equivalent of the geometric transformation just described is the interchange of x and y in the equations defining the functions: If we interchange x and y in $y = \frac{9}{5}x + 32$, the equation determining the function f,

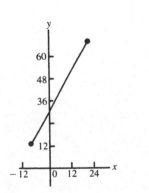

Fig. 1-19

62

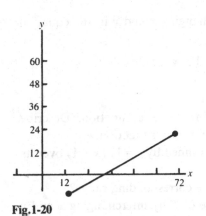

Fig. 1-20

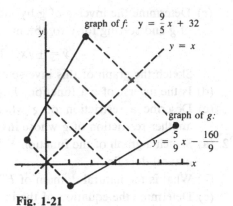

Fig. 1-21

we obtain $x = \frac{9}{5}y + 32$, which, when solved for y, yields $y = \frac{5}{9}x - \frac{160}{9}$, the equation determining g, the inverse of f.

We can proceed to find the inverse of *any* function by the algebraic process just described, but the result is not necessarily a function. For example, suppose F is the function given by

$$F: y = x^2$$

over its natural domain, which is the set of all real numbers: $D_F = (-\infty, \infty)$. Note that R_F is the set of all nonnegative real numbers, $[0, \infty)$. Then G, the inverse of F, has an equation obtained by interchanging x and y in the equation of F and solving for y:

$$x = y^2, \quad \text{or}$$
$$G: y = \pm\sqrt{x}.$$

But G is not a function, because to each positive x correspond two values of y – the graph of G (Figure 1-22) fails "the vertical-line test," as we noted at the end of Section 1.9.

We can, however, choose a restriction of F such that its inverse *is* a function. Suppose we consider

$$p: y = x^2, \quad \text{with } D_p = [0, \infty).$$

Note that in this case R_p is also $[0, \infty)$. Then q, the inverse of p, is given by $q: y = \sqrt{x}$, with $D_q = [0, \infty)$ and $R_q = [0, \infty)$. The graph of q (Figure 1-23) passes the vertical-line test, so it is a function.

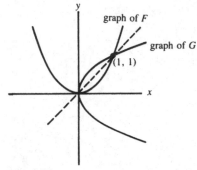

Fig. 1-22

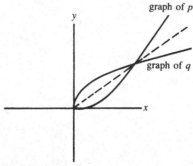

Fig. 1-23

PROBLEMS

1. (a) Sketch a graph of the function g determined by

$$y = x^2 + 1, \quad -2 \le x \le 2.$$

   (b) What is the range of g?

(c) Determine the inverse of g by interchanging x and y in the equation of g and solving for y to obtain

$$y = \pm\sqrt{x-1}, \quad 1 \le x \le 5.$$

Sketch the graph of this inverse of g.

(d) Is the inverse of g a function? Explain.

(e) Describe a restriction of g whose inverse *is* a function. Describe another restriction of g whose inverse is also a function.

2. (a) Sketch a graph of the function F determined by $y = 1/(x-1)$ over its natural domain.

(b) What is the natural domain of F? The corresponding range?

(c) Determine the equation of the inverse of F by interchanging x and y in the equation of F and solving for y. What is the domain of this inverse? Its range? Sketch the graph of this inverse of F.

(d) Is the inverse of F a function? Explain.

3. (a) Sketch a graph of the function f determined by

$$y = 10x - x^2, \quad 0 \le x \le 10.$$

(b) What do you think to be the range of f?

(c) Interchange x and y in the equation determining f, and solve for y to obtain

$$y = 5 \pm \sqrt{25-x}, \quad 0 \le x \le 25.$$

Sketch the graph corresponding to this equation.

(d) Is the inverse of f a function? Explain.

(e) Describe a restriction of f whose inverse is a function. Describe another restriction of f whose inverse is also a function.

4. In Problem 3, **1.9**, you were asked to decide which of the given equations determined v as a function of u. For those that are functions, which have inverses that are functions?

5. Does the linear function $f(x) = ax + b$ always have an inverse that is a function? Discuss fully.

○ 6. What can you say about the graph of $y = f(x)$ if the inverse of f is to be a function?

7. A merchandiser finds that the number of plastic wastebaskets, N, he sells per week is a function of the price he charges. For price ($\$p$) between 1 and 6, inclusive, he finds that $N = 18{,}000/p^2$.

(a) What are the domain and range of this function? Sketch its graph.

(b) The merchandiser wants to know the price at which he can just dispose of a certain number of wastebaskets in a week. Solve his problem by finding a formula for p as a function of N. What are the domain and range of this function? Sketch its graph. (In economics, either of the graphs – in (a) or (b) – is called a "demand curve.")

(c) Now consider the function given by $y = 18{,}000/x^2$. What is its natural domain? What is the corresponding range? Sketch its graph.

(d) Find an expression for the inverse of the function in (c). Sketch a graph of this inverse. Is this inverse a function? If not, describe a restriction of the function in (c) whose inverse *is* a function.

★ 1.12 Absolute values

A notation of considerable usefulness is that of *absolute value*: $|a|$ (read "the absolute value of a") is defined by

$$|a| = \begin{cases} a, & \text{if } a \geq 0, \\ -a, & \text{if } a < 0. \end{cases}$$

The absolute value of a number a is the same as the *numerical value* of a. Thus, $|7| = 7$; $|-3| = 3$; $|-1.56| = 1.56$; $|0| = 0$, and so on. This means that $-5 \leq x \leq 5$ can be written as $|x| \leq 5$. The graph of $y = |x|$ is shown in Figure 1-24.

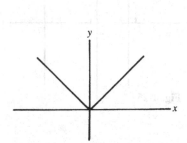

Fig. 1-24

PROBLEMS

1. Write each of the following expressions without use of the absolute-value symbol:
   (a) $|x| < 5$ (b) $0 < |x| \leq 5$ (c) $3 \leq |x| < 5$
2. Sketch the graph of each of the following equations:
   (a) $y = |x - 2|$ (b) $y = |x^3|$ (c) $y = |x^2 + 4|$
   (d) $y = |x^2 - 4|$ (e) $y = |10x - x^2|$
3. Sketch the graph of each of the following equations:
   (a) $|y| = x$ (b) $|x| + |y| = 1$ (c) $|y| = x^2 + 4$
   (d) $|y| = x^2 - 4$ (e) $|y| = |x^2 - 4|$ (f) $|y| = \sqrt{x}$
4. If x and y are any real numbers, find a relation between $|x|$ and $|y|$ and
   (a) $|xy|$ (b) $\left| \dfrac{x}{y} \right|$

   (Hint: Experiment with numbers to guess the relations.)
5. Use the inequalities $-|a| \leq a \leq |a|$, first with $a = x$ and second with $a = y$, to show that $|x + y| \leq |x| + |y|$ for all real numbers x and y.

1.13 Summary

Variation can be exhibited by means of tables, graphs, and formulas. Functions can be extended by linear interpolation in tables and by graphical interpolation. In many cases, formulas are the most convenient form for the analysis of functions. Among the formulas that we have encountered are the following:

   (a) The linear function: $f(x) = ax + b$. The graph of the equation $y = ax + b$ is a straight line passing through the point $(0, b)$ – the "y intercept" of

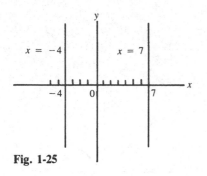

Fig. 1-25

the line is b. For each increase of 1 in x, y changes by a. Conversely, any nonvertical straight line has an equation of the form $y = ax + b$. A vertical straight line 7 units to the right of the y axis has the equation $x = 7$, and 4 units to the left of the y axis it has the equation $x = -4$ (Figure 1-25). On a vertical straight line there are indefinitely many y's corresponding to the same x, so a function is *not* defined.

(b) Other polynomial functions, particularly the quadratic: $f(x) = ax^2 + bx + c$. The graph of a quadratic function is a parabola with a vertical axis of symmetry. If $a > 0$, the parabola opens upward; if $a < 0$, the parabola opens downward.

(c) The exponential function: $f(x) = c \cdot b^x$.

(d) The power function: $f(x) = kx^n$.

An important concept is that of the inverse of a function. We shall develop all these ideas in subsequent chapters.

PROBLEMS

1. An experiment has shown that, within a certain range, y (the number of discharges per hour of an electric knife fish, *Gymnotus carape*) varies with x (the number of milligrams of the tranquilizer chlorpromazine per 20 liters of water in the aquarium) as follows:

$$y = 300 + 20x - x^2.$$

Sketch a graph of this equation, and discuss the function given by the equation as fully as possible, in relation both to mathematics and to the biological experiment (*Medical Research Engineering*, Vol. 6, No. 3, 1967).

2. At the Pizza Hut, the prices of anchovy pizzas are as follows: 10 in., $2.15; 13 in., $3.25; 15 in., $4.35. Make reasonable assumptions to construct a mathematical model of the "real-world" situation:
   (a) Discuss the appropriateness of the price structure. Is it better to buy two 10-in. pizzas or one 15-in. pizza?
   (b) If the middle-size pizza is actually $12\frac{1}{2}$ in., find a formula for the price that fits the (altered) data. By your formula, what would be the price of a very small pizza?
   ★ (c) Find a quadratic formula for the price that fits the data as initially given; with this formula, what would be the price of a 6-in. pizza? Of a 3-in. pizza?
   ★ (d) Find a formula consisting of the sum of two terms, one a constant (representing fixed costs) and the other proportional to the area of the pizza, that gives values as close as feasible to the quoted prices.

3. Extrapolation. An aphorism attributed to Niels Bohr goes as follows: "Prediction is difficult, especially about the future."

Some amusing predictions about the "ultimate of human effort" appear in Chapter 4, "The Sport of Track and Field: Flights of Fancy," in *The Worlds of Brutus Hamilton*, edited by L. J. Baack (Tafnews Press, 1975).

Extremely important examples of extrapolation pervade Barry Commoner's *The Poverty of Power: Energy and the Economic Crisis* (Knopf, 1976), in which, among other things, he compares the expected costs of making electricity by coal and by nuclear power in the next 10 to 20 years.

Annual U.S. electricity consumption (*C* billions of kwh) at various dates is shown in Table 1-24 (*New York Times*, April 18, 1976, News of the Week in Review, p. 4).

(a) From these data, what prediction would you make about *C* in 1970? In 1975? In 1980?

(b) Using only the values of *C* in 1955 and 1960, what prediction would you make about *C* in 1970? In 1975? In 1980?

(c) The actual value of *C* was 1.4 in 1970 and 1.69 in 1975. Using only these values, what prediction would you make about *C* in 1980?

Predictions about demand for energy are changing rapidly. From *Science*, June 20, 1980, "Energy Forecasts: Sinking to New Lows":

> Although some astute energy watchers predicted early in the 1970's that demand would soon level off, the big institutional forecasters have only begun to consider this a real possibility in the last couple of years ... even the most cautious energy forecasters are making revisions today....
>
> A graphic illustration of this behavior has been put together by Amory Lovins, British representative for Friends of the Earth and bête noire of the utility industry. He is one of many who argue that enormous efficiency improvements can and will be made in technology in the next two decades, and that these will reduce energy demand far below the present level of 78 quadrillion British thermal units (quads) per year.
>
> Lovins points out that, no matter what the bias of the forecaster, all energy predictions for the year 2000 have been dropping at about the same speed since the oil embargo [see Table 1-25].

See also the *New York Times*, October 11, 1981, Section 3, p. 1, "Winning the War on Energy: All the Old Predictions Were Wrong."

4. *Curve fitting.* In the summary to his article "Public Support for American Presidents: A Cyclical Model" (*Public Opinion Quarterly*, Vol. 40, Spring 1976), James A. Stimson wrote

The approval accorded to Presidents by the American public is

1.13
Summary

Table 1-24

Date	C
1955	0.54
1956	0.53
1957	0.55
1958	0.58
1959	0.63
1960	0.68
1961	0.73
1962	0.78
1963	0.84
1964	0.90
1965	0.96

Table 1-25

Year of forecast	Beyond the pale	Heresy	Conventional wisdom	Superstition
1972	125 (Lovins)	140 (Sierra)	160 (AEC)	190 (FPC)
1974	100 (Ford zeg)	124 (Ford tf)	140 (ERDA)	160 (EEI)
1976	75 (Lovins)	89–95 (Von Hippel)	124 (ERDA)	140 (EEI)
1977–8	33 (Steinhart)	67–77 (NAS I, II)	96–101 (NAS III, AW)	124 (Lapp)

Abbreviations: Sierra, Sierra Club; AEC, Atomic Energy Commission; FPC, Federal Power Commission; Ford zeg, Ford Foundation zero energy growth scenario; Ford tf, Ford Foundation technical fix scenario; Von Hippel, Frank Von Hippel and Robert Williams of the Princeton Center for Environmental Studies; ERDA, the Energy Research and Development Administration; EEI, Edison Electric Institute; Steinhart, 2050 forecast by John Steinhart of the University of Wisconsin; NAS I, II, III, the spread of the National Academy of Sciences Committee on Nuclear and Alternative Energy Systems (CONAES); AW, Alvin Weinberg study done at the Institute for Energy Analysis, Oak Ridge; Lapp, energy consultant Ralph Lapp.
Source: Amory Lovins put together this table showing the downward drift in forecasts. Figures represent the total U.S. energy demand in year 2000 or 2010.

found to follow a cyclical pattern over time. All Presidents begin their terms with great popularity, experience parabolic declines, steadily lose popular support for about three years, and then recover some at the ends of their terms. These distinctive cycles, it is argued, reflect regular expectation/disillusionment cycles among the less well-informed segments of the public and are tied to the four-year election calendar. The extraordinary fit of parabolic curves to actual presidential approval leads to the suspicion that presidential approval may be almost wholly independent of the President's behavior in office, a function largely of inevitable forces associated with time.

Using "relative approval" to designate the percentage of people queried who express approval of the president, he found that the following equation gave close agreement with the data from public-opinion polls for President Truman's first term:

$$\text{relative approval} = 53.37 + 8.85(2.25 - t)^2,$$

where t is the number of years since the beginning of the term.

(a) What was the relative approval at the start of the term? At its end?

(b) What was the minimum relative approval, and when did it occur?

(c) Write the equation in the form

$$\text{relative approval} = at^2 + bt + c,$$

for appropriate a, b, and c.

5. Samples from the Quinnipiac River at Wallingford gave the relation-ships between specific conductance (x micromhos at $25\,^{\circ}$C) and dis-solved-solids concentration (y mg/liter) shown in Table 1-26. Plot these data and, by eye, draw the straight line that seems to you the best fit. Then find the equation of this line. (Data from Wesleyan University M.A. thesis of Elinor Handman.)

C 6. Find the sum of the first 10 terms of the geometric series, and also G_∞, the sum to infinity, as defined in Section 0.12.

(a) $1 + \frac{1}{4} + \frac{1}{16} + \frac{1}{64} + \cdots$ (b) $1 - \frac{1}{4} + \frac{1}{16} - \frac{1}{64} + \cdots$

C 7. The *harmonic series* is $1 + \frac{1}{2} + \frac{1}{3} + \frac{1}{4} + \frac{1}{5} + \cdots$.

(a) Calculate

$$H_n = 1 + \frac{1}{2} + \frac{1}{3} + \frac{1}{4} + \cdots + \frac{1}{n}$$

for $n = 1, 2, 3, \ldots, 12$.

(b) Do you think that the H_n will ultimately behave like geometric series with $-1 < r < 1$, that is, that H_n will get closer and closer to some limiting value as n increases?

(c) With a programmable calculator or computer, find H_{50}, H_{100}, and H_{200}. Also find how large n must be to make $H_n > 4$, $H_n > 6$, and $H_n > 8$. Do these results reinforce or change your guess about the answer to (b)?

(d) You should have found in (a) that H_{10} is a little more than 2.9. Note that each of the terms $\frac{1}{11}, \frac{1}{12}, \frac{1}{13}, \ldots, \frac{1}{99}$ exceeds $\frac{1}{100}$, so $\frac{1}{11} + \frac{1}{12} + \frac{1}{13} + \cdots + \frac{1}{100} > (\frac{1}{100})90 = 0.9$. Hence, $H_{100} > 2.9 + 0.9 = 3.8$. (Actually, $H_{100} \approx 5.4$.) Similarly, the sum of the next 900 terms is greater than 0.9, so $H_{1000} > 3.8 + 0.9 = 4.7$; and the sum of the next 9000 terms is greater than 0.9, so $H_{10,000} > 4.7 + 0.9 = 5.6$. Use this argument to make a definite statement about H_n as n increases without bound.

8. (a) Let x be the length of a side of a regular polygon inscribed in a circle of radius 1, and let y be the length of a side of a regular polygon with twice the number of sides, inscribed in the same circle. Use geometry and algebra to show that

$$y^2 = 2 - \sqrt{4 - x^2}.$$

(b) What is the perimeter of a regular hexagon (6 sides) inscribed in a circle of radius 1?

Table 1-26

x	y
166	103
170	107
172	108
174	110
180	113
182	118
186	114
188	117
192	118
194	124
196	126
200	122
208	130
210	133
214	136
216	132
220	138
222	141
228	140
232	146
238	146
246	161
254	150
255	153
256	174
257	162

69

(c) Hence, what is the perimeter of a regular dodecagon (12 sides) inscribed in a circle of radius 1?

c   (d) Find the perimeters of regular polygons of 24, 48, and 96 sides inscribed in a circle of radius 1.

(e) Knowing that the circumference of a circle of radius r equals $2\pi r$, use your last result of part (d) to find an approximation to π.

9. *Proof systems.* Discuss the relationships among the scales shown in Table 1-27.

Table 1-27

Britain and Canada	Sykes scale	American	Alcohol (by volume)
75.25 overproof	175	200 proof	100%
50 overproof	150	172 proof	86%
30 overproof	130	149 proof	74.5%
Proof	100	114.2 proof	57.1%
12.5 underproof	87.5	100 proof	50%
30 underproof	70	80 proof	40%
50 underproof	50	57 proof	28.5%
100 underproof	0	0 proof	0%

10. The accompanying graph (Figure 1-26) shows the 10-year cumulative increases in college costs, 1972–3 to 1981–2. Examine the graph and

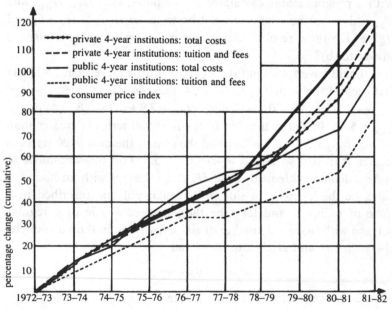

Fig. 1-26

state what interesting and significant information it provides for you. (Graph from *The College Board News*, Fall 1981.)

SAMPLE TEST

In a certain type of mechanism, the pressure (y lb/in.^{2}) in a cylinder is related to the diameter (x in.) of a piston as follows: y varies inversely as x^2, and, when $x = 2$, $y = 900$.

1. Find a formula for y in terms of x.
2. Make a table of values of y corresponding to $x = 2, 4, 6, 8, 10,$ and 12, and sketch the graph showing how y is related to x.
3. Use linear interpolation in your table from Problem 2 to approximate the value of y corresponding to $x = 11$.
4. Explain by reference to your graph whether you expect your answer to Problem 3 to be larger or smaller than the true value.
5. Find an equation for the straight line joining the points on your graph where $x = 6$ and $x = 10$.
6. Let the function given by your formula in Problem 1, with no reference to physical considerations, be called F. What is the natural domain of F?
7. Suppose that, in an actual mechanism, x cannot be less than 1 nor greater than 20. Let the function given by your formula in Problem 1, as restricted in the preceding sentence, be called G. What is the range of G?
8. Sketch a rough graph of the function F. Is the inverse of F a function? If so, write a formula for the inverse function, with x as independent variable and y as dependent variable, and state its domain.
9. Is the inverse of G a function? If so, write a formula for the inverse function, with x as independent variable and y as dependent variable, and state its domain.
★ 10. Draw the graph corresponding to the equation $|x| - |y| = 1$.

71

2 Rate of change

We begin now an investigation of the second question raised in Chapter 1: *How fast* does y vary with t? We shall return to the first question (*How* does y vary with t?) from time to time as we encounter more kinds of functional relationships.

The development of this chapter follows this sequence of topics:
1. *Average speed* as a familiar example of rate of change – of *how fast* distance traveled varies with time
2. *Average velocity* as rate of change of *displacement* with respect to time
3. The intuitive idea of *instantaneous velocity* leading to the concept of *limit*
4. Computations of instantaneous velocity
5. Statements of theorems on limits, to simplify evaluations of limits
6. Proofs of some results on limits
7. Geometrical examples: slope of a line, average slope of a curve in an interval, slope of a curve at a point
8. Tangent to a curve defined in terms of slope
9. Unifying concept: the *derivative* as a *rate of change*, showing *how fast* one quantity varies with another
10. A *specific* formula for the derivative, and several *general* formulas

As you proceed through the chapter, you may find it helpful to refer back to this outline.

2.1 Average speed and average velocity

Probably the most familiar example of a *rate* is found in the concept of *average speed*. If you drive 150 miles between 2 o'clock and 5 o'clock, your average speed is 50 miles per hour (i.e., the distance traveled divided by the time it takes). More formally:

Definition 1
If an object travels a distance $s \geq 0$ from time $t = a$ to $t = b$, then its **average speed** in this interval is $s/(b - a)$.

In the case of our much-traveled rocket, with the height above the ground given by

$$y = -16t^2 + 88t + 48, \quad 0 \le t \le 6,$$

we have, at $t = 0$, $y = 48$, and at $t = 2$, $y = 160$. Thus, the rocket travels 112 ft in the 2-sec interval. We say, then, that the *average speed* of the rocket from $t = 0$ to $t = 2$ is 56 ft/sec.

Problem 1
The average speed in some other 2-sec interval will usually be different. Check that the average speed from $t = \frac{1}{2}$ to $t = 2\frac{1}{2}$ is 40 ft/sec.

Problem 2
Likewise, check that the average speed from $t = 0$ to $t = 1$ is 72 ft/sec and that the average speed from $t = 1$ to $t = 2$ is 40 ft/sec. What is the average of these two averages?

Most dictionaries give velocity as a synonym for speed, but there is a clear mathematical distinction between them: "velocity" takes direction into account; "speed" does not. To account for direction, we use a "directed line," with positive and negative coordinates.

In Figure 2-1(a), the *displacement* of P from the origin is 3, and the displacement of Q is -4. In Figure 2-1(b), the displacement of R from the origin is 2, and the displacement of S is -2.

We define average velocity here only for motion in a straight line:

(a)

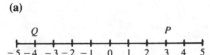

(b)

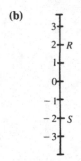

Definition 2
If an object moving on a straight line has displacement c from a fixed point on the line at time $t = a$, and displacement d at time $t = b$, then the **average velocity** in the interval is $(d - c)/(b - a)$.

Fig. 2-1

For the interval $[0, 2]$ in the rocket problem, we get the same value for average velocity as for average speed, but for the interval $[3, 4]$ the average velocity is $\frac{144 - 168}{4 - 3} = -24$ ft/sec, while the average speed is 24 ft/sec. The negative value indicates that the motion in that interval is downward.

Clearly, the average velocity in an interval can be positive, negative, or zero, whereas the average speed is always positive if the object is moving at all. It is the average speed one needs to know in order to find the distance traveled in an interval. By way of contrast, note that if a particle moving on a straight line starts at a certain point and later returns to that point, the average velocity during the interval under consideration is zero, no matter how far and fast the particle has traveled.

PROBLEMS

3. The height (y ft) of an object above the ground t sec after it started is given by $y = 64t - 16t^2$.
   (a) Make a table of values showing y at various values of t, from the start ($t = 0$) until the object hits the ground.
   (b) On the basis of this table (or from an accompanying graph), estimate what t corresponds to the maximum height.
   (c) Find the maximum height.
   (d) Find the average speed and the average velocity of the object for the following intervals: $[1,2]$, $[2,3]$, $[0,3]$, and the whole time interval of its motion.

★ 4. In Problem 2, you probably observed that

   $$\tfrac{1}{2} \,(\text{avg. speed in } [0,1] + \text{avg. speed in } [1,2]) = \text{avg. speed in } [0,2].$$

   Is this true in general, that is, for any time intervals? For any type of variation? Formulate a general result that you think to be true, and prove it if you can. Answer the same question if "speed" is replaced by "velocity."

○ 5. (a) For the rocket problem, with the height above the ground (y ft) varying with elapsed time (t sec) as follows,

   $$y = -16t^2 + 88t + 48, \quad 0 \le t \le 6,$$

   calculate the average velocity from $t = 1$ to $t = 1.1$. From $t = 1$ to $t = 1.01$. From $t = 1$ to $t = 1.001$.
   (b) Formulate a possible definition of what might be *meant* by the phrase "the velocity of a particle *at* $t = 1$."
   (c) In the light of your definition in (b), what do you think is the *value* of the velocity of the rocket at $t = 1$?

6. In Problems 1 and 2, you probably observed that the average speed in $[1,2]$ equals the average speed in $[\tfrac{1}{2}, \tfrac{5}{2}]$. Is there a general principle here? Formulate a general result that you think to be true, and prove it if you can. Can "speed" be replaced by "velocity" in your statement?

★ 7. (a) For the rocket problem, tell *how to find* the average speed from $t = 1$ to $t = 4$. (Be careful!) Actually find this average speed. [Problem 5(c)(iii) of Section 1.9 may be helpful.]
   (b) Likewise, find the average speed from $t = 0$ to $t = \tfrac{11}{4}$. From $t = \tfrac{11}{4}$ to $t = 6$. From $t = 0$ to $t = 6$.
   (c) Find the average velocity in each interval mentioned in (a) and (b).

2.2 Instantaneous velocity and limits

Problem 5(a) in the preceding section may be effectively attacked as follows: For the rocket we can designate y by $f(t)$, so that $f(t) = -16t^2 + 88t + 48$. The subsequent manipulations will turn out to proceed a bit more smoothly if we change the order of terms and write $f(t) = 48 + 88t - 16t^2$. Hence, at $t = 1$, the displacement of the rocket from ground level $= f(1) = 48 + 88 - 16 = 120$. At $t = 1 + h$, for any h, the displacement of the rocket from ground level $= f(1 + h) = 48 + 88(1 + h) - 16(1 + h)^2$.

Problem 1
Show that $f(1 + h)$ simplifies to $120 + 56h - 16h^2$.

Then, by the definition of average velocity (Definition 2, **2.1**), with $a = 1$ and $b = 1 + h$, the average velocity of the rocket in this interval is

$$\frac{f(1 + h) - f(1)}{(1 + h) - 1} = \frac{(120 + 56h - 16h^2) - 120}{h} = \frac{56h - 16h^2}{h}$$

$$= 56 - 16h, \quad \text{for } h \neq 0.$$

Thus, for the rocket, the average velocity

from $t = 1$ to $t = 1.1$	is $56 - 16(0.1) = 54.4$	(ft/sec),
from $t = 1$ to $t = 1.01$	is $56 - 16(0.01) = 55.84$	(ft/sec),
from $t = 1$ to $t = 1.001$	is $56 - 16(0.001) = 55.984$	(ft/sec),
and from $t = 1$ to $t = 1.0001$	is $56 - 16(0.0001) = 55.9984$	(ft/sec).

It seems clear that the closer h is to zero, the closer the average velocity is to 56 ft/sec. It is meaningless to speak of the average velocity or speed when h equals zero, for these terms are defined only over an interval – the phrase "*average* velocity (or speed) *at* a certain instant" has no sense. Put otherwise, we note that the expression we had earlier for average velocity, namely,

$$\frac{f(1 + h) - f(1)}{h},$$

has no meaning when $h = 0$. But intuitively it does make sense to ask "How fast am I going *right* now?" or "What was the velocity of the rocket one second after it started – that is, at the instant when $t = 1$?" And it seems reasonable in the light of our computations to give 56 ft/sec as the answer to the latter question, for it is apparent that we can get an average velocity as close to 56 as we like by taking a sufficiently short interval beginning at $t = 1$. We say in this case that "56 is the *limit* of the average velocity in the interval $[1, 1 + h]$ as h approaches zero." We define the *velocity at the instant* $t = 1$ as the limit of the average velocity in the interval $[1, 1 + h]$ as h approaches zero, and the *speed at* an instant similarly. The crucial word here is "limit," as yet undefined, and we turn our attention now to a discussion of that concept.

The notion of limit is an extremely important one. Later we shall devote considerable attention to the theory of limits, but for the time being we shall get along with a definition in verbal terms. We lead up to such a definition now. In the case treated earlier, $56 - 16h$ will be closer to 56 the closer h is to zero. But more than this: We shall show later that $56-16h$ can be made *arbitrarily close* to 56 by choosing h *sufficiently close* to zero. It is not merely that we can come close to 56 – the important aspect is that we can come as *close as we like* to 56 by choosing h appropriately close to zero.

The symbolism that we use in this case is as follows:

$$\lim_{h \to 0} (56-16h) = 56,$$

which is read "the limit of $56-16h$, as h approaches zero, is 56."

Definition 3 (in verbal terms)

The **limit of $G(u)$ as u approaches a is L** means that the number $G(u)$ is arbitrarily close to L for all u in the domain of G sufficiently close to a, except perhaps for $u = a$.

Notation If the limit of $G(u)$ as u approaches a is L, we write $\lim_{u \to a} G(u) = L$. Thus, if $G(h) = [f(1+h) - f(1)]/h = (56h - 16h^2)/h$, then it can be shown that $\lim_{h \to 0} G(h) = 56$. This example illustrates the significance of the last phrase in the definition of limit, "except perhaps for $u = a$." As we noted before, $[f(1+h) - f(1)]/h$ is not defined for $h = 0$; hence, we cannot speak of the value $G(0)$. In studying the limit of $G(h)$ as h approaches zero, we deal with $G(h)$ for h close to, but not equal to, zero.

The distinctions we are drawing here are easily made in terms of the language of functions. Consider the equations

$$g = \frac{56h - 16h^2}{h}, \quad h \neq 0, \tag{1}$$

$$r = 56 - 16h, \quad h \neq 0, \tag{2}$$

$$t = 56 - 16h. \tag{3}$$

Equations (1) and (2) are alternative ways of defining a function, G, whose natural domain consists of all real numbers except $h = 0$. Equation (3) defines another function, F, say, whose domain is the set of all real numbers. F is an extension of G. $G(0)$ does not exist, but $\lim_{h \to 0} G(h)$ *does* exist, and equals 56. Moreover, $F(0) = 56$, and $\lim_{h \to 0} F(h)$ also equals 56.

You may wonder why we "make things hard for ourselves" by bothering with functions like G – why not consider simply the function F and its value at $h = 0$? The reason is that it is G, rather than F, that expresses the quantity in which we were initially interested – the average velocity of the rocket. Lack of a clear distinction between G and F, and between $G(0)$ and $\lim_{h \to 0} G(h)$, led to a lack of precision in Newton's exposition of the basic concepts of the calculus, and to Bishop Berkeley's biting criticism of how Newton and other mathematicians explained their methods. It took over a hundred years to resolve the matter satisfactorily, through the clear-cut definition of limit by the French mathematician Cauchy early in the nineteenth century.

In the definition of limit, note the phrase "for all u in the domain of the function sufficiently close to a." This means that we should consider values of u somewhat *less* than a, as well as values of u somewhat *greater* than a, provided that the domain of G extends to both sides of a. Consider, for example, $F(h) = 56 - 16h$, with the natural domain suggested by this equation. If we are to show that $\lim_{h \to 0} F(h) = 56$, we must establish that the value of $F(h)$ is arbitrarily close to 56 for all h sufficiently close to zero (i.e., for small positive h and also for small negative h). Notice that although we have used only positive values of h in our computations, there is nowhere any assumption that $h > 0$, and the derivation of the formula,

$$\text{average velocity} = 56 - 16h,$$

holds for negative, as well as positive, numbers h.

Problem 2
For the rocket, with $y = f(t) = 48 + 88t - 16t^2$, compute the average velocity for each of the following intervals: (a) [0.9, 1], (b) [0.99, 1], (c) [1 - h, 1], for any $h > 0$.

Thus, whether we consider an interval beginning at $t = 1$ or an interval ending at $t = 1$, we obtain the same expression for the average velocity, and applying the definition of limit involves consideration of intervals beginning at $t = 1$ and also of intervals ending at $t = 1$. Indeed, this is exactly what we need to define instantaneous velocity:

Definition 4
The **instantaneous velocity** of a particle at $t = a$ is the limit of the average velocity of the particle in an interval of length h, with a as one end point of the interval, as h approaches zero.

For the present we shall merely use our intuition about the values of limits and shall not verify that the definition of limit is indeed fulfilled. Here is a typical argument: If x is close to 3, then $7x - 5$ is close to 16; the closer x is to 3, the closer $7x - 5$ will be to 16. It looks as though we can make the value of $7x - 5$ as close to 16 as we like by choosing x sufficiently close to 3; hence, it seems apparent that $\lim_{x \to 3}(7x - 5) = 16$. Similarly, it looks as though

$$\lim_{x \to 6} \sqrt{10 + x} = 4;$$

$$\lim_{x \to 0} \frac{3x^2 - 2x + 1}{x - 2} = -\frac{1}{2};$$

$$\lim_{h \to 0} (3x^2 + 2xh + h^2) = 3x^2, \quad \text{for each real } x;$$

$$\lim_{u \to -2} \frac{u}{u + 2} \quad \text{does not exist; etc.}$$

77

Example 1

Suppose that for a particle moving in a straight line, the distance (s ft) covered in the first t sec is given by

$$s = 16t^2, \quad \text{for all } t \geq 0.$$

Let us think of this as the function $s(t)$.

Problem 3

"Speed" and "velocity" will be the same for this function. Why?

We can find the instantaneous velocity at $t = 1$ as follows:

At $t = 1$, $s(1) = 16$.

At $t = 1 + h$, $s(1 + h) = 16(1 + h)^2 = 16 + 32h + 16h^2$.

For the interval $[1, 1 + h]$, $h > 0$, the average velocity of the particle is

$$\frac{s(1 + h) - s(1)}{h} = \frac{32h + 16h^2}{h} = 32 + 16h.$$

The same result would be obtained for the average velocity in the interval $[1 + h, 1]$ for $h < 0$. It seems as though $\lim_{h \to 0}(32 + 16h) = 32$. Hence, the instantaneous velocity at $t = 1$ is 32 ft/sec.

Problem 4

Use a similar argument to show that the instantaneous velocity at $t = 2$ is 64 ft/sec and that the instantaneous velocity at $t = 3$ is 96 ft/sec.

We can deal with all these cases at once by finding a formula for the instantaneous velocity for any $t \geq 0$:

For any t, $s(t) = 16t^2$.

At $t + h$, $s(t + h) = 16(t + h)^2 = 16t^2 + 32th + 16h^2$. For the interval $[t, t + h]$, the average velocity of the particle is

$$\frac{s(t + h) - s(t)}{h} + \frac{32th + 16h^2}{h} = 32t + 16h.$$

The same result would be obtained for the average velocity in the interval $[t + h, t]$ for $h < 0$, provided that $t + h$ lies in the domain of this function (i.e., provided that $t + h \geq 0$).

For each t, it seems as though $\lim_{h \to 0}(32t + 16h) = 32t$. Hence, for each $t \geq 0$, the instantaneous velocity of the particle is $32t$ ft/sec. Thus, for $t = 0$, 1, 2, and 3, the instantaneous velocities are 0, 32, 64, and 96 ft/sec, respectively.

Note that the situation at $t = 0$ is slightly different from that for any other t. Because the domain of our function is $[0, \infty)$, when we come to analyze average velocities we consider intervals that begin at zero (i.e., $[0, 0 + h]$, $h > 0$), but we cannot consider intervals that end at zero (i.e., $[0 + h, 0]$ for

$h < 0$). But this is perfectly consistent with Definition 3 – remember the phrase "for all u in the *domain of G* sufficiently close to a."

PROBLEMS

5. State the value of $f(0)$ and what you think to be the value of $\lim_{u \to 0} f(u)$ if $f(u)$ equals

(a) $u^2 + 4$ (b) $\dfrac{3u + 4}{u - 2}$ (c) $(u + 1)(2u - 3)$

(d) $-u$ for all $u < 0$, and 0 for all $u \geq 0$. (A sketch of the graph of this function may be of help.)

6. In each of the following cases, state the natural domain of $F(x)$ and the value of $F(3)$, if it exists. Also state what you think to be the value of $\lim_{x \to 3} F(x)$, or state that it does not exist.

(a) $F(x) = 2x + 4$ (b) $F(x) = \dfrac{x - 3}{x + 3}$ (c) $F(x) = \dfrac{x^2 + 4}{x - 3}$

(d) $F(x) = \dfrac{x^2 - 9}{x - 3}$ (e) $F(x) = (x + 2)(x - 2)$

(f) $F(x) = \begin{cases} 3, & \text{for } x \leq 3 \\ x, & \text{for } x > 3 \end{cases}$

(g) $F(x) = \begin{cases} x^2 + 1, & \text{for } x < 3 \\ 2x + 4, & \text{for } x \geq 3 \end{cases}$ (h) $F(x) = \begin{cases} 14 - 4x, & \text{for } x < 3 \\ 11 - x^2, & \text{for } x > 3 \end{cases}$

(i) $F(x) = \begin{cases} -\frac{2}{3}x, & \text{for } x < 3 \\ 0, & \text{for } x = 3 \\ -2, & \text{for } x > 3 \end{cases}$ (j) $F(x) = \begin{cases} -\frac{2}{3}x, & \text{for } x < 3 \\ 0, & \text{for } x = 3 \\ 11 - x^2, & \text{for } x > 3 \end{cases}$

7. For the function f determined by $f(x) = 3 - 2x + x^2$, $-\infty < x < \infty$, state what you think to be the value of each of the following indicated limits or why you think that the limit does not exist.

(a) $\lim_{x \to 4} f(x)$ (b) $\lim_{h \to 0} [f(4 + h) - f(4)]$ (c) $\lim_{h \to 0} \dfrac{f(4 + h) - f(4)}{h}$

(d) $\lim_{x \to 0} [f(x + h) - f(x)]$ (e) $\lim_{h \to 0} [f(h) - f(0)]$

(f) $\lim_{h \to 0} [f(x + h) - f(x)]$, for each x

(g) $\lim_{h \to 0} \dfrac{f(x + h) - f(x)}{h}$, for each x

8. For the function f determined by

$$f(t) = \begin{cases} 3t + 49, & 0 \leq t \leq 17, \\ 100, & 17 < t \leq 30, \end{cases}$$

state what you think to be the value of the indicated limit or why you think that the limit does not exist.

(a) $\lim_{t \to 5} f(t)$ (b) $\lim_{t \to 17} f(t)$ (c) $\lim_{t \to 0} f(t)$ (d) $\lim_{t \to 30} f(t)$

(e) $\lim_{h \to 0} [f(5 + h) - f(5)]$ (f) $\lim_{h \to 0} \dfrac{f(5 + h) - f(5)}{h}$

(g) $\lim_{h \to 0} [f(15 + h) - f(15)]$ (h) $\lim_{h \to 0} \dfrac{f(15 + h) - f(15)}{h}$

(i) $\lim_{h \to 0} [f(17+h) - f(17)]$ (j) $\lim_{h \to 0} \dfrac{f(17+h) - f(17)}{h}$

(k) $\lim_{h \to 0} \dfrac{f(x+h) - f(x)}{h}$, for each x in $[0, 17]$

9. A certain savings certificate provides interest of 20% per year, compounded semiannually, but no interest is added until a half year has passed. Thus, if you deposit $1000 on January 1, that sum (and only that sum) is available to you until July. On July 1, you get interest of 20% per year (or 10% per half year) on the $1000, so your interest is $100, and your account stands at $1000 + $100 = $1100. It remains at $1100 through December 31, and on the next day, interest of 10% of $1100, or $110, is added to your account, making the total $1210. Continue the computation for another half-year period.

Letting the amount in the account be represented by A (dollars) and time (t years) be measured from the date of initial deposit, we have the following expression for a function with domain $0 \leq t < 2$:

$$A = \begin{cases} 1000, & 0 \leq t < \frac{1}{2}, \\ 1100, & \frac{1}{2} \leq t < 1, \\ 1210, & 1 \leq t < \frac{3}{2}, \\ 1331, & \frac{3}{2} \leq t < 2. \end{cases}$$

Sketch a graph with t on the horizontal axis and A on the vertical axis, showing by heavy dots on your graph the values of A at $t = 0, \frac{1}{2}, 1$, and $\frac{3}{2}$. Because of the appearance of the graph, we have here what is called a "step function."

State what you think to be the value of each of the following limits or why you think that the limit does not exist.

(a) $\lim_{t \to \frac{1}{4}} A$ (b) $\lim_{t \to 1} A$ (c) $\lim_{t \to \frac{5}{3}} A$ (d) $\lim_{t \to \frac{3}{2}} A$

10. For the function f given by

$$f = \begin{cases} 20, & 0 < w \leq 1, \\ 37, & 1 < w \leq 2, \\ 54, & 2 < w \leq 3, \end{cases}$$

sketch a graph with w on the horizontal axis and $f(w)$ on the vertical axis. What real-life situation corresponds to this function? (This was written in 1982.) State what you think to be the value of each of the following indicated limits or why you think that the limit does not exist.

(a) $\lim_{w \to \frac{3}{2}} f(w)$ (b) $\lim_{w \to 1} f(w)$ (c) $\lim_{w \to 2} f(w)$ (d) $\lim_{w \to 3} f(w)$

(e) $\lim_{h \to 0} [f(\frac{3}{2} + h) - f(\frac{3}{2})]$ (f) $\lim_{h \to 0} [f(2+h) - f(2)]$

(g) $\lim_{h \to 0} \dfrac{f(2+h) - f(2)}{h}$

11. A particle moves on a straight line, with the formulas for the distance (s ft) traveled in the first t sec as follows:

$$s = \begin{cases} t^2, & 0 \le t < 3, \\ -18 + 12t - t^2, & 3 \le t \le 6. \end{cases}$$

Sketch a graph showing this variation. State what you think to be the instantaneous velocity of the particle at

(a) $t = 1$ (b) $t = 0$ (c) $t = 4$ (d) $t = 3$ (e) $t = 6$

(f) any t in $[0, 3)$ (g) any t in $[0, 3]$ (h) any t in $[3, 6]$

c 12. Calculate the (approximate) values of each of the following expressions for $h = 0.1, 0.01, 0.001$, and 0.0001, and thus guess the value of the limit of each expression, as $h \to 0$:

(a) $\dfrac{(1+h)^5 - 1}{h}$ (b) $\dfrac{(1+h)^{10} - 1}{h}$ (c) $\dfrac{(2+h)^5 - 32}{h}$

(d) $\dfrac{(2+h)^5 - 31}{h}$ (e) $\left(\dfrac{1}{(1+h)^2} - 1\right) \Big/ h$ (f) $\left(\dfrac{1}{(2+h)^2} - \dfrac{1}{4}\right) \Big/ h$

13. State what you think to be the value of each of the following limits by doing some computing of values. (A table of square roots or a calculator will be helpful.)

(a) $\displaystyle\lim_{h \to 0} \dfrac{\sqrt{4+h} - 2}{h}$ (b) $\displaystyle\lim_{h \to 0} \dfrac{\sqrt{4+h^2} - 2}{h}$

(c) $\displaystyle\lim_{h \to 0} \left(\dfrac{1}{\sqrt{4+h}} - \dfrac{1}{2}\right) \Big/ h$

14. (a) Try, by algebraic manipulations, to rewrite the expression in Problem 13(a) so as to "see" the value of the limit without computation. (Hint: multiply numerator and denominator of the given fraction by $\sqrt{4+h} + 2$.)

(b) Same as (a) for 13(b).

(c) Same as (a) for 13(c).

15. A motorist enters a thruway at 2 p.m. and picks up a toll ticket on which the time is stamped. He leaves the thruway at a point 200 miles away at 5 p.m., where a state policeman examines his toll ticket and arrests him. Why? Explain clearly.

2.3 Theorems on limits

We have been obtaining answers to problems about instantaneous velocity by guessing various limits. To be confident of our answers, we need some theorems on limits. In this section we shall state the principal theorems on limits, all of which should seem entirely reasonable to you. In Section 2.4 we shall indicate something of how the theorems are proved. Even if you do not master the proofs, you should use the theorems as needed.

We begin with two simple results of great importance:

Theorem on the Limit of a Constant Function

If $f(u) = k$, a constant, then $\lim_{u \to a} f(u) = k$, for any number a in the domain of f.

Theorem on the Limit of the Function $g(u) = u$

If $g(u) = u$, then $\lim_{u \to a} g(u) = a$, for any number a in the domain of g.

You should draw a graph of each of these functions, with u as the independent variable, and convince yourself that the conclusions of the theorems are just what you would expect.

As introduction to the next result, we refer to some guesses that we made earlier. We said that if $f(x) = 7x - 5$, then "it seems apparent that" $\lim_{x \to 3} f(x) = \lim_{x \to 3} (7x - 5) = 16$. Likewise, if $g(x) = 2x + 4$, then "it seems apparent that" $\lim_{x \to 3} g(x) = \lim_{x \to 3} (2x + 4) = 10$.

What can we say about the behavior of the *sum* of these functions: $F(x) = f(x) + g(x) = (7x - 5) + (2x + 4) = 9x - 1$? Because $f(x)$ is arbitrarily close to 16 for x sufficiently close to 3, and $g(x)$ is arbitrarily close to 10 for x sufficiently close to 3, then $F(x) = f(x) + g(x)$ should surely be close to $16 + 10 = 26$ for x close to 3. In other words, we guess that $\lim_{x \to 3} F(x) = \lim_{x \to 3} f(x) + \lim_{x \to 3} g(x)$. This is a general result for the *sum* of any two functions:

Theorem on the Limit of the Sum of Two Functions

If $\lim_{u \to a} f(u) = p$ and $\lim_{u \to a} g(u) = q$, then

$$\lim_{u \to a} [f(u) + g(u)] = p + q.$$

"The limit of a sum equals the sum of the limits."

A similar result is true for products:

Theorem on the Limit of the Product of Two Functions

If $\lim_{u \to a} f(u) = p$ and $\lim_{u \to a} g(u) = q$, then

$$\lim_{u \to a} [f(u) \cdot g(u)] = p \cdot q.$$

"The limit of a product equals the product of the limits."

The result for quotients requires the additional hypothesis that the limit of the denominator is not zero:

Theorem on the Limit of the Quotient of Two Functions

If $\lim_{u \to a} f(u) = p$ and $\lim_{u \to a} g(u) = q \neq 0$, then

$$\lim_{u \to a} \left[\frac{f(u)}{g(u)} \right] = \frac{p}{q}.$$

"The limit of a quotient equals the quotient of the limits provided that the limit of the denominator is not zero."

If the limit of the denominator *is* zero, then "anything can happen." For example, because $\lim_{u \to 5}(u-3) = 2$ and $\lim_{u \to 5}(u-5)^2 = 0$, it seems clear that $(u-3)/(u-5)^2$ increases without bound as $u \to 5$. Similarly, $(u-3)/(u-5)$ is a very large positive number if u is a bit larger than 5 and is a very large negative number if u is a bit smaller than 5. Thus, neither $\lim_{u \to 5}[(u-3)/(u-5)^2]$ nor $\lim_{u \to 5}[(u-3)/(u-5)]$ exists. Likewise, because $(u-5)/(u-5)^2 = 1/(u-5)$ for $u \neq 5$, and because $\lim_{u \to 5} 1/(u-5)$ does not exist, $\lim_{u \to 5}[(u-5)/(u-5)^2]$ does not exist either.

On the other hand, because $(u-5)^2/(u-5) = (u-5)$ for $u \neq 5$, and because $\lim_{u \to 5}(u-5) = 0$, we conclude that $\lim_{u \to 5}[(u-5)^2/(u-5)] = 0$. And because $3(u-5)/(u-5) = 3$ for all $u \neq 5$, we know that $\lim_{u \to 5}[3(u-5)/(u-5)] = 3$. (Remember, from the foregoing illustrations, that what happens *at* $u = 5$ is irrelevant to the existence or the value of the limit as u *approaches* 5.)

In colloquial terms, we can say that if the numerator and denominator both approach zero, then what happens to the fraction depends on "how fast" each approaches zero. We can summarize the foregoing results in the following statement.

Supplement to the Theorem on the Limit of the Quotient of Two Functions

If $\lim_{u \to a} f(u) = p$ and $\lim_{u \to a} g(u) = 0$, then,
 (i) if $p \neq 0$, $\lim_{u \to a}[f(u)/g(u)]$ does not exist, and
 (ii) if $p = 0$, then further investigation is required to determine $\lim_{u \to a}[f(u)/g(u)]$, or whether it exists.

The situation in part (ii) of the foregoing supplement has already been met in *every* problem on instantaneous velocity, for example,

$$\lim_{h \to 0} \frac{56h - 16h^2}{h},$$

and will be met again in *all* problems on other instantaneous rates of change.

The theorem on the limit of a constant function and the theorem on the limit of a product give the following result: If $\lim_{u \to a} F(u) = L$, and if k is a constant, then $\lim_{u \to a} kF(u) = (\lim_{u \to a} k)[\lim_{u \to a} F(u)] = k \cdot L$.

The theorem on the limit of the function $g(u) = u$ and the theorem on the limit of a product give the following results: $\lim_{u \to a} u^2 = \lim_{u \to a}(u \cdot u) = (\lim_{u \to a} u)(\lim_{u \to a} u) = a^2$; $\lim_{u \to a} u^3 = a^3$; and, in general, for any positive integer n, $\lim_{u \to a} u^n = a^n$.

The foregoing results imply that

$$\lim_{u \to a} (ku + l) = ka + l,$$

$$\lim_{u \to a} (ku^2 + lu + m) = ka^2 + la + m,$$

$$\lim_{u \to a} (ku^3 + lu^2 + mu + n) = ka^3 + la^2 + ma + n,$$

and so forth.

Thus, for any polynomial $P(u)$ and any real number a, we know that $\lim_{u \to a} P(u) = P(a)$. In words, the *limit* of a polynomial function, as the variable *approaches* some number a, equals the value of the polynomial *at a*.

PROBLEMS

1. In Section 2.2 we had the following formula for the average velocity of a particle in an interval of length h: average velocity $= 56 - 16h$, for $h \neq 0$. Use the theorems of this section to demonstrate that the limit of the average velocity, as $h \to 0$, equals 56.

2. What is wrong with each of the following statements?
   (a) $\lim_{x \to 3} (2x + 4) \to 10$. (b) As $x \to 3$, $(2x + 4) = 10$.

3. (a)–(c). Use the theorems of this section to obtain the limits that you guessed in Problem 5(a)–(c) in Section 2.2.

4. (a)–(f). Use the theorems of this section to obtain the limits that you guessed in Problem 6(a)–(f) in Section 2.2.

5. (a)–(f). Show how the result on the limit of a polynomial, as presented at the end of this section, enables you to state with confidence the answers to each of the parts of Problem 7 in Section 2.2.

○ 6. State the values of each of the following indicated limits, or that they do not exist:

   (a) $\lim_{h \to 0} \dfrac{50h + 0.1h^2}{h}$ (b) $\lim_{h \to 0} \dfrac{50h + 0.1h^2}{h^2}$

   (c) $\lim_{h \to 0} \dfrac{3h^{1/2}}{h^{1/2}}$ (d) $\lim_{h \to 0} \dfrac{-h}{2h^{1/2}}$

   (e) $\lim_{h \to 0} \dfrac{-2}{h^{1/3}}$ (f) $\lim_{h \to 0} \dfrac{h}{3h^{3/2}}$

★ 7. Discuss $\lim_{u \to a} u^n$ if n is a negative integer.

★ 2.4 Proofs of some results on limits

To prove the theorems on limits, or to establish results about limits without appeal to those theorems, we must refer to the *definition* of limit (Definition

2.4
Proofs of some
results on limits

3, **2.2**), repeated here using some symbolism:

Definition 3

$\lim_{u \to a} G(u) = L$ *means that* the number $G(u)$ is arbitrarily close to L for all u in the domain of G sufficiently close to a, except perhaps for $u = a$.

We shall illustrate how to use this definition in establishing a result that we guessed earlier: $\lim_{x \to 3}(7x - 5) = 16$. Can we ensure that $(7x - 5)$ lies within a distance of 0.01, say, of 16? That is, can we make $(7x - 5)$ lie between $(16 - 0.01)$ and $(16 + 0.01)$?

$$\text{If} \quad 16 - 0.01 < 7x - 5 < 16 + 0.01, \tag{4}$$

$$\text{then} \quad 16 - 0.01 + 5 < 7x < 16 + 0.01 + 5, \tag{5}$$

$$\text{or} \quad 21 - 0.01 < 7x < 21 + 0.01, \tag{6}$$

$$\text{or} \quad 3 - \frac{0.01}{7} < x < 3 + \frac{0.01}{7}. \tag{7}$$

Conversely, if (7) is true, then (6) is true, and then, in turn, (5) and (4) are true.

In other words, if x is within a distance of $\frac{0.01}{7}$ of 3, then $(7x - 5)$ is within a distance of 0.01 of 16.

Suppose, now, that we want $(7x - 5)$ to be closer still to 16 – within a distance of 0.0001, say, of 16.

Problem 1

Follow the model of the preceding argument to show that if x is within a distance of $\frac{0.0001}{7}$ of 3, then $(7x - 5)$ is within a distance of 0.0001 of 16.

Specific numbers, even small ones like 0.01 and 0.0001, don't logically fill the bill of being "as close as we like" to 16. For that we should use a letter, like ϵ, to represent an *arbitrary positive number*. Following the model of the preceding argument enables us to prove that $\lim_{x \to 3}(7x - 5) = 16$: Let ϵ be an arbitrary positive number.

$$\text{If} \quad 16 - \epsilon < 7x - 5 < 16 + \epsilon, \tag{8}$$

$$\text{then} \quad 16 - \epsilon + 5 < 7x < 16 + \epsilon + 5, \tag{9}$$

$$\text{or} \quad 21 - \epsilon < 7x < 21 + \epsilon, \tag{10}$$

$$\text{or} \quad 3 - \frac{\epsilon}{7} < x < 3 + \frac{\epsilon}{7}. \tag{11}$$

If we use "$\Rightarrow$" to signify "implies," we have just shown that $(8) \Rightarrow (9) \Rightarrow (10) \Rightarrow (11)$. But each of the steps is reversible, so we can conclude that $(11) \Rightarrow (10) \Rightarrow (9) \Rightarrow (8)$, or (11) implies (8), or if $3 - \epsilon/7 < x < 3 + \epsilon/7$, then $16 - \epsilon < 7x - 5 < 16 + \epsilon$; or, if x is within a distance of $\epsilon/7$ of 3, then $(7x - 5)$ is within a distance of ϵ of 16; or, $(7x - 5)$ is within an arbitrary distance, ϵ, of 16 provided that x is within the distance $\epsilon/7$ of 3. This is just what is required to demonstrate that $\lim_{x \to 3}(7x - 5) = 16$, on the basis of Definition 3.

Note that in the foregoing illustration we *first guessed* the value of the limit and then demonstrated the correctness of the guess by showing that the criteria of the definition of limit are fulfilled. This is a common procedure.

Problem 2

Follow this procedure by guessing the value of $\lim_{x \to 3}(2x + 4)$ and then showing that your guess is correct.

We now know the limits of two functions, given by $(7x - 5)$ and $(2x + 4)$, as $x \to 3$. What can be said about the *limit of their sum*, as $x \to 3$; that is, what is the value of $\lim_{x \to 3}[(7x - 5) + (2x + 4)]$? In Section 2.3 we intuitively argued that if $(7x - 5)$ is close to 16 and $(2x + 4)$ is close to 10, then their sum is close to 26. It would be easy to demonstrate that 26 is the limit of the sum of these functions by noting that $(7x - 5) + (2x + 4) = 9x - 1$ and working with these expressions as we did in the illustration or as you did in Problem 2. It will be more instructive to proceed in a different way:

We wish to ensure that $(7x - 5) + (2x + 4)$ is within a distance of 0.01, say, of 26. Let us assign half of 0.01 to the function $(7x - 5)$, so to speak, and the other half of 0.01 to the function $(2x + 4)$. From the illustration we know that

$$16 - \tfrac{1}{2}(0.01) < 7x - 5 < 16 + \tfrac{1}{2}(0.01), \tag{12}$$

provided that x is within a distance of $\tfrac{1}{2}(0.01)/7$ of 3. Similarly, from Problem 2 we know that

$$10 - \tfrac{1}{2}(0.01) < 2x + 4 < 10 + \tfrac{1}{2}(0.01), \tag{13}$$

provided that x is within a distance of $\tfrac{1}{2}(0.01)/2$ of 3.

Now, $\tfrac{1}{2}(0.01)/7 < \tfrac{1}{2}(0.01)/2$. Hence, if x is within a distance of $\tfrac{1}{2}(0.01)/7$ of 3, it surely is within a distance of $\tfrac{1}{2}(0.01)/2$ of 3. Thus, if x is within a distance of $\tfrac{1}{2}(0.01)/7$ of 3, both the extended inequalities (12) and (13) are valid. By adding (12) and (13) we obtain

$$26 - 0.01 < (7x - 5) + (2x + 4) < 26 + 0.01.$$

This argument could be used equally well if 0.01 were replaced by 0.0001, or by an arbitrary positive number ϵ, thus showing that

$$\lim_{x \to 3}[(7x - 5) + (2x + 4)] = 26,$$

as predicted.

The very same argument can be used for any functions, f and g, to prove the Theorem on the Limit of the Sum of Two Functions (2.3). It is somewhat more complicated, but not different in principle, to establish the

Theorem on the Limit of the Product of Two Functions and the Theorem on the Limit of the Quotient of Two Functions.

A proof of the Theorem on the Limit of a Constant Function is confusing because it is so easy: If $f(u) = k$, a constant, we wish to show that $\lim_{u \to a} f(u) = k$, by using the definition of limit. We say that

$$\text{if} \quad k - \epsilon < f(u) < k + \epsilon, \tag{14}$$

$$\text{then} \quad k - \epsilon < k < k + \epsilon, \tag{15}$$

$$\text{so} \quad -\epsilon < 0 < \epsilon. \tag{16}$$

But, indeed, zero *does* lie between $-\epsilon$ and ϵ for any positive number ϵ, and because the steps in the foregoing sequence are reversible, we know that $(16) \Rightarrow (15) \Rightarrow (14)$. Thus, $f(u)$ lies arbitrarily close to k, not only for values of u close to a, but for *all a*!

A proof of the Theorem on the Limit of the Function $g(u) = u$ is also "too easy": If $g(u) = u$, we wish to show that $\lim_{u \to a} g(u) = a$, by using the definition of limit. We say that

$$\text{if} \quad a - \epsilon < g(u) < a + \epsilon, \tag{17}$$

$$\text{then} \quad a - \epsilon < u < a + \epsilon. \tag{18}$$

The foregoing step being reversible, we know that $(18) \Rightarrow (17)$. That is, if u lies within distance ϵ of a, then $g(u)$ lies within distance ϵ of a. This completes the argument.

PROBLEMS

3. By direct use of the definition of limit, prove that if $\lim_{u \to a} F(u) = L$, and if k is a constant, then $\lim_{u \to a} kF(u) = kL$.

4. Using the definition of limit, prove the Theorem on the Limit of the Sum of Two Functions:

$$\text{if} \quad \lim_{u \to a} f(u) = p \quad \text{and} \quad \lim_{u \to a} g(u) = q,$$

$$\text{then} \quad \lim_{u \to a} [f(u) + g(u)] = p + q.$$

Here is a start on the argument: We wish to show that $(p + q) - \epsilon < f(u) + g(u) < (p + q) + \epsilon$, provided that u is sufficiently close to a. From the definition of limit, we know that the hypothesis $\lim_{u \to a} f(u) = p$ means that $p - \epsilon/2 < f(u) < p + \epsilon/2$, provided that u is sufficiently close to a, that is, provided that $a - d_1 < u < a + d_1$ for some positive number d_1. Similarly, $\lim_{u \to a} g(u) = q$ means....

2.5 Average slope in an interval and slope at a point

In connection with linear interpolation, we have made use of the characteristic property of a straight line – that it rises (or falls) at the same rate

2
Rate of change

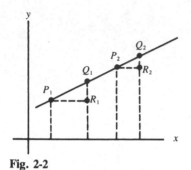

Fig. 2-2

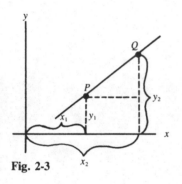

Fig. 2-3

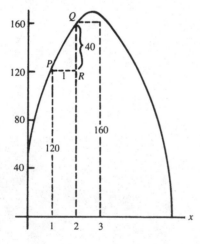

Fig. 2-4

everywhere. That means, in Figure 2-2, that if P_1, Q_1, P_2, and Q_2 are any four points on the line and if R_1Q_1 and R_2Q_2 are parallel to the vertical axis and P_1R_1 and P_2R_2 are parallel to the horizontal axis, then

$$\frac{R_1Q_1}{P_1R_1} = \frac{R_2Q_2}{P_2R_2}.$$

(Note that these are *directed* line segments. If the line is falling to the right, R_1Q_1 and R_2Q_2 are negative.) This ratio is called the *slope of the line*. It may be helpful to remember the definition of slope in colloquial language:

$$\text{slope} = \frac{\text{rise}}{\text{run}}.$$

Subscripts furnish a convenient notation for fixed values of a variable: x_1 will denote a particular value of x, x_2 another value, and so on. If P has coordinates (x_1, y_1) and Q has coordinates (x_2, y_2), then the slope of the line through P and Q equals $(y_2 - y_1)/(x_2 - x_1)$, as seen in Figure 2-3.

A geometric interpretation of the material in Section 2.2 on average and instantaneous velocities leads to results of great importance. In that section we dealt with the relation $y = 48 + 88t - 16t^2$, where y was the height of a rocket and t was the time. Figure 2-4 shows a graph of $y = 48 + 88x - 16x^2$. The height of P above the x axis is 120; the height of Q above the x axis is 160; hence, the line segment RQ has length $\overline{RQ} = 40$. And, of course, the line segment PR has length $\overline{PR} = 1$.

The number that previously we have thought of as measuring the average velocity in the interval $t = 1$ to $t = 2$ appears in the figure as the ratio $\overline{RQ}/\overline{PR} = 40$. This is also the slope of the line through P and Q. We call this the *average slope* of the curve between P and Q. In general, we have the following:

Definition 5
If P has coordinates (x_1, y_1) and Q has coordinates (x_2, y_2), with $x_2 \neq x_1$, then, for any curve passing through P and Q, the **average slope of the curve** between P and Q equals $(y_2 - y_1)/(x_2 - x_1)$. Alternatively, we call this the *average slope of the curve in the interval* $[x_1, x_2]$.

The average slope may be positive or zero or negative.

Problem 1
For the curve that we have been considering, verify that between the points $(4, 144)$ and $(6, 0)$ the average slope is -72. This corresponds to the fact that for the rocket, the average velocity is -72 ft/sec (or 72 ft/sec downward) for the interval $[4, 6]$.

As in the development of the concept of instantaneous velocity, we consider average slopes in smaller and smaller intervals to arrive at the notion of "instantaneous" slope, or slope *at* a point:

Definition 6

The **slope of a curve at** $x = a$ is the limit of the average slope of the curve in an interval of length h, with a as one end point of the interval, as h approaches zero.

It may be that the limit of the average slope does not exist, in which case we say that the slope of the curve at the point in question does not exist.

Example 1

For the curve whose equation is $y = 3 - 2x + x^2$, then

$$\text{when } x = 4, \qquad y = 3 - 8 + 16 = 11;$$

$$\text{when } x = 4 + h, \quad y = 3 - 2(4 + h) + (4 + h)^2$$

$$= 3 - 8 - 2h + 16 + 8h + h^2 = 11 + 6h + h^2.$$

For $h > 0$, the average slope of the curve in the interval $[4, 4 + h]$ is

$$\frac{11 + 6h + h^2 - 11}{h} = \frac{6h + h^2}{h} = 6 + h.$$

For $h < 0$, the average slope of the curve in the interval $[4 + h, 4]$ is

$$\frac{11 - (11 + 6h + h^2)}{-h} = 6 + h,$$

again. As we know from the discussion in Section 2.3, $\lim_{h \to 0}(6 + h) = 6$. Thus, we conclude that the slope of this curve *at* $x = 4$ is 6.

For practice, let us use the same method to find the slope of this curve at $x = -3$:

$$\text{at } x = -3, \qquad y = 3 + 6 + 9 = 18;$$

$$\text{at } x = -3 + h, \quad y = 3 - 2(-3 + h) + (-3 + h)^2$$

$$= 3 + 6 - 2h + 9 - 6h + h^2 = 18 - 8h + h^2.$$

For $h > 0$, the average slope in the interval $[-3, -3 + h]$ is

$$\frac{18 - 8h + h^2 - 18}{h} = \frac{-8h + h^2}{h} = -8 + h.$$

For $h < 0$, the average slope in the interval $[-3 + h, -3]$ is also $-8 + h$. Because $\lim_{h \to 0}(-8 + h) = -8$, we conclude that the slope of the curve *at* $x = -3$ is -8.

Let us find the slope of this curve at an arbitrary point:

$$\text{If } x = x_1, \qquad \text{then } y = 3 - 2x_1 + x_1^2.$$

$$\text{If } x = x_1 + h, \quad \text{then } y = 3 - 2(x_1 + h) + (x_1 + h)^2$$

$$= 3 - 2x_1 - 2h + x_1^2 + 2x_1 h + h^2.$$

89

Thus, the average slope in an interval of length h, beginning or ending at x_1, is

$$\frac{\left(3-2x_1-2h+x_1^2+2x_1h+h^2\right)-\left(3-2x_1+x_1^2\right)}{h}$$

$$=\frac{-2h+2x_1h+h^2}{h}=-2+2x_1+h.$$

As we know from the discussion in Section 2.3, $\lim_{h\to 0}(-2+2x_1+h)=-2+2x_1$. Thus, we conclude that the slope of this curve *at* $x=x_1$ is $-2+2x_1$. Note that if $x_1=4$, we obtain 6 as the slope, and if $x_1=-3$, we obtain -8 as the slope. Both results are in agreement with what we had earlier.

Problem 2
Note also that if $x_1=1$, the slope of the curve equals zero. Sketch the curve and give an interpretation of this result.

Example 2
For the curve whose equation is $y=x^{1/3}$, we have $y=0$ at $x=0$; and $y=h^{1/3}$ at $x=0+h$. Thus, the average slope in an interval of length h beginning or ending at $x=0$ is $(h^{1/3}-0)/h=1/h^{2/3}$. As we know from the discussion in Section 2.3, $\lim_{h\to 0}(1/h^{2/3})$ does not exist. Thus, we conclude that the slope of the curve does not exist at $x=0$.

Problem 3
Sketch the curve and give an interpretation of this result.

PROBLEMS

○ 4. (a) For the curve with equation $y=x^2$, find the slope at $x=2$; at $x=0$; at $x=-2$; at $x=x_1$. Also find the average slope in the interval $[-2,2]$.
  (b) Same as part (a) for the curve with equation $y=x^3$.
  (c) Same as part (a) for the curve with equation $y=x^4$.

○ 5. For the curve with equation $y=\frac{3}{2}x^2$, find the slope at $x=2$; at $x=0$; at $x=-2$; at $x=x_1$. Compare with the results of Problem 4(a). Also find the average slope in the interval $[0,2]$.

○ 6. For the curve with equation $y=x^2+5$, find the slope at $x=2$; at $x=0$; at $x=-2$; at $x=x_1$. Compare with the results of Problem 4(a).

○ 7. For the curve with equation $y=x^2+x^3$, find the slope at $x=2$; at $x=0$; at $x=-2$; at $x=x_1$. Compare with the results of Problems 4(a) and (b).

○ 8. (a) For the curve with equation $y=1/x$, find the slope at $x=2$; at $x=1$; at $x=-1$; at $x=x_1$. Any restriction on x_1?

(b) Same as part (a) for the curve with equation $y = 1/x^2$.

(c) Same as part (a) for the curve with equation $y = 1/x^3$.

9. For the curve with equation $y = 2 + 6x - x^2$, find the slope at $x = 0$; at $x = 2$; at $x = 3$; at $x = 4$; at $x = x_1$. Also find the average slope in the interval $[2, 3]$.

10. For the curve with equation $y = -3 + 8x + 2x^2$, find the slope at $x = -3$; at $x = -2$; at $x = 0$; at $x = x_1$. Also find the average slope in the interval $[-3, 0]$.

11. For the curve with equation $y = 2 + 3x$, find the slope at $x = x_1$.

12. For the curve with equation $y = 2 - 3x$, find the slope at $x = x_1$.

★ 13. For the curve with equations

$$y = \begin{cases} 3x + 49, & 0 \le x \le 17, \\ 100, & 17 < x \le 30, \end{cases}$$

find the slope at $x = 1$; at $x = 0$; at $x = 20$; at $x = 17$.

★ 14. For the curve with equation $y = \sqrt{x}$, find the slope at $x = 4$; at $x = 0$; at $x = x_1$. [Hint: See Problem 13(a), **2.2**.]

★ 15. For the curve with equations

$$y = \begin{cases} x^2, & x < 3, \\ -18 + 12x - x^2, & x \ge 3, \end{cases}$$

find the slope at $x = x_1$.

★ 16. For the curve with equations

$$y = \begin{cases} x^2, & x < 3, \\ 5, & x = 3, \\ -18 + 12x - x^2, & x > 3, \end{cases}$$

find the slope at $x = x_1$.

★ 17. For the curve with equation $y = x^{1/3}$, find the slope at $x = x_1 \ne 0$. [Hint: See Problem 13(c), **2.2**.]

18. Try to express what might be *meant* by "the line tangent to a curve at a point." Would it be satisfactory to say "a line that meets the curve at only one point" or "a line that touches but doesn't cross the curve at the point" or "the line perpendicular to the radius"?

★ 19. For the curve with equation $y = |x|$, find the slope at $x = 2$; at $x = 3$; at $x = -1$; at $x = x_1 \ne 0$. What about the slope at $x = 0$?

2.6 Tangent to a curve

The discussion of the preceding section leads to a satisfactory definition of a tangent line. In Figure 2-5, the secant PQ_2 seems to be a better approximation than the secant PQ_1 to what we intuitively think of as the tangent at P, and the secant PQ_3 is a still better approximation. We might attempt something like this: The tangent at P is the limiting position of secants PQ

2
Rate of change

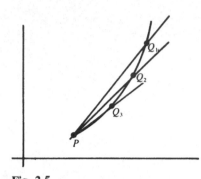

Fig. 2-5

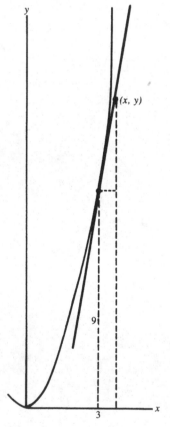

Fig. 2-6

as Q approaches P along the curve. Whereas such a statement seems to incorporate the salient feature with which we are concerned, it is not satisfactory as a definition because of the vagueness of the phrase "limiting position." We have defined limits for the values of functions – real numbers – not for "positions." We can make the shift from the vagueness of our attempted formulation to a satisfactory definition of tangent line by using the concept of *slope*, as follows:

Definition 7

The **tangent to a curve** at a point of the curve is the straight line through that point whose slope equals the slope of the curve at the point. If the slope does not exist because the average slope increases without bound as the length of the interval approaches zero, then the tangent is the vertical line through the point in question; likewise if the slope does not exist because the average slope decreases without bound. If the slope does not exist for any other reason, then the tangent does not exist.

Example 1

Let us investigate the tangent to the curve with equation $y = x^2$ at the point $(3, 9)$.

Problem 1

Verify that at the point $(3, 9)$ the slope of this curve is 6.

Thus, the tangent at the point is the line through $(3, 9)$ with slope 6, as seen in Figure 2-6. We know that the equation of a nonvertical line can be written in the form $y = ax + b$, where a is the slope. Therefore, the equation of this tangent has the form $y = 6x + b$. Because the equation must be satisfied by $x = 3$, $y = 9$, we conclude that $9 = 6 \cdot 3 + b$, or $b = -9$. Thus, the equation of the tangent line at $(3, 9)$ is $y = 6x - 9$.

Problem 2

Verify that the slope of this curve at any point (x_1, x_1^2) is $2x_1$. Thus, the tangent at this point is the line through (x_1, x_1^2) with slope $2x_1$.

Problem 3

Follow the same steps as before to find that the required tangent has equation $y = 2x_1 x - x_1^2$. In particular, if $x_1 = 0$, the tangent has equation $y = 0$. Sketch the curve and interpret this result.

Example 2

As was found in Problem 14, **2.5**, at the point $(4, 2)$ the slope of the curve with equation $y = \sqrt{x}$ is $\frac{1}{4}$.

Problem 4

Use the same steps as in the preceding Example 1 to obtain the following equation for the tangent to this curve at $(4,2)$: $y = \frac{1}{4}x + 1$.

PROBLEMS

In Problems 4–10, **2.5**, you were asked to find the slopes of certain curves at certain points on those curves. As Problems 5–11 for this section, find equations of the corresponding tangent lines.

In Problems 14 and 17, **2.5**, you were asked to find the slopes of certain curves at certain points on those curves. As Problems 12 and 13 for this section, find equations of the corresponding tangent lines.

2.7 The derivative

We have been nibbling away at a problem that concerned many of the best mathematicians of the seventeenth century: Just how do we determine the tangent to a given curve at a given point of the curve? For a circle, the problem is simple: We merely draw the perpendicular to the radius at the given point. Likewise, for certain other curves, ad hoc methods suffice. But Newton, building on the work of Descartes, Fermat, and others, took a great step forward in this "problem of tangents" by developing a method of great generality. This is the core of the differential calculus.

If the equation of a curve is $y = f(x)$, then, as we have seen several times already, the slope of the curve at $x = x_1$ is

$$\lim_{h \to 0} \frac{f(x_1 + h) - f(x_1)}{h}.$$

For each x_1 of a certain domain, the foregoing limit has a definite value. Thus, if there is at least one real number x_1 for which it exists, the limit determines a function, called the *derivative* of f.

Definition 8

Let f be a function. Then the **derivative of f** is a function, f', given by

$$f'(x) = \lim_{h \to 0} \frac{f(x+h) - f(x)}{h},$$

provided this limit exists.

The domain of f' is not larger than the domain of f. Sometimes f' is called the *first derived function* of f. The phrases "derivative of f" and "first derived function of f" have the connotation of a function obtained from f – but obtained in a very definite way. The process by which we obtain f' from f is called *differentiation*.

In Problem 4, **2.5**, you should have obtained the following results:

For the curve with equation $y =$	The slope at the point where $x = x_1$ is
x^2	$2x_1$
x^3	$3x_1^2$
x^4	$4x_1^3$

Problem 1

On the basis of these data, what is your guess as to the slope of the curve $y = x^n$ at $x = x_1$, for any positive integer n?

In order to find the derivatives as given in the foregoing table, we used the expansions of $(x + h)^2$, $(x + h)^3$, and $(x + h)^4$. In the general case, with $f(x) = x^n$, we write $f(x + h) = (x + h)^n$. We do not actually need the details of the binomial theorem; it is sufficient to know that

$$(x + h)^n = x^n + n \cdot x^{n-1} \cdot h + \text{terms involving } h^2, h^3, \ldots, h^n.$$

Then

$$f(x + h) - f(x) = (x + h)^n - x^n$$

$$= n \cdot x^{n-1} \cdot h + \text{terms involving } h^2, h^3, \ldots, h^n.$$

For $h \neq 0$,

$$\frac{f(x + h) - f(x)}{h} = n \cdot x^{n-1} + \text{terms involving } h, h^2, \ldots, h^{n-1}.$$

On the right side of this equation, as $h \to 0$, every term after the first approaches zero. Hence, if n is a positive integer,

$$f'(x) = \lim_{h \to 0} \frac{f(x + h) - f(x)}{h} = nx^{n-1},$$

as you doubtless had guessed. Thus:

If f is given by $f(x) = x^n$, with n being a positive integer, then f' is given by $f'(x) = nx^{n-1}$.

Problem 2

What result does this formula give for $n = 1$? Is this what you expected from other considerations?

Now let us see what happens if the exponent is a *negative* integer. In Problem 8, **2.5**, you should have obtained the following results:

For the curve with equation $y =$	The slope at the point where $x = x_1 \neq 0$ is
$\dfrac{1}{x} = x^{-1}$	$-\dfrac{1}{x_1^2} = -x_1^{-2}$
$\dfrac{1}{x^2} = x^{-2}$	$-\dfrac{2}{x_1^3} = -2x_1^{-3}$
$\dfrac{1}{x^3} = x^{-3}$	$-\dfrac{3}{x_1^4} = -3x_1^{-4}$

Note that these are the same values we would obtain if we used the formula for the derivative that we established with n a *positive* integer. It looks as though that formula applies if n is a *negative* integer, as well.

If we try to follow the preceding proof exactly, we run into a snag, because the expansion of $(x + h)^{-n}$ does not come to an end – it results in what is known as an "infinite series." We get around the difficulty in the following way: Suppose $f(x) = x^n$, where n is a negative integer. Let us set $n = -m$, where m is a *positive* integer. Then

$$f(x) = x^{-m} = \frac{1}{x^m} \quad \text{and} \quad f(x + h) = \frac{1}{(x + h)^m}.$$

(If $x \neq 0$, h can be taken so small that $x + h \neq 0$, so that $x + h$ lies in the domain of f.) Then

$$f(x + h) - f(x) = \frac{1}{(x + h)^m} - \frac{1}{x^m} = \frac{x^m - (x + h)^m}{(x + h)^m \cdot x^m}$$

$$= \frac{x^m - \left[x^m + mx^{m-1}h + \text{terms involving } h^2, h^3, \ldots, h^{m-1} \right]}{(x + h)^m x^m}.$$

Problem 3
Finish the argument by simplifying the numerator of this last fraction, dividing by h, and using the limit theorems, to obtain

$$f'(x) = \lim_{h \to 0} \frac{f(x + h) - f(x)}{h} = \frac{-mx^{m-1}}{x^{2m}} = -mx^{-m-1}.$$

Because $-m = n$, this is equivalent to $f'(x) = nx^{n-1}$, as predicted.

Problem 4
Show that the same formula applies if the exponent is zero, and interpret the result geometrically.

We can combine these results into a single formal statement.

Theorem 1

If $f(x) = x^n$, n any integer (positive, negative, or zero), then $f'(x) = nx^{n-1}$.

Indeed, the method of Problem 14, **2.5**, shows that if $f(x) = \sqrt{x} = x^{1/2}$, then $f'(x) = \frac{1}{2}x^{-1/2}$, and the method of Problem 17, **2.5**, shows that if $f(x) = x^{1/3}$, then $f'(x) = \frac{1}{3}x^{-2/3}$. Thus, the formula of Theorem 1 for the derivative of a function also is valid for $n = \frac{1}{2}$ and $\frac{1}{3}$. We shall later establish that the same result is true for any rational number n, as we shall demonstrate in Chapter 4 when we have more tools. Indeed, the same formula applies for any *real* number n, but we shall not attempt a proof.

We now know that if $f(x) = x^2$ and $g(x) = x^3$, then $f'(x) = 2x$ and $g'(x) = 3x^2$. What can we say about the derivative of $F(x) = f(x) + g(x) = x^2 + x^3$? If we work it out from first principles, that is, investigate $\lim_{h \to 0}\{[F(x+h) - F(x)]/h\}$, we shall find that

$$F'(x) = 2x + 3x^2 = f'(x) + g'(x).$$

This is a general result:

Theorem 2

If F is the function $f + g$, where f has derivative f' and g has derivative g', then F has a derivative, F', and $F' = f' + g'$. ("The derivative of the sum of two functions is the sum of their derivatives.")

The proof of this comes quickly from our result in Section 2.3 that "the limit of the sum equals the sum of the limits." The theorem immediately extends to the sum of three or four or any finite number of functions.

Another important question arises: If we know the derivative of a function f, what can we say about the derivative of the function kf, where k is a constant? For example, if $f(x) = x^2$, then we know $f'(x) = 2x$; what about the derivative of $5x^2$? If we work out the derivative of $5x^2$ from first principles, we shall quickly see that it is 5 times the derivative of x^2. This, too, is a general result:

Theorem 3

If G is the function $k \cdot f$, where k is a constant, and the derivative of f is f', then G has a derivative, G', and $G' = k \cdot f'$. ("The derivative of a constant times a function equals the constant times the derivative of the function.")

The proof of Theorem 3 is an easy consequence of the result of Section 2.3 that "the limit of a constant times a function equals the constant times the limit of the function."

Because any constant function, $f(x) = k$, can be written as $f(x) = k \cdot x^0$, we know that $f'(x) = 0$. ("The derivative of a constant is zero.") We can see this geometrically: The graph of $y = k$ is a horizontal straight line, the slope of which is zero everywhere.

2.7
The derivative

The (specific) formulas embodied in Theorem 1 and the (general) formulas embodied in Theorems 2 and 3 enable us to find the derivative of any polynomial function

$$a_0 + a_1 x + a_2 x^2 + \cdots + a_n x^n,$$

where the a's are constants and n is a positive integer. For example, if $f(x) = 3 + 2x - 4x^2 + x^3/2 - x^4 + \frac{2}{3}x^5$, then

$$f'(x) = 2 - 8x + \tfrac{3}{2}x^2 - 4x^3 + \tfrac{10}{3}x^4.$$

In fact, because Theorem 1 also applies in case the power of x is negative, we can differentiate not only polynomials but also some other functions. We can, for example, differentiate

$$f(x) = 3x^2 + x + 2 - \frac{1}{x} + \frac{2}{x^2}$$

to obtain

$$f'(x) = 6x + 1 + \frac{1}{x^2} - \frac{4}{x^3},$$

and

$$g(x) = \frac{7+x}{x^4} = \frac{7}{x^4} + \frac{1}{x^3}$$

to obtain

$$g'(x) = \frac{-28}{x^5} - \frac{3}{x^4}.$$

Warning! The theorems on limits and the elegant results on derivatives may tempt you to draw some unwarranted conclusions. For example, although "the limit of a product is equal to the product of the limits," the derivative of a product is *not* equal to the product of the derivatives. Consider the function given by $f(x) = r(x) \cdot s(x)$, where $r(x) = x^3$ and $s(x) = x^2$. Then $f(x) = x^5$, of course. Now, $f'(x) = 5x^4$, $r'(x) = 3x^2$, $s'(x) = 2x$. But $5x^4$ is *not* equal to $3x^2 \cdot 2x$. The correct formula for the derivative of a product is obtained in Chapter 4; until then, if we need the derivative of a product, we must multiply out (if possible) and apply Theorems 1, 2, and 3. Similarly, the derivative of a quotient is *not* equal to the quotient of the derivatives – just try $f(x) = x^3/x^2$.

Still another possible mistake occurs in finding the derivative of a function like $G(x) = (x^2 + 1)^2$. It is tempting – but incorrect! – to write $2(x^2 + 1)$ for $G'(x)$. Actually, $G(x) = x^4 + 2x^2 + 1$, by the Binomial Theorem, so $G'(x) = 4x^3 + 4x$, which is *not* equal to $2(x^2 + 1)$. For now, the only method we have for finding $G'(x)$ is that of first "multiplying out" (as described earlier); in Chapter 4 we shall find an alternative method.

97

2

Rate of change

Table 1-1

t	A
0	4.4
1	6.8
2	10.2
3	14.4
4	19.2
5	24.2
6	28.6
7	32.2
8	34.8
9	36.7
10	38.0

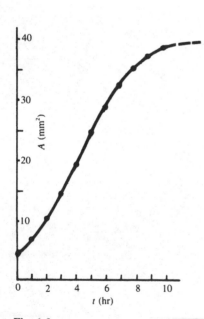

Fig. 1-2

To recapitulate: The derivative of a function $f(x)$, if it exists, is defined for any x as

$$f'(x) = \lim_{h \to 0} \frac{f(x+h)-f(x)}{h}.$$

This is a concept of great generality and many interpretations, of which we have considered two:

(a) If $y = f(x)$ is the equation of a curve, $f'(x_1)$ is the slope of the curve (= the slope of the tangent to the curve) *at* the point where $x = x_1$.

(b) If the displacement, y, from a fixed point at time x of a particle moving on a straight line is given by $y = f(x)$, then the (instantaneous) velocity *at* time x_1 is $f'(x_1)$.

Both foregoing examples are special cases of the general concept of *rate of change*: If x and y are any related quantities, with y a function of x, then the *average* rate of change of y with respect to x in the interval $[x_1, x_1 + h]$ is defined as the change in y divided by the change in x, and the *instantaneous* rate of change of y with respect to x *at* x_1 is defined as the limit of the average rate of change in an interval of length h, with x_1 as one end point, as h approaches zero.

For example, in the case of the problem of the bacterial population in Section 1.2 (Table 1-1), we have at $t = 1$, $A = 6.8$, and at $t = 3$, $A = 14.4$. Hence, the average rate of change of A with respect to t, in the interval $[1, 3]$, is $\frac{14.4-6.8}{3-1} = \frac{7.6}{2} = 3.8$ (mm²/hr).

Problem 5
Check from Table 1-1 that the average rate of change in the interval $[0, 4]$ is 3.7 (mm²/hr).

Assuming the validity of the formulas in Section 1.7 for expressing the variation of $\bar{A}$ with t, namely,

$$\bar{A} = f(t) = \begin{cases} 0.4t^2 + 2.1t + 4.4, & 0 \le t \le 4, \\ -0.4t^2 + 8.7t - 9.2, & 4 < t \le 10, \end{cases}$$

we obtain, for the instantaneous rate of change at any time t such that $0 \le t < 4$, $f'(t) = 0.8t + 2.1$. Hence, at $t = 2$, $f'(t) = 3.7$ (mm²/hr). This can be interpreted as the slope of the tangent to the graph shown in Figure 1-2 at $t = 2$ (Figure 2-7).

Problem 6
Obtain the formula $f'(t) = -0.8t + 8.7$ valid for any t such that $4 < t \le 10$. What, then, is the value of $f'(5)$?

Problem 7
At $t = 4$, $f'(t)$ does not exist. Why? The graph, then, is not smooth at $t = 4$.

The question "How *fast* does one quantity change with another?" is interpreted to mean "What is the rate of change of the first quantity with respect to the second?" and is answered by calculating the *derivative*. Our Theorems 1, 2, and 3 enable us to do this now for quite a broad class of functions, and we shall learn later how to find the derivatives for some others. But before doing that we shall investigate other uses of the derivative.

A closing observation: Keep clearly in mind the distinction between the value of a *function* and the value of the *rate of change* of that function. It is possible for $f(x_1)$ to be large and $f'(x_1)$ to be small, for $f(x_1)$ to be positive and $f'(x_1)$ to be negative, and so forth. Clearly, if $f'(x)$ has large positive values over the interval $[x_1, x_2]$, then $f(x_2)$ will be much larger than $f(x_1)$, but there is no correlation between the value of $f(x)$ at $x = x_1$ and the value of $f'(x)$ there (Figure 2-8).

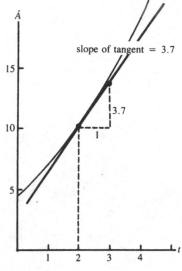

slope of tangent = 3.7

Fig. 2-7

PROBLEMS

8. Make rough sketches of small pieces of graphs that exhibit each of the following at $x = x_1$:
    (a) $f(x_1)$ a small positive number, $f'(x_1)$ a large positive number
    (b) $f(x_1)$ a large negative number, $f'(x_1)$ a small positive number
    (c) $f(x_1)$ zero, $f'(x_1)$ a large negative number
    (d) $f(x_1)$ a large positive number, $f'(x_1)$ zero
    (e) $f(x_1)$ a small negative number, $f'(x_1)$ a large negative number

9. Find f' in each of the following cases:

    (a) $f(x) = 7x$ (b) $f(x) = \dfrac{x^3}{3}$

    (c) $f(x) = \dfrac{2}{x}$ (d) $f(x) = -\dfrac{1}{2x^2}$

    (e) $f(x) = \pi$ (f) $f(t) = -18 + 2t - t^2$

    (g) $f(x) = (x^2 + 3)^2$ (h) $f(x) = (x - \dfrac{1}{x})^2$

    (i) $f(t) = \dfrac{3t^2 + t - 5}{2}$ (j) $f(x) = \dfrac{2}{(3x)^2}$

    (k) $f(t) = a + bt + ct^2$, where a, b, and c are constants   (l) $f(x) = 4 - (\sqrt{2}\,x)^2 + 3(\sqrt{2}\,x)^3$

    (m) $f(t) = (t-1)(2t+3)$ (n) $f(x) = x^2(x^2 - 1)$

    (o) $f(t) = \dfrac{5t^2 - 2t + 1}{t}$ (p) $f(x) = \dfrac{3x - 4}{2x^2}$

10. For each of the curves given by the following equations, find an equation of the tangent line at the point on the curve that has the given value of x:

    (a) $y = x^2 + 3x - 1$, at $x = -1$ (b) $y = \dfrac{1}{x}$, at $x = 2$

    (c) $y = \dfrac{1}{3}x^3 + 5$, at $x = -3$ (d) $y = \dfrac{1}{2x^2} - 3$, at $x = \dfrac{1}{2}$

Fig. 2-8 "I didn't say the prices were tapering off. I said the increases were tapering off." (Drawing by Modell; © 1974, *The New Yorker Magazine, Inc.*)

○ 11. For the curve given by the equation

$$y = f(x) = \tfrac{1}{3}x^3 - 2x^2 + 3x + 1,$$

(a) calculate the height and the slope of the curve at the points where $x = -1, 0, 1, 2, 3$, and 4,

(b) sketch the curve, using the same scale on the vertical axis as on the horizontal axis,

(c) write the equation of the tangent line to the curve at the point where $x = 0$, where $x = 2$, where $x = 3$.

12. A particle moves on a straight line so that its displacement (s ft) from a point on the line varies with the time (t sec) as follows:

$$s = 2t^3 - 15t^2 + 36t + 10.$$

(a) Find a formula for the velocity of the particle at any time.

(b) When is the particle stationary? That is, when is the velocity equal to zero?

★ 13. Find f' in each of the following cases:

(a) $f(x) = |x + 2|$ (b) $f(x) = |x^3|$

(c) $f(x) = \begin{cases} 0, & x = 0 \\ \dfrac{x}{|x|}, & x \neq 0 \end{cases}$ (d) $f(x) = |x^2 + x - 2|$

14. For the example of the flooding river (Problem 3, **1.4**), the formula for the level (L ft) in terms of time (t hr) is as follows:

$$L = 0.5 + 2.1t - 0.1t^2.$$

(a) What is the rate of change of L with respect to t at $t = 1$? At $t = 3$? At $t = 6$?

(b) When is the rate of change of L with respect to t equal to zero? What is the significance of your answer?

★ 15. What is the value of each of the following?

$$\lim_{h \to 0} \frac{(3+h)^{101} - 3^{101}}{h}, \quad \lim_{h \to 0} \frac{(3+h)^{101} - 3^{101}}{\sqrt{h}}, \quad \lim_{h \to 0} \frac{(3+h)^{101} - 3^{101}}{h^2}$$

★ 16. Find the value of

$$\lim_{h \to 0} \frac{1/(2+h)^5 - 1/32}{h}.$$

17. Find a formula for the derivative (including its domain) in each of the following cases, and draw graphs of the functions:

(a) $f(x) = \begin{cases} x^2 + 1, & 0 \leq x \leq 2 \\ -7 + 8x - x^2, & 2 < x \leq 5 \end{cases}$

(b) $g(x) = \begin{cases} x^2 + 1, & 0 \leq x \leq 2 \\ -5 + 8x - x^2, & 2 < x \leq 5 \end{cases}$

(c) $F(x) = \begin{cases} x^2 + 1, & 0 \leq x \leq 2 \\ -9 + 9x - x^2, & 2 < x \leq 5 \end{cases}$

18. A function f is defined as follows:

$$f(x) = \begin{cases} 5, & 0 \le x \le 1, \\ 10, & 1 < x \le 2, \\ 15, & 2 < x \le 3. \end{cases}$$

(a) Draw a graph of $y = f(x)$.

(b) $f(\frac{3}{2}) = \qquad ; f(2) =$

(c) $\lim_{x \to 3/2} f(x) = \qquad ; \lim_{x \to 2} f(x) =$

(d) The average slope of the graph from $x = \frac{3}{2}$ to $x = \frac{5}{2}$ is

(e) The average slope of the graph from $x = 1$ to $x = \frac{5}{2}$ is

(f) The slope of the graph at $x = \frac{3}{2}$ is

(g) The slope of the graph at $x = 2$ is

★ 2.8 Guessing limits with a calculator

Often, in establishing that $\lim_{x \to a} f(x) = L$, we must begin by *guessing* L. Sometimes it is easy to guess this number; but, when it is not, it is usually helpful to do some calculations – to calculate the values of $f(x)$ for x's closer and closer to a, in an effort to discover some "pattern." A calculator or computer can save us a lot of labor. But there are limitations, and it is the purpose of this section to make you a little cautious in using what is, most of the time, an excellent tool.

Generally, the limitations lie in the finiteness of the calculator, which must truncate and round nonterminating decimals. The most common problem is loss of accuracy through cancellation of significant digits by subtraction.

For example, if we return to the *definition* to find $f'(1)$ for $f(x) = x^2$, we need

$$\lim_{h \to 0} \frac{(1+h)^2 - 1}{h}.$$

Of course,

$$\frac{(1+h)^2 - 1}{h} = \frac{1 + 2h + h^2 - 1}{h} = \frac{2h + h^2}{h} = 2 + h,$$

and we recognize that the limit is 2.

If we ask the calculator for the value of $A = [(1+h)^2 - 1]/h$ for some (small) h, it should give us $2 + h$ for that h. The results for one calculator are given in Table 2-1.

Table 2-1

h	$A = \dfrac{(1+h)^2 - 1}{h}$	Calculator's approximation
10^{-4}	2.0001	2.0001
-10^{-4}	1.9999	1.9999
10^{-5}	2.00001	2.0
$8 \cdot 10^{-10}$	2.0000000008	2.5 (!)
$5 \cdot 10^{-10}$	2.0000000005	4.0 (!)

Table 2-2

n (number of sides of the polygon)	x (length of the side of a polygon of n sides)	P (perimeter of a polygon of n sides)	π (approximately equal to $P/2$)	y (length of the side of a polygon of $2n$ sides)
6	1	6	3	0.51764
12	0.51764	6.21166	3.10583	0.26105
⋮	⋮	⋮	⋮	⋮
384	0.01636	6.28311	3.14156	0.00818
768	0.00818	6.28316	3.14158	0.00409
1536	0.00409	6.28316	3.14158	0.00205
3072	0.00205	6.28297	3.14149	0.00102
6144	0.00102	6.28372	3.14186	

Much the same thing can happen in sequential calculations, such as those of Problem 8, **1.13**, in which we found approximations to π by inscribing regular polygons of increasing numbers of sides in a circle of radius 1. We used the fact that if x is the length of the side of a regular polygon of a certain number of sides, then y, the length of the side of a regular polygon of *twice* the number of sides, is given by $y^2 = 2 - \sqrt{4 - x^2}$. The use of a calculator gives rise to Table 2-2.

Actually, $\pi \approx 3.14159$, so that the inscribed polygon of 768 sides gives us the closest result, and later the approximations become *worse* – not because of the approximating method, but because cancellation in $y^2 = 2 - \sqrt{4 - x^2}$ as x gets small means that the calculator approximations do not get better beyond a certain point. Different calculators will produce slightly different numbers in this example because of variations in the numbers of digits used and in their rounding procedures. But all of them will, at some point or another, show the same kind of worsening in their approximations to π.

Archimedes used essentially our method, with circumscribed as well as inscribed polygons, to obtain $3\frac{10}{71} < \pi < 3\frac{1}{7}$; $3\frac{10}{71} \approx 3.141$, and $3\frac{1}{7} \approx 3.143$, so he achieved accuracy of about 0.002.

To find approximations to π correct to *many* decimal places, we need other methods not so sensitive to calculator round-off error.

PROBLEMS

C 1. (a) In Problem 13(a), **2.2**, you were asked to guess the value of $\lim_{h \to 0}[(\sqrt{4+h} - 2)/h]$. Use a calculator to make a table of approximations to $(\sqrt{4+h} - 2)/h$ for small h, using finer gradations when the estimates seem to be deteriorating. (One good calculator gives 0.4 for $h = 2.5 \times 10^{-9}$; another gives 37.5 for $h = 8 \times 10^{-10}$.)

(b) Problem 14(a), **2.2**, suggests rewriting the expression $(\sqrt{4+h} - 2)/h$ by "rationalizing the numerator" to obtain $1/(\sqrt{4+h} + 2)$. The limit, then, equals what?

C 2. Use a calculator with a y^x key to guess $\lim_{x \to 0} x^x$ (x must be positive).

C 3. (a) Perform the following sequence of calculations, and guess what limit is being approached:

(i) take the square root of 6;

(ii) add 6 to the result of (i);

(iii) take the square root of the result of (ii);

(iv) add 6 to the result of (iii);

(v) take the square root of the result of (iv);

$\vdots$

(b) Same as (a), starting with the square root of 20.

(c) Same as (a), starting with the square root of 30.

(d) Demonstrate that if the limits in (a), (b), and (c) exist, they must be the numbers you guessed.

2.9 Review

In reviewing the material in this chapter, the 10 topics mentioned at the start should be helpful to you. You should begin to appreciate the power of differentiation in providing the answer to *how fast* one quantity changes with another.

PROBLEMS

Sample test (Problems 1–7)

1. Find f' in each of the following cases:

   (a) $f(x) = \dfrac{2}{3x^3}$

   (b) $f(t) = \tfrac{1}{3} - 2t + 3t^2$

   (c) $f(x) = (x + \dfrac{1}{x})^2$

2. For the curve given by $y = f(x) = x^3 - 3x^2 + 4$, calculate the height and the slope of the curve at the points where $x = 0$, 1, and 2. Write an equation of the tangent line to the curve at the point where $x = 1$; where $x = 2$.

3. A particle moves on a straight line so that its displacement (s ft) from a point on the line varies thus with the time (t sec): $s = 20t - t^2$.

   (a) Find a formula for the velocity of the particle at any time.

   (b) When is the particle stationary? That is, when is its velocity equal to zero? At that time, what is the displacement of the particle?

   (c) When does the particle return to its starting point?

   (d) How far does the particle travel before it returns to its starting point?

   (e) What *speed* does the particle have when it returns to its starting point?

★ 4. How close does x have to be to 3 in order to ensure that $5x + 2$ differs from 17 by no more than 0.01? By no more than ϵ? Explain clearly how you know your answers to be correct.

5. Find the coordinates of a point on the graph of $y = x^2 + 1$ where the tangent line has the same slope as that of the line joining the points on the graph with $x = 1$ and $x = 5$.

6. Sketch a graph of a function f, given by $y = f(x)$, $3 \leq x \leq 9$, such that $f(3) = 2$, $f(7) = 10$, $f(9) = 5$, and $f'(7) = 0$, and

   (a) $f'(3) = 4$ and the values of f' decrease as x increases from 3 to 7;

   (b) $f'(3) = \tfrac{1}{2}$ and the values of f' increase as x increases from 3 to 5 and decrease as x increases from 5 to 7.

★ 7. Let g be the function given by

$$y = \begin{cases} x^2 + 1, & 0 < x \leq 2, \\ -2 + 8x - x^2, & 2 < x < 4. \end{cases}$$

   (a) What is the domain of g'?

   (b) Express g' by a formula or formulas over its domain.

8. Let f be the function from x to y given by $y = 100/(x-2)^2$ over the natural domain. What is D_f? R_f? Is the inverse of f a function? Explain.

9. Find F' in each of the following:

   (a) $F(x) = \dfrac{3x^2}{2} - 5x + \dfrac{1}{2x^2}$

   (b) $F(t) = t(t-1)^2$ (c) $F(u) = (3u)^2$

10. An ant crawls up the trunk of a tree and then slips back down, its height above the ground (s in.) varying with the elapsed time (t min) as follows: $s = 9t^2 - t^3$.
    (a) Find a formula for the velocity of the ant.
    (b) When is the ant stationary (i.e., velocity = 0)?
    (c) When does the ant return to ground level?
    (d) What is the *speed* of the ant when it returns to ground level?
    (e) How high does the ant get before starting to slip back?
    (f) What is the ant's *average* speed from the start until returning to ground level?

11. For the curve with equation $y = (x^2 - 3x + 28)/x$, find the coordinates of a point P on the curve between the points where $x = 1$ and $x = 4$ such that the tangent line at P has the same slope as the line joining the points on the curve where $x = 1$ and $x = 4$. Find the equation of this tangent line.

★ 12. How close does x have to be to 2 to ensure that $5x - 3$ differs from 7 by no more than 0.05? By no more than ϵ? Explain clearly how you know your answers to be correct.

13. Draw a neat sketch of a function f, given by $y = f(x)$, $-2 \le x \le 4$, such that $f(-2) = 0$, $f(0) = 2$, $f(2) = 0$, $f(3) = -1$, $f'(-1) = 0$, $f'(0) = -\frac{1}{2}$, $f'(x) > 0$ for $3 < x \le 4$.

★ 14. Consider the function defined by

$$F(x) = \begin{cases} 5 + 3x - x^2, & 0 \le x < 3, \\ 20 - 9x + x^2, & 3 \le x \le 5. \end{cases}$$

    (a) Does $\lim_{x \to 3} F'(x)$ exist? Explain.
    (b) Does $F'(3)$ exist? Explain.

15. Consider the function given by the equation $y = 20/(x + 1)$ for $-3 \le x \le 4$.
    (a) For what value or values of x in this interval is y undefined?
    (b) Plot the graph of this function, using the same scale vertically and horizontally.
    (c) Find the average rate of change of y with respect to x from $x = 0$ to $x = 3$.
    (d) Find the average rate of change of y with respect to x from $x = 1$ to $x = 1 + h$.
    (e) Hence, find the instantaneous rate of change of y with respect to x at $x = 1$.
    (f) Exhibit your answer to (c) by showing the slope of a certain chord of your graph, and exhibit your answer to (e) by showing the slope of a certain tangent to your graph. What can you say about this pair of lines?

16. A car is traveling north on High Street (Figure 2-9). (a) What is the *definition* of its average speed between a point P in front of Fisk Hall

2
Rate of change

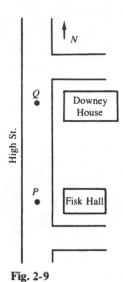

Fig. 2-9

and a point Q in front of Downey House? (b) What is the *definition* of its instantaneous speed at P?

17. If $f(x) = x^2$, what is the value of $f(3)$? Of $f(3+h)$? Of $f(3+h) - f(3)$? Of $[f(3+h) - f(3)]/h$? Of $\lim_{h \to 0} \{[f(3+h) - f(3)]/h\}$?

18. Find g' in each of the following cases:

    (a) $g(x) = -\dfrac{3}{2}x^2 + 5 - \dfrac{1}{3x^3}$

    (b) $g(u) = 10 + 2u - \frac{1}{2}u^2$

    (c) $g(t) = (t+1)(2t-1)$

    (d) $g(t) = \dfrac{t^2 - 1}{t^2}$

19. Obtain the answer to Problem 18(b) by use of the *definition* of the derivative.

20. The height (s ft) above the ground of a vertically fired projectile varies with the time (t sec) after firing as follows: $s = 4 + 96t - 16t^2$.

    (a) From what height was the projectile fired?

    (b) When was the projectile stationary? (That is, when did its velocity equal zero?) What was its height at that time?

    (c) When was the projectile at the 84-ft level, rising? Falling? What was the speed of the projectile at each of these times?

    (d) How far did the projectile travel before hitting the ground?

★ 21. How close does x have to be to 4 to ensure that $6x - 2$ differs from 22 by no more than 0.05? By no more than ϵ? Explain clearly how you know your answers to be correct.

22. (a) Find the coordinates of all points on the curve $y = 2x^3$ where the tangent line to the curve has the same slope as the line joining the point on the curve where $x = 1$ to the point on the curve where $x = 4$.

★   (b) Find the coordinates of all points on the curve $y = cx^3$ (c constant) where the tangent line to the curve has the same slope as the line joining the point on the curve where $x = x_1$ to the point on the curve where $x = x_2$.

23. Find the equation of the tangent line to the curve $y = 6x - x^2$ at the point on the curve where $x = 2$. Where $x = 3$.

24. Find f' in each of the following cases:

    (a) $f(x) = 1 - \dfrac{3}{2}x + 5x^2$ (b) $f(t) = \dfrac{7}{2t^2}$

    (c) $f(u) = (u^2 + 3)^2$

25. Obtain the formula for f' in Problem 24(a) by using the *definition* of derivative.

26. For the curve given by $y = 2x^3 - 3x^2 - 12x + 1$, calculate the height and slope of the curve at the points where $x = -1$, 0, and 1. Write the equation of the tangent line to the curve at the point where $x = -1$. Where $x = 1$.

27. The height above the ground (y ft) of a vertically fired rocket varies with the time (t sec) after firing as follows: $y = 80 + 64t - 16t^2$.
    (a) From what height above the ground was the rocket fired?
    (b) Find a formula for the velocity of the rocket at time t.
    (c) When was the rocket stationary? That is, when was its velocity equal to zero?
    (d) When did the rocket hit the ground?
    (e) What was the total distance traveled by the rocket?
    (f) With what *speed* did the rocket hit the ground?

28. (a) Find the coordinates of all points on the curve $y = 24/x$ where the tangent line to the curve has the same slope as the line joining the point on the curve where $x = 2$ to the point on the curve where $x = 8$.

★ (b) Find the coordinates of all points on the curve $y = a/x$ (a constant) where the tangent line to the curve is parallel to the line joining the point on the curve where $x = x_1$ to the point on the curve where $x = x_2$.

29. Sketch a graph of a function F given by $y = F(x)$, $0 \leq x \leq 10$, such that $F(0) = 3$, $F(2) = 1$, $F(7) = 5$, $F(10) = 1$, $F'(0) = -2$, $F'(2) = 0$, $F'(7) = 0$, $F'(10) = -\frac{1}{2}$, F' increases throughout $0 < x < 4$, F' decreases throughout $4 < x < 8$, and F' increases throughout $8 < x < 10$.

C 30. For the geometric series $1 + r + r^2 + r^3 + \cdots$, find the approximate value of the sum of 5 terms, of 10 terms, and of 15 terms for the following values of r:
    (a) $\frac{1}{3}$ (b) $\frac{1}{4}$ (c) $\frac{1}{5}$ (d) $\frac{1}{6}$ (e) $\frac{1}{13}$ (f) $\frac{2}{5}$
    (g) $\frac{2}{7}$ (h) $\frac{2}{11}$ (i) $\frac{2}{13}$ (j) $-\frac{2}{7}$
    In each case, guess the value of the limit of the sum as the number of terms increases indefinitely. Check your guess using the formula for the "sum to infinity" of a geometric progression (see Section 0.12 if you do not know this).

C 31. For the series that begins $4 - \frac{4}{3} + \frac{4}{5} - \frac{4}{7} + \frac{4}{9} - \frac{4}{11}$, find the approximate value of the sum of the first 20 terms and of the first 30 terms, and *try* to guess the value of the limit of the sum as the number of terms increases indefinitely. (Use a programmable calculator. The trouble here is not cancellation, but slow convergence. It takes close to 200 terms to get accuracy to two decimal places.)

C 32. For the series that begins $1 + 2x + 3x^2 + 4x^3 + 5x^4$, find the approximate value of the sum of the first 10 terms and of the first 20 terms for these values of x:
    (a) $\frac{1}{2}$ (b) $\frac{1}{3}$ (c) $\frac{1}{5}$
    In each case, *try* to guess the value of the limit of the sum as the number of terms increases indefinitely.

33. Let F be the function from x to y given by $y = \sqrt{8 + x^3}$ over the natural domain. What is D_F? R_F? Is the inverse of F a function? Explain.

34. Draw a neat sketch of a function f given by $y = f(x)$, $0 \le x \le 6$, such that $f(0) = 1$; the tangent to the curve at the point where $x = 1$ has equation $y = 1 + x$; $f(2) = 5$; $f'(4) = 0$; $f'(x) < 0$ for $4 < x < 6$; $f(6) = 0$.

★ 35. Sketch a graph of the function given by $y = |x^2 - 7x + 10|$, $0 \le x \le 7$, and find formulas for y', showing clearly the domain of each formula.

36. Find f' in each of the following cases:

(a) $f(x) = \dfrac{3}{4x^4}$ (b) $f(t) = 5 - 3t + \dfrac{1}{2}t^3$ (c) $f(x) = (x - \dfrac{1}{2x})^2$

37. For the curve given by $f(x) = x^3 - 4x^2 + 4x$, calculate the height and the slope of the curve at the points where $x = 0$, 1, and 2. Write an equation of the tangent line to the curve at the point where $x = 1$; where $x = 2$.

38. Let F be the function from x to y given by $y = \sqrt{36 - x^2}$ over the natural domain. What is D_F? R_F? Is the inverse of F a function? Explain.

39. Find f' in each of the following cases:

(a) $f(x) = \dfrac{5x^2}{3} - 4x + \dfrac{3}{x}$ (b) $f(t) = (t^2 - \dfrac{1}{t})^2$ (c) $f(u) = \dfrac{2}{5u^5}$

40. A rock climber is making a difficult ascent, up a vertical chimney, with the height (s ft) above the bottom varying with the time (t min) as follows: $s = 4t^3 - t^4$.

(a) Find a formula for the velocity of the climber.

(b) For what t's was the velocity positive?

(c) When had the climber slid back to the bottom of the chimney?

(d) What is the greatest height attained by the climber?

(e) What is the *speed* with which the climber hits the bottom of the chimney?

(f) What is the *average* speed of the climber from the start until hitting the bottom of the chimney?

41. For the curve with equation $y = x^3 - 3x - 2$, calculate the height and the slope of the curve at the points where $x = -1$, 0, and 1. Write the equation of the tangent line to the curve at the point on the curve where $x = 0$; where $x = 1$.

42. Draw a neat sketch of a function f, given by $y = f(x)$, $1 \le x \le 7$, such that the tangent to the curve at the point where $x = 1$ has equation $y = 6 - x$; $f(3) = 2$; $f'(3) = 0$; $f'(x) > 0$ for $3 < x < 4$; $f(6) = 0$; $f'(6) = 0$; $f(7) = 1$.

43. Find f' in each of the following cases:

(a) $f(x) = \dfrac{4}{3}x^3 - 5x^2 + \dfrac{1}{2} - \dfrac{1}{2x^2}$ (b) $f(t) = \dfrac{3t^2 + 1}{t}$

(c) $f(u) = \left(u - \dfrac{3}{u}\right)^2$

44. A particle moves on a horizontal straight line so that its displacement (x cm) from a certain point A on the line varies thus with elapsed time (t sec): $x = 6t^2 - t^3$. Directions to the right of A are positive; to the left of A, negative.

(a) Find a formula for the velocity of the particle at any time.

(b) When is the particle stationary (i.e., velocity = 0)? How far has it then traveled?

(c) Where is the particle at the start, when $t = 0$?

(d) Locate, relative to point A, the point B at which the particle has velocity $= -63$.

(e) What is the average speed of the particle in traveling from A (at the start) to B?

45. Consider the curve given by the equation $y = x^2 - 3x + 5$.

(a) What values of y correspond to $x = 2$ and $x = 6$?

(b) What is the equation of the line joining the points on the curve where $x = 2$ and $x = 6$?

(c) Find the coordinates of the point on the curve where the tangent line is parallel to the line of (b).

(d) Find the equation of the tangent line of (c).

★ 46. If $Q = 10x - 3$, what is your guess as to the value of $\lim_{x \to 2} Q$? Using the *definition* of limit, demonstrate that your guess is correct.

47. Let G be the function from x to y given by $y = \sqrt[3]{64 - x^2}$ over the natural domain. What is D_G? R_G? Is the inverse of G a function? Explain.

48. Draw a neat sketch of a function f, given by $y = f(x)$, $-1 \le x \le 5$, such that $f(-1) = 0$; $f'(0) > 0$; $f(1) = -2$; $f'(2) = 0$; $f(4) = 0$; $f'(4) = 0$; $f(5) = -1$.

★ 49. By using the definition of derivative, find a formula for $f'(x)$ if $f(x) = 1/(x+1)^2$.

3 Applications of the derivative

In this chapter we shall apply the derivative to the solution of five important types of problems:

1. Determination of regions in which functions increase, or decrease
2. Finding approximations to the changes (increments) in functions corresponding to small changes in the independent variable
3. Study of "marginality," especially marginal cost, in economics
4. Locating the extremes – maxima and minima – of functions
5. Approximate solution of equations

We shall devote most attention to type 4, because although the mathematical ideas are simple, it takes practice to translate the words of a practical problem into the equations to which we can apply the techniques of differentiation.

3.1   The Mean-Value Theorem

We begin with a discussion of the Mean-Value Theorem, which will be applied in the next section to deal with the first of the kinds of problems in this chapter – the determination of regions in which functions increase, or decrease. This theorem is a central result of the calculus, and we shall encounter it again in Chapter 5.

The Mean-Value Theorem is related to a type of problem you met in Chapter 2, in which you found points on a curve where the tangent line was parallel to a chord joining two points of the curve. You will gain a feeling for the theorem by working through the following problems of the same type:

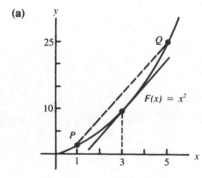

(a)

Fig. 3-1(a)

Problem 1

On the graph of $y = x^2$, let P be the point with abscissa 1 and Q the point with abscissa 5. Sketch the graph, and on the curve between P and Q find any points where the tangent line is parallel to the chord PQ.

Problem 2

Same as Problem 1, for $y = x^3$, with P having abscissa -2 and Q having abscissa 2.

Problem 3

Same as Problem 1, for $y = x^{2/3}$, with P having abscissa -1 and Q having abscissa 1.

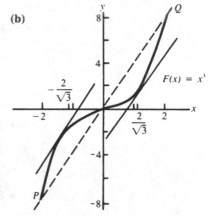

You should have found that, for the graph of $F(x) = x^2$, there is one point (with abscissa 3) between P and Q where the tangent line is parallel to the chord PQ, as seen in Figure 3-1(a); for the graph of $F(x) = x^3$, there are two points (with abscissas $\pm 2/\sqrt{3} \approx \pm 1.15$) between P and Q where the tangent line is parallel to the chord PQ, as seen in Figure 3-1(b); and for the graph of $F(x) = x^{2/3}$, there is *no* point between P and Q where the tangent line is parallel to the chord PQ, because the slope of PQ equals zero, and $F'(x) = 2/3x^{1/3}$, and there is no number x for which $2/3x^{1/3} = 0$; the situation is pictured in Figure 3-1(c). The feature that distinguishes the last case from the other two is that the graph of $F(x) = x^{2/3}$ is not smooth – the derivative, $F'(x)$, does not exist at $x = 0$. Thus, a conjecture might be the following:

If $F(x)$ has a derivative for all numbers in the interval $[a, b]$, then there is *at least* one point on the graph of $F(x)$ between $P\,(a, f(a))$ and $Q\,(b, f(b))$ where the tangent line is parallel to the line PQ.

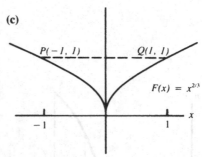

This conjecture is true, but we shall not prove it. In fact, the hypotheses can be weakened a bit: It turns out to suffice to assume that $F(x)$ is continuous in $[a, b]$ – meaning that its graph is unbroken from a to b, both ends included – and has a derivative for all numbers in the open interval (a, b).

Fig. 3-1(b–c)

The valid result just quoted is called the *Mean-Value Theorem*, pictured in Figure 3-2. Its analytic formulation goes as follows:

Mean-Value Theorem

If $F(x)$ is continuous for all x in $[a, b]$ and if $F'(x)$ exists for all x in (a, b), then there is at least one number, z, in (a, b) such that

$$\frac{F(b) - F(a)}{b - a} = F'(z).$$

Note that the conclusion can equally well be written $F(b) - F(a) = F'(z) \cdot (b - a)$. It is often convenient to use this form of the conclusion of the Mean-Value Theorem.

This statement is an example of what is called an *existence theorem*: It asserts the existence of z, but it does not tell us how to find it. In the simple cases we have investigated, we have been able to find the value (or values) of z, but in a complicated case this is unlikely to be possible. We shall discover it to be useful to know that z exists, even if we do not know the value.

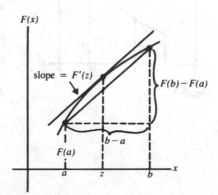

Fig. 3-2 Geometric depiction of the Mean-Value Theorem.

The proof of the Mean-Value Theorem depends on deep properties of the real number system, which we have not investigated, and this is the reason we omit the proof.

PROBLEMS

4. Same as Problem 1 for $y = \sqrt{x}$, with P having abscissa 0 and Q having abscissa 4. What is the natural domain of this function? What is the domain of its derivative? Are the hypotheses of the Mean-Value Theorem satisfied?

★ 5. Same as Problem 1 for $y = x^{2/3}$, with P having abscissa -1 and Q having abscissa 27. Are the hypotheses of the Mean-Value Theorem satisfied? (Note that the conclusion of a theorem may sometimes hold even if the hypotheses fail.)

★ 6. Same as Problem 1 for $y = x^3 + x^2 - 5x + 7$, with P having abscissa -2 and Q having abscissa 3. Does the Mean-Value Theorem apply?

3.2   Increasing and decreasing functions

In the graph shown in Figure 3-3, $f(x)$ increases as x increases – the curve gets higher as we move to the right. Formally, we have these definitions:

(i) A function, $f(x)$, is **increasing over an interval $[a, b]$** if, for *any* two numbers x_1, x_2 in $[a, b]$, with $x_2 > x_1$, it is the case that $f(x_2) \geq f(x_1)$.

(ii) If $f(x_2) > f(x_1)$, we say that $f(x)$ is **strictly increasing**.

The definitions of **decreasing** and **strictly decreasing** are analogous. Thus, in Figure 3-4, we can say that $g(x)$ is strictly increasing over $[a, p]$, $g(x)$ is strictly decreasing over $[p, q]$, and $g(x)$ is strictly increasing over $[q, b]$.

Similarly, in Figure 3-5, we can say that $F(x)$ is strictly increasing over $[a, d]$, despite the horizontal tangent at $x = c$; and $F(x)$ is increasing over $[a, b]$, despite the fact that $F(x)$ is constant over $[d, b]$. Indeed, our definition makes us use language that is slightly peculiar: We are saying that $F(x)$ is an increasing function over $[d, b]$!

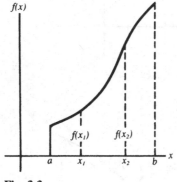

Fig. 3-3

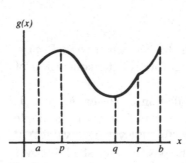

Fig. 3-4

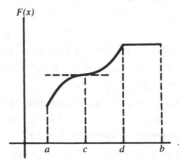

Fig. 3-5

If a function has a derivative at each point of an interval $[p, q]$, then it looks as though the function will be increasing over the interval if its derivative is nonnegative (≥ 0) over the interval, and strictly increasing if its derivative is positive (> 0). We can prove this statement by use of the Mean-Value Theorem:

Theorem

If $f'(x)$ exists over $[p, q]$, and (i) if $f'(x) \geq 0$ over (p, q), then f is increasing over $[p, q]$, and (ii) if $f'(x) > 0$ over (p, q), then f is strictly increasing over $[p, q]$.

Proof of (i) Let x_1 and x_2 be any two numbers in $[p, q]$, with $x_2 > x_1$. Then we know from the Mean-Value Theorem that there is at least one number, z, in (x_1, x_2) such that $f(x_2) - f(x_1) = f'(z) \cdot (x_2 - x_1)$. Now, by our hypothesis, $f' \geq 0$ over (p, q). Hence, surely $f'(z) \geq 0$. Because $x_2 > x_1$, we know that $(x_2 - x_1) > 0$. Therefore, $f'(z) \cdot (x_2 - x_1) \geq 0$. That is, $f(x_2) - f(x_1) \geq 0$, or $f(x_2) \geq f(x_1)$. But this is just what is needed to show that $f(x)$ is increasing.

Problem 1
Prove part (ii) of the theorem.

Problem 2
State an analogous theorem about derivatives and *decreasing* functions.

Example 1
For the curve of Problem 11, **2.6**,

$$y = f(x) = \tfrac{1}{3}x^3 - 2x^2 + 3x + 1,$$

we have $f'(x) = x^2 - 4x + 3 = (x - 1)(x - 3)$. Now,

if $x < 1$, $(x - 1)$ is negative and $(x - 3)$ is negative, so $f'(x) > 0$;
if $x = 1$, $(x - 1)$ is zero, so $f'(x) = 0$;
if $1 < x < 3$, $(x - 1)$ is positive and $(x - 3)$ is negative, so $f'(x) < 0$;
if $x = 3$, $(x - 3)$ is zero, so $f'(x) = 0$;
if $x > 3$, $(x - 1)$ is positive and $(x - 3)$ is positive, so $f'(x) > 0$.

We conclude that f is strictly increasing over $(-\infty, 1]$ and over $[3, \infty)$; f is strictly decreasing over $[1, 3]$.

At $x = 1$, with $f'(1) = 0$, we can say that "instantaneously f is neither increasing nor decreasing." We call f *stationary* at $x = 1$, and also at $x = 3$. The situation is shown in Figure 3-6. The function is strictly increasing to the left of point A $(1, \tfrac{7}{3})$ and to the right of point B $(3, 1)$ – in those regions the slope of the tangent is *positive*. The function is strictly decreasing between points A and B – in that interval the slope of the tangent is

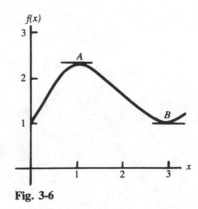

Fig. 3-6

113

negative. The function is *stationary* at points A and B – the tangents are horizontal (slope zero) at these points.

PROBLEMS

3. For each of the following functions, considered over its natural domain, determine where the function is stationary, where it is (strictly) increasing, and where it is (strictly) decreasing:
   (a) $f(x) = x^3 - 3x + 4$ (b) $f(x) = 6 + 12x + 3x^2 - 2x^3$
   (c) $f(x) = x^3 - 6x^2 + 12x + 6$ (d) $f(x) = 2x^3 - 9x^2 + 10$
   (e) $f(x) = x^3 + 9x^2 + 27x - 20$ (f) $f(x) = 3x^4 - 16x^3 + 8$
   ★ (g) $f(x) = 3x^4 - 16x^3 + 30x^2 - 24x + 9$

★ 4. Consider the converse of part (ii) of the theorem presented earlier in this section: "If f is strictly increasing over $[p, q]$, then $f'(x) > 0$ over (p, q)." Is this converse statement valid? Discuss.

5. Let

$$f(x) = \begin{cases} \frac{1}{2}x, & -1 \le x < 0, \\ x, & 0 \le x \le 1. \end{cases}$$

Show that f is strictly increasing over $[-1, 1]$. Comment on this result relative to the theorem in this section.

6. Consider $f(x) = 2x^3 + 9x^2 - 30x + 6$.
   (a) Find where $f'(x) = 0$.
   c (b) Evaluate $f(x)$ where $f'(x) = 0$.
   c (c) Graph $y = f(x)$.

3.3 Approximate increments

If we wish to determine the change in a function $y = f(x)$ as x changes from x_1 to x_2, we can do it directly by reading a graph or substituting in a formula to find the values y_1 and y_2 corresponding to x_1 and x_2 and subtracting to get $y_2 - y_1$, the required increment. If the difference, $x_2 - x_1$, is small, there is a method to find $y_2 - y_1$ *approximately*, which has some advantages over the straightforward method. We shall describe the approximate method first in a graphical context and then show how it applies to a formula.

But, first of all, why not use the straightforward graphical method? Here is the reason: Suppose that we can read the height of a graph to the nearest tenth of a unit, leading to the following results from a hypothetical graph:

$$\text{at } x_1 = 4, \quad \text{we read } y_1 = 15.3 \pm 0.1;$$
$$\text{at } x_2 = 7, \quad \text{we read } y_2 = 22.7 \pm 0.1.$$

A standard notation convenient here is $\Delta x = x_2 - x_1$, $\Delta y = y_2 - y_1$ (read "delta x" and "delta y") for the differences in x and y (Δx and Δy are to be treated as single letters; previously we used h for what we are now calling Δx).

In terms of the new notation, we can say that for $\Delta x = 7 - 4 = 3$, $\Delta y = 22.7 - 15.3 = 7.4$.

But y_1 might be as small as 15.2 and y_2 as large as 22.8, so Δy might be as large as 7.6. Likewise, y_1 might be as large as 15.4 and y_2 as small as 22.6, so Δy might be as small as 7.2. Hence, we really should say that for $\Delta x = 3$, $\Delta y = 7.4 \pm 0.2$. No problem so far.

Suppose, now, that we read from our hypothetical graph:

$$\text{at } x_1 = 4 \quad, \quad y_1 = 15.3 \pm 0.1 \text{ (as before)};$$
$$\text{at } x_2 = 4.2, \quad y_2 = 15.8 \pm 0.1.$$

Then we would conclude that for $\Delta x = 0.2$, $\Delta y = 0.5 \pm 0.2$.

The possible error of 0.2 is such a large fraction of Δy that we can't be happy with the result, so an alternative approach is desirable.

What we can do in the case of a small interval $[x_1, x_2]$ is to draw our best estimate of the tangent line to the graph at $x = x_1$ and argue that the average slope of the curve in the interval $[x_1, x_2]$ is *approximately equal to* the slope of the tangent line at $x = x_1$ – remember the definition of the tangent line (Figure 3-7). Hence, $\Delta y / \Delta x = (y_2 - y_1)/(x_2 - x_1) \approx$ slope of tangent line at x_1 ("$\approx$" means "approximately equal").

Indeed, in a small interval the slope of the tangent does not change much, so we can write

$$\frac{\Delta y}{\Delta x} \approx \text{slope of tangent line at } any \text{ point of } [x_1, x_2] \quad \text{or}$$

$$\Delta y \approx (\text{slope of tangent line at } any \text{ point of } [x_1, x_2]) \cdot \Delta x.$$

Problem 1
On as large a scale as practicable, draw a graph of $y = \frac{1}{3}x^2$ over the interval $[0, 3]$.

(a) From your graph, estimate the values of y corresponding to $x_1 = 2$ and $x_2 = 2.1$, and thus compute Δy.

(b) Draw your best estimate of the tangent line at any point of the interval $[2, 2.1]$, measure its slope, and thus compute Δy.

(c) Compute the *actual* values of y corresponding to $x_1 = 2$ and $x_2 = 2.1$, and thus compute Δy.

Most people who do this find that their answer to (b) is closer to the true value of (c) than is their answer to (a).

If we have a formula for the function $y = f(x)$, as we indeed have in Problem 1, we do not need to face the inaccuracies inherent in graphical measurement. The method we have been describing to approximate Δy can

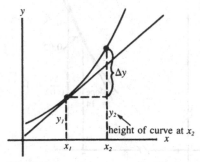

Fig. 3-7

be applied with the formula to avoid some labor in computing the two values of y. We repeat:

For small intervals, the *average* rate of change of y with respect to x is *approximately equal to* the *instantaneous* rate of change of y with respect to x at any point of the interval; that is,

$$\frac{\Delta y}{\Delta x} \approx f'(x),$$

or

$$\Delta y \approx f'(x)\Delta x, \quad \text{for any } x \text{ in } [x_1, x_2].$$

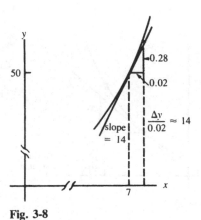

Fig. 3-8

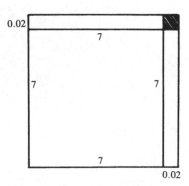

Fig. 3-9

Example 1

What is the approximate change in the square of a number if the number changes from 7 to 7.02?

Here $y = f(x) = x^2$, and we wish to determine Δy corresponding to $x_1 = 7$ and $\Delta x = 0.02$. We note that $f'(x) = 2x$, so that $f'(7) = 14$. Hence, $\Delta y \approx (14)(0.02) = 0.28$. This is shown graphically in Figure 3-8. It is also instructive to look at the example with the help of Figure 3-9. Here, $y_1 = f(7) = 7^2$ is seen as the area of a square of side 7, and $y_2 = y_1 + \Delta y = f(7.02) = (7.02)^2$ is seen as the area of a square of side 7.02; Δy, the difference between these areas, is the area of two rectangles, each of length 7 and width 0.02, plus the area of a square of side 0.02. The approximation we are using is equivalent to the area of these two rectangles and ignores the area of the tiny square of area $(0.02)^2 = 0.0004$. Doing the same thing algebraically, we have $y_2 = y_1 + \Delta y = (7.02)^2 = (7 + 0.02)^2 = 7^2 + 2(7)(0.02) + (0.02)^2$. Hence, $\Delta y = 2(7)(0.02) + (0.02)^2$. Our approximation is just the first term on the right side of this equation.

Example 2

What is the approximate change in the square of a number if the number changes from 4.99 to 5.02?

We are again dealing with $y = f(x) = x^2$, and we wish to determine Δy corresponding to $x_1 = 4.99$ and $\Delta x = 0.03$.

Problem 2

Use these numbers in the approximation to obtain $\Delta y \approx 0.2994$.

Because the derivative is nearly constant throughout this small interval, we may as well calculate the derivative at a point of the interval that will simplify the arithmetic.

Problem 3

Calculate the derivative at $x = 5$ to obtain $\Delta y \approx 0.3$.

Problem 4

Verify that the exact value of Δy is 0.3003.

So we see that, in this case, the second approximation is better than the first.

Problem 5

Draw a large graph of $y = x^2$, and, with the help of two tangent lines, show geometrically the two approximations found in this example.

Example 3

A circular plate has a radius of about 5 in. Approximately how accurately must the radius be measured if the error in the calculated area of the plate shall not be more than 1 in.^{2}?

The formula for the area of a circle is $A = f(r) = \pi r^2$. In this case, then, $f'(r) = 2\pi r$, and $f'(5) = 10\pi$. Hence, $\Delta A \approx 10\pi \cdot \Delta r$. If we give ΔA its maximum permitted value, we see that $\Delta r \approx 1/10\pi \approx 0.032$ in.

The method of approximation explained in this section is so important, both practically and theoretically, that it is worthwhile to summarize and extend slightly what we have presented. As we have seen, replacement of a small piece of a curve by a segment of the tangent line at some point on the piece of the curve leads to

$$\Delta y \approx f'(x) \cdot \Delta x, \quad \text{for any } x \text{ in } [x_1, x_2],$$

or

$$y_2 - y_1 \approx f'(x) \cdot \Delta x, \quad \text{for any } x \text{ in } [x_1, x_2],$$

or

$$f(x_2) - f(x_1) \approx f'(x) \cdot \Delta x, \quad \text{for any } x \text{ in } [x_1, x_2],$$

or

$$f(x_2) \approx f(x_1) + f'(x) \cdot \Delta x, \quad \text{for any } x \text{ in } [x_1, x_2].$$

Writing $x_2 = x_1 + \Delta x$, this last equation becomes

$$f(x_1 + \Delta x) \approx f(x_1) + f'(x) \cdot \Delta x, \quad \text{for any } x \text{ in } [x_1, x_1 + \Delta x].$$

The process used here is called "linearization" of the problem – treating a (perhaps complicated) function as though it were a simple linear function over a short interval. Unfortunately, at this stage we cannot answer a couple of important questions: About how accurate is this approximation? How short an interval must be used to ensure that the approximation has a specified accuracy? Such estimates come later in the calculus.

In linear interpolation, we encountered another example of linearization. In that case, the curve over an interval was replaced by the line segment joining the two points on the curve at the ends of the interval. There, also, we could say little about the accuracy of the approximation.

117

The idea of linearization of problems is extensively used in mathematics and science to cope with situations that could not otherwise be handled, or to simplify the work when an approximate answer is sufficient. The use of differentiation, as in this section, is of great help in such linearization.

It may have occurred to you that each of the approximate increments discussed in this section could be replaced simply by calculating the *exact* increment on a calculator. This observation, however, does not diminish the importance of the approximate-increment technique even in practical applications. There are many practical functions that, unlike the simple examples we use here for illustration, are beyond a calculator's (or a computer's) ability to handle easily.

PROBLEMS

6. If x is a number, $1/x$ is called the *reciprocal* of x. Find an approximate value of the reciprocal of 10.03.

7. (a) Approximately how much greater is $(5.02)^3$ than 5^3? Hence, what is an approximate value of $(5.02)^3$?

   (b) Use a three-dimensional sketch analogous to Figure 3-9, or a model from cardboard, plastic, or wood, to display the geometrical significance of the approximation in part (a), showing the terms ignored in that approximation.

   (c) Use the binomial theorem, as in the last part of Example 1, to display the algebraic significance of the approximation in part (a), showing the terms ignored in that approximation.

   ★ (d) Use the Mean-Value Theorem to yield a new expression for $\Delta y = f(x_1 + \Delta x) - f(x_1)$. What is the difference between this expression and that for the approximate increment? Does either provide information the other does not?

   ★ (e) Show that $3 \cdot 5^2 \cdot (0.02) \le (5.02)^3 - 5^3 \le 3(5.02)^2 \cdot (0.02)$ and calculate these two bounds. What do these inequalities say about your estimation in part (a)?

○ 8. The result $f(x) = x^n \Rightarrow f'(x) = nx^{n-1}$ holds for *all* real values of n, although we have not proved it. Assuming this result is valid, find approximations for the following:

   (a) $\sqrt{4.02}$ (b) $\sqrt{3.98}$ (c) $\dfrac{1}{\sqrt{4.02}}$

9. For ships of a certain type, the cost of operation per mile ($\$C$) varies as the cube of the speed (x knots), and $C = 40$ when $x = 20$. Find a formula for C in terms of x. Approximately how much larger is C for $x = 30.2$ than for $x = 30$?

10. A wooden cylinder has height 15 in. and base radius 5 in. If the height remains constant, by approximately how much must the radius be changed to reduce the volume by 50 in.^{3}?

11. (a) For blood vessels of average size, the rate of flow (F mm^3/sec) varies as the fourth power of the diameter (D mm) of the blood vessel, and $F = 40.5$ at $D = 3$. Approximately how much larger is F for $D = 3.02$ than for $D = 3$?

  (b) If the rate of flow in a blood vessel is constant, then the blood pressure (P units) varies inversely as the fourth power of the diameter, and $P = 100$ at $D = 3$. By approximately how much must D increase to decrease P from 100 to 80 (a 20% decrease in P)? What is the percentage increase in D? (Doctors administer various drugs to dilate blood vessels, permitting the same flow of blood with substantial reduction in blood pressure, thus reducing the load on the heart.)

12. The cost of paint for a large spherical gas storage tank comes to 10 cents per square foot of surface area. Approximately how much less will be the cost of paint for a tank of radius 19.9 ft than for a tank of radius 20 ft?

13. For certain aircraft, the lift (L units) varies as the square of the velocity (v mph), and $L = 54,000$ when $v = 300$. Use differentiation to find approximately how much smaller L is for $v = 499.5$ than for $v = 500$.

14. A growing pile of fine sand constantly has the shape of a cone in which the height equals three-fourths the radius of the base. If the radius is about 6 ft, use differentiation to find approximately how much the radius will increase through the addition of 9π ft^3 to the volume.

15. The water resistance (R units) encountered by a barge varies as the square of the speed (v knots), and $R = 24,000$ when $v = 10$. Use differentiation to find approximately how much larger R is for $v = 8.1$ than for $v = 8$.

16. A metal cylinder contracts so as to keep its height equal to the diameter of its base. If the radius of its base is about 5 in., use differentiation to find by approximately how much the radius must change to reduce the volume by 1 in.^{3}

3.4 Applications to economics: marginal cost and unit cost

The problems we shall analyze here represent interesting applications of the derivative and of the method of approximate increments. The topics, which we shall present under the titles of *marginal cost* and *unit cost*, have far-reaching significance in economics.

Let us study the expense of the sausage manufacturer we met in Problem 8, **1.4**. In that problem and its sequel (Problem 5, **1.7**) we decided that the daily cost (E) of making x lb of sausage was given, with satisfactory accuracy, by the formula $E = 128 + 2.8x - 0.01x^2$.

119

In this formula, the constant, 128, represents the sum of all the "fixed costs" associated with the plant, whether or not any sausage is made – interest on the investment, maintenance of the building, salaries and wages of personnel whose pay goes on in any case, and so forth. The coefficient, 2.8, represents the basic cost of material and labor (i.e., \$2.80 per pound of sausage). The last term, which has a negligible effect for small x, reduces substantially the sum of 128 and $2.8x$ when x is of moderate size. It represents the fact that the manufacturer can operate somewhat more efficiently if his production increases, for he can buy his materials more cheaply in larger quantities, his personnel can be more effectively utilized, and so forth.

But note that if x is large, the term $0.01x^2$ dominates the other two terms in the formula for E, and E will unrealistically become small, or even negative!

Problem 1
Use the formula to calculate E for $x = 300$; for $x = 400$. Also calculate E for $x = 150$ and for $x = 160$.

Thus, this formula cannot possibly be correct for large values of x. Indeed, despite the efficiencies involved in large-scale production, we don't expect the cost of making 160 lb of sausage, say, to be *less* than the cost of making 150 lb. In this illustration, $x = 120$ is about as large a value as it is reasonable to use with this formula, so the domain of the function we are considering is $[0, 120]$:

$$E = 128 + 2.8x - 0.01x^2, \quad 0 \le x \le 120. \tag{1}$$

$$\text{At } x = 50, \quad E = 243;$$
$$\text{at } x = 51, \quad E = 244.79.$$

Thus, for $x = 50$ and $\Delta x = 1$, $\Delta E = 1.79$. In other words, the *incremental cost* of producing 51 lb of sausage per day, rather than 50 lb, is \$1.79. We can readily find an approximation to this incremental cost by use of the derivative: $E' = 2.8 - 0.02x$, so at $x = 50$, $E' = 2.8 - 1 = 1.8$.

Remembering that $\Delta E \approx E' \cdot \Delta x$, we conclude that the incremental cost is approximately $1.80 \times 1 = 1.80$ (dollars).

Problem 2
Use the method of approximate increments to show that

at $x = 6$, E is about 2.7 larger than at $x = 5$;
at $x = 11$, E is about 2.6 larger than at $x = 10$;
at $x = 21$, E is about 2.4 larger than at $x = 20$;
at $x = 80$, E is about 1.2 larger than at $x = 79$;
at $x = 101$, E is about 0.8 larger than at $x = 100$;
at $x = 120$, E is about 0.4 larger than at $x = 119$.

We make the following definitions.

If $E = f(x)$ is the formula for the total cost ($\$E$) of producing x items, then the **incremental cost at $x = n$** is $f(n+1) - f(n)$, and the **marginal cost at $x = n$** is $f'(n)$.

The significant fact is that

the marginal cost is an approximation to the incremental cost.

Problem 3

If the formula for E were $E = 128 + 2.8x$, what could you say about the marginal cost of production?

Problem 4

If the sum of the fixed costs were 1000, rather than 128, what effect would this have on the marginal cost of production?

The ideas of increments and marginality arise in contexts other than cost – for instance, in *value*.

For example, if you are ravenously hungry and have plenty of money, a hamburger might be worth $10 to you. After you have eaten one hamburger, you still might set a high value on a second one, but perhaps something less than $10. After you have eaten five hamburgers, you might set very low the *incremental value* of a sixth hamburger. When you're stuffed, even the thought of a hamburger might be so distasteful that you might give a *negative* incremental value to the tenth hamburger!

If the value ($\$V$) to you of x hamburgers is given by $V = g(x)$, then the *marginal value* of a hamburger to you at $x = n$ is given by $g'(n)$.

We turn now to the concept of *unit cost*. If the total cost of producing x items is E, then the **average total cost** or the **total unit cost** (TUC) is defined by

$$\text{TUC} = \frac{E}{x}.$$

Thus, for the simple cost function of Problem 3, where $E = 128 + 2.8x$,

$$\text{TUC} = \frac{E}{x} = \frac{128 + 2.8x}{x} = \frac{128}{x} + 2.8.$$

Because x appears only in the denominator in this formula, it is clear that we have a *decreasing* function – TUC decreases as x increases. Another way to see this is to examine the derivative of TUC: Because $\text{TUC} = 128x^{-1} + 2.8$,

$$(\text{TUC})' = -128x^{-2} = -\frac{128}{x^2}.$$

Because the derivative is negative for all x in the domain, the function is decreasing. This result is intuitively sensible: Because the total cost consists of a fixed cost of 128 plus a variable cost that amounts to 2.8 per unit produced, the *average total cost* or TUC will decrease with increasing x because the fixed cost will be spread over more units as production increases.

The same result is seen in the original example of sausage manufacture, where $E = 128 + 2.8x - 0.01x^2$. In this case,

$$\text{TUC} = \frac{E}{x} = \frac{128 + 2.8x - 0.01x^2}{x} = \frac{128}{x} + 2.8 - 0.01x,$$

and

$$(\text{TUC})' = -\frac{128}{x^2} - 0.01.$$

Because the derivative is negative for all x in the domain, TUC is again a decreasing function.

In both these cases, we say that the manufacturing process exhibits *economies of scale* for all levels of production – the larger the production, the lower the total unit cost.

A more realistic situation is one in which *diseconomies* appear for large production levels. Suppose, for example, that E is given by $E = 128 + 2.8x + 0.02x^2$. The last term on the right side, being positive, represents some *inefficiencies* associated with large production levels. This does not imply, however, that there are no economies of scale, for as production increases, we still spread the fixed costs over more units. Exactly what happens is discovered by expressing the total unit cost and differentiating:

$$\text{TUC} = \frac{128 + 2.8x + 0.02x^2}{x} = \frac{128}{x} + 2.8 + 0.02x;$$

$$(\text{TUC})' = \frac{-128}{x^2} + 0.02 = \frac{-128 + 0.02x^2}{x^2}.$$

Problem 5

Find the value of x that makes $(\text{TUC})'$ equal to zero. For what values of x, then, is $(\text{TUC})'$ negative? Positive? Sketch a graph of TUC as a function of x to exhibit these results.

You should have discovered that $(\text{TUC})' = 0$ at $x = 80$. For $0 < x < 80$, $(\text{TUC})'$ is *negative*, so TUC is *decreasing*. For $x > 80$, $(\text{TUC})'$ is *positive*, so TUC is *increasing*. Thus, we have *economies* of scale for $0 < x < 80$, and *diseconomies* for $x > 80$.

PROBLEMS

6. The total cost (E) of producing x gal of maple syrup is given by $E = 2500 + 4x - 0.005x^2$, for $0 \le x \le 225$. What is the marginal cost of producing the 10th gallon? The 100th gallon? The 200th gallon?

7. The total cost (E) of producing x lb of cheese is given by $E = 100 + 0.8x - 0.002x^2$, for $0 \le x \le 150$. What is the marginal cost of producing the 10th lb? The 50th lb? The 100th lb?

8. The total cost (C) of obtaining and delivering x cords of firewood is given by $C = 6000 + 35x - 0.1x^2$, for $0 \le x \le 150$. What is the marginal cost of the 10th cord? The 100th cord?

9. The total cost (E) of producing x belts is given by $E = 1000 + 5x - 1/x$, for $1 \le x \le 100$. What is the marginal cost of producing the first belt? The 5th? The 10th?

10. The total cost (E) of distributing x barrels of oil is given by $E = 400 + 24x + 0.01x^2$, valid for $0 \le x \le 500$.
   (a) What is the marginal cost of distributing the 50th barrel? The 100th barrel?
   (b) Write a formula for the total unit cost, E/x.
   (c) Find $(TUC)'$ and the value of x that makes $(TUC)' = 0$.
   (d) For what x's are there economies of scale? Diseconomies?
   (e) For the x found in (c), compute the value of E/x. Of the marginal cost.

3.5 Maxima and minima: the basic idea

There are many situations in which we want to know the maximum or the minimum (the biggest or the smallest) value attained by a varying quantity:
   What is the greatest height reached by the rocket?
   What is the highest level attained by the flooding river?
   What is the lowest cost of manufacturing a certain product?
   What is the greatest profit of operating a certain factory?
   What is the largest rectangular field that can be enclosed with a given amount of fencing?

The basic idea we use is exemplified in Figure 3-10. If y is a function of x, with a graph that has a tangent everywhere, then at a maximum point, as in the figure, the tangent is horizontal. Hence, if $y = f(x)$, and if y has a maximum value for $x = p$, we expect $f'(p)$ to be zero; similarly for a minimum value. Hence, we search for maximum and minimum values of $f(x)$ among the roots of the equation $f'(x) = 0$.

Example 1
For what positive number is the sum of the number and its reciprocal an extreme (maximum or minimum)?

If we call the number x and the sum S, we are to investigate $S = f(x) = x + 1/x$, $x > 0$, for maxima and minima.

We note that if x is a very small positive number, then $1/x$ is very large, so S is very large; and if x is very large, S is again very large.

(a)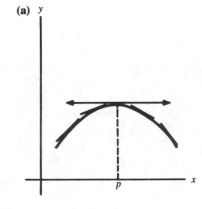

(b)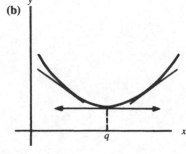

Fig. 3-10

123

Problem 1

Calculate S for $x = 0.1, 0.2, 0.5, 0.8, 1, 2, 4$, and 10, and sketch the graph of S as a function of x.

The graph makes it appear that S has a minimum value for some x.

Problem 2

Find $f'(x)$, set it equal to zero, and solve for x, to find that $x = 1$ is the only positive solution.

Hence, $f(1) = 2$ is the minimum value of S for positive x's. There is no maximum value of S for positive x's, because we can make S as large as desired by taking x sufficiently large (also by taking x sufficiently small).

We shall investigate the whole matter of maxima and minima more carefully in the following sections, but we can solve simple problems now.

PROBLEMS

3. Using the method of this section, find the time at which the river crests (Problem 3, **1.4**), assuming, as in the result of Problem 10, **1.6**, that the formula for the level (L ft) is given by $L = 0.5 + 2.1t - 0.1t^2$. What is the maximum value of L?

4. For the toy rocket (Problem 4, **1.3**), for which the height above the ground (y ft) is given by $y = 48 + 88t - 16t^2$, $0 \le t \le 6$, find the time at which y is a maximum. What is the maximum value of y?

5. For the sausage maker of Problem 8, **1.4**, assume that the formula for the total daily expense ($\$E$) varies with the daily production (x lb) as follows: $E = 128 + 2.8x - 0.01x^2$. Find the value of x that makes E a maximum. Check that it really gives a maximum. What is the maximum value of E? [Note that if the domain of E is limited, as in equation (1) in Section 3.4, there is no maximum of the type we are now seeking. This is consistent with the intuitive notion that a reasonable function for total cost of production should always be *increasing* – the more sausage we produce, the greater will be the total cost.]

6. Find the coordinates of the lowest point on the graph of $y = 8 - 6x + x^2$, over the natural domain. Sketch the graph. Is there a highest point? Where does the graph meet the y axis? The x axis?

7. Find the coordinates of the highest point on the graph of $y = 9 - 8x - x^2$, over the natural domain. Sketch the graph. Is there a lowest point? Where does the graph meet the y axis? The x axis?

8. For the graph of each of the following functions, over the natural domain, find the coordinates of the lowest point. Also find where the

3.6
Is it a maximum
or a minimum?

graph meets the y axis, and where it meets the x axis (if it does). Sketch each graph.

(a) $y = -30 + 4x + 2x^2$ (b) $y = 4x + 2x^2$
(c) $y = 2 + 4x + 2x^2$ (d) $y = 10 + 4x + 2x^2$
(e) $y = 20 - 10x + x^2$ (f) $y = -10x + x^2$
(g) $y = 25 - 10x + x^2$ (h) $y = 30 - 10x + x^2$

9. For the graph of each of the following functions, over the natural domain, find the coordinates of the highest point (if there is one) or of the lowest point (if there is one). Also find where the graph meets the y axis, and where it meets the x axis (if it does). Sketch each graph.

(a) $y = 6 + x - 2x^2$ (b) $y = 3 + x - 2x^2$
(c) $y = 1 + x - 2x^2$ (d) $y = x - 2x^2$
(e) $y = -2 + x - 2x^2$ (f) $y = 10 - 8x + 2x^2$
(g) $y = 8 - 8x + 2x^2$ (h) $y = 6 - 8x + 2x^2$
(i) $y = 4 - 8x + 2x^2$ (j) $y = -8x + 2x^2$

★ 10. (a) Find the coordinates of the lowest point on the graph of $y = 8 - 6x + x^2$, without using calculus. (Hint: See Problem 5, **1.9**.)

(b) Similarly for the graph of $y = 9 - 8x - x^2$.

3.6 How do we know whether we have a maximum or a minimum?

In the case of Example 1 in the previous section, a rough analysis of the behavior of the function $f(x) = x + 1/x$, or a sketch of the associated graph, convinced us that there was one extreme and that it was a minimum. In more complicated problems, where it is not easy to analyze the behavior of the function or to sketch a graph, it is useful to have a test to decide just what it is that the solutions of $f'(x) = 0$ signify. There are, in fact, three standard tests for maxima and minima; we shall consider two of them now and leave the third for later work.

We first introduce some new terms informally through consideration of a function F with domain $1 \leq x \leq 8$. We shall assume that F is *differentiable* (i.e., that it has a derivative in the interval under discussion). The graph of $y = F(x)$ is shown in Figure 3-11. At the points on the curve where $x = 4$ and $x = 6$, the tangent lines are horizontal: $F'(4) = 0$ and $F'(6) = 0$.

We call 4 and 6 *critical numbers* of the function F because $F'(x) = 0$ for those numbers.

$F(4) = 9$ is called a *relative maximum* of F, because in the "neighborhood" of $x = 4$ there is no value of F larger than 9.

$F(8) = 6$ is also called a *relative maximum* of F, because in the neighborhood of $x = 8$ there is no value of F larger than 6. Because the domain of F ends at $x = 8$, we use a "one-sided neighborhood."

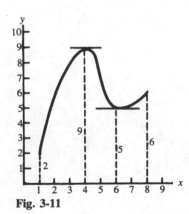
Fig. 3-11

125

$F(6) = 5$ is called a *relative minimum* of F, because in the neighborhood of $x = 6$ there is no value of F smaller than 5.

$F(1) = 2$ is also called a *relative minimum* of F, because in the neighborhood of $x = 1$ there is no value of F smaller than 2. Once again, we use a one-sided neighborhood.

Of the two relative maxima, 9 is greater than 6, and we call 9 the *absolute maximum* of F. Of the two relative minima, 2 is less than 5, and we call 2 the *absolute minimum* of F.

Maxima and minima where the derivative equals zero are called "turning-point extremes." Maxima and minima at the ends of the domain are called "end-point extremes."

To summarize for this example, we say that

at $x = 1$, F has the relative minimum value 2 as an end-point extreme;
at $x = 4$, F has the relative maximum value 9 as a turning-point extreme;
at $x = 6$, F has the relative minimum value 5 as a turning-point extreme;
at $x = 8$, F has the relative maximum value 6 as an end-point extreme.
The critical numbers of F are $x = 4$ and $x = 6$.
The absolute maximum of F equals 9, attained at $x = 4$.
The absolute minimum of F equals 2, attained at $x = 1$.
Thus, the range of F is the interval $2 \le y \le 9$.

We now give formal definitions of our basic terms.

Definition 1

The solutions of $f'(x) = 0$ are called the **critical numbers** of the function f.

Definition 2

A number, c, corresponds to a **relative minimum** of the function f if there is an interval $[p, q]$ of the domain, with $p < c < q$, such that no value of f in $[p, q]$ is smaller than $f(c)$. If c should be the left end of the domain of f, then we compare $f(c)$ with the values of f in an interval $[c, q]$, $q > c$. Similarly for **relative maximum**. Similarly if c should be the right end of the domain.

Definition 3

A number c corresponds to an **absolute minimum** of the function f if there is no x in the domain of f such that $f(x)$ is smaller than $f(c)$. Similarly for **absolute maximum**.

End-point extremes are important because the functions encountered in applications frequently have limited domains. Later we shall consider such problems, but first we shall analyze turning-point extremes more thoroughly.

The first test for relative turning-point extremes is an entirely obvious one, which we shall explain in terms of Example 1, 3.5. In that example we sought extremes of $S = f(x) = x + 1/x$, for $x > 0$. We found $x = 1$ to be a critical number. Let us calculate the values of f on either side of the critical number as well as at the critical number: $f(\frac{1}{2}) = \frac{5}{2}, f(1) = 2$, and $f(\frac{3}{2}) = \frac{13}{6}$, as pictured in Figure 3-12. Because $f(1) < f(\frac{1}{2})$ and $f(1) < f(\frac{3}{2})$, the critical number corresponds to a *minimum*.

Clearly, we must choose values reasonably close to the critical number to make the test reliable. In particular, if in testing the critical number $x = 6$ for the function graphed in Figure 3-11, we were to choose $x = 2$ on one side and $x = 8$ on the other, we would be led to an erroneous conclusion, because $F(2)$ is less than $F(6)$. We should choose an interval small enough not to include any critical numbers other than the one we are testing. We can now state our result formally.

Test 1

Let $x = c$ be a critical number of a differentiable function, f, and let $[p, q]$ be an interval containing c but no other critical number of f. If $f(c) < f(p)$ and $f(c) < f(q)$, then $f(c)$ is a *relative minimum* of f. Similarly for a *relative maximum*.

We also explain the second test in terms of Example 1, 3.5. Let us calculate the values of f' on either side of the critical number: $S = f(x) = x + 1/x$, $f'(x) = 1 - 1/x^2$, $f'(\frac{1}{2}) = -3$, $f'(\frac{3}{2}) = \frac{5}{9}$, and $f'(1) = 0$, of course (Figure 3-13). Putting this in general terms, we have the following.

Test 2

Let $x = c$ be a critical number of a differentiable function, f, and let $[p, q]$ be an interval containing c but no other critical number of f. If $f'(p) < 0$ and $f'(q) > 0$, then $f(c)$ is a *relative minimum* of f. Similarly for a *relative maximum*.

Let us use the foregoing ideas to help sketch the graph of $y = f(x) = 2x^3 - 21x^2 + 60x - 40$ over its natural domain (which is the set of all real numbers), by finding relative extremes of y. We write

$$f'(x) = 6x^2 - 42x + 60 = 6(x^2 - 7x + 10) = 6(x - 2)(x - 5).$$

Thus, $f'(x) = 0$ for $x = 2$ and $x = 5$ – these are the critical numbers.

At $x = 2$, $y = f(2) = 16 - 84 + 120 - 40 = 12$.

At $x = 1$, $y = f(1) = 2 - 21 + 60 - 40 = 1$.

At $x = 3$, $y = f(3) = 54 - 189 + 180 - 40 = 5$.

Because $12 > 1$ and $12 > 5$, we conclude by Test 1 that 12 is a *relative maximum* of f.

3.6
Is it a maximum
or a minimum?

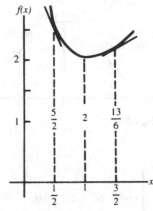

Fig. 3-12

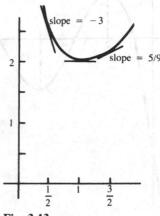

Fig. 3-13

127

3
Applications of the derivative

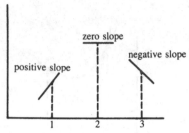

Fig. 3-14

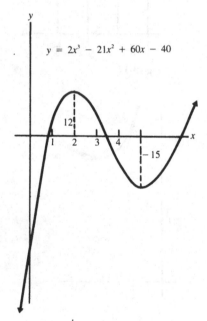

$y = 2x^3 - 21x^2 + 60x - 40$

Fig. 3-15

For practice, let us apply Test 2 for the same critical number (Figure 3-14):

$$\text{At } x = 2, \quad f'(x) = 0.$$
$$\text{At } x = 1, \quad f'(x) \text{ is positive.}$$
$$\text{At } x = 3, \quad f'(x) \text{ is negative.}$$

Thus, Test 2 confirms that 12 is a relative maximum of f. (Often, but not always, Test 2 involves less arithmetic computation than Test 1.)

Problem 1
Check at the other critical number, $x = 5$, that $y = f(5) = -15$, and use both tests to verify that -15 is a relative minimum of f.

Now, if x is a very large positive number, the first term in $f(x)$, $2x^3$, will be dominant, and thus y will be a positive number – as x increases without bound, so does y. Hence, there is *no absolute maximum* for y. If you wish, you can say that y approaches infinity.

If x is a very large *negative* number, the first term, $2x^3$, is again dominant, and thus y will be a negative number (because the cube of a negative number is negative). As x gets larger negatively without bound, so does y. Hence, there is *no absolute minimum* for y, either. We now have enough information to sketch the general form of the curve, as shown in Figure 3-15.

PROBLEMS

2. Use both Test 1 and Test 2 to check your result in
   (a) Problem 3, **3.4**
   (b) Problem 4, **3.4**
   (c) Problem 5, **3.4**

3. It is possible to have $x = c$ as a critical number [i.e., $f'(c) = 0$] without its corresponding to either a relative minimum or a relative maximum. Sketch a graph to illustrate how this can occur.

4. Investigate each following $f(x)$ for maximum and minimum values, and sketch the graph of $y = f(x)$:
   (a) $f(x) = x^3 - 3x^2 - 9x + 12$ (b) $f(x) = 1 + 3x - x^3$
   (c) $f(x) = \frac{1}{3}x^3 - 3x^2 + 9x + 1$

★ 5. Investigate the function determined by

$$y = \tfrac{1}{5}x^5 - x^4 + \tfrac{2}{3}x^3 + 2x^2 - 3x + 2$$

for maximum and minimum values, and sketch the curve. (Hint: Find the derivative and factor it by trial.)

★ 6. Sketch some graphs showing functions that have maxima and minima *without* their derivatives being zero.

3.7 Further questions about maxima and minima

We have seen that it is possible to have a situation like that in Figure 3-16, where there is a horizontal tangent at P (the derivative of the function is zero at P), but P does *not* correspond to a relative maximum or a relative minimum of the function. We call P a *horizontal point of inflection* of the curve, and we shall deal with points of inflection, horizontal and otherwise, in the next chapter.

We now consider the converse question: Is it possible to have a maximum or a minimum *without* having the derivative equal to zero? There are two ways in which this can occur:

1. There may be abrupt extremes, as in Figure 3-17, corresponding to values of x for which the derivative does not exist. In Figure 3-17(a), P corresponds to a relative maximum of the function, but neither $f'(x)$ nor the tangent line to the curve exists at P. In Figure 3-17(b), Q corresponds to a relative minimum of the function; $f'(x)$ does not exist there, but the curve does have a (vertical) tangent line at Q. The points P and Q are called *cusps* on their curves.

2. The second way in which we may get an extreme without having the derivative zero occurs if the domain of the function is limited, as illustrated in Figure 3-11, repeated here. Here the domain of F is the interval $[1, 8]$. The points A and B correspond to end-point extremes of the function. We have relative minima at A and D, and relative maxima at C and B. Because $F(1) < F(6)$, A corresponds to the absolute minimum for this function; and because $F(4) > F(8)$, C corresponds to the absolute maximum.

For functions whose graphs are unbroken over a finite closed interval, $[a, b]$, we can set out the following procedure for the complete analysis of maxima and minima:

(a) Locate any turning-point extremes by finding critical numbers (where f' is zero) and testing for relative maxima and minima.

(b) Locate any "abrupt" extremes by investigating values of x for which f' does not exist and testing for relative maxima and minima.

(c) Locate any end-point extremes.

If the domain of the function is a closed interval, the maximum of the maxima (if there is more than one) is the *absolute maximum*; similarly for the *absolute minimum*.

Most of the functions we encounter lead to nothing more complicated than turning-point extremes, but we shall now work a few problems that illustrate the other types.

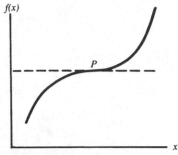

Fig. 3-16

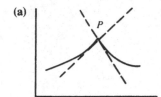

(a)

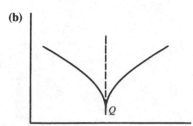

(b)

Fig. 3-17

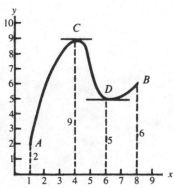

Fig. 3-11

129

PROBLEMS

1. Draw a sketch of a function that has no derivative at some point P, but for which P does not correspond to a relative maximum or a relative minimum.

2. Assuming the validity of the formula for the derivative of x^n for all values of n, investigate the extremes of the functions defined by each of the following equations, over their natural domains:

(a) $y = x^{1/2}$ (b) $y = x^{1/3}$ (c) $y = x^{2/3}$ (d) $y = x^{3/2}$

3. For the function determined by $E = 128 + 2.8x - 0.01x^2$, $0 \le x \le 120$, find minimum and maximum values of E.

4. Find the maxima and minima of the following functions. Be sure to identify absolute extremes.

(a) $y = x^2 - 4x + 6$, $0 \le x \le 3$

(b) $y = -\dfrac{x^3}{3} - x^2 - 1$, $-6 \le x \le 3$

(c) $y = x^3 + 3x^2 + 3x + 7$, $-2 \le x \le 2$

(d) $y = x^3 - 3x^2 + 1$, $-2 \le x \le 3$

(e) $y = \begin{cases} x^2, & -\frac{1}{2} \le x \le 1 \\ \dfrac{1}{x}, & 1 < x \le 2 \end{cases}$

(f) $y = \begin{cases} \frac{1}{2}x^2 + x + 4, & -4 \le x \le -2 \\ -x + 2, & -2 < x \le -1 \\ x^2, & -1 < x \le 1 \\ -2x^2 + 3, & 1 < x \le 2 \end{cases}$

★ 5. Find the maxima and minima of the following functions:

(a) $y = |x|$ (b) $y = |x^2 - 3|$ (c) $y = \left| \dfrac{x^3}{3} - x \right| - 4$

3.8   Applied maxima and minima

When it comes to applications of the basic idea presented in Sec. 3.5, the process of differentiation and the testing of the result often prove to be easier than two other parts of the problem: (a) translating the *words* of the problem into mathematical *symbols* and (b) expressing the quantity to be maximized or minimized in terms of a single variable, so that we have a function to which our method can be applied. There is no set of rules that can be presented to make these matters automatic, but some examples should be of help. The main thing that is needed is *practice*.

Example 1

A farmer has 200 rods of fencing with which he wishes to enclose a rectangular pasture. What are the dimensions of the rectangle that will provide the maximum area? How large an area can be enclosed?

If the dimensions of the rectangular pasture (Figure 3-18) are called x and y, we are trying to maximize the quantity A, given by $A = xy$. Were there no other conditions, there would be no maximum – we could get larger and larger areas by increasing either x or y or both.

But there is a constraint, of course: If we increase x we must decrease y, for $2x + 2y$ cannot be greater than 200. In using all the fencing, we have $2x + 2y = 200$, or $y = 100 - x$. Thus,

$$A = xy = x(100 - x) = 100x - x^2.$$

Problem 1

We now have A expressed as a function of the single variable x. Differentiate to obtain $x = 50$ as a critical number, and test that it corresponds to maximum A.

When $x = 50$, y also equals 50 (i.e., the rectangle is a square, a result in agreement with most people's intuition), and the maximum $A = 2500$ square rods.

Problem 2

Work the same problem if the length of fencing is l rods, to obtain the dimensions $x = l/4$, $y = l/4$, leading to the following general result: "The rectangle of maximum area for a fixed perimeter is a square."

Remark   In Example 1 we found that $x = 50$ gives a relative maximum for the area of the rectangular field. We have not yet checked that it also gives the *absolute* maximum. To accomplish this, we must check for abrupt and end-point extremes, as noted in (b) and (c) at the end of Section 3.7. Let us do that now:

When we wrote $A = 100x - x^2$, the domain was really limited: x could not be less than 0 nor more than 100. Indeed, if $x = 0$ or if $x = 100$, we don't have much of a pasture, for $A = 0$ at each of these ends of the domain. Moreover, the derivative of $100x - x^2$ exists for all numbers x, so there are no abrupt extremes. So, a complete answer to this problem might be worded as follows: At each end point of the domain, A has the relative minimum value zero, and at $x = 50$, A has a (turning-point) relative maximum value. Hence, $x = 50$ corresponds to the absolute maximum of A. Figure 3-19 shows the situation.

Example 2

The load (L lb) that a rectangular beam of a certain length can carry depends as follows on the dimensions of the cross section: L varies as the width (x in.) and the *square* of the depth (y in.). [Does this agree in general with your intuition? In Figure 3-20, would you lay floor joists as in (a) or as

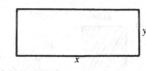

Fig. 3-18

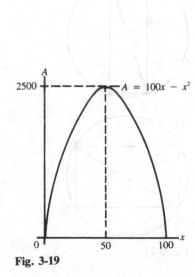

Fig. 3-19

131

3

Applications of the derivative

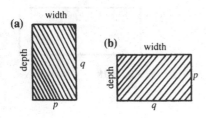
(a)

Fig. 3-20

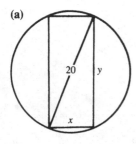
(a)

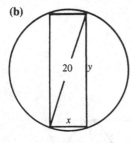
(b)

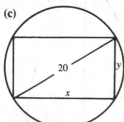
(c)

Fig. 3-21

in (b) for greater strength?] What are the dimensions of the strongest beam (maximum L) that can be cut from a circular log of diameter 20 in.? What is the value of the maximum L?

In this case, $L = kxy^2$, with no information given to determine k. Were there no further conditions, we could obtain a beam of arbitrarily large strength by making x or y or both sufficiently large. But once again there is a constraint: Any rectangle representing the cross section of the beam must be inscribable in a circle of diameter 20, as shown in Figure 3-21(a)–(c). Clearly, then, x is restricted to the interval $(0, 20)$, and y is also restricted to the same interval. The relation between x and y comes from the Pythagorean Theorem, as seen from the diameters drawn in each of the adjoining figures:

$$x^2 + y^2 = 400.$$

In principle, we can solve this equation for x in terms of y and substitute in $L = kxy^2$ to obtain L as a function of y alone; or we can solve for y in terms of x and substitute to obtain L as a function of x alone.

Problem 3

One of these two routes gives a simpler result than the other. Follow the route to the simpler result, obtaining $L = kx(400 - x^2) = k(400x - x^3)$, and differentiate to obtain $L' = k(400 - 3x^2)$.

We find the critical number(s) by setting $L' = 0$.

Problem 4

Remembering that x must be positive, do the algebra to obtain $x = 20/\sqrt{3} = (20\sqrt{3})/3 \approx 11.55$ in. as the critical number. Show that the corresponding y is $(20/\sqrt{3})\sqrt{2} \approx 16.33$ in. and that the corresponding L is approximately $3079k$ units.

Problem 5

Use either Test 1 or Test 2 to check that we have obtained a maximum L.

As with the remark at the end of Example 1, the domain of the function we are concerned with is limited: For $L = kx(400 - x^2)$ we must have $0 \le x \le 20$. Once again, the derivative of the function exists everywhere, so there is no abrupt extreme, and the end points correspond to *minima*. Hence, we have found the desired absolute maximum.

Problem 6

Solve this same problem for a circular log of arbitrary diameter, D, to obtain the dimensions $x = D/\sqrt{3}$, $y = (D/\sqrt{3})\sqrt{2}$, leading to the following general result: "The strongest beam that can be cut from a circular log has a depth that is $\sqrt{2}$ times the width."

Example 3

A rectangular garden is to be plowed in a piece of land in the shape of a right triangle with legs 50 ft and 80 ft, as in Figure 3-22. What are the dimensions of the rectangle of largest area that can be obtained?

If we call the sides of the rectangle x and y, we have $A = xy$ to be maximized. In this case, the relation between x and y comes from similar triangles: Because triangle PQC is similar to triangle ABC, we conclude that

$$\frac{y}{80 - x} = \frac{50}{80} = \frac{5}{8}.$$

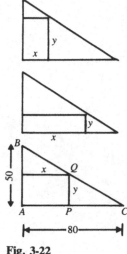

Fig. 3-22

Problem 7

Solve this equation for y in terms of x, and substitute in $A = xy$ to obtain $A = \frac{5}{8}(80x - x^2)$.

Problem 8

Differentiate, and set the derivative equal to zero, to obtain the critical number $x = 40$ (ft). Then get the corresponding $y = 25$ (ft) and $A = 1000$ (ft^2). What fractional part is this area of the area of triangle ABC?

Problem 9

Test that we have obtained a maximum and that it is an *absolute* maximum.

Problem 10

Solve the same problem as in Example 3 for a right triangle with legs a and b, to obtain dimensions $x = b/2$, $y = a/2$, and area $A = ab/4$, leading to the following general result: The rectangle of maximum area that can be "inscribed" in a right triangle has sides that are half the lengths of the legs of the right triangle and has an area half that of the right triangle.

These examples illustrate three of the common ways in which the relationship between two variables appears:

(a) In Example 1, the phrase "a farmer has 200 rods of fencing" leads *directly* to the equation $2x + 2y = 200$.

(b) In Example 2, the phrase "the dimensions of the...beam...that can be cut from a circular log of diameter 20 in." leads, *through the Pythagorean Theorem*, to the equation $x^2 + y^2 = 400$.

(c) In Example 3, the phrase "as in Figure 3-22" leads, *through a proportionality in similar triangles*, to the equation $y/(80 - x) = \frac{5}{8}$.

It will be worthwhile to keep in mind these three methods of finding the relationship between two variables, for they arise frequently.

We can summarize the procedure for finding turning-point extremes in applied problems as follows:

1. Express the quantity to be maximized or minimized, perhaps in terms of more than one variable.

2. Use the data of the problem, and/or your knowledge of geometry, to express the quantity to be maximized or minimized as a function of one variable.
3. Differentiate this function.
4. Set the derivative equal to zero, and solve to find the critical number(s).
5. Test to see whether you have a maximum or a minimum.
6. To find the extreme value(s) of the quantity, substitute the critical number(s) in the formula for the quantity.

Then, to find *absolute* extremes, check also for abrupt extremes and end-point extremes.

PROBLEMS

11. What are the dimensions of a rectangle of minimum perimeter if the area is fixed at 2500 square units?

★ 12. What is the shape of a rectangle of minimum perimeter for fixed area?

13. A farmer has 300 yards of fencing to be used to enclose three sides of a rectangular field bordering a straight stream – no fencing is needed along the stream. What are the dimensions of the field of maximum area that can be so enclosed? Check that you have obtained a maximum.

   Now visualize the situation if the farmer has another 300 yards of fencing that he uses for a second field, directly across the stream from the first. What are the dimensions of the two fields taken together? Explain why the result is to be expected.

★ 14. What is the shape of a rectangle of maximum area if the sum of the lengths of three sides is fixed?

15. A rectangular corral of width x yards and length y yards is to be built with materials costing $4 per yard for the width and $2 per yard for the length.

   (a) What are the dimensions of the corral of maximum area that can be built for $80? Check that you have obtained a maximum.

   (b) What are the dimensions of the corral of minimum cost that can be built to enclose 50 square yards? Check that you have obtained a minimum.

★ 16. A rectangular corral is to be built with materials costing $a per yard for the width and $b per yard for the length. What is the optimal shape of the corral?

17. A rectangular gutter is to be made by bending up x in. from each end of a sheet of aluminum 20 in. wide, as shown in Figure 3-23. What are the dimensions of the gutter of maximum carrying capacity (maximum cross-sectional area) that can be so constructed? Check that you have obtained a maximum.

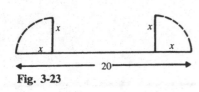

Fig. 3-23

134

18. (a) A closed cylindrical container (i.e., a cylinder with both top and bottom) is to hold 250π in.^{3} What dimensions will require the least material? How much material?

    (b) A closed cylindrical container is to be made from 150π in.^{2} of plastic. Assuming no waste, what are the dimensions of the container of maximum volume that can be made? What is the maximum volume?

★ (c) What is the optimal shape for a closed cylindrical container?

19. (a) A cylindrical container with a bottom but no top is to hold 125π in.^{3} What dimensions will require the least material? How much material?

    (b) A cylindrical container with a bottom but no top is to be made from 75π in.^{2} of plastic. Assuming no waste, what are the dimensions of the container of maximum volume that can be made? What is the maximum volume?

    (c) If you were to take two open-topped cylinders of the dimensions obtained as the solutions of parts (a) and (b) and put them together with the open tops coinciding, what would you have? (No wisecracks, please.)

★ (d) What is the optimal shape for a cylindrical container with a bottom but no top?

20. What are the dimensions of the largest cylinder (cylinder of maximum volume) that can be inscribed in a sphere (Figure 3-24) of radius 10 in.?

21. What are the dimensions of the largest cone (cone of maximum volume) that can be inscribed in a sphere (Figure 3-25) of radius 10 in.?

★ 22. As in Figure 3-26, a spring is at point S, at distances a and b from two perpendicular paths, p_1 and p_2. It is desired to make a path AB from p_1 to p_2 and passing through S. Find the lengths x and y so that the length of AB is a minimum. You may assume that this is equivalent to finding x and y so that the *square* of the length of AB is a minimum.

23. A potato crop, if dug now, would yield 50 bushels worth \$2 per bushel. In every week from now on, the crop will grow by 5 bushels, and the price will drop by 10 cents per bushel. When should the potatoes be dug for maximum total value?

24. Today a farmer has 1000 boxes of apples in storage, worth \$4 per box. In every week from now, he will lose 20 boxes through spoilage, and the value of the apples will increase by 10 cents per box. Express the number of boxes he will have x weeks from now and the value of each box at that time. When should he sell the apples for maximum return?

25. A closed cylindrical can is to be made of metal costing 5 cents per square inch for the bottom, the same for the top, and 4 cents per square inch for the lateral surface. What are the dimensions of a can containing 1280 in.^{3} if the cost of the metal is to be a minimum?

26. (a) The material for the bottom of a rectangular box with a square base costs 10 cents per square inch; the material for the sides costs 2

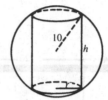

Fig. 3-24

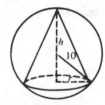

Fig. 3-25

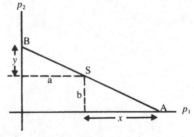

Fig. 3-26

135

cents per square inch; there is no top. Find the dimensions of the box of maximum volume that can be made for $30.

(b) Same as (a), if the material for the sides costs 5 cents per square inch.

(c) Same as (a), if the material for the sides costs 10 cents per square inch.

27. An apple grower has a yield of 30 bushels per tree when there are 20 trees per acre. Each additional tree per acre decreases the yield per tree by 1 bushel because of crowding. How many trees per acre should there be for maximum total yield? What is the maximum total yield?

28. A calf weighing 300 lb could now be sold for 80 cents per pound. If in every week from now on, it is estimated that the weight of the calf will increase by 10 lb and that the market price will decrease by 2 cents per pound, when should the calf be sold for maximum return? (Note that we neglect the expense of care and feeding of the calf over the next weeks.)

29. A rectangular corral is to be built using 300 yards of fencing for three sides and using part or all of a 120-yard wall for the fourth side, as in Figure 3-27.

(a) Express the area (A square yards) of the corral as a function of x, specifying the domain of this function. Find the dimensions of the corral of maximum area that can be so constructed, and demonstrate conclusively that you have obtained a maximum. Illustrate your result by drawing a sketch of A as a function of x.

(b) Also express A as a function of y, specifying the domain of this function, and illustrate your result by drawing a sketch of A as a function of y.

30. (a) You are to build a rectangular box with square base to hold 36,000 cm³. If material for the base of the box costs 10 cents/cm², material for the top costs 6 cents/cm², and material for the four sides also costs 6 cents/cm², what are the dimensions of the box for minimum cost?

(b) Same as (a), if the side of the base must not exceed 25 cm.

(c) Same as (a), if the height of the box must not exceed 30 cm.

(d) Same as (a), if no dimension of the box may exceed 36 cm.

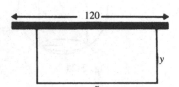

Fig. 3-27

3.9 Maxima and minima in some problems in economics

The problems we shall consider here represent an important application of our methods for determining maxima and minima. The mathematics is quite simple–we have only to translate each problem into appropriate mathematical symbolism.

Let us study the profit made by the sausage manufacturer we first met in Problem 8, **1.4**, and then again in Problem 5, **1.7**, and in Section 3.4. In

Section 3.4 we decided that an adequate formula for the total daily expense (E) of making x lb of sausage per day is

$$E = 128 + 2.8x - 0.01x^2, \quad 0 \le x \le 120. \tag{1}$$

If p is the selling price of a pound of sausage, then the total daily revenue (R) from selling x lb of sausage is $R = px$, and the total daily profit (P) from making and selling x lb of sausage is

$$P = R - E = px - (128 + 2.8x - 0.01x^2).$$

But it is unlikely that an arbitrarily large number of pounds can be sold per day at some fixed price, p. Usually we find that p is a decreasing function of x, as with the example of the wastebaskets in Problem 7, **1.11**. Let us assume that a market study has shown that to sell the output of the manufacturer and to operate steadily, p is related to x by the following "demand function":

$$p = 6.4 - 0.03x. \tag{2}$$

Problem 1
The natural domain of the function given by equation (2), if we think only in mathematical terms, is $(-\infty, \infty)$. But, if x and p have their economic significance, what is the domain of the function, and what is its range?

Problem 2
Do the algebra to show that the daily profit is given by $P = -128 + 3.6x - 0.02x^2$. What is P if $x = 0$? Is this result expected?

Problem 3
Use differentiation to show that the maximum daily profit is obtained for $x = 90$. Check that it gives a maximum. What is the maximum value of P?

Problem 4
Find the maximum daily profit if the cost function is given by equation (1) as before, but the demand function is given by $p = 4.6 - 0.02x$, rather than by equation (2). What's the trouble?

Problem 5
If the daily profit is P on the sale of x lb, the daily profit per pound is, of course, P/x.
  (a) Compute the profit per pound for the values of x and P in Problem 3.
  (b) If the cost function is given by equation (1) and the demand function by equation (2), find the formula for the profit per pound, and show that it is a maximum for $x = 80$. What, then, is the maximum profit per pound, and what is the corresponding total profit?

137

As you see, the problems of this section are approached in the standard manner: Translate the words into mathematical symbols, and find the extreme by differentiating and setting the derivative equal to zero.

PROBLEMS

6. An oil company, selling x gal of gasoline per year at y cents per gallon, finds from experience that the relation between x and y for steady operation is

$$x = 10^7(150 - y)$$

and that their total annual expense (E cents) for producing and marketing x gal is $E = 4 \cdot 10^9 + 90x$.
   (a) Express the annual profit, P cents, as a function of y, and find y, x, and P corresponding to maximum P.
   (b) Check your result in (a) by expressing P as a function of x and by finding x, y, and P corresponding to maximum P.
   (c) How much is the profit per gallon in (a)?
   (d) Express the profit per gallon as a function of x, and find x, y, and P corresponding to maximum profit per gallon. How much is the maximum profit per gallon?

★ 7. The total cost, C, of producing x items is given by

$$C = a + bx,$$

and the relation between the number x that can be sold and the selling price, y, is

$$x = p - qy.$$

   (a) Show that for maximum total profit, $y = (p + bq)/2q$.
   (b) Show that for maximum unit profit (profit per item), $x = \sqrt{aq}$, and hence that $y = (p - \sqrt{aq})/q$.

★ 8. The total cost, C, of producing x items is given by

$$C = a + bx - cx^2,$$

and the relation between the number x that can be sold and the selling price, y, is

$$x = p - qy.$$

   (a) Show that for maximum total profit, $x = (p - bq)/[2(1 - cq)]$.
   (b) Show that for maximum unit profit (profit per item), $x = \sqrt{aq/(1 - cq)}$.

9. Suppose that the total cost ($\$E$) of manufacturing x units of a certain commodity is given by

$$E = 40 + 1.6x + 0.001x^2, \quad 0 \le x \le 250.$$

Note that the "fixed cost" is 40 and the "variable cost" is $1.6x + 0.001x^2$.
   (a) Find the minimum and maximum values of E.

(b) Find the marginal cost of manufacturing the first unit. The 250th unit.

(c) If the "average total cost" (or "total unit cost") is defined as E/x, write a formula for the average total cost in terms of x. What is the domain of this function?

(d) Find the minimum value of the average total cost. Is there a maximum value? Explain.

(e) If the "average variable cost" is defined as (variable cost)$/x$, find the minimum and maximum values of the average variable cost.

★ 10. In Problem 9(d), you should have found that the minimum value of the average total cost is $2, attained at $x = 200$.

(a) Verify that the derivative, E', is also 2 at $x = 200$. In other words, for the cost function of Problem 9,

$$E' = \frac{E}{x}$$

at the value of x that makes $(E/x)' = 0$.

(b) Show that the result stated in (a) is also valid for the cost function $E = a + bx + cx^2$, where a, b, and c are any positive constants. (In Chapter 4 we shall find that this is a quite general result, not restricted to quadratic cost functions.)

11. In a small shoe factory (this doesn't mean that they manufacture only Cinderella slippers), the expense ($\$E$) of producing x pairs per week is as follows: $E = 1{,}000 + 12.5x + 0.001x^2$.

(a) Find a formula for the marginal cost. What, then, is the marginal cost of producing 1 pair per week? The 100th pair? The 1000th pair?

(b) Find a formula for the total unit cost, E/x.

(c) What x makes the total unit cost an extreme? Maximum or minimum?

(d) What is the total unit cost at the x found in (c)?

(e) What is the marginal cost at the x found in (c)?

3.10 Approximate solution of equations: the Newton–Raphson method and the bisection method

It is simple to solve any linear equation $ax + b = 0$, with a and b being constants, and $a \neq 0$.

For any quadratic equation, $ax^2 + bx + c = 0$, with a, b, and c being constants, and $a \neq 0$, we have the "quadratic formula":

$$x = \left(-b \pm \sqrt{b^2 - 4ac}\right)/2a.$$

(In some cases, the roots are not real numbers, of course.)

There is a formula for the (three) roots of the general cubic equation,

$$ax^3 + bx^2 + cx^2 + d = 0,$$

and a formula for the (four) roots of the general quartic equation,

$$ax^4 + bx^3 + cx^2 + dx + e = 0.$$

There are no comparable formulas for the roots of the general polynomial equation of degree greater than four, and even the formulas for the roots of the cubic and quartic are not convenient to use if we want a numerical approximation to the roots.

A method essentially due to Newton uses the derivative to find such approximations. The particular *form* that is now common appears in the work of Raphson, a contemporary of Newton. We shall illustrate the Newton–Raphson method by working an example that Newton himself used in his *Methods of Fluxions*:

Example 1

The problem is to find the real root(s) of the equation $x^3 - 2x - 5 = 0$. For short, we shall designate $x^3 - 2x - 5$ as $f(x)$.

It is easy to make the following table:

x	-2	-1	0	1	2	3
$f(x)$	-9	-4	-5	-6	-1	16

By the rule stated at the end of Section 0.11 we know that there are no roots less than -2, and none greater than 3.

Now, $f'(x) = 3x^2 - 2$, so critical numbers of this function are $\pm\sqrt{2/3}$, or ± 0.8165, approximately. We can quickly check that $x = \sqrt{2/3}$ corresponds to a relative minimum of f, and $x = -\sqrt{2/3}$ corresponds to a relative maximum. We calculate that $f(\sqrt{2/3}) \approx -6.09$, and $f(-\sqrt{2/3}) \approx -3.91$.

The foregoing calculations enable us to draw Figure 3-28 as the graph of $y = f(x)$, indicating a root between $x = 2$ and $x = 3$, with no other real roots. We choose $x_0 = 2$ as an initial approximation to the root, and we get a better approximation through "linearization": We approximate the curve near $x = 2$ by a segment of the tangent line to the curve at $x = 2$ – as shown in Figure 3-29, we use the straight-line segment PT as an approximation to the arc PQ. Of course, the root that we seek is the abscissa of point Q. As our next approximation we use x_1, the abscissa of point T.

Now the slope of $PT = [0 - (-1)]/(x_1 - 2)$. Also, the slope of $PT = f'(x_0) = f'(2) = 10$.

Thus, $1/(x_1 - 2) = 10$; so $x_1 - 2 = \frac{1}{10}$, or $x_1 = 2 + \frac{1}{10} = 2.1$. Substituting 2.1 for x in $f(x)$ gives $f(2.1) = (2.1)^3 - 2(2.1) - 5 = 0.061$, a small number; so $x_1 = 2.1$ is a reasonably good approximation to the desired root.

To obtain a better approximation we repeat the process, as shown in Figure 3-30, which is not drawn to scale: We approximate the arc UQ by the segment of the tangent UV. As our next approximation we use x_2, the abscissa of point V. We see that the slope of $UV = (0 - 0.061)/(x_2 - 2.1)$. Also, the slope of $UV = f'(x_1) = f'(2.1) = 3 \cdot (2.1)^2 - 2 = 11.23$.

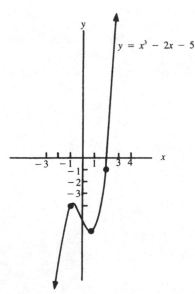

$y = x^3 - 2x - 5$

Fig. 3-28

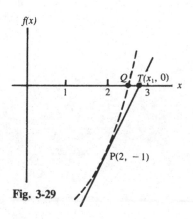

$f(x)$

Q $T(x_1, 0)$

$P(2, -1)$

Fig. 3-29

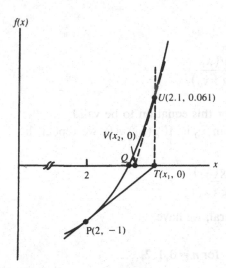

Fig. 3-30

Thus,

$$-\frac{0.061}{x_2 - 2.1} = 11.23;$$

so

$$11.23(x_2 - 2.1) = -0.061, \quad \text{or} \quad x_2 - 2.1 = \frac{-0.061}{11.23},$$

$$\text{or } x_2 = 2.1 - \frac{0.061}{11.23} \approx 2.095.$$

We calculate $f(x_2) = f(2.095) = (2.095)^3 - 2(2.095) - 5 = 0.005$. Because $f(x_2)$ is such a small number, $x_2 = 2.095$ is a good approximation to the root. But we could repeat the process yet again for a still better approximation.

Problem 1
Starting with $x_2 = 2.095$, and $f(x_2) = 0.005$, repeat the process to find x_3 and $f(x_3)$.

We now generalize the method of this example. Let r be a root of the equation $g(x) = 0$. Suppose that x_0 is an approximation to r. Let the tangent line to the curve at the point with abscissa x_0 meet the x axis at x_1, as in Figure 3-31. Then the slope of this tangent line is $[0 - g(x_0)]/(x_1 - x_0)$. But the slope also equals $g'(x_0)$. Equating these two expressions for the slope gives

$$\frac{0 - g(x_0)}{x_1 - x_0} = g'(x_0).$$

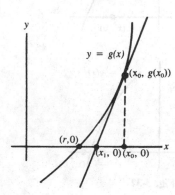

Fig. 3-31

141

Problem 2

Solve this equation for x_1 to obtain

$$x_1 = x_0 - \frac{g(x_0)}{g'(x_0)}.$$

Note that $g'(x_0)$ must not equal zero for this equation to be valid.

If all is well, x_1 will be closer to r than x_0 is. If $g(x_1) \neq 0$, we repeat the process, obtaining

$$x_2 = x_1 - \frac{g(x_1)}{g'(x_1)}$$

as the next approximation to r. In general, we have

$$x_{n+1} = x_n - \frac{g(x_n)}{g'(x_n)}, \quad \text{for } n = 0, 1, 2, \ldots.$$

Problem 3

With each of the sketches in Figure 3-32, determine x_{n+1} geometrically, and verify that you are led to the preceding formula in each case.

Example 2

Let us apply the method to approximate $\sqrt{7}$.

We take $g(x) = x^2 - 7$. Then $g'(x) = 2x$. Because $2^2 = 4$ and $3^2 = 9$, it would seem reasonable to start at $x_0 = 3$. The formula for successive approximations is

$$x_{n+1} = x_n - \frac{x_n^2 - 7}{2x_n}.$$

Thus, $x_1 = 3 - \frac{9-7}{6} = \frac{8}{3} = 2.666\cdots$.

Problem 4

Show that $x_2 = \frac{127}{48} = 2.6458333\cdots$, and that $x_3 \approx 2.6457513$, the square of which is approximately 6.9999999.

A good check on accuracy is obtained by comparing each approximation with a quite accurate approximation to $\sqrt{7}$ from a calculator or a large table. We would find that the error in x_0 is about 0.35, the error in x_2 is about 8×10^{-5}, and the error in x_3 is about 1×10^{-6}.

This pattern of increasing accuracy is typical. When Newton's method is working, the error at any step is roughly the square of that at the preceding step. Thus, if the error is smaller than 1, the succeeding errors decrease rapidly.

It is hard to predict exactly where the method does work. Certainly it is unlikely to do so when $g'(x_0)$ is very close to zero, for then x_1 may be far

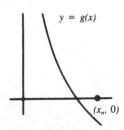

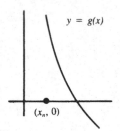

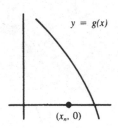

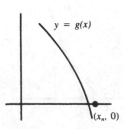

Fig. 3-32

away (Figure 3-33). Another difficulty will be indicated later in Problem 9. For these reasons, the method tends to be more useful in calculator work, where one can watch what is happening, than in computer work.

The important thing is to start close to the root r. We can ensure this by using the following simple, sure-fire method first: It is called *bisection*, and it always works so long as the graph of f is unbroken (*continuous* is the term we shall use shortly) and crosses the x axis at an isolated root r. What this amounts to is that there is an interval $[a, b]$ surrounding r such that $f(a)$ and $f(b)$ are of different signs – one positive, the other negative.

The method is simply to bisect $[a, b]$ successively, each time narrowing down on r. Let $x_1 = (a + b)/2$. There are three cases to consider.

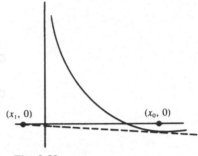

$(x_1, 0)$ $(x_0, 0)$

Fig. 3-33

Case 1. $f(x_1) = 0$. We were lucky enough to land on r immediately, and we are done.

Case 2. $f(x_1) \cdot f(a) < 0$. That is, $f(x_1)$ and $f(a)$ are of different signs. Then we know that r lies between a and x_1, as shown in Figure 3-34. The next step is to let x_1 take the place of b to determine $x_2 = (a + x_1)/2$.

Case 3. If neither Case 1 nor Case 2 obtains, $f(x_1) \cdot f(b) < 0$, and r lies between x_1 and b. We set $x_2 = (x_1 + b)/2$.

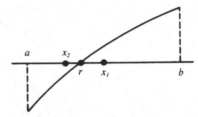

Fig. 3-34

And we keep going, forming $x_3, x_4, \ldots, x_n$, stopping if we land on r and otherwise isolating r in ever-diminishing intervals. How close are we to r at each step? Well, we know that x_1 can be no further from r than $(b - a)/2$. And x_2 can be no further from r than $(b - a)/2^2$. In general, x_n is certain to be within $(b - a)/2^n$ of r.

Let us take as an example $f(x) = x^2 - 7$ and find $r = \sqrt{7}$ to an accuracy of two decimal places. Because $f(2) = -3$ and $f(3) = 2$ are of opposite sign, we have r isolated in $[a, b]$, with $a = 2$ and $b = 3$. Then $x_1 = \frac{5}{2}$ and $f(x_1) = -\frac{3}{4}$ < 0, so that r lies between x_1 and b (Case 3). Thus, $x_2 = (\frac{5}{2} + 3)/2 = \frac{11}{4}$. Then $f(x_2) = f(\frac{11}{4}) = \frac{9}{16} > 0$, which means that r lies between x_1 and x_2 (Case 2). Thus, we set $x_3 = (x_1 + x_2)/2 = \frac{21}{8}$. Can we stop? Hardly. To be certain of the desired accuracy, we must find x_n so that $(b - a)/2^n = 2^{-n}$ is less than $5 \cdot 10^{-3}$.

Problem 5
Show that n must be at least 8. Then verify that $x_4 = \frac{43}{16}$, $x_5 = \frac{85}{32}$, $x_6 = \frac{169}{64}$, $x_7 = \frac{339}{128}$, $x_8 = \frac{677}{256}$. Check to see that x_8 is sufficiently close to $\sqrt{7}$.

It is clear that bisection is a very tedious process. However, its simplicity and reliability make it a handy device for use with computers, as well as for narrowing down on a root with a calculator to a point where the Newton–Raphson method can find the root quickly and to high accuracy.

PROBLEMS

C 6. Use the Newton–Raphson method to approximate

(a) $\sqrt{17}$ (b) $\sqrt[3]{120}$ (c) $\sqrt[4]{85}$

7. Let A be a positive number. Show that the Newton–Raphson formula for finding $\sqrt{A}$ can be written in the form

$$x_{n+1} = \frac{1}{2}\left(x_n + \frac{A}{x_n}\right).$$

C 8. Use three steps of the Newton–Raphson method to approximate $\sqrt[3]{7}$. Determine the error at each step.

C 9. Let $f(x) = x^3 - 3.3x^2 - 7.7x + 12.4$. Make a rough graph of f to determine that there are roots of $f(x) = 0$ near -2, 1, and 4. Then use the Newton–Raphson method to get close to these roots. If you have programmed the calculation, use the program with various starting values not very close to the roots, in particular with $x_0 = 0.8$ and $x_0 = 3$, where the derivative is numerically small. The problem is not one of convergence, but rather of how long it takes and where it ends up.

C 10. If your calculator is programmable and permits branching, program it to bisect. Use your program on Problems 8 and 9.

3.11 Review

In reviewing, note the five types of problems mentioned at the start of this chapter. Remember the definitions of *functions increasing or decreasing over an interval* (Section 3.2). We have had two tests for "turning-point" maxima and minima (Section 3.6). Translating the words of a maximum and minimum problem into a function of a single variable may require ingenuity; three helpful observations are given in Section 3.8. A procedure for a complete analysis of maxima and minima is given at the end of Section 3.7, and a procedure for applied maxima and minima is given at the end of Section 3.8.

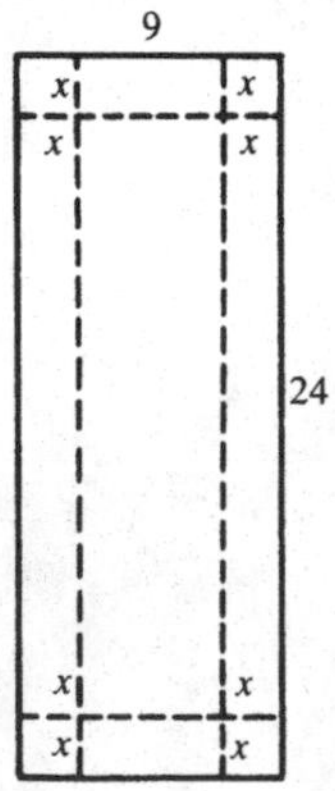

PROBLEMS

1. A rectangular sheet of aluminum 9 in. by 24 in. is to be formed into an open rectangular box by cutting out squares x in. on a side from the corners and bending up the aluminum along the dotted lines (Figure 3-35).

(a) Express the volume (V in.^{3}) of the box as a function of x. What is the domain of this function?

(b) What x gives maximum V?

(c) Discuss the question of extremes (maximum or minimum) for the area of the bottom of the box.

Fig. 3-35

2. You wish to make a rectangular box with a square base and no lid, of volume 80 ft.^{3} The material for the sides will cost 20 cents per square foot, and that for the base 50 cents per square foot. Find a formula for the total cost (C cents) in terms of the length (x ft) of the side of the base. Determine whether or not, with $20 to spend, you can afford to make such a box.

3. (a) A quarter-mile (440-yard) running track consists of two sides of a rectangle and two semicircles erected on the ends of the rectangle, as in Figure 3-36. Show that if the area of the rectangle is to be as large as possible, the two "straightaways" of the track represent half the length of the track. (These are the standard dimensions in the rule book.)

★ (b) Show that the result in (a) applies to a track of any length, l. (In particular, then, it will apply to a 400-m track.)

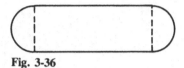

Fig. 3-36

4. For the function $f(x) = x^3 - 3x^2 - 9x + 2$ over its natural domain, what are the values of x for which f is stationary? Increasing? Decreasing?

5. The cost (C) of producing x tons of fertilizer is given by $C = 100 + 13x - 0.01x^2$. The selling price *per ton* (S) at which x tons can be sold is given by $S = 25 - 0.02x$. What x gives maximum profit? How much is the maximum profit? Check that you have a maximum.

6. For C as given in Problem 5, what is the marginal cost of producing the 1st ton? The 10th? The 100th?

7. The force of attraction (F dynes) between two particles varies inversely as the square of the distance between them (x cm), and $F = 40$ when $x = 5$. Find the formula for F in terms of x. Use differentiation to find the approximate change in F if x decreases from 10.2 to 9.9.

8. A closed cylindrical can is to be made of metal costing 4¢ per square inch for the bottom, the same for the top, and 3¢ per square inch for the lateral surface. What are the dimensions of a can containing 72π in.^{3} if the cost of the metal is to be a minimum?

9. Sketch a graph of the function given by

$$y = \begin{cases} x + 3, & -5 \le x < -3, \\ 9 - x^2, & -3 \le x \le 3, \\ (x - 3)^2, & 3 < x \le 5. \end{cases}$$

For what values of x in the domain of f does f' not exist? Where do relative minima of f occur? Relative maxima? What is the absolute minimum of f? The absolute maximum?

★ 10. If the total expense (E) of producing x units of a commodity is given by $E = a + bx + cx^3$, where a, b, and c are positive constants, show that $E' = E/x$ at that x which makes E/x an extreme.

11. The cost (C) of producing x lb of grass seed is given by $C = 100 + 4x - 0.01x^2$, for $0 \le x \le 180$. The selling price *per pound* (S) at which x lb can be sold is given by $S = 7 - 0.02x$.

(a) Express the revenue (i.e., gross income), R, as a function of x.

(b) Express the profit (P) as a function of x.

(c) What x gives maximum profit?

(d) How much is the maximum profit? Check that you have a maximum.

(e) Find a formula for the marginal cost. Hence, what is the marginal cost for $x = 10$? For $x = 100$?

12. The resistance (R ohms) of a wire varies inversely as the square of its diameter (x mm), and $R = 20$ at $x = 5$. Find a formula for R in terms of x. Use differentiation to approximate the change in R if x decreases from 10.2 to 9.9, giving the sign and units of your answer.

13. A voyage of 1000 miles is made at some speed, v mph.

(a) How many hours does the voyage take?

(b) What is the cost of the voyage if the *hourly* cost (y per hour) is given by $y = 160 + 0.01v^3$?

(c) What speed makes the cost of the voyage a minimum?

(d) What is the minimum cost?

14. Figure 3-37 shows the graph of a function, $y = f(x)$.

(a) Find a formula (or formulas) for f.

(b) For what values of x in the domain of f does f' not exist?

(c) What are the relative maxima of f, and where do they occur?

(d) What is the absolute maximum of f, and where does it occur?

15. For the function $f(x) = 2x^3 - 9x^2 - 24x + 2$, defined over its natural domain, find those x's for which f is stationary. Is increasing. Is decreasing.

16. As in Figure 3-38, a rectangular field, $ABCD$, is to be fenced, with no fencing required along the river, but with a fence down the middle. If 1200 m of fencing is available, what are the dimensions of the field for maximum area? What is the maximum area? Give the units of your answer.

17. The intensity (I calories/min) of heat from a lamp varies *inversely* as the *square* of the distance (x cm) from the lamp, and $I = 1000$ for $x = 2$. Find a formula for I in terms of x. Use differentiation to approximate the change in I if x increases from 4.9 to 5.2. Give the sign and units of your answer.

★ 18. Use the *definition* of derivative to demonstrate that if $f(x) = 1/4x^2$, then $f'(x) = -1/2x^3$.

19. (a) A closed cylindrical can is to contain 54π in.^{3} What are the dimensions of the can if the metal needed to make the can is to be minimized?

(b) Same as (a), with the diameter of the base of the can not to exceed 5 in.

(c) Same as (a), with the height of the can to be at least three times the diameter of the base.

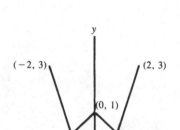

Fig. 3-37

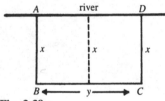

Fig. 3-38

20. A peanut butter jar consists of an open glass cylinder with a metal screw-on top. Make a reasonable model of the cost of making the jar, and find the most economical shape for a jar to contain a given quantity of peanut butter.

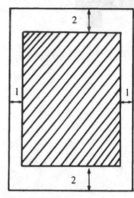

Fig. 3-39

SAMPLE TEST

1. For the function $f(x) = x^3 + 3x^2 + 3x - 5$, considered over its natural domain, determine where the function is stationary, is increasing, and is decreasing.

2. For certain aircraft, the lift (L units) varies as the square of the velocity (v mph), and $L = 64,000$ when $v = 400$. Use differentiation to find approximately how much smaller L is for $v = 498$ than for $v = 500$.

3. A sheet of paper of area A in.^{2} is to have printing x in. by y in., with margins of 2 in. top and bottom and margins of 1 in. on each side (Figure 3-39). If the printed area is to be 128 in.^{2}, what are the dimensions of the paper for minimum A?

4. The total cost ($\$E$) of producing x tons of fertilizer is given by $E = 5000 + 45x - 0.02x^2$, $5 \le x \le 100$. What is the marginal cost of producing the 10th ton? The 50th ton? The 100th ton?

5. Figure 3-40 shows the graph of a function, $y = f(x)$.
   (a) Find a formula (or formulas) for f.
   (b) For what values of x in the domain of f does f' not exist?
   (c) What are the relative minima of f, and where do they occur?
   (d) What is the absolute minimum of f, and where does it occur?

6. The material for the bottom of a rectangular box with a square base costs 2¢ per square inch; the material for the sides costs 1¢ per square inch; there is no top.
   (a) Find the dimensions of the box of maximum volume that can be made for $6.
   ★ (b) Can you explain intuitively why the box turns out to be a cube in this case?

★ 7. (a) A rectangular corral is to be built with 100 yards of fencing. An existing 20-yard wall is to be used as *part or all* of one side, or the wall can be extended by fencing, if desired (Figure 3-41). What are the dimensions of the corral for maximum area?
   (b) Discuss the problem of part (a) if the length of fencing is some constant l and the length of the existing wall is some constant w.

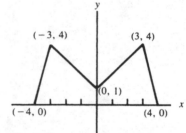

Fig. 3-40

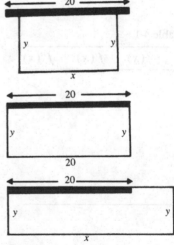

Fig. 3-41

4 Further differentiation

In this chapter we shall extend the theory and technique of differentiation begun in Chapter 3, enabling us to apply our methods to problems that we couldn't handle earlier.

4.1 Repeated differentiation and derived curves

If a function, $f(x)$, has a derivative, $f'(x)$, at all points of its domain, D_f, then f' is also a function of x, with a domain $D_{f'} = D_f$. Hence, we can ask whether or not f' has a derivative – the rate of change of f'. If it exists, the derivative of f' is called the *second derivative* of f and is denoted by f''.

Problem 1
State the definition of $f''(x)$ in terms of f' and the concept of limit.

For example, if $f(x) = x^2$, then $f'(x) = 2x$, and $f''(x) = 2$.

Problem 2
Find $g''(x)$ if $g(x) = x^4 - 2x^3 + x - 1 + 1/x$.

Problem 3
If $f(x) = \frac{1}{3}x^3 - 2x^2 + 3x + 1$, find $f'(x)$ and $f''(x)$, and fill in Table 4-1.

Geometrically, $f'(x)$ is interpreted as the slope of the curve $y = f(x)$. The second derivative, $f''(x)$, is called the **flexion** of the curve $y = f(x)$. Of course, $f''(x)$ can also be interpreted as the slope of the curve $y = f'(x)$. Let us now show this graphically: If we use the f of Problem 3 and plot the graph of $y = f(x)$ as in Figure 4-1(a), we recognize that for each x, the value of $f'(x)$ is the slope of this curve. Hence, if we plot $y = f'(x)$ as in Figure 4-1(b), the *height* of this curve for each x equals the *slope* of the original curve at the same x. The curve $y = f'(x)$ is called the **first derived curve** of $y = f(x)$.

Table 4-1

x	$f(x)$	$f'(x)$	$f''(x)$
0			
$\frac{1}{2}$			
1			
$\frac{3}{2}$			
2			
3			
4			

148

Similarly, if we plot $y = f''(x)$ as in Figure 4-1(c), the *height* of this curve for each x equals the *slope* of the first derived curve at the same x, and it also equals the *rate of change of slope*, or *flexion*, of the original curve at the same x. The curve $y = f''(x)$ is called the **second derived curve** of $y = f(x)$.

Problem 4
Various relationships are set out in Table 4-2. Check them by referring to the appropriate graphs, and complete the table where there are blanks.

Table 4-2

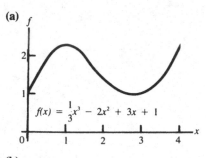

(a)

x	$f(x) = \frac{1}{3}x^3 - 2x^2 + 3x + 1$	$f'(x) = x^2 - 4x + 3$ $= (x-1)(x-3)$	$f''(x) = 2(x-2)$
Interval $(0,1)$	Increasing Slope of graph is decreasing	Positive Decreasing	Negative
1	Maximum value Slope of graph is decreasing	0 Decreasing	Negative
Interval $(1,2)$	Decreasing Slope of graph is decreasing	Negative Decreasing	Negative
2	Decreasing Slope of graph is minimum	Negative Minimum value	0
Interval $(2,3)$			
3			
Interval $(3,4)$			

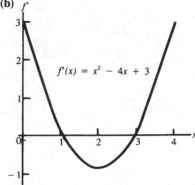

(b)

In the equation $y = F(x)$, if y represents displacement from some starting point and x represents elapsed time, for a particle moving in a straight line, then, as we know, $F'(x)$ represents *instantaneous velocity*, the rate of change of displacement with respect to time. The rate of change of velocity with respect to time, $F''(x)$, has a special name – it is called **acceleration**.

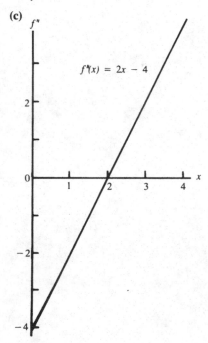

(c)

Problem 5
A projectile thrown vertically has height (s ft) above the ground at time (t sec), according to the following formula:

$$s = -16t^2 + 80t + 96, \quad 0 \le t \le 6.$$

(a) Verify that the velocity (v ft/sec) is given by $v = -32t + 80$ and that the acceleration (a ft/sec/sec) is constantly -32. In other words, the velocity decreases by 32 ft/sec in each second.

Fig. 4-1

(b) Verify that the projectile was launched ($t = 0$) from a height of 96 ft above the ground with an initial upward speed of 80 ft/sec. What was the initial acceleration?

(c) Verify that the projectile achieves its maximum height at $t = \frac{5}{2}$. What is the maximum height, and what is the acceleration at that time?

(d) Verify that the projectile hits the ground ($s = 0$) at $t = 6$. What are the velocity and acceleration at that time?

Problem 6

A particle moves on the x axis so that its displacement (x ft) from the origin varies with the elapsed time (t sec) as follows: $x = t^3 - 9t^2 + 6t + 10$. Find a formula for the acceleration at any time. For what values of t, then, is the velocity increasing? Decreasing? When is the velocity an extreme? Is it then a maximum or a minimum?

We sometimes deal with derivatives higher than the second. For example, in considering the graph of $y = f(x)$, the rate of change of flexion is given by the third derivative of y with respect to x, $f'''(x)$. Similarly, in the equation $y = F(x)$, if y represents displacement from some starting point and x represents elapsed time, for a particle moving in a straight line, then the rate of change of acceleration is given by $F'''(x)$. We do not have any special names (comparable to slope, velocity, flexion, acceleration) for these rates of change.

Note that one may be led astray by memorizing a shorthand statement like "acceleration is the second derivative, and rate of change of acceleration is the third derivative." If we should start with a formula for velocity, $v = f(t)$, then acceleration is $f'(t)$ and rate of change of acceleration is $f''(t)$. The correct statement reads "*acceleration is the second derivative of displacement with respect to time*," and so forth.

PROBLEMS

○ 7. The displacement (s cm) from a fixed point on a line of a particle moving on that line varies with elapsed time (t sec) as follows:

$$s = t^3 - 3t^2 - 9t + 4, \quad -2 \le t \le 4.$$

(a) Find turning-point extreme values of s, and test for maximum and minimum.

(b) Find turning-point extreme values of the velocity, and test for maximum and minimum.

(c) Find intervals of increasing s and of decreasing s.

(d) Find intervals of increasing velocity and of decreasing velocity.

(e) Plot the graph of s as a function of t, and relate your answers to (a)–(d) to geometrical features of this graph.

8. Look again at Problem 7.
   (a) What are the end-point extremes of s? Hence, what is the absolute maximum of s? The absolute minimum?
   (b) What are the end-point extremes of v? Hence, what is the absolute maximum of v? The absolute minimum?

9. In Problem 10, **2.9**, the height (s in.) above the ground of an ant crawling up the trunk of a tree varied with the elapsed time (t sec) as follows: $s = 9t^2 - t^3$, for $0 \le t \le 9$.
   (a) Find the turning-point and end-point extremes of s, and thus find the absolute minimum of s; find also the absolute maximum. (Give units of your answers.)
   (b) Find the turning-point and end-point extremes of the velocity, v, and thus find the absolute minimum of v; find also the absolute maximum. (Give units of your answers.)
   (c) Find the absolute minimum of the acceleration, a; find also the absolute maximum.
   (d) Find intervals of increasing s and of decreasing s.
   (e) Find intervals of increasing v and of decreasing v.

10. In Problem 40, **2.9**, the height (s ft) of a rock climber above the bottom of a chimney varied with elapsed time (t min) as follows: $s = 4t^3 - t^4$, for $0 \le t \le 4$.
    (a) Find the turning-point and end-point extremes of s, and thus find the absolute minimum of s; find also the absolute maximum. (Give units of your answers.)
    (b) Find the turning-point and end-point extremes of the velocity, v, and thus find the absolute minimum of v; find also the absolute maximum. (Give units of your answers.)
    (c) Find the turning-point and end-point extremes of the acceleration, a, and thus find the absolute minimum of a; find also the absolute maximum. (Give units of your answers.)
    (d) Find the absolute minimum of the rate of change of acceleration; find also the absolute maximum. (Give units of your answers.)
    (e) Find intervals of increasing s and of decreasing s.
    (f) Find intervals of increasing v and of decreasing v.
    (g) Find intervals of increasing a and of decreasing a.

11. Consider the curve given by $y = 2x^3 - 3x^2 - 12x + 1$, for $-2 \le x \le 4$.
    (a) Find the turning-point and end-point extremes of the height of the curve, and thus find the absolute minimum of the height; find also the absolute maximum.
    (b) Find the turning-point and end-point extremes of the slope of the curve, and thus find the absolute minimum of the slope; find also the absolute maximum.
    (c) Find the absolute minimum of the flexion and the absolute maximum.

4.2 Points of inflection and third test for maxima and minima

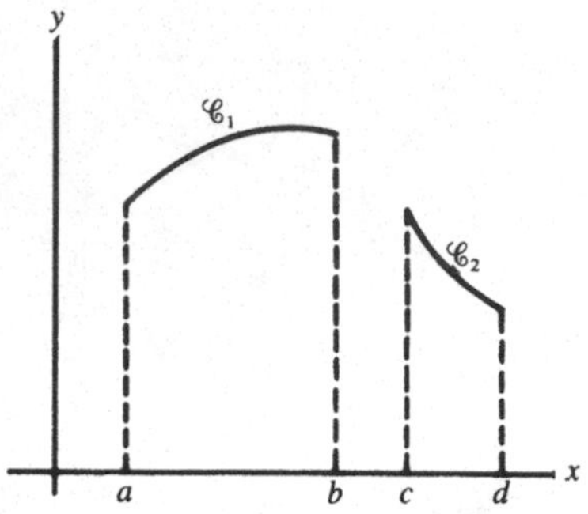

Fig. 4-2

We begin with some terminology. In Figure 4-2, it seems reasonable to call the curve $\mathscr{C}_1$ *concave down* over the interval (a, b) and the curve $\mathscr{C}_2$ *concave up* over the interval (c, d). Here is a formal definition (refer to Figure 4-3):

> Let P and Q be any two points on the curve in the interval, with abscissas p and q. Let $\bar{x}$ be any number between p and q. If the height of the line segment PQ at $\bar{x}$ is less than the height of $\mathscr{C}_1$ at $\bar{x}$, for *all* choices of P, Q, and $\bar{x}$, the curve $\mathscr{C}_1$ is **concave down** over (a, b).

Problem 1
Show by a sketch why it would be unsatisfactory to consider only *one* choice of P and Q, say at the ends of the interval.

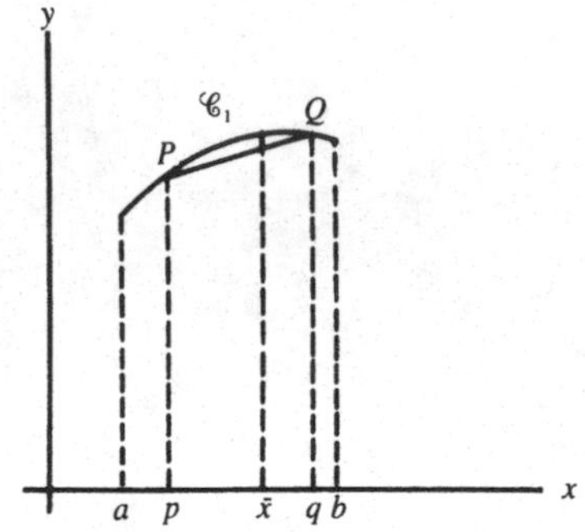

Fig. 4-3

Problem 2
Give the definition of **concave up** over an interval.

In Figure 4-4, the curve is concave down between H and J, concave up between J and K, and concave down again between K and L. The point J where the curve changes its direction of bending from concave down to concave up is called a *point of inflection*. The point K is also a point of inflection:

> A point where a curve changes its direction of bending from concave down to concave up (or, from concave up to concave down) is called a **point of inflection** of the curve.

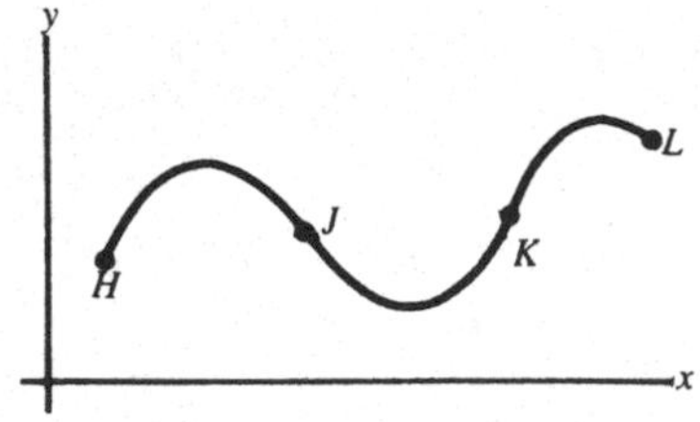

Fig. 4-4

In Figure 4-1(a), reproduced here, we note that the slope of $y = f(x)$ decreases from A to C (where the curve is concave down) and increases from C to E (where the curve is concave up). At the point of inflection C, then, the slope of $y = f(x)$ is a minimum – a result that is also clearly seen in Figure 4-1(b). By taking the second derivative of $f(x)$, which is graphed in Figure 4-1(c), we can seen that for functions like $f(x)$, which have first and second derivatives, the slope is decreasing and the curve is concave down where the second derivative is negative. Likewise, the slope is increasing and the curve is concave up where the second derivative is positive. At C, the point of inflection where the slope is a minimum, $f''(x) = 0$. The point P in Figure 4-5(a) is also a point of inflection.

Problem 3
Estimate the slopes at various points of the curve in Figure 4-5(a) and sketch its first derived curve. What can you say about the value of $g'(x)$ at P?

As suggested by the tangents at C in Figure 4-1(a) and at P in Figure 4-5(a), the tangent line at a point of inflection crosses the curve there. This is seen also in Figure 4-5(b) at Q, which is also a point of inflection – in this case with a horizontal tangent, that is, $F'(x) = 0$ at Q, or the corresponding x is a critical number of $F(x)$. But Q is not a maximum or minimum point of the curve, of course. As we approach Q from the left, $F'(x)$, the slope of the tangent, is decreasing toward zero; and as we proceed on to the right from Q, $F'(x)$ increases from zero to greater and greater positive values. In other words, $F'(x)$ has a minimum value (which happens to be zero) at the point of inflection, Q. We encountered a similar situation in Problem 4(c), 3.6, where we studied the curve $f(x) = \frac{1}{3}x^3 - 3x^2 + 9x + 1$.

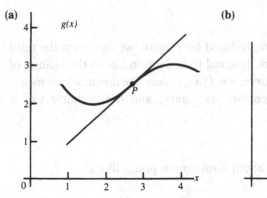

(a)   **(b)**

Fig. 4-5

In summary, for functions that have first and second derivatives, we can state that *at a point of inflection, the slope of the curve is an extreme (maximum or minimum), and hence the second derivative of the original function equals zero.* However, the converse is not true: If the second derivative is zero at some x, the corresponding point on the curve is not necessarily a point of inflection, just as, if the first derivative is zero at some x, the corresponding point on the curve is not necessarily a maximum or minimum point. In searching for maximum and minimum points on a curve, we find critical numbers by setting the first derivative equal to zero, and then *test* to make sure we have an extreme, in addition to checking any points where the derivative does not exist. Likewise, in searching for points of inflection on a curve, we set the second derivative equal to zero, and then *test* to make sure that we have an extreme (maximum or minimum) of the slope. Again we must check any points where the derivative being used (this time the second derivative) does not exist, by finding the direction of concavity of the curve (up or down) on each side of such points.

Problem 4

For the curve given by $y = f(x) = x^4 - 4x^3 + 6x^2 + 8x + 4$, check that $f''(x) = 12(x-1)^2$. Hence, if there is a point of inflection, it occurs for what

4.2

Points of inflection
and third test for
maxima and minima

(a)

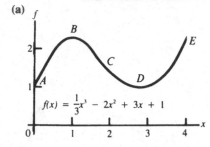

$$f(x) = \frac{1}{3}x^3 - 2x^2 + 3x + 1$$

(b)

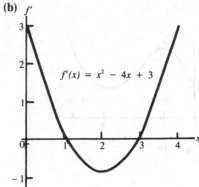

$$f'(x) = x^2 - 4x + 3$$

(c)

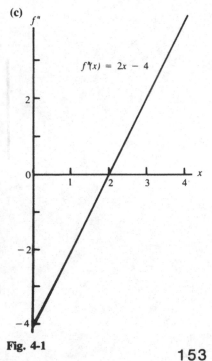

$$f''(x) = 2x - 4$$

Fig. 4-1

153

x? Find $f'(x)$ at this x, at a smaller x, and at a larger x. Do we have a point of inflection?

The following example shows that it is necessary, as in finding extrema, to worry about points at which derivatives do not exist.

Problem 5

Sketch a graph of the function g given by

$$g(x) = \begin{cases} -x^2, & -1 \le x < 0, \\ x^2, & 0 \le x \le 1. \end{cases}$$

Show that the graph has a horizontal point of inflection at $x = 0$. Show that $g''(0)$ does not exist.

Returning to Figure 4-1(a), reproduced here again, we can learn the third test for maximum and minimum, referred to in Section 3.6. In the vicinity of a *maximum* point, like B, the curve, $y = f(x)$, is concave down, which means that $f'(x)$ is decreasing (or possibly stationary) and hence that $f''(x)$ is *negative* (or possibly zero).

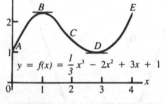

$$y = f(x) = \frac{1}{3}x^3 - 2x^2 + 3x + 1$$

Fig. 4-1a

Problem 6

Make a comparable statement about a minimum point, like D.

In light of the foregoing observations, the following test for extrema appears valid. A proof of its validity follows directly from an extended version of the Mean-Value Theorem:

Test 3

If $f'(c) = 0$ and $f''(c) < 0$, then $f(x)$ has a relative maximum value for $x = c$.

If $f'(c) = 0$ and $f''(c) > 0$, then $f(x)$ has a relative minimum value for $x = c$.

If $f'(c) = 0$ and $f''(c) = 0$, we must test further to find if we have a maximum or a minimum or a horizontal point of inflection or none of these.

Note that

in Test 1 we check the *values* of f *on both sides of* the critical number;
in Test 2 we check the *sign* of f' *on both sides of* the critical number;
in Test 3 we check the *sign* of f'' *at* the critical number.

PROBLEMS

7. Use Test 3 to check for maxima and minima in the following problems in Section 3.5:
   (a) Problem 3, in which the relevant equation is $L = 0.5 + 2.1t - 0.1t^2$

4.2
Points of inflection
and third test for
maxima and minima

(b) Problem 4, in which the relevant equation is $y = 48 + 88t - 16t^2$

(c) Problem 5, in which the relevant equation is $E = 128 + 2.8x - 0.01x^2$

(d) Problem 6, in which the relevant equation is $y = 8 - 6x + x^2$

(e) Problem 7, in which the relevant equation is $y = 9 - 8x - x^2$

8. Use Test 3 to check for maxima and minima in the following problems in Section 3.8:

   (a) Problem 18(a) (b) Problem 20 (c) Problem 21

9. Recall Problem 7, **4.1**.

   (a) Use Test 3 to check your results in part (a).

   (b) Use Test 3 to check your results in part (b).

   (c) For the graph of part (e), locate the point of inflection and the intervals for which the curve is concave up and concave down.

10. For Problem 9, **4.1**, in which the relevant equation is $s = 9t^2 - t^3$, use Test 3 to test for turning-point maxima and minima on the graph of this equation, and also locate point(s) of inflection.

11. Same as the foregoing Problem 10 for Problem 10, **4.1**, in which the relevant equation is $s = 4t^3 - t^4$.

12. Same as the foregoing Problem 10 for Problem 11, **4.1**, in which the relevant equation is $y = 2x^3 - 3x^2 - 12x + 1$.

13. For the curve determined by $y = x^4 - 8x^3 + 18x^2 + 1$, find maximum and minimum points and points of inflection, and sketch the curve over the interval $-1 \le x \le 4$. (Over this interval, y has too great a range to permit drawing the curve with the same scale on the vertical and horizontal axes unless you use an enormous sheet of paper.)

14. (a) Analyze the curves $y = x^2$, $y = x^3$, $y = x^4$, $y = x^5$, and $y = x^6$, for maximum and minimum points and points of inflection.

    (b) Using a large scale, sketch each of these curves on the same axes over the interval $-1 \le x \le 1$.

    (c) On the basis of your results in part (a), guess a result about maximum and minimum points, and points of inflection, for the curve $y = x^n$, where n is an integer greater than or equal to 2.

  ★ (d) Prove your conjecture of part (c). What can you say about maximum and minimum points, and points of inflection, for the curves $y = x^n$, where $n = 1, 0, -1, -2, \ldots$?

15. $Q = ax^2 + bx + c$.

    (a) Find the second derivative of Q with respect to x.

    (b) Show that the graph of Q as a function of x is concave up if $a > 0$ and concave down if $a < 0$.

  ★ (c) Discuss the extreme value (or values) of Q. Compare with Problem 5(e), **1.9**.

★ 16. In Section 1.7 we approximated Figure 1-2 (see Section 1.3 or Section 1.7) by the formula

$$\bar{A} = \begin{cases} 4.4 + 2.1t + 0.4t^2, & 0 \le t \le 4, \\ -9.2 + 8.7t - 0.4t^2, & 4 < t \le 10. \end{cases}$$

155

To use our present terminology, we argued that $t = 4$ seemed to correspond to a point of inflection and that the first of the quadratic expressions in $\bar{A}$ represented adequately the concave-up portion of the curve to the left of $t = 4$, whereas the second of the quadratic expressions did likewise for the concave-down portion of the curve to the right of $t = 4$.

(a) Check that both the quadratic expressions give the same value for $\bar{A}$ at $t = 4$.

(b) Check that the two quadratic expressions have *different* derivatives at $t = 4$. This means that the left-hand tangent at $t = 4$ is not the same as the right-hand tangent there – our approximating curve is not smooth at $t = 4$.

(c) It is hardly believable that the function A, giving the size of the bacterial colony, does not have a derivative (i.e., a growth rate) at $t = 4$. Yet we have just seen that such is the case for the approximating function $\bar{A}$. Explain this anomaly.

17. For each of the following functions, defined for all real numbers, 0 is a critical number. Try to apply Test 3 in each case – you should find that it fails. Use some other test to decide if $x = 0$ corresponds to a maximum of the function or a minimum or a horizontal point of inflection or none of these.

(a) $f_1(x) = x^4$　(b) $f_2(x) = 2 - x^4$　(c) $f_3(x) = x^3$　(d) $f_4(x) = 2$

4.3　Extreme rates

Just as we may wish to find the maximum *height* attained by a rocket, or the maximum *velocity* reached by a car, so likewise we may seek the maximum (or minimum) *acceleration* of a moving particle. The method is straightforward: We first find the acceleration itself (which doubtless will involve some differentiation) and then find the derivative of the acceleration and set it equal to zero to obtain critical numbers corresponding to extreme values of the acceleration. The same idea applies to finding extreme values of any quantity that itself is a rate. Here is an example:

For what values of x is the rate of change of flexion of the curve given by $y = f(x) = x^5 - \frac{1}{12}x^6$ an extreme? Maximum or minimum? In this case we have

$$\text{height of curve} = f(x) = x^5 - \tfrac{1}{12}x^6,$$
$$\text{slope of curve} = f'(x) = 5x^4 - \tfrac{1}{2}x^5,$$
$$\text{flexion of curve} = f''(x) = 20x^3 - \tfrac{5}{2}x^4,$$
$$R = \text{rate of change of flexion} = f'''(x) = 60x^2 - 10x^3.$$

Because the rate of change of flexion is the quantity we are to investigate for maxima and minima, we have designated it by the special letter R. For the rest of our work it is irrelevant that we had to do a lot of differentiating to find R – we simply now proceed to find the maximum or minimum of R given by $R = 60x^2 - 10x^3$.

Problem 1
Verify that the critical numbers of the function determined by the equation for R are $x = 0$ and $x = 4$.

Problem 2
For the practice, use all three tests to determine that $x = 0$ corresponds to minimum R and that $x = 4$ corresponds to maximum R.

In problems of this sort we have to be careful in translating the words into the correct rate to be maximized or minimized; from then on, it's usually simple.

PROBLEMS

3. The distance (y ft) traveled by an object in t min is $y = 15t^4 - t^5$. Find when the velocity is a maximum; likewise the acceleration. Test all the critical numbers.
4. In Problem 3, find when the acceleration is increasing most rapidly.
5. If the distance (y ft) traveled by an object in t min is given by $y = 60t^3 - t^4$, find the acceleration, a, at $t = 20$. Is a then increasing or decreasing, and how rapidly? Find the maximum acceleration.
6. The equation of a curve is $y = 12x^3 - x^4$. Find the flexion of the curve at $x = 2$. At $x = 2$, is the flexion increasing or decreasing, and how rapidly? Find the maximum flexion.
7. If the displacement (s ft) of a particle traveling on a straight line is related to the time (t min) by $s = 80 + 7t + 4t^2 + 10t^3 - t^4$, find the acceleration, a, at $t = 5$. Is a then increasing or decreasing, and how rapidly? Find the maximum acceleration.

4.4 Derivative of a function of a function: the Chain Rule

It is common to have a "chain" of functional dependence. For example, in any year the price per bushel ($\$ p$) of corn is a function of the size of the crop (u bushels), and u, in turn, is a function of the amount of rainfall (x in.); so, finally, p is a function of x. In symbols, this might be put as follows: $p = f(u)$ and $u = g(x)$, so $p = f[g(x)]$. This is read "p equals f of g of x" – note that this is *not* a case of multiplying functions f and g. The idea of a "function of a function" was introduced in Section 1.10. We shall have occasion to work with this concept extensively.

157

4

Further differentiation

Problem 1

For practice, check that if $y = f(u) = u^2 - u + 1$, and $u = g(v) = 3v - 1$, then $y = f[g(v)] = 3(3v^2 - 3v + 1)$.

Problem 2

If $y = g(u) = 3u - 1$ and $u = f(v) = v^2 - v + 1$, what is the expression for $y = g[f(v)]$?

Problem 3

If $f(w) = 1/w$, what is $f[f(w)]$?

Problem 4

If $f(x) = 1 - x$, $g(x) = 1/x$, $h(x) = 1 - 1/x$, what is $h\{g[f(x)]\}$?

In our initial example in this section,

$$p = f(u), \quad u = g(x); \quad \text{so} \quad p = f[g(x)].$$

We can designate this last equation by $p = F(x)$.

We pose the question: How is the rate of change of price with respect to rainfall related to the rate of change of price with respect to size of crop and the rate of change of size of crop with respect to rainfall? That is:

How is $F'(x)$ related to $f'(u)$ and $g'(x)$?

The answer to this question turns out to be simple and one of the most powerful results in the calculus. A couple of problems help lead to the answer.

Problem 5

Suppose that $y = f(u) = u^4$ and $u = g(x) = 3x^2$. Then $y = F(x) = f[g(x)] = (3x^2)^4 = 81x^8$. Verify that $F'(x) = 648x^7$, and also check that this can be written as $F'(x) = 4(3x^2)^3 \cdot 6x$. Can you see why we have chosen to write the answer in this complicated form?

Problem 6

Suppose that $y = F(x) = (2x + 1)^3$. Multiply this out, differentiate, and collect terms to obtain $F'(x) = 3(2x + 1)^2 \cdot 2$. Can you see why we have chosen to write the answer in this particular form?

In Problem 5, the answer, $F'(x) = 4(3x^2)^3 \cdot 6x$ can also be written as $F'(x) = 4u^3 \cdot 6x$. Now $4u^3 = f'(u)$, and $6x = g'(x)$, so in this case, if $F(x) = f(u)$ with $u = g(x)$, then $F'(x) = f'(u) \cdot g'(x)$. The right side of this last equation *is* a product. As we shall see, the result we have here is quite general. It applies also to Problem 6 in the following sense: Suppose that we set $2x + 1 = u$, $= g(x)$, say. Then we have $F(x) = f[g(x)]$, with $f(u) = u^3$, and our answer can be written as $F'(x) = 3u^2 \cdot 2$. Because $f'(u) = 3u^2$ and $g'(x) = 2$, once again we see that $F'(x) = f'(u) \cdot g'(x)$. The statement of the general result reads as follows:

Theorem: the Chain Rule

If $f(u)$ and $g(x)$ have derivatives, then $F(x) = f[g(x)]$ has a derivative, and $F'(x) = f'(u) \cdot g'(x)$, where $u = g(x)$.

The two illustrations contained in Problems 5 and 6 do not *prove* the Chain Rule, of course. A plausible argument, applied to Problem 5, goes as follows: If $u = g(x) = 3x^2$, then our method of approximate increments (Section 3.3) shows that

$$\Delta u \approx g'(x) \cdot \Delta x = 6x \, \Delta x. \qquad (1)$$

Likewise, if $y = g(u) = u^4$, then the same method shows that

$$\Delta y \approx f'(u) \cdot \Delta u = 4u^3 \cdot \Delta u. \qquad (2)$$

Substituting for Δu from (1) into (2), we have

$$\Delta y \approx 4u^3 \cdot 6x \, \Delta x. \qquad (3)$$

But if we write $y = f[g(x)] = F(x)$, as we did in Problem 5, the same method of approximate increments gives

$$\Delta y \approx F'(x) \, \Delta x. \qquad (4)$$

From (3) and (4), it seems reasonable to expect that

$$F'(x) = 4u^3 \cdot 6x,$$

which is what the Chain Rule gives in this case.

Because a valid proof of the Chain Rule is a bit involved, we shall defer it to Section 4.6, and we shall use the result without proof, comforting ourselves with its reasonableness as seen in simple examples and in the foregoing plausibility argument.

The foregoing discussion of Problem 6 illustrates an extremely important special case of the Chain Rule:

Special Case of the Chain Rule

If $F(x) = [g(x)]^n$, then $F'(x) = n[g(x)]^{n-1} \cdot g'(x)$.

In words, this can be put as follows: "The derivative of the nth power of a function equals n, times the $(n-1)$st power of the function, times the derivative of the function."

The justification of the statement comes from setting $u = g(x)$, so that $F(x) = u^n$, $= f(u)$, say. Then $f'(u) = nu^{n-1} = n[g(x)]^{n-1}$, and the Chain Rule leads to the stated result.

159

Problem 7

It is true that Problem 6 was solved without too much effort by expanding $F(x)$ into a polynomial before differentiating. In principle, this could also be done for the function $F(x) = (x^2 - 2x + 5)^{12}$, but it wouldn't be much fun. Use the Special Case of the Chain Rule to find the derivative of this F. Also, find maximum and minimum values of $F(x)$.

Problem 8

If $G(x) = 1/(2x + 1)^3$, we could find $G'(x)$ by going back to the definition of the derivative. It is much easier to find $G'(x)$ through use of the Chain Rule. Do so.

So far we have seen the Chain Rule as a device for finding derivatives more simply than we otherwise could or for finding derivatives that we could not obtain without it. In subsequent sections there will be still other uses for the Chain Rule.

PROBLEMS

○ 9. The cost ($\$C$) of producing n sport shirts in a certain factory is given by $C = 1000 + 4n - 0.001n^2$. The number (n shirts) that can be sold at $\$x$ each is given by $n = 2500 - x^2$.
   (a) Find the rate of change of C with respect to n and the rate of change of n with respect to x.
   (b) Hence, what is an expression for the rate of change of C with respect to x? Find the value of this rate of change at $x = 40$.
   (c) What is the approximate incremental cost of producing the 901st shirt?

10. Consider the curve determined by the equation $y = \sqrt{25 - x^2}$, $= F(x)$, say.
   (a) What is the natural domain of F? What is its range?
   (b) Find an expression for $F'(x)$, the slope of the tangent to this curve.
   (c) What is the domain of F'? What is its range?
   (d) What is the slope of the tangent to the curve F at $x = 4$? At $x = -3$?
   (e) Draw the curve with some care, and show the tangents at the points where $x = -3, 0, 4$, and 5.
   (f) Do you know what kind of curve this is?

11. Same as Problem 10 for the curve determined by the equation $y = 12/(x-1)$.

12. Find derivatives of each of the functions given by the following expressions:
   (a) $(3x + 1)^{50}$ (b) $\dfrac{5}{\sqrt{9 - x^2}}$ (c) $t^2 + \dfrac{3}{4-t}$ (d) $\left(w + \dfrac{1}{w}\right)^4$

(e) $\left(2x + \dfrac{8}{x^2}\right)^3$ (f) $\sqrt[3]{c^2 - x^2}$, where c is a constant

(g) $\sqrt{1 - x^2} + \dfrac{1}{\sqrt{1 - x^2}}$ (simplify your result as much as possible)

(h) $\sqrt[3]{1 - \dfrac{x^2}{36}}$ (i) $\sqrt{x^2 - 6x + 1}$

(j) $\left(\sqrt{10 + x^2} + x - 1\right)^5$ (Hint: Use the Special Case of the Chain Rule twice.)

(k) $\sqrt{1 + \sqrt{x}}$ [Same hint as in (j)]

13. (a)–(k) For each of the parts of Problem 12, try to find the critical number(s) of the function.

 14. The length of an edge (x cm) of a metal cube varies with the temperature ($T°$ above zero Celsius) as follows: $x = 5 + 0.0002T$. Find how fast the volume (V cm^3) changes, per degree, at $T = 1000$. At $x = 5$.

4.5 Continuity

Nearly all the graphs of functions we have examined in this and earlier chapters have been "continuous" curves. In informal language this means merely that the graphs are unbroken. Because variation is often continuous in situations arising in the physical, life, and social sciences, continuity can often be expected in our work. But there are exceptions, one such appearing in Problem 10, **2.2**, where the postage required for a letter of w oz, $0 < w \le 3$, is given by

$$f = \begin{cases} 20, & 0 < w \le 1, \\ 37, & 1 < w \le 2, \\ 54, & 2 < w \le 3. \end{cases}$$

The graph of f (Figure 4-6) has breaks at the points corresponding to $w = 1$ and $w = 2$, and hence we say that the function is *discontinuous* for these values of w.

In general, what we expect of a function not having a break at some number p is that if x is close to p, then $f(x)$ is close to $f(p)$. Note that this is not always the case with the postage function: We can choose numbers as close as we like to 1, say 1.00001, for which $f(1.00001) = 37$, but $f(1) = 20$. Our formal definition reads as follows:

Definition

A function f is said to be **continuous** at p in the domain of f if $\lim_{u \to p} f(u) = f(p)$.

Note that this definition really has three parts to it:
1. $\lim_{u \to p} f(u)$ must exist;

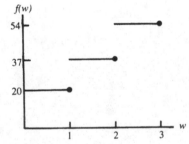

Fig. 4-6

161

2. $f(p)$ must exist;

3. the two numbers, $\lim_{u \to p} f(u)$ and $f(p)$, must be the same.

Condition 1 fails in the postage function at each integral (whole number) w. For example, as w approaches 1 from the left (through weights *less than* 1), $f(w)$ approaches 20; but as w approaches 1 from the right (through weights *greater than* 1), $f(w)$ approaches 37 – the "left-hand limit" is not the same as the "right-hand limit," so $\lim_{w \to 1} f(w)$ does not exist.

Condition 2 fails at $p = 0$ for any function expressing an average rate of change. For example, suppose that $F(p)$ represents the average rate of change of some function in the interval $[a, a + p]$. Then $F(0)$ does not exist – we cannot have an *average* rate of change in an interval of *zero* length – so 0 is not in the domain of F, and we do not have continuity of F at $p = 0$. Remember that this was basically the reason for the final phrase in the definition of limit in Section 2.2.

Condition 3 can fail for "pathological" functions. For example, suppose

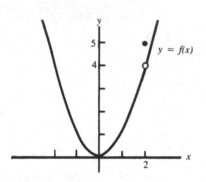

Fig. 4-7

$$f(x) = \begin{cases} x^2, & \text{for } x \neq 2, \quad \text{and} \\ 5, & \text{for } x = 2. \end{cases}$$

The graph of $y = f(x)$ is shown in Figure 4-7. This function is continuous at all $x \neq 2$. But, because $\lim_{x \to 2} f(x) = 4$, while $f(2) = 5$, the function is *not* continuous at $x = 2$.

We have here a *removable* discontinuity – we can easily "fix things up" by dealing with a different function:

$$g(x) = \begin{cases} x^2, & \text{for } x \neq 2, \quad \text{and} \\ 4, & \text{for } x = 2, \end{cases}$$

or, more simply put, $g(x) = x^2$ for all x.

At the end of Section 2.3 we had the following important result about polynomials: For any polynomial,

$$P(u) = c_0 + c_1 u + c_2 u^2 + \cdots + c_n u^n,$$

and for any real number a,

$$\lim_{u \to a} P(u) = P(a).$$

In terms of our definition of continuity, this result is equivalent to the following statement:

Every polynomial, $P(u)$, is a continuous function for all real numbers, u.

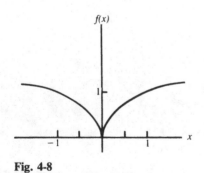

Fig. 4-8

It would be pleasant if every continuous function were also differentiable, but such is not the case. For example, the function given by $f(x) = x^{2/3}$ is continuous for all x (Figure 4-8). However, $f'(x)$ does not exist at $x = 0$, although the curve does have a tangent (which is vertical) at $x = 0$. A sketch of the graph of a continuous function with a cusp where there is *no* tangent is shown in Figure 3-17(a). On the other hand, if a function is differentiable, then it *is* continuous. This will be proved in the next section.

PROBLEMS

1. For what numbers is each of the following functions continuous? Consider the domain to be the *natural domain* unless otherwise specified.

   (a) $f(x) = 3 + \frac{1}{2}x - 5x^2$ (b) $g(x) = \frac{1}{x}$

   (c) $f(u) = u^2, \quad -2 \le u < 2$ (d) $g(t) = t + \frac{3}{1-t}$

   (e) $f(w) = \frac{w-2}{w^2 - 9}$ (f) $g(t) = \begin{cases} 3t + 49, & 0 \le t \le 17, \quad \text{and} \\ 100, & 17 < t \le 30 \end{cases}$

   (g) $F(h) = \frac{(3+h)^2 - 9}{h}$ (h) $G(x) = \begin{cases} x^2 - 2x + 3, & x \le 2, \quad \text{and} \\ -x^2 + 3x + 1, & x > 2 \end{cases}$

   (i) $f(x) = \begin{cases} x^2 - 2x + 3, & x \le 0, \quad \text{and} \\ 4, & x > 0 \end{cases}$

   (j) $f(x) = \begin{cases} 2x + 4, & x \ne 3, \quad \text{and} \\ 8, & x = 3 \end{cases}$

   (k) $G(x) = \begin{cases} \frac{1}{n}, & \text{for } x = \frac{1}{n}, \quad n = 1, 2, 3, \dots, \quad \text{and} \\ 0, & \text{otherwise} \end{cases}$

   (l) $F(h) = \frac{(a+h)^5 - a^5}{h}$, where a is a constant

2. (a) Find an extension of F in Problem 1(g) that is continuous for all h.
   (b) Same as (a), for Problem 1(l).
   (c) Find a modification of f in Problem 1(i) that is continuous for all x.
   (d) Same as (c), for Problem 1(j).

★ 3. Suppose that $f(x)$ and $g(x)$ are both continuous functions at $x = p$. Use one of the limit theorems to show that the function $F = f + g$ is also continuous at p. What can be said about the continuity of $f \cdot g$, of f/g, and of kf, where k is a constant?

★ 4. (a) Write out the proof that the function given by $f(x) = x$ is continuous at all real numbers.
   (b) Write out the proof that the polynomial $P(x) = c_0 + c_1 x + c_2 x^2 + \cdots + c_n x^n$, where the c's are constants, is continuous at all real numbers.

★ 5. Show that the statement $\lim_{u \to p} f(u) = L$ is equivalent to the statement $\lim_{h \to 0} f(p + h) = L$. [Hence, the definition of continuity of function f at p could be written $\lim_{h \to 0} f(p + h) = f(p)$.]

★ 6. Suppose $f(x)$ is a differentiable function everywhere (i.e., for $-\infty < x < \infty$). Let $F(h) = [f(x+h) - f(x)]/h$, $h \ne 0$. Using the fact that a differentiable function is continuous, show that $F(h)$ is continuous for all $h \ne 0$ and has a removable discontinuity at $h = 0$. Remove the discontinuity.

⋆ **4.6   Proof that differentiability implies continuity and proof of the Chain Rule**

Let f be a function, and suppose that f' exists at p in D_f. We wish to demonstrate that f is continuous at p. We express the difference between an average rate of change and the instantaneous rate of change, and we then extend this function to make the extension continuous at zero: Let E be the function defined by the formula

$$E(k) = \begin{cases} \dfrac{f(p+k)-f(p)}{k} - f'(p), & k \neq 0, \quad \text{and} \\ 0, & k = 0. \end{cases}$$

Problem 1

Show that E is continuous at zero.

Problem 2

Show that for all k (including 0) such that $(p+k)$ is in D_f,

$$f(p+k) = f(p) + f'(p) \cdot k + E(k) \cdot k, \tag{5}$$

where $\lim_{k \to 0} E(k) = 0$.

Problem 3

Use equation (5), Problem 5 in the previous section, and the existence of $f'(p)$ to show that f is continuous at p.

This is what we set out to prove. The immediately preceding work leads to a proof of the Chain Rule.

Suppose that g is a function that is differentiable at a number r and that f is a function that is differentiable at the number $p = g(r)$. Because in equation (5), k can be any number, we can let $k = g(r+h) - g(r) = g(r+h) - p$, where h is such that $r+h$ is in D_g.

Problem 4

Make this substitution in equation (5) and whatever other changes you need to produce the equation

$$\frac{f[g(r+h)] - f[g(r)]}{h} = f'[g(r)] \cdot \frac{g(r+h) - g(r)}{h}$$

$$+ E[g(r+h) - g(r)] \cdot \frac{g(r+h) - g(r)}{h}, \quad h \neq 0.$$

Problem 5

Show that $\lim_{h \to 0} [g(r+h) - g(r)] = 0$ and hence that

$$\lim_{h \to 0} E[g(r+h) - g(r)] = 0.$$

Problem 6
Complete the proof.

A more straightforward approach to this result would seem to lie in the following manipulation:

$$\frac{f[g(r+h)]-f[g(r)]}{h} = \frac{f[g(r+h)]-f[g(r)]}{g(r+h)-g(r)} \cdot \frac{g(r+h)-g(r)}{h}$$

$$= \frac{f(p+k)-f(p)}{k} \cdot \frac{g(r+h)-g(r)}{h}.$$

Problem 7
How would the rest of the argument go? What is wrong with it? (Hint: Consider what would happen if g were constant, for example.)

4.7 Notation

The calculus was invented in the latter half of the seventeenth century almost simultaneously by Newton in England and Leibniz in what is now Germany. They worked quite independently, and it is not surprising that they created different notations for the concepts they developed. So intense was the rivalry for priority in the creation of the calculus – more a competition between the friends and supporters of Newton and Leibniz than between the principals themselves – that the calculus developed with one notation in England and another notation on the Continent. Over the years, still other notations have appeared, for specialized purposes.

The notation we have been using for the successive derivatives of $f(x)$ with respect to x – $f'(x)$, $f''(x)$,... – is similar to that introduced by Newton; if displacement is represented by s, he wrote its derivative with respect to time (i.e., velocity) as $\dot{s}$, its second derivative with respect to time (i.e., acceleration) as $\ddot{s}$, and so forth. We shall now consider Leibniz's notation for derivatives, because it is so widely used that you will undoubtedly encounter it in your reading, and, moreover, it is especially convenient for some purposes, so that we shall use it whenever it is simpler than our original notation.

To explain Leibniz's notation, it is easiest if we go back to the development of the derivative: Starting with the function $y = f(x)$, we give x an *increment* that we called h before and that we now designate by Δx (read "delta x"). So, in Figure 4-9, $PR = h = \Delta x$. We consider the change in the value of the function, which previously we wrote as $f(x+h)-f(x)$, and which we shall now think of as the increment in y, designated by Δy. So, in the figure,

$$RQ = f(x+h)-f(x) = \Delta y.$$

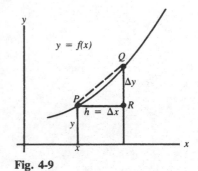

Fig. 4-9

165

Next we compute the *average rate of change* of the function over the interval in question:

$$\frac{RQ}{PR} = \frac{f(x+h)-f(x)}{h} = \frac{\Delta y}{\Delta x}.$$

Finally, we define the derivative as the *limit* of this average rate of change, as the length of the interval approaches zero:

$$\lim_{h \to 0} \frac{f(x+h)-f(x)}{h} = f'(x),$$

or

$$\lim_{\Delta x \to 0} \frac{\Delta y}{\Delta x} = \frac{dy}{dx} \quad \text{(read "dee } y, \text{ dee } x \text{ ")}.$$

Although $\Delta y/\Delta x$ is actually the ratio of two quantities, Δy and Δx, and although we shall soon encounter some new quantities, dy and dx, called "differentials," the new notation for the derivative, dy/dx, is *not defined* as the ratio of two quantities – it is the *limit* of a ratio. For this reason, avoid saying "dee *y over* dee *x*."

We practice with the new notation:

$$\text{if } y = x^3, \quad \text{then } \frac{dy}{dx} = 3x^2;$$

$$\text{if } s = (t^2-1)^{3/2}, \quad \text{then } \frac{ds}{dt} = \frac{3}{2}(t^2-1)^{1/2} \cdot 2t = 3t(t^2-1)^{1/2};$$

$$\text{if } z = u + \frac{1}{u}, \quad \text{then } \frac{dz}{du} = 1 - \frac{1}{u^2}; \quad \text{and so forth.}$$

These results can also be written as follows:

$$\frac{d}{dx}(x^3) = 3x^2;$$

$$\frac{d}{dt}(t^2-1)^{3/2} = 3t(t^2-1)^{1/2};$$

$$\frac{d}{du}\left(u + \frac{1}{u}\right) = 1 - \frac{1}{u^2}.$$

The first of these equations is read "the derivative, with respect to x, of x^3 equals $3x^2$," and similarly for the others.

Now, if we wish the second derivative of y with respect to x, we recognize it as "the derivative, with respect to x, of dy/dx"; and, in keeping with the notation introduced earlier, we might write it as

$$\frac{d}{dx}\left(\frac{dy}{dx}\right).$$

This is the motivation for the new notation for the second derivative, d^2y/dx^2 (note where the superscript 2's appear). It is read "dee second y, dee x second." We have, then $d^2y/dx^2 = f''(x)$. Similarly, $d^3y/dx^3 = f'''(x)$, and so forth.

In the new notation, the Chain Rule becomes particularly easy to remember. Recall that it reads as follows: If $y = f(u)$ and $u = g(x)$, so that $y = f[g(x)] = F(x)$, then $F'(x) = f'(u) \cdot g'(x)$. In the new notation, $F'(x) = dy/dx$, $f'(u) = dy/du$, $g'(x) = du/dx$, so the Chain Rule becomes

$$\frac{dy}{dx} = \frac{dy}{du} \cdot \frac{du}{dx}.$$

It is just as though we could "cancel the du's" on the right side of the displayed equation. After all the emphasis that expressions like dy/dx are not quotients of two quantities, they behave as though they were!

Let us see how the new notation applies to a problem involving the Chain Rule. We shall work through it using both notations. You will probably find the new notation easier to understand and apply. Consider Problem 9(b), **4.4**. We have $C = f(n) = 1000 + 4n - 0.001n^2$, and $n = g(x) = 2500 - x^2$. This implies that C is a function of x – $C = F(x)$, say – and we are to find the rate of change of C with respect to x – we are to find $F'(x)$, or dC/dx. By the Chain Rule,

$$F'(x) = f'(n) \cdot g'(x) \qquad \text{or} \qquad \frac{dC}{dx} = \frac{dC}{dn} \cdot \frac{dn}{dx}.$$

Now,

$$f'(n) = \qquad 4 - 0.002n \qquad = \frac{dC}{dn},$$

and

$$g'(x) = \qquad -2x \qquad = \frac{dn}{dx}.$$

So

$$F'(x) = \qquad (4 - 0.002n)(-2x) \qquad = \frac{dC}{dx}.$$

We can now substitute $x = 40$, and the corresponding value of n, into this equation for the derivative to complete the problem.

Observe that there is an annoying ambiguity in one aspect of our earlier notation: Starting with $C = f(n) = 1000 + 4n - 0.001n^2$, we have sometimes written $C' = f'(n) = 4 - 0.002n$. This gives rise to no difficulty if no other variable is involved. But, if $n = g(x) = 2500 - x^2$, so that we also have $C = F(x) = f[g(x)] = 1000 + 4(2500 - x^2) - 0.001(2500 - x^2)^2$, we might write $C' = F'(x) = \cdots$, and we would then be using the same symbol C' for two different derivatives – the first being the rate of change of C *with respect to n*, and the second being the rate of change of C *with respect to x*. The symbols dC/dn and dC/dx make the distinction clear. So do the symbols $f'(n)$ and $F'(x)$, but we sometimes omit the functional notation and write $C = 1000 + 4n - 0.001n^2$. Then it may be safer to write $dC/dn = 4 - 0.002n$, rather than $C' = 4 - 0.002n$. We shall practice working with *both* notations in the next several sections.

4.8   Related rates

We introduce the topic of *related rates* by using the Leibniz notation to solve Problem 14, **4.4**: The length of an edge (x cm) of a metal cube varies with the temperature ($T°$ Celsius) as follows: $x = 5 + 0.0002T$. Find how fast the volume (V cm^3) changes, per degree, at $T = 1000$. At $x = 5$.

We are given, explicitly, one variable, x, as a function of T: $x = 5 + 0.0002T$, $= g(T)$, say. We need V expressed as a function of T. Because we are dealing with a cube, we know that $V = x^3$, $= f(x)$, say.

We can put these results together to read

$$V = (5 + 0.0002T)^3, \quad = F(T),$$

say. We can apply the Special Case of the Chain Rule to $F(T)$ to obtain

$$\frac{dV}{dT} = F'(T) = 3(5 + 0.0002T)^2(0.0002).$$

Problem 1
Show that $F'(1000) \approx 0.0162$ cm^3/degree.

Problem 2
What T corresponds to $x = 5$? Show that for this T, $F'(T) = 0.015$ cm^3/degree.

But, in fact, there was no need to write the explicit expression for $F(T)$; we could simply have differentiated the formulas for $f(x)$ and $g(T)$ to get $f'(x) = 3x^2$ and $g'(T) = 0.0002$. Then, from the Chain Rule itself, $F'(T) = f'(x) \cdot g'(T) = 3x^2(0.0002) = 0.0006x^2$. Now, at $T = 1000$, $x = 5.2$, so that $F'(1000) = 0.0006(5.2)^2$, the same result, of course, as obtained earlier. Likewise, at $x = 5$, $F'(T$ corresponding to $x = 5)$ $= 0.0006(5)^2$, also as before. Note an interesting aspect of this last method of treating the problem: To find $F'(T)$ at $x = 5$ we do not need to know that particular formula for $g(T)$ as given in the problem; all we need to know is that $g'(T) = 0.0002$.

In the other notation we would write:

$$V = x^3, \quad \text{so} \quad \frac{dV}{dx} = 3x^2;$$

$$x = 5 + 0.0002T, \quad \text{so} \quad \frac{dx}{dT} = 0.0002.$$

Hence,

$$\frac{dV}{dT} = \frac{dV}{dx} \cdot \frac{dx}{dT} = 3x^2(0.0002) = 0.0006x^2.$$

As observed earlier, we don't need the explicit formula for x in terms of T; all we need is the formula for dx/dT. A pair of examples will bring out two further points in such problems.

Example 1

A growing pile of sawdust has the shape of a right circular cone whose height always equals two-thirds the radius of its base. How fast is the volume of the pile increasing when $r = 6$ ft, if r is then increasing at the rate of 0.5 ft/min?

First we need the formula for the volume of a cone: $V = \frac{1}{3}\pi r^2 h$. In this case, $h = \frac{2}{3}r$, so V can be expressed as a function of r alone:

$$V = \frac{1}{3}\pi r^2\left(\frac{2}{3}r\right) = \frac{2}{9}\pi r^3, \quad = f(r),$$

say. Because r varies with the time, t, we know that r is some function of t: $r = g(t)$, say. We do not know an expression for g, but we are told that at the time in question the value of $g'(t) = 0.5$.

Because V varies with r and r varies with t, V is a function of t: $V = F(t)$, say. This is a typical example of a "function of a function": V is a function of t through the medium of r, we might say. Because we do not have an expression for r as a function of t, we also do not have an expression for V as a function of t. But we don't need $F(t)$; what we want is $F'(t)$, which the Chain Rule provides as

$$F'(t) = f'(r) \cdot g'(t), \quad \text{or} \quad \frac{dV}{dt} = \frac{dV}{dr} \cdot \frac{dr}{dt}. \tag{6}$$

Problem 3

Complete the problem by filling in the blanks:

Because $V = \frac{2}{9}\pi r^3$, $\dfrac{dV}{dr} = \underline{\qquad}$. We know that $\dfrac{dr}{dt} = \underline{\qquad}$.

Hence, $\dfrac{dV}{dt} = \underline{\qquad}$ and, at $r = 6$, $\dfrac{dV}{dt} = 12\pi \ (\text{ft}^3/\text{min})$.

The only new feature in this example is that we have to use the data to obtain the equation expressing V as a function of the single variable r.

Example 2

As in Example 1, a growing pile of sawdust has the shape of a right circular cone whose height always equals two-thirds the radius of its base. If sawdust is added to the pile at the constant rate of 50 ft^3/min, how fast is the radius increasing when it equals 5 ft?

Using the same notation, we have $V = f(r)$ just as before; we are given that $dV/dt = F'(t) = 50$, and we are asked to find $dr/dt = g'(t)$ at $r = 5$. In Example 1, we used equation (6) to find the required value of dV/dt by using the values of the two factors on the right. Now we can use the same equation to find dr/dt, knowing the values of the other two terms in the equation.

Problem 4

Follow this method to obtain the result that at $r = 5$, $dr/dt = 3/\pi$ (ft/min).

169

In summary, we can say that the relation among rates of change, as set out in equation (6), permits us to find $F'(t)$ if we know $f'(r)$ and $g'(t)$, or to find $g'(t)$ if we know $F'(t)$ and $f'(r)$. We shall usually need an expression for $f(r)$, perhaps from a geometrical relation, in order to obtain $f'(r)$.

Alternatively expressed in the Leibniz notation, we can say the following: By differentiating to obtain dV/dr, we can use the equation

$$\frac{dV}{dt} = \frac{dV}{dr} \cdot \frac{dr}{dt}$$

to obtain dV/dt if we know dr/dt, and dr/dt if we know dV/dt.

PROBLEMS

5. A cylinder contracts so that its height always equals four times its radius. If the volume is decreasing at the rate of 2 in.^{3}/hr, how fast is r decreasing when $r = 10$?

6. A right circular cone expands so that its height always equals three times the radius of its base. If the volume increases at the rate of 10 in.^{3}/min, how fast is the radius increasing when it is 5 in.?

7. A cylinder expands in such a way that its height always equals twice the radius of its base. If the volume increases at the rate of 30 in.^{3}/hr, how fast does r increase when $r = 5$?

8. If the volume of a spherical balloon increases at the rate of 10 in.^{3}/min, how fast does the radius increase when it is 5 in.? How fast is the surface area of the balloon then changing? (Give the units of your answer.)

9. (a) At the moment that the radius of an expanding sphere equals 5 cm, the surface area of the sphere is increasing at the rate of 8 cm^2/min. How fast is the volume of the sphere then changing? (Give the units of your answer.) By approximately how much does the volume change in the next 10 sec?

   ★ (b) Show that if the surface area of a sphere is increasing at A cm^2/min when the radius of the sphere equals a cm, the volume of the sphere is then increasing at $(A \cdot a)/2$ cm^3/min.

10. (a) The volume of a metal cube increases with temperature at 1.2 cm^3/degree. If the cube contains 216 cm^3 at a certain temperature, how fast are the edge and the surface area of the cube increasing at that temperature? (Give the units of your answer.) By approximately how much does the surface area change if the temperature increases 0.5 degree from that temperature?

   ★ (b) Show that if the volume of a cube increases at the rate of a cm^3/degree when the volume equals b cm^3, the surface area of the cube then increases at the rate of $4a/\sqrt[3]{b}$ cm^2/degree.

4.9 Functions in implicit form and implicit differentiation

Equations like $y = x^2 - 3x + 1$ and $y = \sqrt{25 - x^2}$ define functions *explicitly*: In each case it is immediately clear what value of y corresponds to a value of x. The equation has already been "solved for y in terms of x." Unlike the situation just described, the equation

$$x^3 + y^3 = 9 \tag{7}$$

is not a formula stating the value of y corresponding to a value of x, although it is the case that there is one, and only one, such value of y – and we can rather easily write the formula:

$$y = \sqrt[3]{9 - x^3}. \tag{8}$$

We say that equation (7) defines y as a function of x *implicitly*, or that the function is in *implicit form*.

Sometimes it is difficult or impossible to solve an equation like (7) explicitly for y, and we therefore are led to this question: How do we find the rate of change of y with respect to x from equation (7)? Or, equivalently, how do we find the slope of the tangent at an arbitrary point of the curve given by equation (7)?

We proceed as follows: Assume that equation (7) determines one or more differentiable functions for y in terms of x. Let one of them be given by $y = f(x)$. Then

$$x^3 + [f(x)]^3 = 9. \tag{9}$$

The left side of equation (9) is a function of x; call it $F(x)$, so that

$$F(x) = x^3 + [f(x)]^3. \tag{10}$$

Problem 1
Use the Special Case of the Chain Rule to obtain

$$F'(x) = 3x^2 + 3[f(x)]^2 \cdot f'(x). \tag{11}$$

But equation (9) also implies that $F(x)$ is constantly equal to 9, so that $F'(x)$ equals zero. That is,

$$3x^2 + 3[f(x)]^2 \cdot f'(x) = 0.$$

Problem 2
Solve this equation to obtain

$$f'(x) = -\frac{x^2}{y^2}. \tag{12}$$

Thus, at the point $(1,2)$ on the curve given by equation (7), the slope of the tangent $= -\frac{1}{4}$.

The process we have exemplified is called **implicit differentiation**.

Problem 3

Use the Special Case of the Chain Rule with equation (8) to show that the slope of the tangent to the curve $= -x^2/(9-x^3)^{2/3}$. Reconcile this result with equation (12).

As we know, an equation in x and y sometimes does not define y as a function of x. For example, the equation $x^2 + y^2 = 25$ is equivalent to $y = \pm\sqrt{25 - x^2}$, and because there are *two* values of y corresponding to each x in $(-5, 5)$, this is not a function. However, there are two continuous functions determined by

$$x^2 + y^2 = 25, \tag{13}$$

namely,

$$y = \sqrt{25 - x^2} \tag{14}$$

and

$$y = -\sqrt{25 - x^2}, \tag{15}$$

and the derivatives of these functions are correctly obtained by the process of implicit differentiation.

Problem 4

(a) Use implicit differentiation with equation (13) to show that the slope of the tangent $= -x/y$.

(b) Use equation (14) to show that the slope of the tangent $= -x/\sqrt{25 - x^2}$. Find the slope at $x = 3$, reconcile with the result in part (a), and show the tangent on a sketch of equation (14).

(c) Use equation (15) to show that the slope of the tangent $= x/\sqrt{25 - x^2}$. Complete as in part (b).

Your results in Problem 4 should look like the sketches in Figure 4-10.

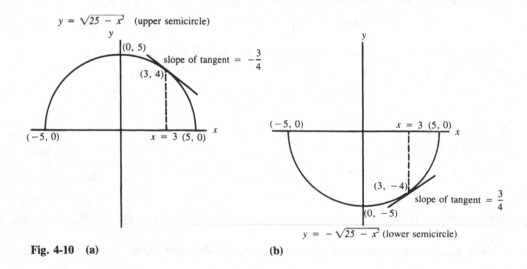

Fig. 4-10 (a)   (b)

4.10 Derivatives of fractional powers

In Section 2.7 we established the formula for the derivative of x^n when n is an integer – positive, negative, or zero – and we checked the validity of the same formula for $n = \frac{1}{2}$ and $\frac{1}{3}$. But, although we have been blithely using the formula for many fractional powers, we haven't yet proved it correct. The Chain Rule and the method of implicit differentiation now permit us to do this.

Theorem
If $f(x) = x^{p/q}$, where p and q are integers and $q \neq 0$, then

$$f'(x) = \frac{p}{q} \cdot x^{(p/q)-1}.$$

We illustrate the method of proof in the numerical case of $p = 3$, $q = 5$, that is, for $f(x) = x^{3/5}$, which we shall assume is a differentiable function. We raise both sides of the equation to the 5th power, in order to eliminate fractional exponents:

$$[f(x)]^5 = x^3.$$

By the Chain Rule, the derivative of the left side is $5[f(x)]^4 \cdot f'(x)$, and that of the right is $3x^2$. Setting them equal and solving for $f'(x)$ gives

$$f'(x) = \frac{3x^2}{5[f(x)]^4}.$$

Problem 1
Substitute the expression for $f(x)$ into the right side of this equation to obtain

$$f'(x) = \tfrac{3}{5}x^{-2/5}.$$

Note that this is nx^{n-1} for $n = \frac{3}{5}$.

★ Problem 2
Work through the same method for the general power, p/q, p and q integers and $q \neq 0$, to prove the theorem.

PROBLEMS

3. Here are routine problems for practice in differentiation, including the use of the Chain Rule. Find the derivative in each case.

   (a) $y = 2x^{1/2} + 6x^{1/3} - 2x^{3/2}$ (b) $y = \dfrac{2}{x^{1/2}} + \dfrac{6}{x^{1/3}} - \dfrac{2}{x^{3/2}} - \dfrac{4}{x^{3/4}}$

   (c) $y = \sqrt[3]{3x^2} - \dfrac{1}{\sqrt{5x}}$ (d) $s = (t^2 - 3)^4$

173

(e) $y = \dfrac{1}{2x^2} + \dfrac{4}{\sqrt{x}}$ (f) $y = \sqrt{2x} + 2\sqrt{x}$

(g) $Q = \dfrac{2}{\sqrt{t}} + \dfrac{6}{\sqrt[3]{t}}$ (h) $y = (1 - 5\sqrt{x})^2$

(i) $u = \left(\dfrac{\sqrt{v}}{2} - \dfrac{1}{\sqrt{v}}\right)^2$ (j) $z = \sqrt{1 + \sqrt{1 - x^2}}$

4. Consider the curve with equation $x^{1/2} + y^{1/2} = 2$.

   (a) Find dy/dx by implicit differentiation. What, then, is the slope of the curve at $x = 4$? At $x = 1$? What happens at $x = 0$?

   (b) Check all your results in (a) by first solving the given equation for y and then differentiating explicitly.

   (c) Write the equation of the tangent line at the point on the curve where $x = 1$.

   (d) Sketch the curve.

5. Consider the curve with equation $x^2 + 4y^2 = 25$.

   (a) Find dy/dx by implicit differentiation. What, then, is the slope of the curve at $x = 0$, $y = \frac{5}{2}$? At $x = 3$, $y = 2$? What happens at $x = 5$?

   (b) Check all your results in (a) by solving the given equation for y, choosing the positive square root, and differentiating explicitly.

   (c) Write the equation of the tangent line at the point $(3, 2)$ on the curve.

   (d) Sketch the entire curve.

4.11   Implicit differentiation applied to related rates

Here is a problem that is neatly handled by our method of implicit differentiation: Jeep A leaves a camp at noon and heads due east at 20 mph. At two o'clock, jeep B leaves the same camp and heads due north at 30 mph. How fast are the jeeps separating at four o'clock?

At four o'clock, jeep A is 80 miles from the camp; and at that time jeep B is 60 miles from the camp (Figure 4-11). By the Pythagorean Theorem, A and B are then 100 miles apart. But to find the rate at which they are separating, we have to consider not merely the *static* situation at the end of 4 hr but also the *dynamic* situation at a general time. Suppose that at time t hr after jeep A starts, it is x miles from the camp, that jeep B is then y miles from the camp, and that the jeeps are then z miles apart (Figure 4-12). Then $z^2 = x^2 + y^2$, by the Pythagorean Theorem. Instead of solving for z, we leave the equation in this form and observe that x, y, and z are all functions of t: $x = f(t)$, $y = g(t)$, $z = h(t)$, say. Then $[h(t)]^2 = [f(t)]^2 + [g(t)]^2$. Differentiating each side, implicitly, gives

$$2h(t) \cdot h'(t) = 2f(t) \cdot f'(t) + 2g(t) \cdot g'(t).$$

Now $h'(t)$ is the quantity desired at $t = 4$; $f(t)$ is then 80; $g(t)$ is then 60; $h(t)$ is then 100; $f'(t)$ is constantly 20; $g'(t)$ is constantly 30.

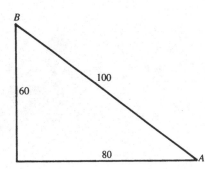

Fig. 4-11

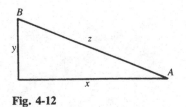

Fig. 4-12

Problem 1

Use these values to obtain 34 mph as the answer to our problem.

We can use the new notation in this problem as follows: We have $z^2 = x^2 + y^2$, with x, y, and z all functions of t. Differentiating with respect to t, we write:

$$2z\frac{dz}{dt} = 2x\frac{dx}{dt} + 2y\frac{dy}{dt}.$$

[If you have difficulty in seeing this, think of the left side, z^2, as being a quantity, Q, say. Then $Q = z^2$, and the Chain Rule gives $dQ/dt = (dQ/dz)\cdot(dz/dt) = 2z(dz/dt)$. Similarly for the derivative of the right side.]

Now, $dx/dt = 20$ and $dy/dt = 30$. At $t = 4$, $x = 80$, $y = 60$, and $z = 100$. So we obtain $dz/dt = 34$.

Problem 2

Work the same problem by showing that $x = 20t$ and $y = 30(t - 2)$. Express z^2 in terms of t, and differentiate implicitly.

Problem 3

Work the same problem by proceeding as in Problem 2 and solving for z before differentiating.

In some cases we may have a right triangle (as in this example), but with one of the legs or the hypotenuse fixed in length. Then the rate of change of that side is zero, of course.

PROBLEMS

4. A 25-ft ladder leans against a wall, and the foot of the ladder is pulled away from the base of the wall at 2 ft/sec. How fast is the top of the ladder descending when the foot of the ladder is 15 ft from the base of the wall?

5. The baseball "diamond" is a square 90 ft on a side. A ball was tapped along the third-base line at a speed of 45 ft/sec. How fast was the distance of the ball from first base changing 1 sec after starting?

6. An airplane flying horizontally at the rate of 200 ft/sec passes straight over a pool, at an elevation of 6000 ft. How fast is its distance from the pool increasing 40 sec later?

7. Jogger A leaves the corner of Church Street and High Street at the constant rate of 300 yards/min. When she is 100 yards from the start along High Street, jogger B leaves the corner of Church and High, going along Church Street, also at 300 yards/min (Figure 4-13). How fast is the straight-line distance between A and B changing 1 min after B starts? (Give the units of your answer.)

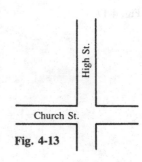

High St.

Church St.

Fig. 4-13

175

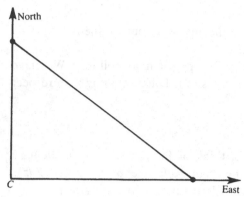

Fig. 4-14

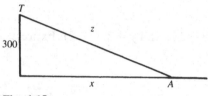

Fig. 4-15

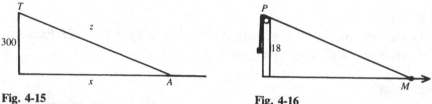

Fig. 4-16

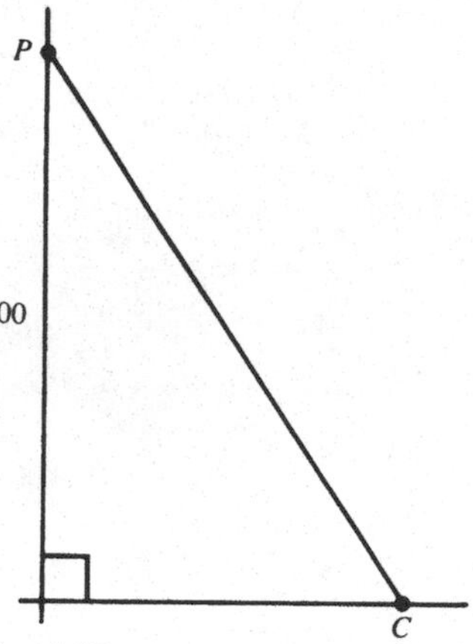

Fig. 4-17

8. Two trains leave a city C, one going east at 40 mph and the other going north at 50 mph. They leave at such times that at a certain instant the eastbound train is 80 miles from C, and the northbound train is 60 miles from C (Figure 4-14). At that instant, how fast is the distance between the trains changing?

9. An auto leaves the base of a 300-ft-tall tower, traveling in a straight line at 40 ft/sec (Figure 4-15). How fast is the distance between the auto and the top of the tower changing 10 sec later?

10. A man, M, raises a weight by means of a rope run over a pulley, P, that is 18 ft above the level of his hands (Figure 4-16). He walks away at the rate of 5 ft/sec. How fast is the distance PM changing when the man is 24 ft from the point directly below P?

11. As in Figure 4-17, a policeman in a stationary helicopter (P) observes a car (C), using radar to determine that the straight-line distance PC is increasing at the rate of 54 ft/sec at the instant that $PC = 750$ ft. If the helicopter is 600 ft up, is the car exceeding the 55-mph speed limit?

4.12 Differentials

Although we have presented the *derivative* as a fundamental concept, Newton emphasized *differentials*, and they still are useful. If $y = f(x)$, the differentials, dy and dx, are defined as any two quantities whose ratio equals the derivative:

$$dy \div dx = f'(x), \tag{16}$$

or

$$\frac{dy}{dx} = \left(\frac{dy}{dx}\right). \tag{17}$$

The long fraction line on the left side of equation (17) calls attention to the fact that we have *dy divided by dx*, and the parentheses on the right serve to emphasize that this is the derivative. Obviously, the defining equations (16) and (17) are equivalent to

$$dy = \left(\frac{dy}{dx}\right) \cdot dx = f'(x) \cdot dx.$$

(In England, what we call the *derivative* was long called the "differential coefficient" – it is the quantity by which *dx* is multiplied, just as 3 is the coefficient in the expression 3*x*.) Because it is only the ratio of *dy* to *dx* that is defined, we can choose *dx* at will, and then *dy* will be determined.

In practice we frequently choose $dx = \Delta x$, an increment in *x*. Then *dy* has the following familiar interpretation: In Figure 4-18, $PR = \Delta x = dx$, and $RT/PR = $ slope of tangent $= f'(x) = dy/dx$. By taking the equality $RT/PR = dy/dx$ and multiplying both sides by *PR*, we obtain

$$RT = \left(\frac{dy}{dx}\right) PR.$$

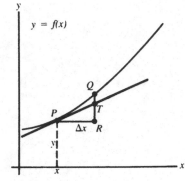

Fig. 4-18

That is,

$$RT = \left(\frac{dy}{dx}\right) \cdot dx, \quad \text{or} \quad RT = dy.$$

Now $RQ = \Delta y$. If the interval is small, the tangent does not deviate much from the curve, so

$$RQ \approx RT, \quad \text{or} \quad \Delta y \approx dy.$$

This is the familiar approximation that we have been using (without the notation for "differential") ever since Section 3.3, Approximate Increments. Because of the interpretation shown here, it is often called *the differential approximation*.

PROBLEMS

1. Simplify each of the following expressions:

   (a) $\dfrac{dV}{dx} \cdot \dfrac{dx}{dt}$ (b) $\dfrac{dS}{dt} \cdot \dfrac{dt}{dx}$

   (c) $\dfrac{dy}{du} \cdot \dfrac{du}{dt} \cdot dt$ (d) $\dfrac{dy}{dx} \cdot \dfrac{dx}{du} \cdot du$

2. (a) If $V = x^3$, find dV/dx at $x = 5$.

   (b) If $x = V^{1/3}$, find dx/dV at $x = 5$.

177

3. (a) If $y = u^3 + 5$, express dy in terms of u and du.

(b) If y is as in (a), and $u = x^2 + 1$, express dy in terms of x and dx.

(c) Are the answers to (a) and (b) equivalent? Explain.

4.13 Formulas for derivatives of products and quotients

One of the most basic of results in calculus is expressed as "the derivative of the sum of two functions equals the sum of the derivatives of the functions," or, in symbols,

$$(f + g)' = f' + g',$$

or

$$\frac{d}{dx}(u + v) = \frac{du}{dx} + \frac{dv}{dx}.$$

According to legend, Leibniz first thought that an analogous formula applied to the *product* of two functions – "the derivative of the product of two functions equals the product of their derivatives" – but he must quickly have seen this to be fallacious from a simple example: If $f(x) = x$ and $g(x) = x^2$, then $G(x) = f(x) \cdot g(x) = x^3$. We know that $f'(x) = 1$, $g'(x) = 2x$, and $G'(x) = 3x^2$. But clearly, $1 \cdot 2x \neq 3x^2$.

The legend goes on to state that after realizing his mistake, Leibniz took 10 days to derive the correct formula. Here is how it goes: Suppose that

$$y = u \cdot v, \tag{18}$$

where u and v are functions of x. If x takes on an increment, Δx, then u changes to $u + \Delta u$, v to $v + \Delta v$, and, finally, y to $y + \Delta y$. Thus, corresponding to $x + \Delta x$, we have

$$y + \Delta y = (u + \Delta u)(v + \Delta v). \tag{19}$$

It may happen that $\Delta u = 0$, or $\Delta v = 0$, or $\Delta y = 0$, for a nonzero change in x, but that doesn't make us any trouble. What is important is that $\lim_{\Delta x \to 0} \Delta u = 0$; that is, as the new value, $x + \Delta x$, approaches the original value, x, we know that the new value of the function, $u + \Delta u$, approaches the original value, u. This will be true if u is a continuous function of x. Likewise, $\lim_{\Delta x \to 0} \Delta v = 0$, if v is continuous.

Now, if we subtract equation (18) from equation (19), we have

$$\Delta y = (u + \Delta u)(v + \Delta v) - uv.$$

Problem 1

Show that this reduces to

$$\Delta y = u \Delta v + v \Delta u + (\Delta u)(\Delta v).$$

The average rate of change of y with respect to x is given by

$$\frac{\Delta y}{\Delta x} = \frac{u \Delta v + v \Delta u + (\Delta u)(\Delta v)}{\Delta x}$$

$$= u \frac{\Delta v}{\Delta x} + v \frac{\Delta u}{\Delta x} + \frac{\Delta u}{\Delta x} \Delta v.$$

By definition,

$$\frac{dy}{dx} = \lim_{\Delta x \to 0} \frac{\Delta y}{\Delta x},$$

so

$$\frac{dy}{dx} = \lim_{\Delta x \to 0} \left(u \frac{\Delta v}{\Delta x} + v \frac{\Delta u}{\Delta x} + \frac{\Delta u}{\Delta x} \Delta v \right). \tag{20}$$

Problem 2

(a) What is an expression for $\lim_{\Delta x \to 0} \Delta v / \Delta x$? For $\lim_{\Delta x \to 0} \Delta u / \Delta x$?
(b) What is the value of $\lim_{\Delta x \to 0} u$? Of $\lim_{\Delta x \to 0} v$? (Trick question)
(c) What is the value of $\lim_{\Delta x \to 0} \Delta v$?
(d) State the basic limit theorems needed to conclude from equation (20) that

$$\frac{dy}{dx} = u \frac{dv}{dx} + v \frac{du}{dx}.$$

In words, we have this Rule for the Derivative of a Product:

The derivative of the product of two differentiable functions equals the first function times the derivative of the second plus the second function times the derivative of the first.

Using our old symbolism, we write:

$$\text{If } G = f \cdot g, \quad \text{then } G' = f \cdot g' + g \cdot f'.$$

If we have the quotient of two functions,

$$y = \frac{u}{v},$$

with u and v functions of x, and $v \neq 0$, we can obtain a formula for the derivative of y with respect to x by rewriting,

$$y = u \cdot \frac{1}{v},$$

and using the formula for the derivative of a product found earlier. If we set $w = 1/v$, we have $y = u \cdot w$ and $dy/dx = u(dw/dx) + w(du/dx)$.

Problem 3

If $w = 1/v = v^{-1}$, use the Special Case of the Chain Rule to obtain $dw/dx = -v^{-2}(dv/dx) = -(dv/dx)/v^2$. Thus, we have $dy/dx = u[-(dv/dx)/v^2] + (1/v)(du/dx)$.

Problem 4

Do the algebra to reduce this to

$$\frac{dy}{dx} = \left(v\frac{du}{dx} - u\frac{dv}{dx} \right) \Big/ v^2.$$

In words, we have the Rule for the Derivative of a Quotient:

> The derivative of the quotient of two differentiable functions, in which the denominator is not zero, equals the denominator times the derivative of the numerator minus the numerator times the derivative of the denominator, all divided by the square of the denominator.

In our old symbolism, we write:

$$\text{If } F = \frac{f}{g}, \quad \text{then } F' = \frac{g \cdot f' - f \cdot g'}{g^2}.$$

We can use these differentiation formulas to solve more general *related rate* problems than we handled in Section 4.8. In that section we solved problems like this:

> A growing pile of sawdust has the shape of a right circular cone whose height always equals two-thirds the radius of its base. How fast is the volume of the pile increasing when $r = 6$ ft, if r is then increasing at the rate of 0.5 ft/min?

The given relation, $h = \frac{2}{3}r$, means that the volume, V, can be expressed as a function of the *single* variable r.

We shall now consider an example in which we do *not* have this kind of simplifying relation.

Example 1

If the radius of a cone increases at the rate of 0.5 in./hr and the height decreases at the rate of 0.2 in./hr, how fast does the volume change when $r = 10$ in. and $h = 6$ in.? We have

$$V = \tfrac{1}{3}\pi r^2 h, \tag{21}$$

where r and h are functions of time, t. Hence, V is also a function of t. Indeed, the problem can be put in symbols as follows:

If $dr/dt = 0.5$ and $dh/dt = -0.2$, what is the value of dV/dt when $r = 10$ and $h = 6$?

So we take equation (21) and differentiate to find dV/dt. The constant, $\frac{1}{3}\pi$, remains as a multiplier, and we have the product of two quantities, r^2 and h, to differentiate with respect to t. In accordance with the Rule for the Derivative of a Product,

$$\frac{dV}{dt} = \frac{1}{3}\pi \left[r^2 \cdot \frac{dh}{dt} + h \cdot \frac{d(r^2)}{dt} \right].$$

By the Chain Rule,

$$\frac{d(r^2)}{dt} = \frac{d(r^2)}{dr} \cdot \frac{dr}{dt} = 2r\frac{dr}{dt}.$$

Thus,

$$\frac{dV}{dt} = \frac{1}{3}\pi\left[r^2 \cdot \frac{dh}{dt} + h \cdot 2r\frac{dr}{dt}\right].$$

Substituting the values given in the problem results in

$$\frac{dV}{dt} = \frac{1}{3}\pi\left[100(-0.2) + 6 \cdot 20 \cdot (0.5)\right]$$

$$= \frac{1}{3}\pi(-20 + 60) = \frac{40\pi}{3} \approx 42 \ (\text{in.}^3/\text{hr}).$$

PROBLEMS

5. Find the derivative for the function given by each of the following expressions. In each case, specify the natural domain of the function and of its derivative.

(a) $x\sqrt{16 - x^2}$

(b) $(x^2 + 1)^2(1 - x)^3$

(c) $\dfrac{2x}{3(x - 2)}$

(d) $\dfrac{2x - 1}{\sqrt{4 - x}}$

(e) $\sqrt{x}(1 - x^2)$

(f) $\dfrac{1 - x^2}{\sqrt{x}}$

(g) $\dfrac{(2x - 5)(x + 1)}{x^2 - 9}$

(h) $\dfrac{x}{\sqrt{9 - x^2}}$

(i) $\sqrt{\dfrac{25 - x^2}{36 - x^2}}$

(j) $\sqrt{\dfrac{9 - x^2}{x^2 - 25}}$

6. If $y = (x - 1)/(x + 1)$, find dy/dx. Does this function, over its natural domain, have any maxima or minima? Does its graph have any points of inflection?

7. Same as Problem 6 for $y = (x^2 - 1)/(x^2 + 1)$.

8. Same as Problem 6 for $y = \sqrt{1 + x^2}/x$.

9. Establish (for instance, by cross-multiplying) the identity

$$1 + x + x^2 + x^3 + \cdots + x^7 = \frac{1 - x^8}{1 - x}.$$

For what values of x is it valid? Calculate the derivatives of each side of this identity and hence obtain a formula for the sum $1 + 2x + 3x^2 + \cdots + 7x^6$. Check it by putting $x = -1$.

10. If the radius of a cylinder decreases at the rate of 0.2 in./min and the height of the cylinder increases at the rate of 0.1 in./min, how fast does the lateral surface area of the cylinder change when $r = 10$ in. and $h = 5$ in.? (Give the units of your answer.)

11. If the radius of a cone increases at the rate of 0.2 in./min and the height of the cone decreases at the rate of 0.3 in./min, how fast does the volume change when $r = 5$ in. and $h = 15$ in.? (Give the units of your answer.)

12. If the radius of a cylinder increases at the rate of 0.1 in./min and the height of the cylinder decreases at the rate of 0.2 in./min, how fast does the total surface area (lateral surface plus top and bottom) change when $r = 10$ in. and $h = 5$ in.? (Give the units of your answer.)

13. (a) If f and g are differentiable functions, and $G = f \cdot g$, find G'', G''', and G''''.

   ★ (b) Conjecture the formula for $G^{(n)}$, the nth derivative, and prove it correct. The formula for $G^{(n)}$ is known as *Leibniz's rule*.

★ 14. In Problem 9, **3.9**, we defined the average total cost of production as E/x, where E is the total cost of producing x units. If E is a differentiable function of x, show that, at the values of x that make the average total cost a turning-point extreme, E' equals that average total cost (cf. Problem 10, **3.9**).

4.14 Marginal cost, marginal revenue, and optimal production levels

The concept of marginal cost was introduced in Section 3.4, where it was noted that the idea of marginality applied also to quantities other than cost (e.g., to *value* or *utility*). We shall now push further the concepts introduced in Section 3.4.

If the cost (C) of producing n items is given by $C = f(n)$, and if f is a differentiable function, defined for all real numbers (not just for integers) over a certain interval, then the *marginal cost* for any n is defined as

$$MC(n) = f'(n).$$

Likewise, if the gross revenue (R) from selling n items is given by $R = g(n)$, then the *marginal revenue* at the nth item is defined as

$$MR(n) = g'(n).$$

Now, the profit (P) obtained from producing and selling n items is clearly

$$P = R - C \quad \text{or} \quad P = g(n) - f(n).$$

To maximize P, we set $P' = 0$:

$$P' = g'(n) - f'(n) = 0.$$

Critical numbers, then, are given by solutions of

$$g'(n) = f'(n).$$

In words, this can be put as follows:

"Profit is maximized for production at which marginal revenue equals marginal cost."

This makes intuitive sense: If the marginal revenue from selling the 500th item is *greater* than the marginal cost of producing that item, it pays the manufacturer to produce and sell the 500th item; if the marginal revenue from selling the 700th item is *less* than the marginal cost of producing that item, the manufacturer will be inclined to stop short of producing and selling the 700th item.

The distinction between *marginal* costs (and benefits) and *total* costs (and benefits) is forcefully brought out in the following letter (*Science*, November 22, 1974):

Emission Standards: Costs and Benefits

It is reported by Constance Holden (News and Comment, 27 Sept., p. 1142) that the National Academy of Sciences (NAS) has okayed auto emission standards. Indeed, the recent NAS study (1) prepared for the Senate Public Works Committee endorses the numerical emission standards set out in the 1970 Clean Air Act and sees "no substantial basis for changing the standards." It claims that the standards are justifiable in cost-benefit terms. It reaches this conclusion by finding "that the benefits in monetary terms ... are commensurate with the expected cost" of about $5 billion to $8 billion per year.

Unfortunately this conclusion is not justified: the optimum point of operation is not one at which the dollar benefits are equal to the dollar costs, but one at which the *marginal* (or incremental) benefits are equal to the *marginal* costs [Figure 4-19]. This optimum point generally occurs where the costs are much lower than the benefits. At the optimum, a $1 increase in cost would buy an additional $1 of benefits; at the point where costs equal benefits (which is well past the optimum), it would buy substantially less. The summary report only hints at this possibility. But the detailed results of the study itself can be used directly to support the following contrary conclusions: Relaxing the emission standards, or reducing their geographic coverage to cities with serious pollution problems, or delaying the implementation of the standards would lower the costs drastically without an important reduction in benefits.

S. FRED SINGER

Department of Environmental Sciences,
University of Virginia,
Charlottesville 22903

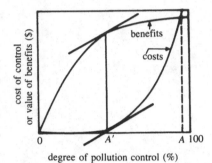

Fig. 4-19 Schematic diagram of costs and benefits versus degree of pollution control (2, p. 949). The optimum level of pollution is not at point *A*, where costs equals benefits, but at point *A'*, where the marginal quantities (slopes) are equal.

References

1. Coordinating Committee on Air Quality Studies, National Academy of Sciences – National Academy of Engineering, *Air Quality and Automobile Emission Control*, vol. 1, *Summary Report* (Government Printing Office, Washington, D.C., 1974).
2. Modified from S. F. Singer, *Eos* (*Trans. Am. Geophys. Union*) 55, 948 (1974).

PROBLEMS

For a certain manufacturing process, the cost (C) of producing n items is given by $C = 200 + 5n - 0.001n^2$, and the revenue (R) obtained from selling n items is given by $R = 7n - 0.002n^2$.

1. Discuss the significance of each of the terms in these cost and revenue functions.
2. What are the formulas for MC and MR?
3. For what values of n is C decreasing (MC negative)?
4. For what values of n is R decreasing (MR negative)?

   (Because it is most unlikely in a real situation that the total cost of producing 3000 items is less than that of producing 2999 items – or that the cost of producing the 3000th item is negative – the domain of the cost function is doubtless an interval $[0, b]$, where b is less than 2500. Likewise, because it is unlikely that the total revenue obtained from selling 2000 items is less than that from selling 1999 items – or that the selling price of the 2000th item is negative – the domain of the revenue function is doubtless an interval $[0, c]$, where c is less than 1750. To be safe, let us assume that the cost and revenue functions both have $[0, 1500]$ as domains.)

5. Find a formula for the profit, P. What n gives maximum P? What is the maximum value of P?
6. Obtain the value of n for maximum P by solving the equation $MR = MC$.
7. Using a large scale, draw graphs of C, R, and P on the same axes over $[0, 1500]$.
8. Immediately below the graphs of Problem 7, draw the graphs of the derived curves MC, MR, and MP on the same axes over $[0, 1500]$.
9. What is the formula for average total cost (or total unit cost, TUC)?
10. For what production levels is $(TUC)'$ negative? Positive? Hence, for what production levels are there economies of scale? Diseconomies?
11. Answer the foregoing set of questions for a different manufacturing process in which $C = 200 + 5n + 0.0002n^2$ and $R = 8n - 0.001n^2$. Also verify for this cost function that at the n that makes average total cost a minimum, the value of the marginal cost equals the value of the average total cost.
12. With reference to the letter in *Science* reprinted in this section, draw various hypothetical "cost-of-control" and "value-of-benefits" curves, and observe how the position of the optimum point of operation changes.

4.15
Maxima and minima
using implicit differentiation

⋆ **4.15   Maxima and minima using implicit differentiation**

Implicit differentiation provides an elegant method for treating some problems of finding maxima and minima, as the following example shows.

A rectangular area is to be fenced with materials costing $2 per yard for the length and $3 per yard for the width. What is the shape of the rectangle for (a) maximum area for given cost and (b) minimum cost for given area?

If the length is x and the width is y (Figure 4-20) then

$$A = xy, \qquad (22)$$

and

$$C = 4x + 6y. \qquad (23)$$

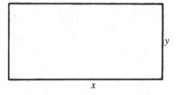

Fig. 4-20

To solve part (a), we observe that because the cost is fixed, equation (23) could be solved for y in terms of x: $y = f(x)$, say. Substituting $f(x)$ for y in equations (22) and (23) gives

$$A = x \cdot f(x)$$

and

$$C = 4x + 6f(x).$$

For A to be a maximum, its derivative must equal zero; if C is constant, its derivative also equals zero.

Problem 1
Use the content of the preceding sentence to obtain $x \cdot f'(x) + f(x) = 0$, $4 + 6f'(x) = 0$.

Problem 2
Eliminate $f'(x)$ between these two equations, and replace $f(x)$ by y to obtain $2x = 3y$, or $x/y = \frac{3}{2}$.

In other words, we have maximum area for given cost if the ratio of length to width of the rectangle equals $\frac{3}{2}$.

To solve part (b), we argue that because the area is fixed, equation (22) could be solved for y in terms of x: $y = g(x)$, say. Substituting $g(x)$ for y in equations (22) and (23) gives

$$A = x \cdot g(x) \quad \text{and} \quad C = 4x + 6g(x).$$

If A is constant, its derivative equals zero; for C to be a minimum, its derivative also must equal zero.

Problem 3
Complete part (b) to obtain the same result as before, namely, $x/y = \frac{3}{2}$.

In other words, the problems are equivalent, and we might phrase either of them as "What is the *optimal* shape of the rectangle?"

PROBLEMS

Use the method of this section to solve each of the following problems from Section 3.8:

4. Problem 6
5. Problem 12
6. Problem 14
7. Problem 18(c)
8. Problem 19(d)
9. Problem 20
10. Problem 21
11. What are the radii of two spheres if the sum of their volumes is fixed and the sum of their surface areas is an extreme? If the sum of their surface areas is fixed and the sum of their volumes is an extreme? Discuss fully.
12. Rectangles are inscribed in the circle $x^2 + y^2 = r^2$, with their sides parallel to the coordinate axes. What are the dimensions and area of the rectangle of maximum area?
13. Same as Problem 12 for the ellipse $(x^2/a^2) + (y^2/b^2) = 1$.

4.16 Summary

Our results in differentiation can be brought together under the following headings. Write out for yourself the complete statement corresponding to each of them.

General results
1. Derivative of the sum of two functions.
2. Derivative of a constant times a function.
3. Chain Rule.
4. Derivative of the product of two functions.
5. Derivative of the quotient of two functions.

Specific results
6. Derivative of x^n, for n a rational number.
7. Derivative of $[f(x)]^n$, for n a rational number (consequence of 3 and 6).

There is a surprisingly large number of problems that can be handled with these rather limited results. Later we shall continue with our question "How *fast* does a quantity change?" by studying the rates of change of the exponential, the logarithmic, and the trigonometric functions. But before we do so, we shall consider the reverse of the rate problem.

PROBLEMS

4.16
Summary

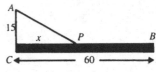

Fig. 4-21

1. (a) A trench is to be dug from point A, 15 yards off a road in a field, to point B, 60 yards down the road from the point C, closest to A (Figure 4-21). If the cost of digging in the field is $5 per yard, and along the road $4 per yard, find the minimum cost of the trench. (Hint: Assume that the trench goes through the field from A to P, x yards along the road from C, and then from P to B.)

   (b) Same as (a), if $\overline{CB} = 20$.

   (c) Same as (a), if $\overline{CB} = 18$.

★ (d) Discuss this problem in general, if $\overline{AC} = a$, $\overline{CB} = b$, cost of digging in the field is $p per yard, and cost of digging along the road is $q per yard.

2. A first approximation to a learning model is one in which the time (t sec) it takes to memorize n nonsense syllables is proportional to n: $t = f(n) = kn$.

   A better model is that in which the formula for t is $t = g(n) = kn\sqrt{n-l}$, where k and l are empirically determined constants, and D_g is the set of integers greater than or equal to $l+1$. Assume that $k = 7$, that $l = 5$, and that we can treat g as though D_g consisted of all real numbers greater than or equal to 6.

   (a) Find $g'(n)$ and show that it is positive for all $n \geq 6$.

   (b) Find $g''(n)$ and thus find a point of inflection on the graph of g.

★ (c) What is a power-law approximation to g' for large n?

3. (a) The hourly cost ($\$ p$/hr) of operating an oil tanker varies with its speed (v knots, i.e., v nautical miles per hour) as follows: $p = 250 + \frac{1}{8}v^3$. What is the minimum cost of a voyage of 500 nautical miles? Check that you have obtained a minimum.

   (b) Does the optimal speed change if the voyage has a length of 1000 nautical miles? Does the minimum cost change?

   (c) What is the optimal speed if competition reduces the cost of leasing an oil tanker so that the formula for the cost function becomes $p = 128 + \frac{1}{8}v^3$?

4. In a table of squares of reciprocals, the entry opposite 0.52 is 3.69822. Find $(0.52001)^{-2}$ approximately.

5. A certain article costs $1 to manufacture. The number, x, sold per annum is related to the selling price, $\$p$, over a certain price range by the equation $x = k/p^2$, where k is a constant; and 2500 are sold when the selling price is $1.

   (a) Express the annual profit as a function of x, and find x and p for maximum profit. What is the maximum annual profit?

   (b) Express the annual profit as a function of p, and find p and x for maximum profit.

6. You wish to make a rectangular box with a square base and no lid, of

187

volume 80 ft³. The material for the sides costs 20 cents/ft², and that for the base 50 cents/ft². Calculate whether, with $20 to spend, you can afford to make such a box.

7. Water is running out of a conical funnel at the constant rate of 10 in.³/min. The axis of the funnel is vertical, the angle of the funnel is 60°, and the hole at the vertex is extremely small.

  (a) Find the rate at which the depth (x in.) of water in the funnel is falling when $x = 5$.

  (b) Sketch the form of the graph of dx/dt as a function of x.

8. A private swimming pool is made in the shape of an equilateral triangle with sides 30 ft (Figure 4-22). A reluctant bather stands at B and waggles his toes in the water, causing circular ripples to spread out from B at a speed of 5 ft/sec. How rapidly is the *area* affected by ripples increasing as the first ripple reaches the side AC?

9. A boy in a skiff is hauling on a painter (i.e., pulling on a rope), one end of which is attached to a dock 8 ft above the level of the boy's hands. If he pulls in rope at the uniform rate of 3 ft/sec, how fast is the skiff moving when 17 ft of rope are still out?

10. The cost (C) of producing n lb of a commodity is given by $C = 500 + 5n - 0.01n^2$. The number, n, of pounds that can be sold at x per pound is given by $n = 250 - 10x$, $0 < x < 20$. Find the approximate selling price that maximizes the profit, and find the maximum profit.

11. U.S. postal regulations specify that, to go by parcel post, the length plus girth of a package must not exceed 100 in.

  (a) What are the dimensions and volume of the largest rectangular box with square cross section that can be sent by parcel post? Discuss fully.

  ★ (b) Can you show that if the condition "with square cross section" were omitted, the rectangular box of maximum volume would nevertheless have square cross section?

  (c) What are the dimensions and volume of the largest cylindrical package that can be sent?

  ★ (d) Solve (a) and (c) of this problem by implicit differentiation, as suggested in Section 4.15.

12. The cost functions we have considered thus far have been associated either with efficiencies in production over the whole domain of the function or with inefficiencies of production over the whole domain. Probably neither type is as realistic as a cost function for which there are efficiencies for small and medium production and inefficiencies for very large production. We turn to such a case now.

  Suppose that the total cost (C) of producing x units of a commodity per week is given by

$$C = f(x) = \begin{cases} 500 + 5x - 0.01x^2, & 0 \le x \le 100, \\ 625 + 2.5x + 0.0025x^2, & 100 < x \le 600. \end{cases}$$

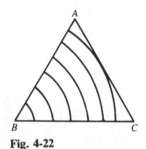

Fig. 4-22

(a) Is f continuous over its domain? Differentiable? Sketch a graph of this function.

(b) If the marginal cost is represented as the function g, write formulas for $g(x)$, indicating the domain. Is g continuous over its domain? Differentiable? Sketch a graph of this function.

(c) If the average total cost is represented as the function F, write formulas for $F(x)$, indicating the domain. Is F continuous over its domain? Differentiable? Sketch a graph of this function.

(d) What is the fixed cost in this operation? Write formulas for the variable costs.

(e) For what x is the average total cost a minimum? What is the minimum average total cost? Verify that this minimum average total cost equals the value of f' (the marginal cost) at the value of x corresponding to the minimum (cf. Problem 14, **4.13**).

(f) If the total revenue ($\$R$) obtained by selling the production of x units is given by $R = 9x - 0.01x^2$, write formulas to express the profit ($\$P$) as a function of x. Sketch graphs of R and P on the same axis as your graph of C. For what x is the profit a maximum? Show on your graph that $MC = MR$ for this x. What is the maximum profit?

13. Find the point(s), P, on the circle $x^2 + y^2 = 1$ such that the sum of the distances from P to the points $(1,0)$ and $(-1,0)$ is an extreme.

★ 14. What are the dimensions of the cone of minimum volume that can be circumscribed about a given sphere?

15. Water is running out of the conical funnel shown in Figure 4-23 at the constant rate of 2 in.^{3}/sec, until $h = \frac{1}{2}$ (in.).

(a) What does your intuition tell you about the rate at which the water level decreases?

(b) Find the rate at which h decreases at $h = 4, 3, 2$, and 1.

(c) What is the maximum rate of decrease of h?

★ 16. By consideration of the first and second derivatives, discuss fully the function given by $g(x) = kx/(1 - x + kx)$, k constant. [From Peter Newman, "Some Properties of Concave Functions," *Journal of Economic Theory* (1969), pp. 291–314.]

17. The *stiffness* (T units) of a rectangular beam of fixed length varies as the width (x in.) and the *cube* of the depth (y in.), as designated in Figure 4-24(a).

(a) Find the dimensions of the stiffest beam that can be cut from a circular log of diameter 20 in. (For interest, compare with Example 2, **3.8**.)

★ (b) Solve this problem by implicit differentiation, as suggested in Section 4.15.

18. Let f be the function defined by $y = \sqrt{4 + x^2}$.

(a) What is the natural domain of f?

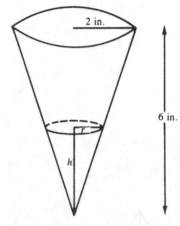

Fig. 4-23

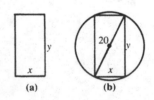

(a)   (b)

Fig. 4-24

189

(b) Find $f'(x)$ and $f''(x)$.

(c) Find whatever numbers x correspond to maximum and minimum values of y, and find the corresponding values of y.

(d) Show that there are no points of inflection on the graph of f and that the graph is everywhere concave up.

(e) Sketch the graph of the equation.

19. The formula for the slope, s, of a certain curve is $s = 5 - 3x + 2x^2 - 8x^3 + x^4$.

    (a) Find a formula for R, the rate of change of flexion of this curve.

    (b) For what x is R an extreme? Maximum or minimum?

20. If the radius of a cylinder increases at the rate of 0.5 in./min and the height decreases at the rate of 0.8 in./min, how fast does the total surface area change when $r = 6$ in. and $h = 10$ in.? (Give the units of
    · your answer.)

21. The material for constructing a rectangular box with square cross section costs 3¢ per in.^{2} for the bottom, the same for the top, and 2¢ per in.^{2} for the sides. If the edge of the bottom is x in. and the height of the box is y in., express the cost of material for the box. What are the dimensions of the box for minimum cost if the box is to contain 1500 in.^{3} and if no dimension of the box can exceed 12 in.?

★ 22. Consider $f(x) = \sqrt{x^2 - 6x}$.

    (a) State the natural domain of f.

    (b) Find any maximum points, minimum points, and points of inflection on the graph of $y = f(x)$.

    (c) Find regions where the graph is concave up. Concave down.

    (d) Sketch the graph of $y = f(x)$, showing where the tangent to the curve is vertical.

★ 23. Same as Problem 22 for $g(x) = \sqrt{x^2 + ax}$, for any constant a.

★ 24. The lateral surface area (L ft^2) of a cone of radius r ft, height h ft, and slant height s ft (Figure 4-25) is given by $L = \pi r s$. If r increases by 0.3 ft/hr and h decreases by 0.2 ft/hr, how fast does L change when $r = 12$ and $h = 9$? Also, try to *derive* the given formula for L.

25. Let $f(x)$ be the function defined by $f(x) = (3 - x^2)/(x + 2)$.

    (a) Find $f'(x)$ and $f''(x)$.

    (b) Find whatever numbers x correspond to maximum and minimum values of $f(x)$ and to points of inflection on the graph of f.

    (c) Find maximum and minimum values of f in $[-1.5, 0]$.

26. If the length (x ft) of a rectangular plate increases at 0.1 ft/min and the width (y ft) decreases at the rate of 0.2 ft/min, how fast does the area change when $x = 5$ and $y = 10$?

27. Consider the curve with equation $y = \frac{1}{2}x^4 - 2x^3$, over the natural domain.

    (a) Find the coordinates of any turning-point maximum and minimum points on the curve.

Fig. 4-25

(b) Find the coordinates of any points of inflection on the curve.

(c) Sketch the curve.

(d) Find the maximum and minimum values of the flexion over the interval $[0, 3]$.

28. Let f be the function given by $f(x) = (2x - 5)/x^2$.

    (a) What is the natural domain of f?

    (b) Find $f'(x)$. (Hint: Simplify *before* differentiating.)

    (c) Find any turning-point extremes of f.

    (d) Find the coordinates of any points of inflection on the graph of f.

    (e) Use two tests to determine whether, in (c), you found maxima or minima.

★ 29. For the curve given by $y = (x - 1) \cdot \sqrt[3]{x^2}$, find any maximum or minimum points, any points of inflection, and any points where dy/dx is not defined. Sketch a neat graph.

30. A rectangular box with a square base and no top is to hold 4000 in.^{3} What dimensions require the least material to build the box, if the edge of the base cannot exceed 16 in.?

31. The volume (V ft^3) of a cone is given by $V = \frac{1}{3}\pi r^2 h$. If the volume increases by $\frac{2}{3}\pi$ ft^3/min and the height increases by 0.03 ft/min, find the rate of change of the radius when $r = 10$ ft and $h = 5$ ft. (Give the units of your answer.)

32. The equation of a curve is $y = x^4 - 8x^3 + 24x^2 - 24x + 10$.

    (a) Find the flexion (y'') at $x = 1$.

    (b) At $x = 1$, is the flexion increasing or decreasing, and how rapidly?

    (c) Find the minimum flexion, and test to show that it is a minimum.

33. The material for constructing a rectangular box with square top and bottom costs \$2 per square meter for the bottom, the same for the top, and \$$\frac{1}{2}$ per square meter for the sides. What are the dimensions and cost of the box of minimum cost that can be constructed to contain 32 m^3 if the height of the box cannot exceed 2 m?

★ 34. Find maximum and minimum points and points of inflection on the graph of $y = 4x/(x^2 + 4)$, and sketch the curve.

35. The material for the bottom of a cylindrical can costs 6¢ per square inch, and for the lateral surface, 2¢ per square inch; there is no top. Express the cost of material for a cylinder of base radius r in. and height h in. If the can is to contain 3000π in.^{3}, find the dimensions of the can of minimum cost if neither r nor h can exceed 15 in.

36. Consider the function g given by $g(x) = (x^2 - 4)/\sqrt{36 - x^2}$.

    (a) What is the natural domain?

    (b) Find the coordinates of any maximum or minimum points.

    (c) What is the range of the function g?

    (d) Where does the graph of $y = g(x)$ cross the x axis? Sketch the graph.

37. An island I is 8 miles off a straight coastline. A town T is 6 miles along the coast from the point A on the coast nearest the island (Figure 4-26).

191

4
Further differentiation

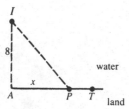

Fig. 4-26

If you can row at 3 mph and jog at 5 mph, where should you hit the beach (point P, x miles from A) so that the total time for rowing the distance IP and jogging the distance PT is a minimum?

★ 38. You are asked to divide a given length of wire into two pieces, one of which you will form into a circle and the other into a square, so as to maximize the total area obtained. What should you do? Explain fully.

★ 39. Same as Problem 38, if you are to form a circle from each portion of the wire.

★ 40. Find maximum and minimum points and points of inflection on the curve $y = x^3/(3x^2 - 3x + 1)$ for $0 \le x \le 1$. Sketch the curve. (Some political scientists have studied the apparent validity of this formula with the interpretation that x is the fraction of the total votes obtained by one of the parties and y is the fraction of seats won in parliament by that party.)

★ 41. A function f is defined by

$$f(x) = \begin{cases} -5 + \sqrt{25 - x^2}, & -5 \le x < 0, \\ 5 - \sqrt{25 - x^2}, & 0 \le x \le 5. \end{cases}$$

(a) For what x's in the domain of f is f continuous?
(b) Using the *definition* of derivative, determine whether $f'(0)$ exists. If it does, state the value of $f'(0)$.
(c) Find $f'(x)$ and state its domain.
(d) For what x's in the domain of f' is f' continuous?
(e) Using the *definition* of derivative, determine whether $f''(0)$ exists. If it does, state the value of $f''(0)$.
(f) Find $f''(x)$ and state its domain.
(g) For what x's in the domain of f'' is f'' continuous?
(h) Sketch graphs of f and its first and second derived curves.
(i) Show that $(0,0)$ is a horizontal point of inflection.

SAMPLE TEST

1. Let $f(x)$ be the function defined by $y = (x^2 + 5)/(x + 2)$.
   (a) What is the natural domain of f?
   (b) Find $f'(x)$ and $f''(x)$.
   (c) Find whatever numbers x correspond to maximum and minimum values of $f(x)$ and to points of inflection on the graph of f.
   (d) Find the values of y corresponding to the values of x found in (c).
   (e) Check whether you have maxima or minima in (c), using two tests in each case.

2. The volume of a cone is given by $V = \frac{1}{3}\pi r^2 h$. If the radius decreases at the rate of 0.1 in./min and the height increases at the rate of 0.2 in./min, how fast does the volume change when $r = 10$ in. and $h = 12$ in.? (Give the units of your answer.)

192

3. A ladder 20 ft long, leaning against a wall, has its foot pulled out from the bottom of the wall at the rate of 2 ft/sec. How fast is the top of the ladder descending when the foot of the ladder is 12 ft from the bottom of the wall?

4. The equation of a curve is $y = 8x^3 - 2x^4$. Find the flexion of the curve at $x = 2$. At $x = 2$, is the flexion increasing or decreasing, and how rapidly? Find the maximum flexion.

5. A rectangular box with square top and bottom is to contain 250 ft³. If the top and bottom are made of material costing $2 per square foot and the sides are made of material costing $1 per square foot, find the dimensions of the box for minimum cost, if no dimension can be greater than 8 ft. Demonstrate conclusively that you have a minimum.

★ 6. Demonstrate that whatever be the values of a and b, the curve $y = (x^2 + a)/(x + b)$ has no points of inflection. Discuss the existence of maximum and minimum points on this curve.

7. A lock-keeper on a bridge 10 ft above a canal is pulling a canal boat by means of a rope attached to the boat at water level. If he hauls in rope at the uniform rate of 6 ft/sec, how fast is the boat moving along the canal when the length of the rope from boat to bridge is 26 ft?

8. (a) Part or all of a concrete slab 15×15 ft is to be used as the base of a rectangular storage container with a square cross section (horizontal), to hold 750 ft³. Material for the four sides of the container costs $2 per square foot; material for the square flat top costs $3 per square foot; no material is needed for the bottom because the concrete slab serves that purpose. What are the dimensions of the container for minimum cost?

   (b) Same as (a), if the concrete slab is $7\frac{1}{2} \times 7\frac{1}{2}$ ft.

★ 9. What is the shape of the container in Problem 8 for any fixed volume?

5 Antidifferentiation and integration

5.1 The reverse of differentiation

Knowing *how* a quantity varies, we have been learning methods of finding *how fast* it varies, that is, how to find f' from f. We now turn to the reverse problem: If we know the rate of change of a function, can we reconstruct the function itself? Often the rate of change is known, from experiment or theory, and the process of "antidifferentiation" turns out to be as important as that of differentiation itself. The ideas of antidifferentiation can best be introduced through a simple example:

Suppose that we know that the *slope* of a curve is

$$\frac{dy}{dx} = x, \quad \text{for all } x,$$

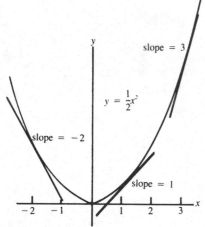

Fig. 5-1

and we wish to find an equation of the curve, that is, an expression for the *height*, y, in terms of x.

Remembering that if $f(x) = x^2$, then $f'(x) = 2x$, we see that a slight modification will do the trick: If $g(x) = \frac{1}{2}x^2$, then $g'(x) = x$, and now we have exactly the given derivative. Thus,

$$y = g(x) = \tfrac{1}{2}x^2$$

is a solution of our problem – for each x, the slope of the curve $y = \frac{1}{2}x^2$ is, indeed, equal to x, as seen in Figure 5-1. But is this the *only* solution?

Problem 1
Before reading further, try to find other solutions.

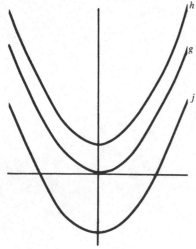

Fig. 5-2

We can quickly see that $y = h(x) = \frac{1}{2}x^2 + 1$, $y = j(x) = \frac{1}{2}x^2 - 2$, and, in fact, $y = F(x) = \frac{1}{2}x^2 + c$, for any constant c, are also solutions:

$$h'(x) = x = j'(x) = F'(x).$$

The geometrical significance of this fact is seen in Figure 5-2: For each x, the slopes of the various curves are all equal. The curves are all the same except for vertical displacement.

194

We are still left with a question: Are there *still other* curves, besides those of the "family" given by $y = F(x) = \frac{1}{2}x^2 + c$, for any constant c, such that $dy/dx = x$? The answer is *no*, as we shall see in the next section. Before proceeding with it, however, there is one more thing to be said about our problem of $dy/dx = x$. Suppose that we assume that $y = 1$ at $x = 0$. Because $dy/dx = 0$ at $x = 0$, we can approximate the curve by a short horizontal line segment in the neighborhood of $x = 0$, as seen in Figure 5-3(a). Near $x = \frac{1}{4}$, the curve can be approximated by a short line segment of slope $\frac{1}{4}$, as shown in Figure 5-3(b).

Problem 2
Check the coordinates shown in Fig. 5-3(b).

Near $x = \frac{1}{2}$, the curve can be approximated by a short line segment of slope $\frac{1}{2}$, as shown in Figure 5-3(c), and so forth.

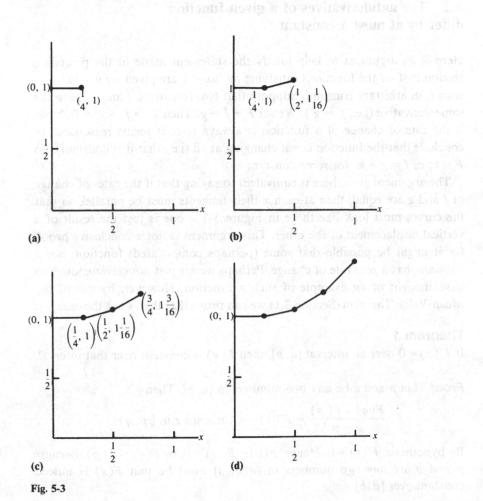

Fig. 5-3

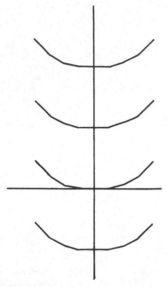

Fig. 5-4

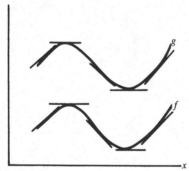

Fig. 5-5

Problem 3
Check the coordinates shown in Figure 5-3(c).

The general effect, then, is seen in Figure 5-3(d). If we did not make the initial assumption that $y = 1$ at $x = 0$, we would have to consider many possibilities: $y = 2$ at $x = 0$, $y = 0$ at $x = 0$, $y = -1$ at $x = 0$, and so forth, and the method exemplified in Figure 5-3 would be translated into the result shown in Figure 5-4.

In our problem, of course, we have found the general formula, $y = \frac{1}{2}x^2 + c$, corresponding to the given equation $dy/dx = x$, so that we do not really need the approximation shown in Figures 5-3 and 5-4. But such approximations are useful in cases in which we cannot find a formula for the antiderivative, and even in our simple example, Figure 5-4 is instructive in helping us to see "what's going on."

5.2 The antiderivatives of a given function differ by at most a constant

Here is an argument to help justify the statement made in the preceding section that *all* the functions satisfying $dy/dx = x$ are given by $y = \frac{1}{2}x^2 + c$, with c an arbitrary constant. Suppose that two functions, f and g, have the same derivative (i.e., $f' = g'$). We set $F = f - g$. Then $F' = f' - g' = 0$. Now, if the rate of change of a function is always zero, it seems reasonable to conclude that the function is not changing at all (i.e., that it is constant). So $F = k$, or $f = g + k$, for some constant k.

The argument given here is equivalent to saying that if the rates of change of f and g are equal, then at each x their tangents must be parallel, so that the curves must look like those in Figure 5-5 – one is just the result of a vertical displacement of the other. This argument is not a conclusive proof, for it might be possible that some (perhaps complicated) function, not a constant, has a zero rate of change. Perhaps we are just not clever enough to have thought of an example of such a function. However, by use of the Mean-Value Theorem (Section 3.1) we can prove that this is not the case:

Theorem 1
If $F'(x) = 0$ over an interval $[a, b]$, then $F(x)$ is constant over that interval.

Proof Let p and q be any two numbers in $[a, b]$. Then

$$\frac{F(q) - F(p)}{q - p} = F'(z) \qquad \text{for some } z \text{ in } (p, q).$$

By hypothesis, $F'(z) = 0$. Hence, $F(q) - F(p) = 0$, or $F(q) = F(p)$. Because p and q are *any* two numbers in $[a, b]$, it must be that $F(x)$ is indeed constant over $[a, b]$.

It now follows that if two functions have the same derivative over an interval, they differ by at most a constant; that is:

Formulas for antiderivatives

Theorem 2

If $f'(x) = g'(x)$ over an interval $[a, b]$, then $f(x) = g(x) + k$, for all x in $[a, b]$, where k is a constant.

5.3 Formulas for antiderivatives

Corresponding to the formulas for differentiation, as summarized in Section 4.16, there are analogous formulas for antidifferentiation. We can develop most of them quickly now. We begin with a specific formula: If $dy/dx = x^n$, what is the expression for y in terms of x? In Section 5.1, as a result of a trial, we found that if $dy/dx = x$, then $y = \frac{1}{2}x^2 + c$.

Problem 1

Use a similar trial method to find a formula for y if $dy/dx = x^2$, if $dy/dx = x^3$, if $dy/dx = x^4$, if $dy/dx = x^{-2}$, if $dy/dx = x^{2/3}$, if $dy/dx = x^{-1/2}$. Conjecture a general result.

The correct conjecture is verified simply by checking that if $y = x^{x+1}/(n+1)$, then $dy/dx = x^n$, and using the result of Section 5.2 that any other function satisfying $dy/dx = x^n$ differs from this one only by a constant.

Problem 2

What happens if you apply the formula stated here to the case $dy/dx = 1/x = x^{-1}$?

Clearly, the formula fails in this case. In a sense, this is not surprising, for we might at first think that $dy/dx = x^{-1}$ comes from differentiating $y = x^0$. But, of course, if $y = x^0$, then $dy/dx = 0 \cdot x^{-1} = 0$.

In Chapter 7 we shall learn that the equation $dy/dx = x^{-1}$ implies that y is a *logarithmic* function of x. For now, we make the following statement:

$$\text{If} \quad \frac{dy}{dx} = x^n, \quad \text{with} \quad n \neq -1, \quad \text{then} \quad y = \frac{1}{n+1}x^{n+1} + c. \quad (1)$$

This is our basic *specific* formula. We also have several *general* formulas for antidifferentiation, as we did for differentiation. We first put into symbols the result of Section 5.2:

$$\text{If} \quad \frac{dy}{dx} = f'(x), \quad \text{then} \quad y = f(x) + c, \quad \text{for an arbitrary constant } c. \quad (2)$$

197

Next, we have

$$\text{If } \frac{dy}{dx} = f'(x) + g'(x), \text{ then } y = f(x) + g(x) + c. \qquad (3)$$

"An antiderivative of the sum of two functions
equals the sum of their antiderivatives."

and

$$\text{If } \frac{dy}{dx} = k \cdot f'(x), \text{ then } y = k \cdot f(x) + c. \qquad (4)$$

"An antiderivative of a constant times a function equals
the constant times an antiderivative of the function."

The analogue of the Chain Rule is straightforward:

$$\text{If } \frac{dy}{dx} = f'(u) \cdot g'(x), \text{ where } u = g(x), \text{ then } y = f(u) + c. \quad (5)$$

The Special Case of the Chain Rule leads to an important formula:

$$\text{If } \frac{dy}{dx} = u^n \cdot g'(x), \text{ where } u = g(x) \text{ and } n \neq -1,$$

$$\text{then } y = \frac{1}{n+1} u^{n+1} + c. \qquad (6)$$

Here are some examples that utilize our antidifferentiation formulas.

Example 1

If $dy/dx = 5$, what is an expression for y? A straightforward approach is to note that the given equation could be written as

$$\frac{dy}{dx} = 5x^0,$$

so that an application of our basic specific formula (1) tells us that

$$y = 5x + c.$$

Alternatively, we could have observed that dy/dx can be interpreted as the slope of a curve, and that one curve that has the slope constantly 5 is the straight line $y = 5x$. Hence, the antiderivatives are $y = 5x + c$.

Example 2

The rate of absorption (R mg/min) by the liver of a certain chemical in the bloodstream of an animal was found to vary with the time (t min) as follows:

$$R = 5 + 6t - 3t^2.$$

How much (Q mg) was absorbed in the first 2 min?

The central feature here is that $R = dQ/dt$, so that Q is an antiderivative of R.

Problem 3

Perform the antidifferentiation, and thus obtain $Q = 5t + 3t^2 - t^3 + c$.

At the start ($t = 0$), none of the chemical had yet been absorbed – Q was then equal to zero. This implies that $c = 0$ for this example, and hence the formula for Q at any time is $Q = 5t + 3t^2 - t^3$.

Problem 4

Verify that at $t = 2$, $Q = 14$. This is the answer.

Example 3

The slope of a certain curve is given by

$$\frac{dy}{dx} = (x^2 - 1)^2 \cdot 2x. \qquad (7)$$

Find an expression for the height of the curve for any x. We shall do the antidifferentiation in two different ways.

Problem 5

Multiply out the right side of equation (7) and perform the antidifferentiation to obtain

$$y = \frac{x^6}{3} - x^4 + x^2 + c. \qquad (8)$$

An alternative approach is to use formula (6) of this section: If we set $u = g(x) = x^2 - 1$, then $du/dx = g'(x) = 2x$. Thus, equation (7) can be written

$$\frac{dy}{dx} = u^2 \cdot g'(x), \quad \text{where} \quad u = g(x) = x^2 - 1. \qquad (9)$$

Applying formula (6) to equation (9) gives

$$y = \tfrac{1}{3}u^3 + k = \tfrac{1}{3}(x^2 - 1)^3 + k. \qquad (10)$$

The letter k is used here because it is not necessarily the same as the c of equation (8).

Problem 6

Expand the right side of equation (10) to get

$$y = \tfrac{1}{3}(x^6 - 3x^4 + 3x^2 - 1) + k. \qquad (11)$$

Equations (8) and (11) are two forms of the solution of the problem. Reconcile them.

Example 4

Formula (6) was not really needed in Example 2, and it was not even much of a help, but here is a case where it is essential. The slope of a certain curve is given by

$$\frac{dy}{dx} = \sqrt{x^2 - 9} \cdot 2x. \tag{12}$$

Find an expression for the height of the curve for any x.

In this case, unlike what was done in Problem 4, we cannot "multiply out" the right side of equation (12). Rather, as was done in the alternative approach to Example 3, we set $u = g(x) = x^2 - 9$, so that $du/dx = g'(x) = 2x$. Hence, equation (12) can be written

$$\frac{dy}{dx} = u^{1/2} \cdot g'(x), \quad \text{where} \quad u = g(x) = x^2 - 9. \tag{13}$$

Problem 7
Verify that applying formula (6) to equation (13) results in $y = \frac{2}{3}(x^2 - 9)^{3/2} + C$.

You probably will have noticed that equation (12) was carefully tailored to make formula (6) applicable to it. Equations like

$$\frac{dy}{dx} = \sqrt{x^2 - 9} \cdot 3x, \tag{14}$$

and

$$\frac{dy}{dx} = \sqrt{x^2 - 9} \cdot x^2 \tag{15}$$

are another story. We shall see later (Section 5.9) how to use formula (6) on equation (14). Equation (15) requires quite a different technique.

Example 5
Remember that the derivative of a product does *not* equal the product of the derivatives. Likewise, we *cannot* multiply the antiderivatives of two functions to obtain antiderivatives of their product. For example, suppose $dy/dx = (x^2 + 1) \cdot x^3 = x^5 + x^3$. Now,

$$\text{an antiderivative of } (x^2 + 1) \text{ is } \quad \frac{x^3}{3} + x,$$

$$\text{an antiderivative of } x^3 \text{ is } \quad \frac{x^4}{4}, \text{ and}$$

$$\text{an antiderivative of } x^5 + x^3 \text{ is } \quad \frac{x^6}{6} + \frac{x^4}{4}.$$

$$\text{But } \left(\frac{x^3}{3} + x\right)\frac{x^4}{4} \neq \frac{x^6}{6} + \frac{x^4}{4}.$$

The only way we can now treat antiderivatives of products is *first* to "multiply out" (if we can), unless formula (5) or (6) applies. (In analogy to the product rule for differentiation, there is a procedure in antidifferentiation called "integration by parts," but we shall not deal with it.) Similarly, for quotients, we must *first* "divide out" (if we can), unless formula (5) or (6) applies.

PROBLEMS

8. Find the antiderivatives of each of the following functions:

   (a) $\frac{2}{3} + \frac{1}{2}x - 3x^2$ (b) $\frac{-2}{x^2} + x + \frac{2}{x^3} - 1$ (c) $3\sqrt{x} - \frac{1}{2\sqrt{x}} + \pi$

   (d) $A + Bx + Cx^2 + Dx^3$, where A, B, C, and D are constants

   (e) $\frac{x^{1/3}}{3} - \frac{2}{x^{2/3}} - \frac{x^{-1/3}}{2}$ (f) $(x-1)(2x-3)$

   (g) $\frac{x^2 - 5}{x^2}$ (h) $\left(\sqrt{x} + \frac{1}{\sqrt[3]{x}}\right)^2$ (i) $x^2\left(1 - \frac{2}{x}\right)$ (j) $\sqrt{x}\,(3x-2)^2$

9. Find the antiderivatives of each of the following functions:

   (a) $4(4x+1)^3$ (b) $\sqrt{25 + x^2} \cdot 2x$

   (c) $(25 + x^2) \cdot 2x$ (Use two methods, and reconcile results.)

   (d) $\frac{2(x-1)}{\sqrt{10 - 2x + x^2}}$ (e) $2x\left(x - \sqrt{9 - x^2}\right)$ (f) $\frac{3(1 - x^2)}{(5 + 3x - x^3)^2}$

   (g) $\frac{(\sqrt{x} - \sqrt{2})^2}{2\sqrt{x}}$ (Use two methods, and reconcile results.)

   (h) $\frac{\sqrt{3x^{1/3} + 1}}{x^{2/3}}$ (i) $2(\sqrt{x^2 + 4x - 3} + 2)(x + 2)$ (j) $\frac{-2x}{\sqrt{16 - x^2}}$

10. If the slope of a curve is given by $dy/dx = x^2 + 1/x^2$, and if $y = 0$ at $x = 1$, find the formula for y in terms of x.

11. Same as Problem 10, with $dy/dx = \frac{1}{2}x^{1/2} + 1/2x^{1/2} + 1$, and $y = 1$ at $x = 9$.

12. For the curve of Example 3 in this section, determine maximum and minimum points and inflectional points.

13. Same as Problem 12 for the curve of Example 4 in this section. Remember to check the domain of the function!

14. The rate of domestic use of coal in a small city (R tons/year) varied with the time (t years after 1930) as follows: $R = 48,000 - 300t^2$. How much coal (G tons) was used between 1930 and 1940?

15. The rate of flow (R gal/day) of water in a stream varied with the time (t days after June 1) as follows: $R = t^3 - 3t^2 - 24t + 1000$. Find the total quantity (Q gal) that flowed between $t = 1$ and $t = 5$.

○ 16. Careful accounting in a small factory manufacturing x items per week shows that the marginal cost ($\$MC$) is given by $MC = 100 - 0.2x$, for $0 \le x \le 150$, and that the marginal revenue ($\$MR$) is given by $MR = 120 - 0.4x$, for $40 \le x \le 150$.

   (a) Find x for maximum weekly profit.

201

(b) If the fixed costs of operation come to $400 per week, find a formula for the total cost of producing x items per week.

(c) Find a formula for the total revenue obtained from selling x items per week.

(d) Express the profit made by manufacturing and selling x items per week.

(e) Use (d) to find x for maximum profit, and thus check your answer to (a). Find the maximum profit.

5.4 Repeated antidifferentiation: projectiles thrown vertically

If we know an expression for the second derivative of a function, one antidifferentiation will give the first derivative of the function, and another antidifferentiation will give the function itself.

Example 1
Suppose that the flexion of a curve is given by $d^2y/dx^2 = 1 - 1/\sqrt{x}$ and that at $x = 4$, $dy/dx = 1$ and $y = 3$. What is the formula for y in terms of x?

Problem 1
Perform one antidifferentiation and use some of the information given to obtain $dy/dx = x - 2x^{1/2} + 1$. Then antidifferentiate again to get $y = \frac{1}{2}x^2 - \frac{4}{3}x^{3/2} + x + \frac{5}{3}$.

Example 2
Suppose that the acceleration of a particle moving on a straight line is given by $d^2s/dt^2 = \frac{5}{4}\sqrt{t}(1-t)$, where t is elapsed time in seconds, and s represents the displacement from a fixed point, in feet. Moreover, suppose that $s = 5$ at $t = 0$ and that $s = 8$ at $t = 1$. What is the formula for s in terms of t?

Problem 2
Perform one antidifferentiation to obtain $ds/dt = \frac{5}{6}t^{3/2} - \frac{1}{2}t^{5/2} + c_1$ and a second antidifferentiation to get $s = \frac{1}{3}t^{5/2} - \frac{1}{7}t^{7/2} + c_1t + c_2$. Use the other given information to find that $c_2 = 5$, $c_1 = \frac{59}{21}$, so that the answer is $s = \frac{1}{3}t^{5/2} - \frac{1}{7}t^{7/2} + \frac{59}{21}t + 5$.

An object not too far above the surface of the earth is attracted so that it has an acceleration of about 32 ft/sec², directed downward. (This is an approximation, neglecting the effects of air friction, varying distance from the surface, etc.) Thus, if the velocity of the object, at $t = 5$, say, is 40 ft/sec (i.e., a speed of 40 ft/sec upward), its velocity at $t = 6$ will be $40 - 32 = 8$ ft/sec in the absence of any forces other than that of gravitational attraction. Similarly, if the velocity of an object at $t = 3$ is 10 ft/sec, its velocity at

$t = 4$ will be $10 - 32 = -22$ ft/sec (i.e., a speed of 22 ft/sec downward); and if the velocity of an object at $t = 12$ is -21 ft/sec, its velocity at $t = 13$ will be $-21 - 32 = -53$ ft/sec. All problems about projectiles thrown vertically can be easily treated by a single method, using antidifferentiation, as shown in the following examples.

Example 3
A rocket is shot straight up from the edge of the top of a building 112 ft above ground level with an initial speed of 96 ft/sec. Find expressions for the velocity (v ft/sec) and the height (y ft) of the rocket above the ground t sec later.

We know that the velocity of the rocket is given by dy/dt and the acceleration by d^2y/dt^2. It is simpler to choose the upward direction as positive. Hence, $d^2y/dt^2 = -32$.

Problem 3
Antidifferentiate twice and use the given information to obtain $y = -16t^2 + 96t + 112$.

Example 4
A stone is hurled straight down from the edge of a cliff 160 ft above a beach, with an initial speed of 48 ft/sec. Find expressions for the velocity (v ft/sec) and the height (y ft) of the stone above the beach t sec later.

Once again, $d^2y/dt^2 = -32$.

Problem 4
Follow the antidifferentiation procedure to obtain $v = -32t - 48$ and $y = -16t^2 - 48t + 160$.

Example 5
We can work out a general result: If a projectile is thrown vertically from a point h ft above ground level with an initial velocity of v_0 ft/sec (v_0 is a constant that might be positive, negative, or zero), then the height (y ft) above the ground t sec later is given by

$$y = h + v_0 \cdot t - 16t^2. \tag{16}$$

Problem 5
Derive this equation.

Each of the terms on the right of equation (16) has a simple interpretation: h, of course, is the *initial* height, $v_0 \cdot t$ is the effect on the height of the *initial* velocity alone (a velocity of 10 ft/sec acting for 3 sec results in a distance covered of $10 \cdot 3 = 30$ ft), and $-16t^2$ gives the effect on the height of the

203

gravitational acceleration alone. The height at any time is the sum of these three components.

PROBLEMS

6. In Example 3 in this section, when is the rocket highest, and what is the maximum height? When does the rocket reach the level of the top of the building in its downward flight? If the rocket just misses the building in its downward flight, when does it hit the ground? With what speed?

7. In Example 4 in this section, when does the stone hit the beach? With what speed?

8. (a) A stone is dropped ($v_0 = 0$) from the edge of a cliff and hits the beach 5 sec later. How high is the cliff? With what speed does the stone hit the beach?

   (b) A stone is hurled straight up from ground level and hits the ground 5 sec later. What was the initial speed of the stone? What was its maximum height?

9. For the curve of Example 1 in this section, find maximum, minimum, and inflectional points. [Hint: $x - 2x^{1/2} + 1 = (x^{1/2} - 1)^2$.]

10. For the motion of the particle in Example 2 in this section, find when the velocity has a maximum value. A minimum value. (Check the domain of the function!)

11. A rocket is shot straight up with an initial speed of 192 ft/sec from an airplane at an altitude of 7500 ft. Find its height above the ground t sec later. When does it pass the 7820-ft level rising? Falling? What is its speed at each of these times? When does it reach its maximum height? What is its maximum height?

12. An object moving on an inclined plane is constantly subject to an acceleration of 15 ft/sec^2, directed downward along the plane.

    (a) If it starts with an initial speed of 10 ft/sec downward along the plane, express the distance (s ft) that it covers in t sec.

    (b) If it starts from the bottom of the plane with an initial speed of 24 ft/sec upward along the plane, how far does it move up the plane?

    ★ (c) If it starts from the bottom of the plane with an initial speed of v_0 ft/sec upward along the plane, what is the value of v_0 if the object just reaches a point 120 ft up the plane?

13. An object sliding on a rough horizontal surface is subject to a constant retarding acceleration of 10 ft/sec^2.

    (a) If it starts with an initial speed of 60 ft/sec, how far does it slide before its speed equals zero?

    ★ (b) What initial speed must it be given to slide 125 ft before its speed equals zero?

★ 14. An object moving on an inclined plane is constantly subject to an acceleration of a ft/sec^2, directed downward along the plane.

(a) If it starts from the bottom of the plane with an initial speed of v_0 ft/sec upward along the plane, how far does it move up the plane?

(b) If it starts from the bottom of the plane with an initial speed upward along the plane, what must that initial speed be if the object is to move b ft up the plane?

15. A block of wood sliding on an inclined plane is subject to gravitational acceleration constantly equal to 16 ft/sec^2, directed *downward* along the plane (Figure 5-6). The block starts 56 ft from the bottom of the plane with a shove that gives it an initial speed of 48 ft/sec *upward* along the plane.

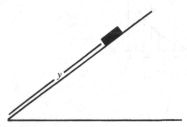

Fig. 5-6

(a) Express the distance (y ft) of the block from the bottom of the plane at time t sec.

(b) What maximum distance from the bottom is reached by the block?

(c) When is the block 96 ft from the bottom, going up? Going down?

(d) When does the block reach the bottom of the plane? With what speed?

5.5   The limit of a sum

We turn now to a completely different type of problem. Then we shall learn that these new problems – those of the integral calculus – have a remarkably close relationship to those we have already encountered. The discovery of this relationship was Newton's major contribution in this field, and the articulation of the relationship – the Fundamental Theorem – links the differential calculus and the integral calculus to form a magnificently unified subject.

We begin by considering a problem that was solved over 2000 years ago by Archimedes, who phrased the result as follows: "The area under a parabolic arch is two-thirds the area of the circumscribing rectangle." Relative to Figure 5-7(a), Archimedes' statement means that the shaded area equals two-thirds the area of rectangle $ABCD$. Because of the obvious symmetry of the figure, this is equivalent to stating that the vertically shaded area in Figure 5-7(b) equals one-third the area of rectangle $ABOO'$.

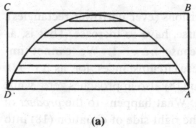
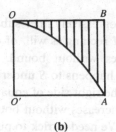

(a)   (b)

Fig. 5-7

5

Antidifferentiation;
integration

Fig. 5-7(c)

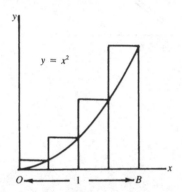

Fig. 5-8

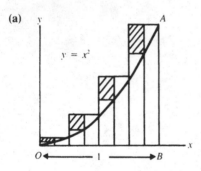

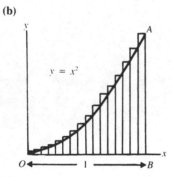

Fig. 5-9

Let us turn the figure upside down, as in Figure 5-7(c), and consider the special case in which the equation of the parabola is $y = x^2$ and the length of OB is 1 unit.

Problem 1
What does this imply about the length of BA, and what is the magnitude of the vertically shaded area, according to Archimedes?

We shall find the vertically shaded area by a method similar in spirit to that of Archimedes. (The method works equally well for parabolas with other equations and for lengths OB different from 1 unit.) We first approximate the desired area by dividing the segment OB into equal parts (four of them in Figure 5-8), erecting verticals up to the curve at the division points, and completing rectangles, as shown. The sum of the areas of these four rectangles is an approximation to the desired area, but it is obviously too large. If we double the number of division points, the sum of the areas of the eight rectangles is also larger than the desired area, but it is a closer approximation than the preceding one – the shaded areas in Figure 5-9(a) were included in the previous approximation, but they have been trimmed away from this one. It appears that we can come as *close as we like* to the desired area by summing the areas of the rectangles of a *sufficiently fine* subdivision, as in Figure 5-9(b). This is reminiscent of the language of limits (see Chapter 2), and we shall return to the idea later. Now let us do some computation. If we divide the unit interval, OB, into n equal parts, each of length Δx, then we have, of course,

$$n \cdot \Delta x = 1. \tag{17}$$

The abscissas of the points of subdivision are $\Delta x, 2\Delta x, 3\Delta x, \ldots, n\Delta x$, as in Figure 5-10; and because the equation of the parabola is $y = x^2$, the heights of the curve at the subdivision points are $(\Delta x)^2, (2\Delta x)^2, (3\Delta x)^2, \ldots, (n\Delta x)^2$.

Problem 2
Show that S, the sum of the areas of the n rectangles, can be written as

$$S = (\Delta x)^3 (1^2 + 2^2 + 3^2 + \cdots + n^2). \tag{18}$$

Now, if we take progressively finer subdivisions (ever narrower rectangles), then the number of rectangles will, of course, have to increase. That is, as $\Delta x \to 0$, n increases without bound, because $n\Delta x = 1$. We cannot immediately tell what happens to S under these circumstances, for, as $\Delta x \to 0$, the first factor on the right side of equation (18) also approaches zero, while the second factor increases without bound. What happens to the *product* of these two factors? We need a trick to put the right side of equation (18) into a form in which we can determine its limit as $\Delta x \to 0$.

Digression A similar situation occurs in every determination of a derivative. For instance, in example 1, **2.2**, in which distance (s ft) varied with time (t sec) according to $s = 16t^2$, the velocity at $t = 1$ is given by

$$v = \lim_{h \to 0} \frac{16(1+h)^2 - 16}{h} = \lim_{h \to 0} \frac{16 + 32h + 16h^2 - 16}{h} = \lim_{h \to 0} \frac{32h + 16h^2}{h}$$

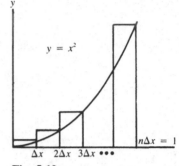

Fig. 5-10

Now, as $h \to 0$, both the numerator and the denominator of the fraction approach zero, and we cannot immediately tell what the limit of the fraction is. (This will always happen for a continuous function, for the derivative of $f(x)$ is defined as $\lim_{h \to 0}\{[f(x+h) - f(x)]/h\}$; and, as $h \to 0$, $f(x+h) \to f(x)$ if f is continuous, so that both numerator and denominator approach zero. Indeed, if the numerator did *not* approach zero as the denominator approached zero, the derivative would not exist.) In the case of $\lim_{h \to 0}[(32h + 16h^2)/h]$, we use an entirely obvious trick – we divide numerator and denominator by h, which is legitimate so long as $h \neq 0$.

Now we return to equation (18). The trick here involves the use of a formula for $1^2 + 2^2 + 3^2 + \cdots + n^2$. The formula is

$$1^2 + 2^2 + 3^2 + \cdots + n^2 = \frac{n(n+1)(2n+1)}{6}. \qquad (19)$$

This is usually derived in algebra courses, often with the help of mathematical induction. We shall assume its validity.

Problem 3

To help make equation (19) seem reasonable, verify its correctness for $n = 1$, 2, 3, 4, and 5.

Problem 4

Multiply out the right side of equation (19) to obtain

$$1^2 + 2^2 + 3^2 + \cdots + n^2 = \tfrac{1}{3}n^3 + \tfrac{1}{2}n^2 + \tfrac{1}{6}n. \qquad (20)$$

Because, from equation (17), $n = 1/\Delta x$, we can write equation (20) as

$$1^2 + 2^2 + 3^2 + \cdots + n^2 = \frac{1}{3} \cdot \frac{1}{(\Delta x)^3} + \frac{1}{2} \cdot \frac{1}{(\Delta x)^2} + \frac{1}{6} \cdot \frac{1}{\Delta x}. \qquad (21)$$

Problem 5

Use equation (21) to put equation (18) into the form

$$S = \tfrac{1}{3} + \tfrac{1}{2}\Delta x + \tfrac{1}{6}(\Delta x)^2. \qquad (22)$$

Now we have an expression for the sum of the areas of the rectangles that permits us to see what happens:

As Δx gets close to zero, S gets close to $\tfrac{1}{3}$. In fact, we can make S arbitrarily close to $\tfrac{1}{3}$ by making Δx sufficiently close to zero.

207

We say, then, that the area bounded by OB, BA, and the parabolic arc OA is $\frac{1}{3}$. Archimedes was right! – at least for the particular parabola we have examined. For the general case, see Problem 7 at the end of this section. Two remarks are in order:

1. Our result could seemingly be abbreviated as $\lim_{\Delta x \to 0} S = \frac{1}{3}$. But there is one difference from the limit situations with which we have dealt before. Previously, in dealing with a derivative, when we have written that $\lim_{h \to 0}(32+16h)=32$ we have thought of the domain of the function, $32+16h$, as being *all real numbers* in the neighborhood of zero. Now, in our area problem, if we write

$$\lim_{\Delta x \to 0} S = \lim_{\Delta x \to 0}\left[\tfrac{1}{3}+\tfrac{1}{2}\Delta x+\tfrac{1}{6}(\Delta x)^2\right]=\tfrac{1}{3}, \tag{23}$$

we must remember that $n\Delta x=1$, so that $\Delta x=1/n$. Because n is a positive integer, the values of Δx are reciprocals of integers, like $\frac{1}{4}$, $\frac{1}{8}$, $\frac{1}{100}$, $\frac{1}{500}$, $\frac{1}{1,000,000}$, and so forth. It still is true that we can make S as close as we like to $\frac{1}{3}$ for all sufficiently small reciprocals of integers, but we should realize that the idea is conceptually a bit different from the previous definition of limit. We shall, however, use the old notation, as in equation (23).

2. The ingenious method of finding the area bounded by OB, BA, and the parabolic arc OA has a serious drawback: It depends on knowing the formula contained in equation (19), and that formula would not help if the curve OA of Figure 5-11 were something other than a parabola. The Newtonian method, which will be presented in Section 5.7, has the tremendous advantage of *generality*.

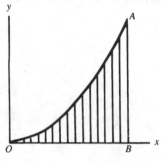

Fig. 5-11

PROBLEMS

6. (a) Compute the values of $1^3+2^3+3^3+\cdots+n^3$ for $n=1, 2, 3, 4$, and 5. Your results should make the following formula seem reasonable:

$$1^3+2^3+3^3+\cdots+n^3=\left[\frac{n(n+1)}{2}\right]^2.$$

  (b) Use the formula of part (a) and the method of this section to find the area (Figure 5-11) bounded by OB, BA, and the arc OA whose equation is $y=x^3$. (As before, $OB=1$.)

★ 7. Use the method of this section to find the area (Figure 5-11) bounded by $OB=c$, BA, and the arc OA whose equation is $y=k\cdot x^2$, where c and k are constants. Check that this area equals one-third the area of the circumscribing rectangle.

★ 8. In deriving the area bounded by OB, BA, and the parabolic arc OA whose equation is $y=x^2$, we constructed rectangles by drawing vertical segments up to the curve *at the right end of each subinterval*.

  (a) Show that if *the left end* were used instead, then equation (18) would become $\bar{S}=(\Delta x)^3[0^2+1^2+2^2+\cdots+(n-1)^2]$. Complete the argument to obtain the same value for $\lim_{\Delta x \to 0} \bar{S}$ as for $\lim_{\Delta x \to 0} S$, namely, $\frac{1}{3}$.

(b) Show that if the *midpoint* of each subinterval were used instead (Figure 5-12), then equation (18) would become

$$\tilde{S} = \Delta x \left[\left(\frac{\Delta x}{2}\right)^2 + \left(\frac{3\Delta x}{2}\right)^2 + \left(\frac{5\Delta x}{2}\right)^2 + \cdots + \left(\frac{(2n-1)\Delta x}{2}\right)^2 \right],$$

or $\quad \tilde{S} = \dfrac{(\Delta x)^3}{4}\left[1^2 + 3^2 + 5^2 + \cdots + (2n-1)^2\right].$

Here is a formula for the sum within square brackets:

$$1^2 + 3^2 + 5^2 + \cdots + (2n-1)^2 = \frac{n(2n-1)(2n+1)}{3}.$$

Check that this formula is valid for $n = 1$, 2, 3, and 4, and use the formula to obtain $\tilde{S} = \frac{1}{3} - \frac{1}{12}(\Delta x)^2$. Hence, $\lim_{\Delta x \to 0} \tilde{S} = ?$

C 9. (a) Refer to Figure 5-9(b). Find an approximate value for the area "under" the parabolic arc by calculating *directly* the sum of the areas of the rectangles if there are 10 of them. Compare your answer with the value given by equation (22) in this case.

  (b) With a programmable calculator or a computer, do the same as (a) for 100 rectangles. For 1000 rectangles.

C 10. (a) With reference to Figure 5-13, calculate directly the sum of the areas of the rectangles, if $\Delta x = 0.1$, and the equation of the curve is $y = x^4$.

  (b) With a programmable calculator or a computer, do the same as (a) if $\Delta x = 0.01$. If $\Delta x = 0.001$.

C 11. (a) and (b) Same as Problem 10, if the equation of the curve is $y = 3x^4$.

C 12. (a) and (b) Same as Problem 10, if the equation of the curve is $y = x^4 + 1$.

C 13. (a) and (b) Same as Problem 10, if the equation of the curve is $y = \sqrt{x}$.

C 14. (a) and (b) Same as Problem 10, if the equation of the curve is $y = \sqrt[3]{x}$.

C 15. (a) With reference to Figure 5-14, calculate directly the sum of the areas of the rectangles, if $a = 1$, $b = 3$, $\Delta x = 0.1$, and the equation of the curve is $y = 10 - x^2$.

  (b) With a programmable calculator or a computer, do the same as (a) if $\Delta x = 0.01$. If $\Delta x = 0.001$.

5.6 Further limits of sums

To find the area bounded by a closed curve, like that in Figure 5-15, we might set up x and y axes and add the areas of a number of component parts of the type shaded. Therefore, we shall consider in detail the area "under" a curve – the area bounded by the x axis, two vertical line segments, and the arc of some curve, whose equation we shall call $y = f(x)$,

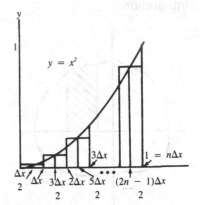

Fig. 5-12

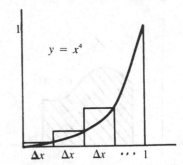

Fig. 5-13

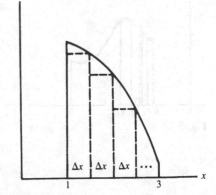

Fig. 5-14

209

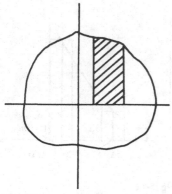

Fig. 5-15

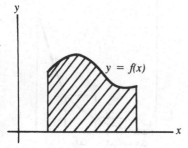

Fig. 5-16

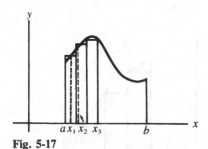

Fig. 5-17

as in Figure 5-16. We shall assume that the curve does not fall below the x axis.

We proceed as we did in the case of the area under the parabola (i.e., by first approximating the desired area as the sum of areas of rectangles). As before, we divide the interval $[a, b]$ on the x axis into n equal subintervals, each of length Δx. Then we construct rectangles whose heights are vertical segments drawn up to the curve, but with a greater generality than before: Instead of drawing the vertical *at the right end of each subinterval*, we draw the vertical *anywhere* in the subinterval. We designate by x_1 the abscissa of the point in the first subinterval at which we draw the vertical, by x_2 the abscissa of the point in the second subinterval, and so forth. In Figure 5-17, x_1 has been chosen close to the midpoint of the first subinterval, x_2 has been chosen quite far to the left in the second subinterval, x_3 has been chosen at the right end of the third subinterval, and so forth. Because the equation of the curve is $y = f(x)$, the heights of the rectangles are $f(x_1), f(x_2), \ldots, f(x_n)$, and the sum, S, of the areas of the rectangles is

$$S = f(x_1) \cdot \Delta x + f(x_2) \cdot \Delta x + f(x_3) \cdot \Delta x + \cdots + f(x_n) \cdot \Delta x,$$

or

$$S = [f(x_1) + f(x_2) + f(x_3) + \cdots + f(x_n)] \Delta x. \qquad (24)$$

As the subdivision gets finer (n increases, and Δx decreases), it appears that S gets closer to what we think of as the *area under the curve* and that, moreover, we approach the same value as $\Delta x \to 0$, no matter where within the respective subintervals we choose $x_1, x_2, \ldots, x_n$. We assume then, that $\lim_{\Delta x \to 0} S$ exists for any choice of the x's, and we define the area under the curve, A, as the value of this limit:

$$A = \lim_{\Delta x \to 0} S = \lim_{\Delta x \to 0} [f(x_1) + f(x_2) + \cdots + f(x_n)] \cdot \Delta x. \qquad (25)$$

As $\Delta x \to 0$, we know that n increases without bound, so that $[f(x_1) + f(x_2) + \cdots + f(x_n)]$ also increases without bound. Thus, we are interested in the product of two factors, one of which is "blowing up," and the other is approaching zero. Another way of thinking of the situation is to observe that we are adding the areas of more and more rectangles, each of which is getting thinner and thinner. We still have no general method of finding the limit.

Before we find the limit, we examine another problem that serves to illustrate that a limit of the sort appearing in equation (25) is encountered not just in defining areas under curves: As in Figure 5-18, consider the line segment $y = x$ from the origin to the point $(1,1)$, and revolve it about the x axis, generating a cone (Figure 5-19). We proceed to find the volume within this cone by dividing its altitude (along the x axis) into n equal parts, each of length Δx. Thus, the abscissas of the division points are $\Delta x, 2\Delta x, 3\Delta x, \ldots, n\Delta x \ (=1)$. At each division point we take a plane section perpendicular to the x axis, cutting a circle from the cone. On each circle we erect a cylinder of height Δx, as in Figure 5-20. The sum of the volumes within these cylinders is an approximation to the volume within the cone, and we assume that the volume within a cylinder is known.

For $x_1, x_2, \ldots, x_n$, we are choosing the right ends of the various subintervals. Because it was the line $y = x$ we revolved to generate the cone, the various radii of the bases of the cylinders are also $x_1, x_2, \ldots, x_n$. Hence, the areas of the bases of the cylinders are $\pi x_1^2, \pi x_2^2, \ldots, \pi x_n^2$, and the volumes within these cylinders are $\pi x_1^2 \cdot \Delta x, \pi x_2^2 \cdot \Delta x, \ldots, \pi x_n^2 \cdot \Delta x$. Therefore, $\mathscr{S}$, the sum of the volumes within these cylinders, is given by

$$\mathscr{S} = \pi x_1^2 \Delta x + \pi x_2^2 \Delta x + \cdots + \pi x_n^2 \Delta x$$

$$= \left(\pi x_1^2 + \pi x_2^2 + \cdots + \pi x_n^2 \right) \Delta x.$$

Note that this is a sum of the sort expressed by equation (24).

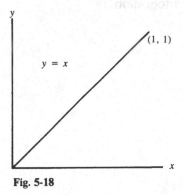

Fig. 5-18

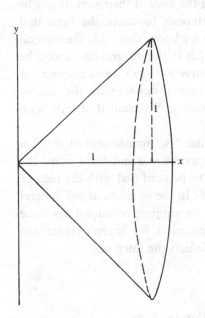

Fig. 5-19   Fig. 5-20

Problem 1
What is $f(x)$ in this case?

As $\Delta x \to 0$, we obtain progressively shorter cylinders, but more and more of them, and we appear to be getting closer to what we think of as the volume within the cone. Indeed, we define V, the volume within the cone, as the limit of the sum of the volumes within the cylinders, as $\Delta x \to 0$:

$$V = \lim_{\Delta x \to 0} \mathscr{S} = \lim_{\Delta x \to 0} \left[\pi x_1^2 + \pi x_2^2 + \cdots + \pi x_n^2 \right] \cdot \Delta x.$$

211

Note that this is very similar to equation (25). Thus, we have an example of *a limit of a sum* of the sort appearing in equation (25), not only in the context of an *area under a curve* but also in the context of a *volume*. We shall find many other examples of the same sort of limit of a sum – so many that it has been given a name, *the integral*. Here is the formal statement:

> Let $f(x)$ be defined over the interval $[a, b]$, which is divided into n subintervals, each of length Δx. Choose $x_1, x_2, \ldots, x_n$ arbitrarily within the first, second, $\ldots, n$th subintervals. Then, $\lim_{\Delta x \to 0} [f(x_1) + f(x_2) + \cdots + f(x_n)] \Delta x$, provided this limit exists, is symbolized as $\int_a^b f(x)\, dx$, which is read, "the integral, from a to b, of $f(x)\, dx$."

Note that this statement serves just as a definition of the *integral*, a word with the connotation of "wholeness" – we make the whole by summing up lots of constituent bits, or, rather, by taking the *limit* of their sum. It can be shown that, at least when $f(x)$ is a continuous function, the limit that appears in the foregoing statement exists, so we know, then, that the integral of a continuous function exists. Because, if $f \geq 0$, the integral can always be *interpreted* as an area – the area under the curve $y = f(x)$ – and because our intuition tells us that the area under a continuous curve exists, the conclusion of the preceding sentence is not surprising. We shall deal later with integrals of functions that are negative.

The integral sign, $\int$, is an elongated capital "S," reminiscent of the *sum* involved in the definition of integral. The quantities a and b are called the *lower* and *upper limits of integration* – not to be confused with the idea of the limit of a function. The appearance of dx in the notation of the integral is a vestige of the historical development of the subject; because it has some usefulness, as we shall see, we preserve the notation. We haven't yet learned how to evaluate, in general, the limit that defines the integral.

PROBLEMS

2. Return to the example of the volume within the cone:

$$V = \lim_{\Delta x \to 0} \mathscr{S} = \lim_{\Delta x \to 0} \left[\pi x_1^2 + \pi x_2^2 + \cdots + \pi x_n^2 \right] \Delta x.$$

(a) Find expressions for $x_1, x_2, \ldots, x_n$ in terms of Δx.

(b) Use the result of part (a) to show that $V = \lim_{\Delta x \to 0} \mathscr{S} = \lim_{\Delta x \to 0} \pi \cdot S$, where S is the sum of the areas of rectangles – equation (18), **5.5**.

(c) Knowing, as in Section 5.5, that $\lim_{\Delta x \to 0} S = \frac{1}{3}$, find V, and check it by use of the formula for the volume of a cone: "one-third the area of the base times the altitude."

★ 3. Use the method of this section to show that the volume, V, within a cone of base radius r and altitude h is given by $V = \frac{1}{3}\pi r^2 h$. [Hint: Revolve the line segment $y = (r/h)x$ from $(0,0)$ to (h, r) about the x axis to obtain the cone, and divide the altitude into n equal parts, each of length Δx, so that $n\Delta x = h, \ldots .$]

C 4. (a) Refer to Figure 5-11. Find an approximate value of the area "under" the parabolic arc, assuming that $OB = 2$ and that the equation of the curve is $y = 3x^2$, by calculating directly the sum of the areas of the rectangles if there are 10 of them.

(b) With a programmable calculator or a computer, make the calculation of (a) if there are 100 rectangles. If there are 1000 rectangles.

C 5. Refer to Figure 5-20. Find an approximate value of the volume of the cone by calculating the sum of the volumes of the cylinders if there are 10 of them. If there are 20 of them.

6. Consider the area bounded by $y = 5 + 6x - x^2$, the x axis, $x = 0$, and $x = 6$. Divide the interval into six equal parts (i.e., $\Delta x = 1$) and erect ordinates to the curve at the division points. What is the sum of the areas of rectangles whose heights are the heights of the curve at

(a) the right end points of the subintervals?

(b) the left end points of the subintervals?

(c) the midpoints of the subintervals?

C 7. (a)–(c) Same as Problem 6, if the interval $[0,6]$ is divided into 60 equal parts (i.e., $\Delta x = 0.1$).

5.7 The Fundamental Theorem

In the preceding section we made definitions leading to

$$A = \lim_{\Delta x \to 0} \left[f(x_1) + f(x_2) + \cdots + f(x_n) \right] \Delta x = \int_a^b f(x)\, dx, \quad (26)$$

where A is the area under the curve $y = f(x)$ between $x = a$ and $x = b$, with $f \geq 0$ (Figure 5-21). We shall now approach the evaluation of A in a different way, thus showing the reasonableness of a result known as the Fundamental Theorem of the calculus. We restrict our attention to non-negative functions. Suppose, then, in evaluating the area A mentioned above, we think first of the area under the curve $y = f(x)$ from $x = a$ to some arbitrary x: Let us call this $A(x)$, recognizing that to each value of x corresponds just one value of the area under the curve from a to that x – hence, $A(x)$ is a function of x, as the notation suggests (Figure 5-22). Because $A(b)$ is the area under the curve from $x = a$ to $x = b$, we know that $A(b) = A$, as defined earlier.

Problem 1

What is the value of $A(a)$?

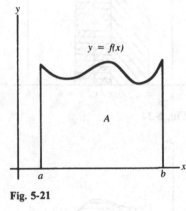

Fig. 5-21

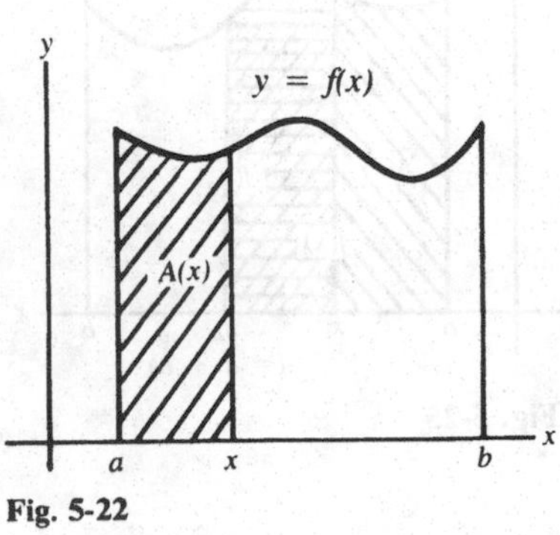

Fig. 5-22

213

5
Antidifferentiation; integration

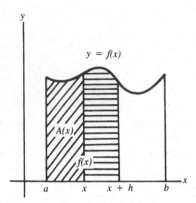

Fig. 5-23

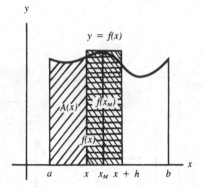

Fig. 5-24

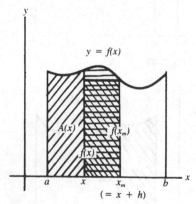

Fig. 5-25

214

Now let us increase x by an amount h. The expression $A(x+h) - A(x)$ is seen as the horizontally shaded area in Figure 5-23. We need bounds on the value of $A(x+h) - A(x)$.

In the closed interval between x and $x+h$, the continuous function $f(x)$ has a maximum value, attained, say, at x_M. Then the maximum value of $f(x)$ in the interval is $f(x_M)$. Clearly, the horizontally shaded area of Figure 5-24 is not greater than the area of the rectangle of height $f(x_M)$ and width h:

$$A(x+h) - A(x) \le f(x_M) \cdot h. \tag{27}$$

Likewise, in the same closed interval the continuous function $f(x)$ has a minimum value, attained, say, at x_m. (In Figure 5-25, x_m happens to coincide with $x+h$, the right end of the interval $[x, x+h]$, but that is not significant.) Then the minimum value of $f(x)$ in the interval is $f(x_m)$.

Problem 2
Write a relation involving $f(x_m)$ analogous to the inequality (27).

Putting (27) together with the result of Problem 2, we have

$$f(x_m) \cdot h \le A(x+h) - A(x) \le f(x_M) \cdot h.$$

We divide through by the positive quantity h, obtaining

$$f(x_m) \le \frac{A(x+h) - A(x)}{h} \le f(x_M). \tag{28}$$

As h gets smaller, so long as $h \ne 0$, the continued inequality (28) remains valid. Now look at Figure 5-24 and imagine h approaching zero. It should be clear, because f is a continuous function, that $\lim_{h \to 0} f(x_M) = f(x)$. Likewise, as we see from Figure 5-25, $\lim_{h \to 0} f(x_m) = f(x)$. Thus, as $h \to 0$, the two end terms of the continued inequality (28) approach the *same* quantity, $f(x)$. But note that the middle term of the continued inequality is always pinned between the end terms. Hence, we conclude that

$$\lim_{h \to 0} \frac{A(x+h) - A(x)}{h} \quad \text{also equals} \quad f(x).$$

But remember the definition of derivative: $\lim_{h \to 0} [A(x+h) - A(x)]/h = A'(x)$. We have arrived at an important conclusion:

$$A'(x) = f(x). \tag{29}$$

We phrase this equation as follows: "If $A(x)$ is the area under the graph of $y = f(x)$ from $x = a$ to an arbitrary value of x, the rate of change of $A(x)$ with respect to x equals $f(x)$."

As we learned in Section 5.1, we can find $A(x)$ from equation (29) by *antidifferentiation*. Suppose that $F(x)$ is an antiderivative of $f(x)$. Then,

$$A(x) = F(x) + c. \tag{30}$$

From Problem 1 we know that $A(a) = 0$.

Problem 3

Use this fact and equation (30) to show that $A(x) = F(x) - F(a)$.

Putting $x = b$, we have

$$A(b) = A = F(b) - F(a). \qquad (31)$$

Here we have an expression for A. Equation (26) provided an alternative expression for A. Equating the two gives us the following:

The Fundamental Theorem of the calculus

If $f(x)$ is a continuous function over the interval $[a, b]$, then

$$\int_a^b f(x)\, dx = F(b) - F(a), \text{ where } F(x) \text{ is an antiderivative of } f(x).$$

This beautiful result, linking the limit of a sum to antidifferentiation, is a tool of extraordinary power. No longer are we required to develop special formulas, like that for $1^2 + 2^2 + 3^2 + \cdots + n^2$, in order to find an integral – rather, the whole kit of formulas resulting from our experience with differentiation is at our command to use in integration problems.

We illustrate the power of the Fundamental Theorem by applying it to find the area under the parabola $y = x^2$ from $x = 0$ to $x = 1$. We have

$$A = \int_0^1 x^2\, dx.$$

We know that if $f(x) = x^2$, then an antiderivative is $F(x) = x^3/3$.

Problem 4

Evaluate $F(1)$ and $F(0)$ to obtain $A = \frac{1}{3}$.

Problem 5

Another antiderivative is $G(x) = x^3/3 + 7$. Evaluate $G(1)$ and $G(0)$ to verify that $A = G(1) - G(0)$.

We see that, in this case, it does not matter whether we use the antiderivative $F(x)$ or the antiderivative $G(x)$. In general, if $H(x) = F(x) + c$, then $H(b) = F(b) + c$, and $H(a) = F(a) + c$, so

$$H(b) - H(a) = F(b) + c - [F(a) + c] = F(b) - F(a).$$

Hence, in applying the Fundamental Theorem, we may as well use the simplest antiderivative we can find.

The discussion in this section has not been called a *proof* of the Fundamental Theorem – rather, it has been called an "argument" showing the "reasonableness" of the statement of that theorem. For a *proof*, it would be necessary to demonstrate that the concept of area defined as a limit of a sum is actually identical with the "growing" area that led to $A'(x) = f(x)$.

Likewise, it would be necessary to prove the "Pinching Theorem": If $r(z) \le s(z) \le t(z)$, and if $\lim_{z \to a} r(z) = l$, $\lim_{z \to a} t(z) = l$, then $\lim_{z \to a} s(z)$ also equals l. This was implicitly used in the last step leading to $A'(x) = f(x)$. There are still other gaps that would have to be filled to have a genuine proof, but the heuristic argument presented should be sufficient for our needs.

PROBLEMS

6. Find the area under $y = x^3$ from $x = 0$ to $x = 1$, and compare your result with what you found in Problem 6(b), **5.5**. Compare the amount of work, too!

7. Find the area under $y = 4 - 2x + x^2$ between the given limits:
   (a) $x = 0$ and $x = 3$ (b) $x = 1$ and $x = 4$ (c) $x = -2$ and $x = 1$

8. Find the area under $y = x^3 - 6x^2 + 9x$ between the given limits:
   (a) $x = 0$ and $x = 2$ (b) $x = 2$ and $x = 4$ (c) $x = 1$ and $x = 3$

9. (a)–(c) Same as Problem 8, for $y = x^3 - 6x^2 + 9x + 5$.

10. Find the area under $y = 2/x^2$ between the given limits:
   (a) $x = 1$ and $x = 5$ (b) $x = 1$ and $x = 10$
   (c) $x = 0.1$ and $x = 1$ (d) $x = -3$ and $x = -1$
   (e) $x = -2$ and $x = -0.1$

11. Find the area under $y = \sqrt{9 + x^2} \cdot 2x$ between $x = 0$ and $x = 4$.

12. Find the area under $y = 2x/\sqrt{9 + x^2}$ between $x = 0$ and $x = 4$.

○ 13. (a) Use the Fundamental Theorem to show that if $a < c < b$, then $\int_a^c f(x)\,dx + \int_c^b f(x)\,dx = \int_a^b f(x)\,dx$. Give a geometric interpretation of this result.

   (b) Because, in the definition of the integral, $[a, b]$ is an interval, which implies that $a < b$, there is no significance to $\int_a^a f(x)\,dx$ or to $\int_b^a f(x)\,dx$. Show, however, that a blind application of the Fundamental Theorem yields

$$\int_a^a f(x)\,dx = 0 \quad \text{and} \quad \int_b^a f(x)\,dx = -\int_a^b f(x)\,dx.$$

   We shall choose these equations as defining the hitherto undefined symbols $\int_a^a f(x)\,dx$ and $\int_b^a f(x)\,dx$.

   (c) Use part (b) to show that in part (a) the condition $a < c < b$ can be ignored.

14. Sketch a graph of $y = 5 + 6x - x^2$, $0 \le x \le 6$, and approximate the area under this curve by adding up the areas of six rectangles, each of base 1, with heights being
   (a) the maximum value of y in each subinterval;
   (b) the minimum value of y in each subinterval;
   (c) the value of y at the midpoint of each subinterval.

c 15. (a)–(c) Same as Problem 14, with 12 rectangles, each of base $\frac{1}{2}$.

C 16. (a)–(c) With a programmable calculator or a computer, same as Problem 14, with 60 rectangles, each of base 0.1.

C 17. (a)–(c) With a programmable calculator or computer, same as Problem 14, with 600 rectangles, each of base 0.01.

18. Find the exact area under the curve of Problem 14.

19. Draw figures like Figure 5-24 and Figure 5-25 for the case in which $h < 0$ to deduce that

$$f(x_m) \cdot (-h) \le A(x) - A(x+h) \le f(x_M)(-h)$$

and hence that the inequalities (28) hold for *all* small $h \ne 0$.

5.8 Applications of the Fundamental Theorem

Appreciation of the Fundamental Theorem follows from an understanding of how it really applies to a host of diverse cases, and this understanding comes only with lots of practice. We shall start that practice now.

Example 1

Find the area under the curve $y = 5 + 6x - 3x^2$ from $x = 0$ to $x = 2$.

We have already solved problems more complicated than this one, but we want to review the ideas and to introduce some convenient notation. The desired area can be approximated by the sum of the areas of rectangles, one of which is shown in Figure 5-26. The area of the shaded rectangle equals $y \cdot \Delta x = (5 + 6x - 3x^2)\Delta x$. The desired area is actually the *limit* of the sum of such rectangles, as $\Delta x \to 0$; that is,

$$A = \int_0^2 (5 + 6x - 3x^2)\, dx.$$

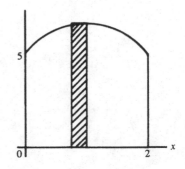

Fig. 5-26

An antiderivative of $5 + 6x - 3x^2$ is $5x + 3x^2 - x^3$. A useful notation is shown now:

$$A = \int_0^2 (5 + 6x - 3x^2)\, dx = (5x + 3x^2 - x^3)\big|_0^2$$

$$= (5 \cdot 2 + 3 \cdot 2^2 - 2^3) - (5 \cdot 0 + 3 \cdot 0^2 - 0^3) = 14 \text{ square units.}$$

The symbol "$|_0^2$" means that the function that precedes it is to be evaluated at 2, and then at 0, and the latter value is to be subtracted from the former. Thus, if $F(x)$ is an antiderivative of $f(x)$, we have

$$\int_a^b f(x)\, dx = F(x)\big|_a^b = F(b) - F(a).$$

Compare this example with Example 2, 5.3.

217

Problem 1

Follow the method and symbolism of the foregoing Example 1 to find the area under the curve $y = 6x - 3x^2$ from $x = 0$ to $x = 3$.

You should have found that $A = 0$ square units. This doesn't seem to make sense – let's investigate it. The graph of $y = 6x - 3x^2$ over the interval $[0, 3]$ appears in Figure 5-27. Because in these discussions Δx is always positive, and because y has negative values between $x = 2$ and $x = 3$, the expression $y \cdot \Delta x$ will likewise be negative there, and the same is true of the sum of such expressions and, finally, of the limit of their sum. In other words, $\int_2^3 (6x - 3x^2)\, dx$ has a negative value.

Problem 2

Verify that $\int_2^3 (6x - 3x^2)\, dx = -4$. Also verify that $\int_0^2 (6x - 3x^2)\, dx = 4$.

We see that the integral from $x = 2$ to $x = 3$ just "cancels out" the integral from $x = 0$ to $x = 2$. If we want the shaded area in Figure 5-28, we observe that the portion above the x axis equals 4 square units, and the portion below the x axis also equals 4 square units, so that the total area is 8 square units.

Whenever we want the area between a curve $y = f(x)$ and the x axis, we should check whether $f(x)$ assumes negative values in the interval under discussion; if it does, we must proceed as we did above.

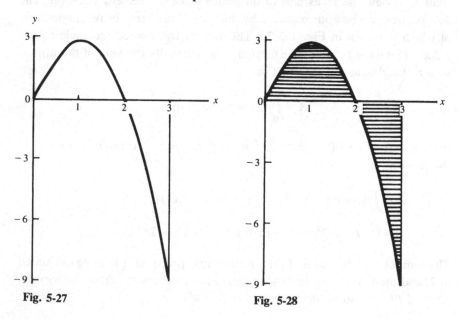

Fig. 5-27   Fig. 5-28

Example 2

Find the area bounded by the curves $y = f(x) = 30 - 14x + 2x^2$ and $y = g(x) = 18 + x - x^2$.

Problem 3

Verify that $f(x)$ has a minimum of $\frac{11}{2}$ at $x=\frac{7}{2}$ and that $g(x)$ has a maximum of $\frac{73}{4}$ at $x=\frac{1}{2}$.

A rough sketch of the curves appears in Figure 5-29. The desired area can be approximated by the sum of the areas of rectangles, one of which is shown in the sketch. The area of the shaded rectangle equals $[g(x)-f(x)]\Delta x=[18+x-x^2-(30-14x+2x^2)]\Delta x$. The desired area is actually the *limit* of the sum of the areas of such rectangles, as $\Delta x \to 0$.

Problem 4

Verify that $A=(-12x+\frac{15}{2}x^2-x^3)|_a^b$. To complete the problem, we must find where the curves intersect, so that we will know a and b.

Problem 5

Solve the equations of the curves simultaneously to find that the abscissas of the points of intersection are 1 and 4.

Problem 6

Verify that $A=\frac{27}{2}$ square units.

Example 2 could also be solved by finding the area under $y=g(x)$ from $x=1$ to $x=4$ and subtracting from it the area under $y=f(x)$ between the same limits.

Problem 7

Compute $\int_1^4 g(x)\,dx$ and $\int_1^4 f(x)\,dx$, and subtract to check the value of A found in Problem 6.

Example 3

Every horizontal section of a steeple is a rectangle with dimensions y ft and z ft given in terms of the vertical distance (x ft) of the section below the top, as follows: $y=\frac{1}{3}x+1$, $z=x^{1/2}$. What is the volume within the steeple from the top down to $x=36$? The steeple is roughly sketched in Figure 5-30(a). If we consider the portion of the steeple between sections taken at distances x and $(x+\Delta x)$ below the top, we have something that looks like Figure 5-30(b). The volume of this portion is approximated by a rectangular slab, as in Figure 5-30(c). The volume, ΔV, of the rectangular slab is given by

$$\Delta V = y \cdot z \cdot \Delta x = \left(\tfrac{1}{3}x+1\right) \cdot x^{1/2} \cdot \Delta x.$$

The volume of the entire steeple is approximated by the sum of the volumes of such slabs, and the volume *equals* the *limit* of the sum of the volumes of such slabs, as their thickness, Δx, approaches zero. In other words,

$$V = \int_0^{36} \left(\tfrac{1}{3}x+1\right)x^{1/2}\,dx.$$

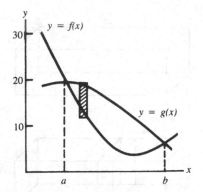

Fig. 5-29

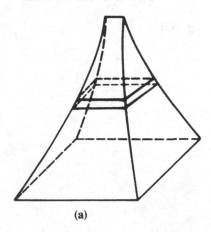

(a)

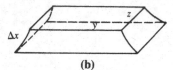

(b)

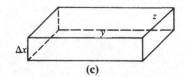

(c)

Fig. 5-30

219

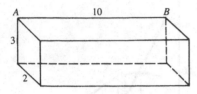

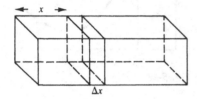

Fig. 5-31

Problem 8

Multiply out the factors of the *integrand* (the function to be integrated), and perform the integration to obtain $V = \frac{5904}{5} = 1180.8 \ (\text{ft}^3)$.

Example 4

A beam of length 10 in. has a uniform cross section of 2 in. by 3 in. (Figure 5-31). The beam has a density (ρ oz/in.^{3}) that varies with the distance (x in.) from the left end of the bar, as follows: $\rho = 2 + \frac{1}{3}x$. What is the weight of the bar?

If we consider the portion of the bar contained between sections at distances x and $(x + \Delta x)$ from the left end, the volume, ΔV, of this slab is given exactly by

$$\Delta V = 2 \cdot 3 \cdot \Delta x = 6\Delta x \ (\text{in.}^3),$$

and the weight, ΔW, of this slab is given *approximately* by

$$\Delta W = \rho \cdot \Delta V = \left(2 + \tfrac{1}{3}x\right) \cdot 6\Delta x \ (\text{oz}).$$

This is not exact because the density varies somewhat within the slab – although the density at the left face of the slab is $2 + \frac{1}{3}x$, by the time we get to the right face the density is $2 + \frac{1}{3}(x + \Delta x)$. But, as usual, we are going to take thinner and thinner slabs, and we shall get the exact weight by taking the *limit* of the sum of the approximate weights of the slabs, as $\Delta x \to 0$. In other words, the weight (W oz) of the bar is given by

$$W = \int_0^{10} \left(2 + \tfrac{1}{3}x\right) 6 \, dx.$$

Problem 9

Verify that $W = 220$ (oz).

Problem 10

What is the density of the bar at the left end? At the right end? Because the formula for the density is $\rho = 2 + \frac{1}{3}x$, we know that the density increases at a *uniform rate* from left to right. What, then, do you think to be the *average density* of the bar? What is the volume of the bar? Using your value of average density, what do you get for the weight of the bar?

Example 5

In a "nitrogen-washout" experiment with a patient, it is found that the concentration of nitrogen ($y\%$) in the gas that he expires from his lungs varies with the total volume (x ml) of gas that he has expired since he started to breathe a nitrogen-free atmosphere, as follows: $y = 80 - 1000x + 30{,}000x^2$. What volume, V_N, of nitrogen does he expire when x goes from 0 to 0.01?

If the concentration of nitrogen were 60%, say, in an expired volume of gas of 0.001 ml, the volume of nitrogen expired would be $\frac{1}{100}(60)(0.001)$ ml.

Similarly, if the volume of gas expired were to increase by a small Δx from x to $(x+\Delta x)$, the volume of nitrogen expired would be approximately $\frac{1}{100}(80-1000x+30{,}000x^2)\Delta x$.

The entire volume of nitrogen expired as x goes from 0 to 0.01 would be the *limit* of the sum of such volumes, as $\Delta x \to 0$; that is,

$$V_N = \int_0^{0.01}(0.80-10x+300x^2)\,dx.$$

Problem 11
Verify that $V_N = 0.0076$ (ml).

In all these cases, the method is essentially the same: We first approximate a small part of the desired quantity (area, volume, weight, volume of nitrogen), and then express the whole of the desired quantity as the *limit* of a *sum* of the approximations of the parts; this is an *integral*, which we evaluate by antidifferentiation. In subsequent chapters we shall encounter further applications of the integral.

PROBLEMS

12. Find the area under the curve $y = x^3 - 8x^2 + 15x + 6$, above the x axis, between $x = 0$ and $x = 4$.

13. Horizontal sections of a steeple are rectangles whose lengths (l ft) and widths (w ft) vary thus with the distance (x ft) below the top: $l = \frac{2}{7}(x + 6)$; $w = x^{1/2}$. Find the volume within the steeple if it is 25 ft high.

14. The density (ρ g/cm^3) of a bar whose cross section is a rectangle 2 cm $\times$ 3 cm is proportional to the distance (x cm) from one end of the bar, and $\rho = 6$ at $x = 2$. Find the weight of the bar if it is 5 cm long.

15. Sketch the curves $y = x^2$ and $y = x^3$ on the same set of axes, and find the area of the region bounded by these curves. (The curves meet in only two points, and there is only one region that is bounded by the curves.)

★ 16. Every horizontal section of a solid is a ring between two concentric circles whose radii (R, r ft) vary thus with the distance (x ft) above the lowest point: $R = \sqrt{x}$, $r = x^2$ (Figure 5-32). Find the volume between $x = 0$ and $x = 1$.

17. The region bounded by $y = x^2$, the x axis, $x = 0$, and $x = 2$ is revolved about the x axis. Find the volume of the solid so generated.

★ 18. The horizontal base of a solid is a circle of radius 5 in. Every vertical section perpendicular to one diameter of the base is an isosceles triangle whose altitude equals $\frac{3}{2}$ its base. Find the volume of the solid.

19. Horizontal sections of an observatory tower 36 ft tall are squares whose sides (s ft) vary thus with the distance (x ft) of the section below the top of the tower: $s = 5 + \frac{1}{2}\sqrt{x}$. Find the volume within the tower.

Fig. 5-32

221

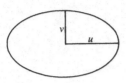

Fig. 5-33

20. The density (ρ g/cm³) of a bar whose cross section is 1 cm × 2 cm varies with the distance (x cm) from one end of the bar as follows: ρ varies as x^2, and $\rho = 12$ at $x = 2$. Find the weight of the bar if it is 5 cm long.

21. Make a rough sketch of the curve $y = (x+1)(x-2)$. Find the area between this curve and the x axis over the interval $x = -2$ to $x = 2$.

22. An ellipse is a squashed circle that looks like Figure 5-33. The area within such an ellipse, with "semiaxes" u and v, is $\pi \cdot u \cdot v$. Find the volume within an elliptical cone of height 6 in. – a surface whose every horizontal section x in. below the vertex is an ellipse with semiaxes given by $u = 2x$, $v = \frac{1}{2}x + \frac{1}{6}x^2$.

23. The area under the curve $y = 3/x$ between $x = 1$ and $x = 2$ is rotated through 360° about the x axis. Find the volume so generated.

24. Evaluate $\int_1^3 [(1-x^2)^2/x^2]\,dx$.

5.9 Use of the Chain Rule in integration (antidifferentiation)

In Example 4, **5.3**, we began with

$$\frac{dy}{dx} = \sqrt{x^2 - 9} \cdot 2x \qquad (32)$$

and antidifferentiated to obtain y in the following way: We set $u = g(x) = x^2 - 9$, so that $du/dx = g'(x) = 2x$. Then equation (32) can be written

$$\frac{dy}{dx} = u^{1/2} \cdot \frac{du}{dx}. \qquad (33)$$

The general expression of the Chain Rule can be put

$$\frac{dy}{dx} = \frac{dy}{du} \cdot \frac{du}{dx}. \qquad (34)$$

Comparison of equations (33) and (34), term by term, gives

$$\frac{dy}{du} = u^{1/2}.$$

Straightforward antidifferentiation of this last equation leads to

$$y = \tfrac{2}{3}u^{3/2} + c,$$

and substitution of $(x^2 - 9)$ for u provides the answer:

$$y = \tfrac{2}{3}(x^2 - 9)^{3/2} + c.$$

You have already encountered many problems of this type, notably all ten parts of Problem 9, **5.3**.

Example 1
Let us now do essentially the problem of equation (32) in the context of integration by evaluating $\int_3^5 \sqrt{x^2 - 9} \cdot 2x\,dx$.

As before, we set $u = x^2 - 9$, which implies that $du/dx = 2x$, and $du = 2x\,dx$. In writing the integral (which we shall abbreviate as $\mathscr{I}$) in terms of u, we must remember that the given limits of integration refer to x. One way to make this clear goes thus:

$$\mathscr{I} = \int_3^5 \sqrt{x^2 - 9} \cdot 2x\,dx = \int_{x=3}^{x=5} u^{1/2}\,du = \left.\tfrac{2}{3} u^{3/2}\right|_{x=3}^{x=5}$$

$$= \left.\tfrac{2}{3}(x^2 - 9)^{3/2}\right|_3^5 = \tfrac{2}{3}\left[(16)^{3/2} - 0\right] = \tfrac{2}{3} \cdot 64 = \tfrac{128}{3}.$$

A slightly different approach is to observe that at $x = 3$, $u = 0$; and at $x = 5$, $u = 16$. Hence,

$$\mathscr{I} = \int_3^5 \sqrt{x^2 - 9} \cdot 2x\,dx = \int_0^{16} u^{1/2}\,du = \left.\tfrac{2}{3} u^{3/2}\right|_0^{16} = \tfrac{2}{3}(16)^{3/2} - 0 = \tfrac{128}{3}.$$

Example 2

We now evaluate an integral that is not quite in the "perfect" form of Example 1, by considering $\int_3^5 \sqrt{x^2 - 9}\, x\,dx$. As before, we set $u = x^2 - 9$, so that $du = 2x\,dx$. We are missing the needed factor 2 in the given integral. But we can get what we need by writing

$$J = \int_3^5 \sqrt{x^2 - 9}\, x\,dx = \int_3^5 \sqrt{x^2 - 9}\, \tfrac{1}{2} \cdot 2x\,dx$$

$$= \tfrac{1}{2} \int_3^5 \sqrt{x^2 - 9} \cdot 2x\,dx.$$

Multiplying by $\tfrac{1}{2}$ *and* by 2 does not change the value of the integral, of course; and bringing the factor out in front of the integral sign is valid because

$$\int_a^b k \cdot f(x)\,dx = k \int_a^b f(x)\,dx, \quad \text{for any } constant\ k.$$

Thus,

$$J = \tfrac{1}{2} \int_3^5 \sqrt{x^2 - 9}\, 2x\,dx = \tfrac{1}{2} \int_0^{16} u^{1/2}\,du$$

$$= \left.\tfrac{1}{2} \cdot \tfrac{2}{3} u^{3/2}\right|_0^{16} = \tfrac{1}{3}\left[(16)^{3/2} - 0\right] = \tfrac{64}{3}.$$

Example 3

For the integral $\int_3^5 \sqrt{x^2 - 9}\, 3x\,dx$, we begin by bringing the unneeded factor 3 out in front:

$$\int_3^5 \sqrt{x^2 - 9}\, 3x\,dx = 3\int_3^5 \sqrt{x^2 - 9}\, x\,dx,$$

and then we proceed as in Example 2.

PROBLEMS

Evaluate each of the following:

1. $\int_0^4 \frac{\sqrt{x^2+9}\,x\,dx}{2}$ 2. $\int_0^4 \frac{-x\,dx}{(x^2+9)^3}$ 3. $\int_0^4 \frac{6x\,dx}{\sqrt{x^2+9}}$

4. $\int_1^5 \frac{t+10}{\sqrt{t^2+20t+4}}\,dt$ 5. $\int_1^5 \frac{3t+30}{2(t^2+20t+4)^2}\,dt$

6. $\int_1^5 (t^2+20t+4)^{-3/2}\left(\frac{t}{5}+2\right)dt$ 7. $\int_0^2 (x^3+1)^{1/2}\frac{x^2}{2}\,dx$

8. $\int_2^3 \frac{4x\,dx}{(1-x^2)^2}$ 9. $\int_{-3}^{-2} \frac{x\,dx}{(1-x^2)^{1/3}}$ 10. $\int_0^1 \frac{x(x+2)}{(x^3+3x^2+1)^2}\,dx$

5.10 The indefinite integral

The symbol $\int_a^b f(x)\,dx$, as we have defined it, is sometimes called the "definite integral," to distinguish it from $\int f(x)\,dx$, called the "indefinite integral." The indefinite integral is synonymous with the antiderivative:

> To say that $\int f(x)\,dx = F(x)$ is equivalent to saying that an anti-derivative of $f(x)$ is $F(x)$, and both statements are equivalent, of course, to $f(x) = F'(x)$.

We have gotten along perfectly well up to this point without the notation of the indefinite integral. It is introduced now only because you may encounter it in other mathematics readings.

In terms of the new notation, we have

$$\int x^3 \, dx = \frac{x^4}{4} + c.$$

If we take the derivatives of both sides, we obtain

$$\frac{d}{dx}\int x^3 \, dx = x^3.$$

This should not seem at all surprising. It is a specific case of the general result:

$$\frac{d}{dx}\int f(x)\,dx = f(x).$$

That is: "The derivative of the integral of $f(x)$ equals $f(x)$." Or, even more obviously: "The derivative of any antiderivative of $f(x)$ equals $f(x)$."

Problem 1

Is it true that $\int f'(x)\,dx = f(x)$? That is, is the integral of the derivative of $f(x)$ equal to $f(x)$? Try several simple functions.

5.11 Summary

Whenever we know a *rate of change* and want to find the function whose rate we know, we perform an antidifferentiation. Our set of results on differentiation leads to an analogous set on antidifferentiation, except that we have as yet no method of finding the antiderivative of one function that we might have thought we could deal with: If $dy/dx = 1/x$, we do not yet know a formula for y. Two observations bring a broad class of problems under the heading of antidifferentiation problems:

1. Any integral, like $\int_a^b f(x)\,dx$, whatever the context in which it arises – a volume, a weight, or whatever – can be interpreted in terms of *areas*, that is, in terms of the areas bounded by the curve $y = f(x)$ and the x axis, from $x = a$ to $x = b$.

2. If $A(x)$ is the area under the curve $y = f(x)$ from $x = a$ to any x, then $dA/dx = f(x)$, so $A(x)$ is an antiderivative of $f(x)$.

PROBLEMS

1. Find y if dy/dx is given as follows:

   (a) $x^2 - \dfrac{1}{x^2}$ (b) $\sqrt{x} + \dfrac{1}{\sqrt{x}}$ (c) $x^{3/2} - x^{1/2} + x^{-2/3}$ (d) $(x+1)^3$

   [Do part (d) two ways, and reconcile your answers.]

2. (a) If $dy/dx = (2-3x)^5$, is $y = (2-3x)^6/6 + C$? (Differentiate the suggested y to see if you get the given dy/dx.) What *is* a correct expression for y?

   (b) If $dy/dx = (ax+b)^k$, with a, b, and k constants, and $k \neq -1$, find y.

   (c) If $dy/dx = \sqrt{x^2+3} \cdot x$, find y.

3. If the slope of a curve is given by $dy/dx = 6x^2 + 1/x^2 - 2$, and if $y = 3$ when $x = 1$, find an equation for y in terms of x.

4. Shortly after violent exercise, the rate of consumption of oxygen (R in.^{3}/min) varies with the time (t min) since termination of exercise as follows: $R = 1000/\sqrt{t}$. Find the number of cubic inches of oxygen consumed between $t = 1$ and $t = 9$.

5. The rate of growth of bacteria in a culture (R bacteria per hour) varies with the time (t hr) as follows: $R = 10^5 \cdot t^{3/2}$. Find the number of bacteria added to the culture between $t = 1$ and $t = 4$.

6. An elevator, rising with a speed of 32 ft/sec, is at a height of 128 ft above ground level when the cable breaks, thus transforming the elevator into a projectile subject to the acceleration due to gravity (because of the lack of any safety equipment).

   (a) What is the maximum height achieved by the elevator?

   (b) When does it pass the 128-ft level on the way down?

   (c) When does it hit the ground?

   (d) With what speed does it hit the ground?

225

7. Sketch the curves $y = 2x^2$ and $y = 12 - x^2$ on the same set of axes, and find the area of the region bounded by the curves.

8. A ball is thrown straight upward from the ground with a speed of 40 ft/sec at the same instant that another ball is dropped (from rest) from a height of 100 ft. Show that they strike the ground simultaneously.

9. A certain function, given by $y = f(x)$, has the property that $d^3y/dx^3 = 6$ for all values of x. The graph of the function passes through the origin with slope -1, and it has a point of inflection there. Find the formula for y, and make a rough sketch of the function.

10. A man driving an automobile in a straight line at a speed of 80 ft/sec applies the brakes at a certain instant, which we take to be $t = 0$. If the brakes furnish a constant acceleration of -20 ft/sec^2, how far will he go before he stops?

★ 11. Let C be a curve with equation $y = 9 - 9x + 6x^2 - x^3$.
   (a) Find relative maximum, relative minimum, and inflection points on C.
   (b) Sketch C and the line L, with equation $y = 9 - x$.
   (c) Find the area bounded by C and L.
   (d) Verify that C and the line M, with equation $y = 19 - 6x$, intersect at points with abscissas -1, 2, and 5.
   (e) Find the area bounded by C and M.
   (f) Use your results in (c) and (e) to conjecture a general result.

12. A curve satisfies the equation $d^2y/dx^2 = -6x$. It passes through the origin in a direction that makes an angle of 45° with the positive direction of the x axis. Find the equation of the curve, and sketch it.

★ 13. Assuming that the area under the curve $y = x^3$ between $x = 0$ and $x = r$ is $\frac{1}{4}r^4$, *deduce* that the area under the curve $y = x^{1/3}$ between the same values of x is $\frac{3}{4}r^{4/3}$. (Hint: Consider the area between the second curve and the y axis.)

14. Thomas Whiteside, in an article entitled "Tomatoes" (*The New Yorker*, January 24, 1977), remarks on the deleterious effects of breeding the fruit for mass shipment, gassed ripening, ability to stand the abuse of self-service markets, and so forth. A solid, tough tomato of the new breed can sustain a 6-ft fall onto a hard tiled floor.

> Out of curiosity, I telephoned Dr. William Haddon, Jr., an auto safety expert, who is president of the Insurance Institute for Highway Safety, and asked him if one of his technical people could compute the approximate impact speed of the Florida MH-1 [tomato] in the six-foot fall I had witnessed in ratio to the minimum federal requirements for impact resistance in the bumpers of cars sold in this country. Dr. Haddon obliged, and on the basis of the figures he provided I concluded that Dr. Bryan's MH-1 was able to survive its fall to the floor at an impact speed of 13.4 miles

per hour, more than two and a half times the speed which federal auto-bumper safety standards provide for the minimum safety of current-model cars. This undoubtedly represents a great step forward in tomato safety. Yet.....

Show that the stated result (13.4 mph) is correct. We don't have to depend on "technical people" for solution of problems of this sort.

★ 15. Find the area bounded by the curve $y = x^3 - 8x^2 + 15x$, the x axis, $x = 0$, and $x = 4$. Find the abscissa(s) of the point(s) on this curve where the tangent line has the same slope as the line joining the points on the curve where $x = 0$ and $x = 8$. Sketch the curve and the tangent line(s).

16. The flexion of a curve is given by $d^2y/dx^2 = x^2/(x^2 + 9)^{1/2}$, and the slope is -2 at $x = 0$.

    (a) Find those x's corresponding to maximum and minimum values of y on the curve, identifying which is maximum and which is minimum.

    (b) Find those x's corresponding to points of inflection on the curve.

★ 17. The horizontal base of a solid is a circle of radius 5 in. Every vertical section perpendicular to one diameter of the base is an equilateral triangle. Find the volume of the solid.

18. Sketch the parabola $y = 72 - 2x^2$ from $x = -6$ to $x = 6$, and verify that the area bounded by this portion of the curve and the x axis equals two-thirds the area of the circumscribing rectangle.

19. If the slope of a curve is given by $dy/dx = 1/x^{3/2} - 1 + 6x^2$, and if $y = 120$ when $x = 4$, find an equation for y in terms of x.

20. The rate of conversion of a certain substance subject to chemical change (R g/hr) varies with the time (t hr) as follows: $R = 10 \cdot t^{1/2}$. Find the number of grams of the substance converted in the first 9 hr.

21. An ellipse is an oval that looks like that in Figure 5-34. The area within this ellipse is πab square units. Find the volume of an "elliptical cone" of height 25 in. – a solid every horizontal section of which x in. below the top is an ellipse in which $a = 2x$ and $b = \frac{1}{5}x^{1/2}$.

22. Consider the curve with equation $y = x^3 - 6x^2 + 9x$.

    (a) Find the coordinates of maximum and minimum points and points of inflection on the curve, and sketch its graph.

    (b) Find the area under the curve from $x = 0$ to $x = 3$.

23. The area (A square miles) covered by an oil slick increases with the time (t days) after an accident in such a way that the rate of increase (R square miles/day) is given by $R = 4 + \frac{1}{2}t - 0.03t^2$. Find A when $t = 10$.

24. A wound heals at such a rate that the area (A mm^2) of the wound decreases at $2\sqrt{t}$ mm^2/day, where t (days) is the time since the wound was inflicted. At the start (when $t = 0$), $A = 36$.

    (a) Find a formula for A at any time.

    (b) When has the wound healed entirely ($A = 0$)?

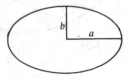

Fig. 5-34

227

5

Antidifferentiation;
integration

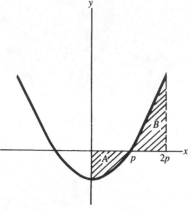

Fig. 5-35

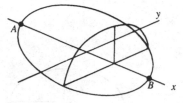

Fig. 5-36

25. A water bag is hurled straight *down* with an initial speed of 32 ft/sec from a fifth-floor window, 54 ft above ground level.
    (a) Find a formula for the height of the bag at time t sec.
    (b) When does it hit the head of a 6-ft-tall professor standing below?
    (c) With what speed does it hit the victim? (Give your answer in miles per hour; 1 mile = 5280 ft.)

26. Consider the curve with equation $y = x^3 - 9x^2 + 15x + 25$.
    (a) Find the coordinates of maximum and minimum points and points of inflection on the curve, and sketch the curve.
    (b) Find the area bounded by the curve, the x axis, $x = 1$, and $x = 3$.

27. The rate of flow (R cc/sec) of blood through a vein varies thus with elapsed time (t sec): $R = 10 - \sqrt{t}$. Find the total quantity (Q cc) of blood that flows through the vein between $t = 0$ and $t = 36$.

★ 28. Figure 5-35 shows a parabola with a vertical axis, symmetrical about the y axis. Demonstrate that the area marked B is twice the area marked A.

29. If $dy/dx = x\sqrt{x^2 - 16}$, and $y = 4$ at $x = -4$, find a formula for y in terms of x.

30. Find the volume of the solid generated by revolving about the x axis the area bounded by
    (a) $y = \sqrt{x}$, the x axis, $x = 2$, and $x = 6$;
    (b) $y^2 = 4ax$, and $x = b$, where a and b are constants.

31. Find the area between the curve $y = 2 - x^2$ and the curve $y = x^2 - 6$.

32. The region under the parabola $y = 9 - x^2$ above the x axis is revolved about the x axis. Find the volume of the solid so generated.

33. Sketch the parabola $y = 12 - 6x + x^2$ and the line $y = 2x$ on the same set of axes. Find the coordinates of their points of intersection. Find the area between the curve and the line.

★ 34. The horizontal base of a solid is the ellipse shown in Figure 5-36. The equation of this ellipse is $x^2/36 + y^2/25 = 1$. Every vertical section of the solid perpendicular to AB is a semicircle. One such section is shown in the sketch. Find the volume of the solid.

★ 35. The arc in Figure 5-37 is part of the parabola $y = kx^2$.

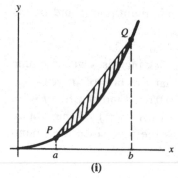

Fig. 5-37

228

(a) Find the shaded area.

(b) In Figure 5-37(ii) the line RS is tangent to the parabola, and the figure $PQRS$ is a rectangle. Find the area of the rectangle.

(c) Verify that your answer to (a) equals two-thirds of your answer to (b). This is the general result obtained by Archimedes.

★ 36. Let C_1 be the curve with equation $y = x^3 - 9x^2 + 25x - 10$, and C_2 the curve with equation $y = 2 + 6x - x^2$.

(a) Find the abscissas of the maximum and minimum points on C_1.

(b) Find the coordinates of the point of inflection on C_1, and the slope of C_1 at that point.

(c) Sketch C_1.

(d) Sketch C_2 on the same axes used in (c).

(e) Find the area bounded by C_1 and C_2.

★ 37. Like Problem 36, if C_1 has equation $y = x^3 - 5x^2 + 2x + 8$, and C_2 has equation $y = 3x^2 - 13x + 8$.

SAMPLE TEST

1. Find the area under the curve $y = x^3 - 6x^2 + 8x + 5$, above the x axis, between $x = 0$ and $x = 4$.

2. Horizontal sections of a modernistic steeple are isosceles triangles whose bases (b ft) and altitudes (h ft) vary thus with the distance (x ft) of the section below the top of the steeple: $b = \frac{1}{30}x^2$; $h = 1 + x$. Find the volume within the steeple from the top down to $x = 30$.

3. A projectile is shot straight up from a point 144 ft above the ground with an initial upward speed of 128 ft/sec.

(a) Find a formula for the height of the projectile above the ground at time t (sec).

(b) When does the projectile hit the ground?

(c) When does the projectile reach its highest point? What is its maximum height above the ground?

(d) When is the projectile at the 336-ft level rising? Falling?

(e) When, and with what initial speed, could the projectile have been fired from ground level so that its behavior for $t \geq 0$ would have been just as in the problem as stated?

4. The slope of a curve is given by $-x/\sqrt{25 - x^2}$, and $y = 8$ at $x = 4$. Find an equation of the curve.

5. The density (ρ oz/in.^{3}) of a bar whose cross section is a square 1 in. on a side varies with the distance x in. from one end of the bar as follows: ρ is proportional to x^2, and $\rho = 12$ at $x = 2$. Find the weight of the bar if it is 10 in. long.

6. Find the area bounded by the curve $y = x^3 - 6x^2 + 8x$, the x axis, and the lines $x = 0$ and $x = 4$. At what point or points is the tangent to this curve

229

parallel to the chord joining the points on the curve where $x = 0$ and $x = 6$?

Here is another sample test on Chapter 5:

1'. A point moved in such a way that $d^3y/dt^3 = 72$, y being the distance traveled. At $t = 0$, the velocity was 200 and the acceleration -6. Find y at any time.

2'. A stone was dropped from a helicopter 2400 ft high when an auto running 80 ft/sec passed directly beneath. How far apart were the stone and the auto 10 sec later, and how fast was the distance between them changing?

3'. Sketch the curves $y = x^3$ and $y = x^4$, and find the area bounded by them in the first quadrant.

4'. The acceleration of a particle is given by $t/\sqrt{10 - t^2}$, and the velocity, v, is 14 at $t = 1$. Find a formula for v at any time.

5'. Sketch the curve $y = x^2$ from $x = 0$ to $x = 2$. Find the volume of the solid within the surface obtained by revolving this portion of the curve about the x axis.

6'. The ellipse in Figure 5-38 has equation $x^2 + 2y^2 = 36$. What are the coordinates of the points A and B? Find the volume of a solid whose base is the region within this ellipse in a horizontal plane and whose every vertical section of the solid perpendicular to the segment AB is a square.

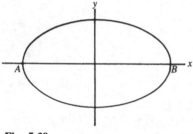

Fig. 5-38

Exponential functions

6

6.1 Introduction to exponential functions

According to the *New York Times* (June 1, 1975), United States consumption of water (W billions of gallons per day) has increased dramatically with time, as shown in Table 6-1 – by over 100-fold in 160 years!

Problem 1

Use these data to plot a reasonably careful graph of the function determined by Table 6-1, extend the function by sketching a smooth curve joining the points you have plotted, and use graphical interpolation to estimate the value of W in 1930. Also, extend the function further by using graphical extrapolation to predict the value of W in the year 2000.

Using the data in Table 6-1, we construct Table 6-2, representing by t the number of 20-year periods after 1800 and computing to the nearest hundredth the ratio of each value of W to the preceding one (after the first, of course). The striking feature is that the ratios are all nearly the same, so we might approximate W by a function, $\overline{W}$, for which the ratios are constant, say 1.8. Then we would have the following:

$$\text{at } t = 0, \quad \overline{W} = 2,$$
$$\text{at } t = 1, \quad \overline{W} = 2(1.8) = 3.6,$$
$$\text{at } t = 2, \quad \overline{W} = (3.6)(1.8) = 2(1.8)^2,$$
$$\text{at } t = 3, \quad \overline{W} = 2(1.8)^2(1.8) = 2(1.8)^3,$$

and, in general,

$$\overline{W} = 2(1.8)^t.$$

Problem 2

Compute the values and graph the function $\overline{W} = 2(1.8)^t$ for $t = 1, 2, \ldots, 10$. (A calculator will be helpful.) Does this formula for $\overline{W}$ give the value of W for the first entry in the table (i.e., for $t = 0$)?

Table 6-1

Date	W
1800	2.0
1820	3.5
1840	6.2
1860	11.2
1880	20.2
1900	36.4
1920	65.5
1940	118.6
1960	214.7

Table 6-2

t	W	Ratio
0	2.0	
1	3.5	1.75
2	6.2	1.78
3	11.2	1.80
4	20.2	1.80
5	36.4	1.80
6	65.5	1.80
7	118.6	1.81
8	214.7	1.81

By the formula, in the year 2000 ($t = 10$), $\overline{W}$ will equal about 713, which should check reasonably closely with your answer to the last part of Problem 1. (This is well over three times as large as the consumption in 1960, a rate that was already straining our resources, so even greater reuse of water is called for in the future – especially when we realize that the factor 1.8 may be low.) We can also use the formula to interpolate for $\overline{W}$ in 1930 ($t = 6.5$): $\overline{W} = 2(1.8)^{6.5}$. With a calculator, we find this to be approximately 91.3, which should be in good agreement with your graphical interpolation in Problem 1.

Problem 3

Use linear interpolation with Table 6-1 (or Table 6-2) to approximate W in 1930. Compare the size of your result with the value found just above, and explain the reasonableness of the difference.

We met a function very like this one in the case of the number of overseas telephone calls at various times between 1950 and 1965 (Example 3, **1.6**). Both functions are of the the type

$$y = c \cdot b^x, \tag{1}$$

where c and b are constants. Note that the variable x appears here as the exponent, so this is not the same as the power law,

$$y = k \cdot x^n, \tag{2}$$

where k and n are constants, which we encountered in Example 5, **1.6**. The distinction between equation (1) and equation (2) was emphasized in Problem 7, **1.6**, where you were asked to plot the graphs of $y = x^2$ and $y = 2^x$ on the same set of axes. You should have obtained a result similar to that of Figure 6-1.

Problem 4

Plot on the same set of axes smooth graphs of

$$y = x^{1/2}, \quad \text{for } 0 \le x \le 9, \quad \text{and} \quad y = \left(\tfrac{1}{2}\right)^x, \quad \text{for } -3 \le x \le 6.$$

Note that for $x = 0, 1, 2, 3, \ldots$, the successive values of 2^x, and of $\left(\tfrac{1}{2}\right)^x$, form the terms of geometric progressions. For this reason, an exponential function is sometimes said to increase, or to decrease, "geometrically."

Problem 5

In equation (1), what limitations are there on the constants c and b if the function is not to be uninterestingly simple?

There is another limitation on b, in addition to those referred to in Problem 5: If $b = -8$ and $c = 1$, say, we would have $y = (-8)^x$. Then for

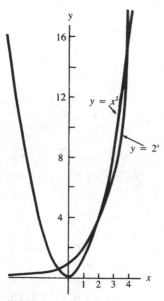

Fig. 6-1

$x = \frac{1}{2}$, y would be $(-8)^{1/2} = \sqrt{-8}$, which is not a real number. On the other hand, for $x = \frac{1}{3}$, we would have $y = (-8)^{1/3} = \sqrt[3]{-8} = -\sqrt[3]{8} = -2$. No difficulty here. In order not to have excessively wild domains, we restrict our attention to positive values of b.

Definition

An **exponential function** is one given by an equation of the form $y = c \cdot b^x$, where c and b are constants, with c nonzero and b positive and not equal to 1.

An implication of the foregoing paragraphs is that the meaning of the symbol b^x is clear for positive values of b and for all x. However, the meaning is *not* simple for some x. If x is a positive integer, there's no problem: b^x means that b is taken as a factor x times. If x is a negative integer, then b^x is defined as $1/b^{-x}$. For $x = 0$, $b^x = b^0$ is defined as 1.

Suppose that x is the reciprocal of a positive integer: $x = 1/q$, say. Then $b^x = b^{1/q}$, which is defined as $\sqrt[q]{b}$, that is, as the number that when taken as a factor q times equals b. Does such a number always exist? For example, is there a number y such that $y^3 = 15$? The answer to this question is somewhat lengthy: Because the cube of a negative number is negative, y cannot be negative. Clearly, y is not an integer, for 2 is too small and 3 is too big, and there is no integer between. Suppose, then, that there is a rational number (i.e., a fraction), $y = r/s$, not an integer, such that $y^3 = (r/s)^3 = 15$. Suppose further that r/s is the fractional form "in lowest terms" (i.e., r and s have no common factor other than 1). Because y is not an integer, $s \neq 1$. Because r and s have no common factor, r^3 and s^3 also have no common factor. Hence, r^3/s^3 is not an integer; in particular, $r^3/s^3 \neq 15$. Thus, there is no integer or other rational number y satisfying $y^3 = 15$.

Problem 6

By generalizing this argument, what can you say about the roots of integers?

We can, however, find by trial successive rational numbers whose cubes get closer and closer to 15, as follows: Find consecutive integers between which y lies:

$$2^3 = 8 < 15; \quad 3^3 = 27 > 15; \quad \text{hence}, 2 < y < 3.$$

Then find values that differ by 0.1; then by 0.01:

$$(2.4)^3 = 13.824 < 15; \qquad (2.5)^3 = 15.635 > 15; \quad \text{hence}, 2.4 < y < 2.5;$$

$$(2.46)^3 = 14.886936 < 15; \quad (2.47)^3 = 15.069223 > 15;$$

$$\text{hence}, 2.46 < y < 2.47.$$

In this way we could find a number whose cube is arbitrarily close to 15. Continuing this process would give us a never-ending sequence of rational numbers: 2, 2.4, 2.46,.... This sequence has a limit that we take to be the number $\sqrt[3]{15}$ or $15^{1/3}$. The same process could be used for any positive integer q and any positive real number b to define $(b)^{1/q}$.

Next, if $x = p/q$, where q is a positive integer and p is any integer, we define $b^{p/q}$ as $(b^{1/q})^p$. By a succession of steps, then, we have defined b^x for all rational values of x.

It is implied in what was said earlier that the real numbers may be defined as the set of all decimals. The set of rational numbers corresponds to exactly the subset made up of decimals that are terminating or periodic from some point on.

★ **Problem 7**
Prove the preceding statement.

Every irrational number, then, has a nonterminating, nonperiodic decimal representation. For example, $\sqrt{2} = 1.4142...$; that is, $\sqrt{2}$ is the limit of the unending sequence of rational numbers:

$$1, 1.4, 1.41, 1.414, 1.4142, \ldots.$$

We then define $b^{\sqrt{2}}$ as the limit approached by

$$b^1, b^{1.4}, b^{1.41}, b^{1.414}, b^{1.4142}, \ldots,$$

as the number of decimal places in our approximation to $\sqrt{2}$ increases indefinitely. Because we earlier defined the meaning of b^x for rational x, each of the numbers in this sequence has a meaning, and we shall assume that a limit always exists. It turns out that the laws of exponents, such as $b^x \cdot b^y = b^{x+y}$, hold for all real numbers x and y.

Finally, then, in this way we have an interpretation of b^x for any positive b and for all real x. We *assume* such an interpretation when we draw a smooth graph joining the points corresponding to $\overline{W} = 2(1.8)^t$ for $t = 0, 1, 2, 3, 4, 5, 6, 7$, and 8.

PROBLEMS

8. By the method described earlier for $\sqrt[3]{15}$, find the first three terms in the sequence for $(21)^{1/2}$ (i.e., find $\sqrt{21}$ to two decimal places).

C 9. Same as Problem 8, to find $\sqrt[3]{100}$ to three decimal places.

C 10. Same as Problem 8, to find $\sqrt{1000}$ to three decimal places.

11. Simplify the following expressions in which e is a constant:

(a) $(e^{-x})^2$ (b) $\sqrt{e^{-0.2x}}$ (c) $(e^{\sqrt{x}})^2$ (d) $\dfrac{3}{e^{-1/x}}$

12. Solve for x, or explain why no solution exists (e is a constant):

   (a) $e^{x^2-4}=1$ (b) $e^{1/x}\cdot e^{-x}=1$ (c) $e^{x/2-1}=0$

   (d) $e^{-x}\cdot x\cdot(x-3)=0$ (e) $e^{2x}-1=0$

13. On the same axes, sketch rough graphs corresponding to each of the following exponential functions. In each case, specify the natural domain and the range.

   (a) $y=3\cdot 2^x$ (b) $y=-3\cdot 2^x$ (c) $y=3\cdot(\frac{1}{2})^x$

   (d) $y=\frac{1}{3}\cdot 2^x$ (e) $y=-3\cdot 5^x$

14. Same as Problem 13 for the following:

   (a) $y=2\cdot 3^x$ (b) $y=-2\cdot 3^x$ (c) $y=2\cdot 3^{-x}$

   (d) $y=-2\cdot 3^{-x}$ (e) $y=2\cdot(\frac{1}{3})^x$ (f) $y=2\cdot(\frac{1}{3})^{-x}$

15. Show that each of the following equations defines an exponential function by stating in each case the values of c and b in the standard form $y=c\cdot b^x$. On the same axes, sketch rough graphs corresponding to each equation.

   (a) $y=2^{-x}$ (b) $y=3\cdot 2^{2x}$ (c) $y=\frac{1}{3}(\frac{1}{2})^{-2x}$ (d) $y=-4^{(1/2)x}$

   (e) $y=10(8)^{(-1/3)x}$ (f) $y=-3(\frac{1}{8})^{(-1/3)x}$

   (g) $y=5\cdot 2^{(1/2)x}$ (h) $y=2^{-0.1x}$ (i) $y=(\frac{1}{3})^{-2x}$

16. (a) What are the natural domain and the range of the function determined by $y=2^{x^2}$? Sketch a graph corresponding to this equation. Is this an instance of an exponential function as defined in this section?

   (b) Same as (a) for $y=2^{-x^2}$.

6.2 The rate of change: preliminary remarks

Having learned, in the previous section, the principal features of *how* an exponential function behaves, we turn next to the question of *how fast* such a function changes. This turns out to be a matter of particularly far-reaching import, but the problem is more difficult than for the functions studied thus far. Happily, we shall eventually discover that the answer is simple and elegant, which is one of the reasons for the significance of the result.

A direct attempt to find the derivative of the exponential function would go as follows:

$$\text{if } f(x)=c\cdot b^x,$$
$$\text{then } f(x+h)=c\cdot b^{x+h}.$$

Problem 1
Show that the average rate of change of f with respect to x in the interval $[x, x+h]$ can be written as $c\cdot b^x(b^h-1)/h$.

Then $f'(x)=\lim_{h\to 0}[c\cdot b^x(b^h-1)/h]$. Because $c\cdot b^x$ does not change with

h, this can be written as

$$f'(x) = c \cdot b^x \lim_{h \to 0} \frac{b^h - 1}{h}.$$

As $h \to 0$, $b^h \to 1$, so that both numerator and denominator of $(b^h - 1)/h$ approach zero. Of course, we *always* have this situation in applying the definition of the derivative, and we need some trick to help us find the value of the limit, if it exists.

But first some calculations will be instructive. Let us start with the case $b = 2$, so that we are trying to find $f'(x)$ if $f(x) = c \cdot 2^x$. We have just seen that we need to know the value of $\lim_{h \to 0}[(2^h - 1)/h]$, if it exists. With a calculator having nothing more sophisticated than a square-root key, we can make the following table:

h	1	$\frac{1}{2}$	$\frac{1}{4}$	$\frac{1}{8}$	$\frac{1}{16}$	$\cdots$	$\frac{1}{256}$	$\cdots$	$\frac{1}{1024}$	$\frac{1}{2048}$	$\frac{1}{4096}$
$\dfrac{2^h - 1}{h}$	1	0.8284	0.7568	0.7241							

Problem 2

Use a calculator to find that for $h = \frac{1}{4096}$, $(2^h - 1)/h \approx 0.6926$.

It appears, then, that if $y = c \cdot 2^x$, $dy/dx = c \cdot 2^x \cdot k$, where k is a constant approximately equal to 0.6926.

Problem 3

Repeat this process for $c \cdot 5^x$ to find that if $y = c \cdot 5^x$, $dy/dx = c \cdot 5^x \cdot l$, where l is a constant approximately equal to 1.6093.

Likewise, do the calculation to show that if $y = c \cdot 10^x$, dy/dx seems to be $c \cdot 10^x \cdot m$, where m is a constant approximately equal to 2.3028.

Problem 4

Do you see any relationship among the three numbers we have found: 0.6926, 1.6093, and 2.3028?

Observe that as the "base" increases from 2 to 5 to 10, the constants appearing in the derivative increase from (approximately) 0.69 to 1.61 to 2.30. We might hope that there is some "base" for which the constant appearing in the derivative equals 1, exactly. If there is such a base (call it B), then for

$$y = c \cdot B^x, \quad \frac{dy}{dx} = c \cdot B^x \cdot 1 = y.$$

That would be a marvelously simple result.

The relationship suggested in Problem 4 and the observation in the preceding paragraph will reappear later. Now we look for the trick needed to find the value of $\lim_{h \to 0}[(b^h - 1)/h]$. The needed trick is much more complicated than for the functions we have met previously. We are forced to use a roundabout approach, but we shall gain some side benefits in the process. An analysis of compound interest in the next two sections is the roundabout approach that will help us find the derivative of the exponential function.

6.3 Compound interest

An investment at *simple interest* pays interest only on the principal. For principal P at simple annual interest rate r, the amount, A, after n years is P plus the annual interest multiplied by the number of years; that is,

$$A = P + nrP = P(1 + nr).$$

For example, after 5 years, $100 at 10% simple annual interest would amount to $150.

Compound interest pays interest on the interest. Compounding annually means that at the end of each year the interest for that year is added to the amount at the beginning of the year, with interest subsequently computed on the total. Compounding semiannually means that the interest for each half year is added to the amount; and, in general, compounding k times per year means that at the end of each interest period $(1/k$ year$)$ the interest is added to the amount at the beginning of that period.

Problem 1
Assuming that no interest is paid for part of an interest period, fill in the values of the amounts in Table 6-3, beginning in each case with $100.

Table 6-3

	Simple interest, 10% per year	Compounding annually, 10% per year	Compounding semiannually, 10% per year
At start	$100	$100	$100
At end of 6 months			
At end of first year			
At end of 18 months			
At end of 2 years			

Problem 10, **1.6**, concerns an investment at compound interest. There you found, or should have, that the formula for the value of an investment of $P at 10% compounded annually for n years is

$$A = P(1+0.1)^n = P(1.1)^n,$$

an exponential formula.

An analysis in the language of investments will help to clarify the behavior of exponential functions in general. Proceeding as in that problem, we can get a formula for the amount after n years of a principal of P dollars at interest rate r, compounded annually. "Compounding annually" means that in each year the amount gained equals the amount at the beginning of that year multiplied by r.

n	A
1	$P(1+r)$
2	$P(1+r)+P(1+r)r = P(1+r)(1+r) = P(1+r)^2$
3	$P(1+r)^2 + P(1+r)^2 r = P(1+r)^2(1+r) = P(1+r)^3$
$\vdots$	$\vdots$
n	$P(1+r)^n$

This table shows how it goes. The formula we are looking for, then, is

$$A = P(1+r)^n.$$

For future reference, we observe that

the amount gained in the following year will be $A \cdot r$.

Clearly, if we dealt with *interest periods* instead of *years*, we could get, in just the same way, the amount, Q, after m interest periods, of a principal of P dollars at interest rate i per interest period, compounded each interest period, as

$$Q = P(1+i)^m.$$

Now suppose that instead of compounding annually we compounded quarterly (i.e., four times per year). Then the number of interest periods would be four times the number of years ($m = 4n$), and the interest rate per interest period would be $\frac{1}{4}$ of the nominal annual rate ($i = r/4$). (Note that 6% per year can equally well be expressed as 3% per half year, or $1\frac{1}{2}$% per quarter, or $\frac{1}{2}$% per month, just as the same velocity can be expressed in miles per hour or miles per minute. To distinguish it from other rates, we shall use the term "nominal" for the annual rate in terms of which the investment is described.)

Thus, the amount after n years ($4n$ quarters), compounding quarterly, is $P(1 + r/4)^{4n}$. Hereafter we shall use the notation A_k for the amount after n years, compounding k times per year. For example, A_1 means the amount after n years, compounding annually; A_2 means the amount after n years, compounding semiannually. We have, then,

$$A_4 = P\left(1 + \frac{r}{4}\right)^{4n};$$

and, in general, we have the amount after n years, at nominal annual rate r, compounded k times per year:

$$A_k = P\left(1 + \frac{r}{k}\right)^{kn}.$$

Problem 2

(a) For a principal of $100 at nominal annual rate of 20%, tabulate the values of each of A_1, A_2, and A_8 at the end of each of its interest periods from $n = 0$ to $n = 2$. (Use the interest table – Table A – at the end of this book to find A_8.)

(b) Assuming that simple interest is paid for a partial period during which a withdrawal is made, draw large-scale graphs of these functions on the same set of axes. (Use 2 in. to represent $\frac{1}{2}$ year on the horizontal axis and $10 on the vertical axis, and make $100 the lowest point on the vertical axis.)

(c) For each, what is the amount gained during the first interest period? The last?

(d) For each, what percentage of the initial amount ($100) is gained during the first interest period? For each, what percentage of the amount at the start of the last interest period is gained during the last interest period?

(e) For each, what is the amount gained during the first year? The last?

(f) For each, what percentage of the initial amount ($100) is gained during the first year? For each, what percentage of the amount at the start of the last year is gained during the last year? (This is called the *yield rate*. It can be defined as the rate that, compounded annually, will give the same amount as actually obtained at the end of each year.)

(g) How should the graphs be drawn if no interest is paid for partial interest periods during which withdrawal is made?

We have been discussing cases in which something (like the amount in a savings account) *appreciates* through the accrual of interest. Mathematically, a similar situation exists when the value of something (like a car) *depreciates* with time. Suppose that a car initially worth $10,000 depreciates each year by 15% of its value at the beginning of the year. Then,

at the start, the car is worth $10,000;
at the end of 1 year, the car is worth $10,000(1 - 0.15) = 10,000(0.85) = \$8,500$;
at the end of 2 years, the car is worth $10,000(0.85)^2 = \$7,225$;
at the end of 3 years, the car is worth $10,000(0.85)^3 = \$6,141.25$; etc.

Note that the loss during the first year is $1500; during the second year, $1275; during the third year, $1083.75. The annual losses, in dollars, decrease with time, because the depreciation percentage is successively applied to smaller amounts. Under our assumption of constant annual percentage loss, the car never becomes valueless.

Problem 3

Check that if the value of something *depreciates* at a nominal annual rate r, all the work of this section carries through, with the sole change that r is a negative number, and we must speak of "loss" rather than "gain."

Problem 4

"CREF's net total rate of return for 1975, reflecting dividend earnings and changes in the market value of CREF's common stocks, was plus 32.1%. In 1974, CREF's net total rate of return was minus 31.0%." What was the 2-year percentage change?

Problem 5

From an editorial, "Reflections on a 'Birthday'" (*Saturday Review*, December 13, 1975): "The value of a citizen's currency would shrink 10 per cent when he or she crossed a state line. Thus a citizen who started out from New Hampshire with $100 in his pocket would have $20.24 left by the time he arrived in Georgia – without having spent a cent." Is the arithmetic of the preceding statement correct?

Problem 6

(a) If your stock portfolio gained 30% in 1979 and then *lost* 30% in 1980, what was your 2-year percentage change?

(b) If your stock portfolio lost 30% in 1980 and then *gained* 30% in 1981, what was your 2-year percentage change?

6.4 Continuous compounding

If we let k increase indefinitely (denoted by $k \to \infty$) so that the length of the interest period approaches zero as a limit, we have a situation that may reasonably be called "continuous compounding," with the amount after n years given by

$$A = \lim_{k \to \infty} A_k = \lim_{k \to \infty} \left[P \left(1 + \frac{r}{k} \right)^{kn} \right],$$

if the limit exists. The crucial question is the existence of this limit. To simplify matters at the start, we shall take $P = 1$, $r = 1$ (i.e., an annual interest rate of 100% – how that prime rate has exploded!), and $n = 1$ (i.e., we see what happens at the end of 1 year). Then the formula for the amount, A_k, to which a principal of \$1 at 100%, compounded k times per year, grows in 1 year is

$$A_k = \left(1 + \frac{1}{k} \right)^k.$$

Problem 1

Verify that $A_1 = 2$; $A_2 = 2.25$; $A_3 \approx 2.37$; $A_4 \approx 2.44$. What is your guess about what happens to A_k for large k?

Problem 2

With a calculator, check the values in the following table:

k	10	100	1000	10,000	100,000	1,000,000
A_k	2.59374	2.70481	2.71692	2.71815	2.71828	2.71828

It appears as though A_k increases as k increases, but that the rate of increase slows down, so we can guess that A_k does approach a limit as k increases indefinitely. This can be proved, although we shall not do so here. The limit of A_k as k increases indefinitely is called e; its value is about 2.718. Like $\sqrt{2}$, π, and many other numbers with which we deal, e is irrational and hence cannot be expressed exactly in terminating decimal form, nor in any other fractional form, nor as a periodic decimal (see the last sentence of the text just prior to Problem 7, **6.1**). (Although the idea of e arose in connection with the development of logarithms in the early seventeenth century, the proof that e is irrational – indeed, "transcendental" – dates only from the work of the French mathematician Hermite in 1873.) To summarize:

> e is defined as the limit of $(1+1/k)^k$ as k increases indefinitely; its value is approximately 2.718.

This implies that $1 invested for a year at an annual rate of 100%, with extremely frequent compounding, amounts to about $2.71 and that, with *continuous compounding*, it amounts to e dollars.

Now we can take the general case:

$$A = \lim_{k \to \infty} A_k = \lim_{k \to \infty} \left[P\left(1 + \frac{r}{k}\right)^{kn} \right] = P\left[\lim_{k \to \infty} \left(1 + \frac{r}{k}\right)^k \right]^n,$$

because P and n are constant throughout the discussion. For convenience, we set $r/k = 1/u$, so that $k = ru$. Clearly, as k increases indefinitely, so does u.

Problem 3

Do the algebra to show that A_k can be written as $A_k = P[(1+1/u)^u]^{rn}$.

Then $A = P[\lim_{u \to \infty} (1+1/u)^u]^{rn}$. We know that $(1+1/u)^u$ approaches e as u increases indefinitely. Assuming that the function we have here is continuous, we conclude that A equals Pe^{rn}. Thus,

> $A = Pe^{rn}$ is the formula for the amount of an initial P after n years at a rate r, compounded continuously.

Depreciation, figured on a continuous compounding basis, is covered by the same formula, with negative r.

Table B at the end of this book, giving approximate numerical values of e^x and e^{-x} for various values of x, will be helpful in many of our problems. When necessary, we can use linear interpolation in this table. Later we shall be able to do more accurate work using logarithms.

Example 1

In how long will any sum be quadrupled at 10% nominal annual interest compounded continuously?

An initial sum of P will amount to $A = Pe^{0.1n}$ at the end of n years. The question amounts to this: For what value of n does $A = 4P$? We have to solve, then, $Pe^{0.1n} = 4P$, or $e^{0.1n} = 4$. (It doesn't matter if we are talking about $1 growing to $4, or $25 growing to $100, or whatever.) The relevant entries from Table B are shown here:

x	e^x
1.3	3.669
1.4	4.055

Problem 4

Interpolate in this table to find that $e^x = 4$ for $x = 1.39$ approximately.

Thus, $e^{0.1n} = 4 = e^{1.39}$. Hence, $0.1n = 1.39$ and $n = 13.9$ years.

Example 2

What is the equivalent, as a rate compounded semiannually, of a rate of 10% per year, compounded continuously?

In n years, a sum of P will amount to $Pe^{0.1n}$, at a rate of 10% compounded continuously. In the same time, at a rate r, compounded semiannually, the same sum will amount to $P(1 + r/2)^{2n}$. The question asks for the value of r that will make these amounts equal: $P(1 + r/2)^{2n} = Pe^{0.1n}$. We divide both sides by P to obtain $(1 + r/2)^{2n} = e^{0.1n}$, and we take the nth root of both sides to obtain

$$\left(1 + \frac{r}{2}\right)^2 = e^{0.1}. \tag{3}$$

(It doesn't matter if we start with $1 or $10 or whatever, and it doesn't matter how many years we consider – if the amounts are equal at the end of 1 year, they will also be equal at the end of 2 years, at the end of 3 years, etc.)

To solve equation (3) for r, it is easiest first to take the square roots of both sides.

Problem 5

Do this, and use Table B to obtain $1 + r/2 = 1.051$.

Hence, $r/2 = 0.051$, and $r = 0.102$, or 10.2%.

6. Use Table B to solve the following equations for x (rough approximations will suffice; interpolation is not required):

(a) $e^{2x} = 3$ (b) $e^{(1/2)x} = 0.5$ (c) $e^{-(1/3)x} = 0.1$

(d) $x = e^9$ (e) $x = \dfrac{3}{e^3}$ (f) $x = \sqrt[4]{0.2}$

7. A risky investment provides a return of 20% per year, compounded continuously. To what rate is this equivalent, compounded (a) annually, (b) semiannually, and (c) quarterly?

8. In how long a time will any sum be doubled at 8% interest compounded continuously?

9. A bank advertises that on certain savings accounts it pays interest of 6%, compounded continuously. To what rate, compounded annually, is this equivalent?

10. How long does it take a sum of money to triple at 11% per year, compounded continuously?

11. If a car depreciates by 35% per year, compounded continuously, about how long does it take to lose half its value? Three-quarters of its value?

12. What can you say about $\lim_{v \to 0}(1 + v)^{1/v}$ if v approaches zero through the sequence $1, \frac{1}{2}, \frac{1}{3}, \frac{1}{4}, \frac{1}{5}, \dots$?

13. Use a calculator to guess the value of $\lim_{k \to \infty}(1 + k)^{1/k}$.

6.5   The derivative of the exponential function

We can now use ideas from compound interest to obtain the formula for the derivative of the exponential function. For the case of annual compounding, at rate $r\%$, an amount A at time t years will become $A + A \cdot r$ at time $(t + 1)$. Hence, the gain is $A \cdot r$, and the average rate of increase in the year is $A \cdot r / 1 = A \cdot r$. Thus, in Figure 6-2, the slope of line UV equals $A \cdot r$.

For the case of semiannual compounding, at the same annual rate of $r\%$, an amount A at time t years will become $A + A \cdot r / 2$ at time $t + \frac{1}{2}$. Thus, as in Figure 6-3, the average rate of increase in the first half year is $(A \cdot r / 2) / \frac{1}{2}$

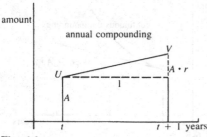

Fig. 6-2

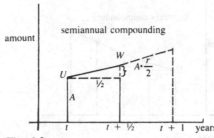

Fig. 6-3

$= A \cdot r =$ the slope of line UW. (Of course, the average rate of increase during the second half year will be somewhat larger, but that is not of concern to us.)

Consider one more case – that of quarterly compounding (Figure 6-4). The gain during the first quarter year is $A \cdot r/4$, so the average rate of increase during the quarter is $(A \cdot r/4)/\frac{1}{4}$, which again reduces to $A \cdot r$. This equals the slope of UX.

Thus, the lines UV, UW, and UX all have the same slope, and we can see that *no matter how often we compound, we always leave point U along a line with slope A·r.* It seems plausible, then, to conclude that for continuous compounding, our initial direction, the tangent at U, will also have slope $A \cdot r$ (Figure 6-5). This argument leads to the result that

$$\text{if} \quad A = Pe^{rt}, \quad \text{then} \quad \frac{dA}{dt} = A \cdot r = Pe^{rt} \cdot r.$$

We no longer need to restrict ourselves to the language of "principal," "interest rate," and "amount," and we write the basic formula for the derivative of an exponential function:

$$\text{If} \quad y = ce^{rx}, \quad \text{where } c \text{ and } r \text{ are constants, then} \quad \frac{dy}{dx} = y \cdot r = ce^{rx} \cdot r. \quad (4)$$

Note that this fits the definition of the exponential function, $y = c \cdot b^x$, with $b = e^r$. Here is a special case:

$$\text{If} \quad y = e^x, \quad \text{then} \quad \frac{dy}{dx} = e^x. \quad (5)$$

What could be more beautifully simple than this?

Admittedly, we have not yet quite done what we set out to do, for we began by looking for the rate of change of the function given by $f(x) = c \cdot b^x$. We have, however, answered a question raised in Section 6.2: There, for $f(x) = c \cdot b^x$, it appeared that $f'(x) = c \cdot b^x \cdot$ (a number depending on b), and we asked "For what b does this number equal 1?" We now have the answer: for $b = e$. That is, if $f(x) = c \cdot e^x$, then $f'(x) = c \cdot e^x \cdot 1$.

In the next chapter we shall discuss the rate of change of an exponential function for an arbitrary "base," b. In the remainder of this chapter we

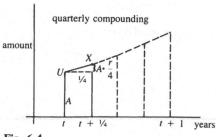

Fig. 6-4

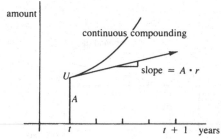

Fig. 6-5

shall practice with equation (4) and show the power of that equation in solving many central problems.

Two further comments are in order:

1. In equation (4), note that c, the analogue of the "principal" or the "original amount," is the value of y at $x = 0$; it is often called the *initial value* of y and is designated by y_0.

2. Our development, culminating in equation (4), was based on the idea of compound interest, so that r was thought of as being a positive number. But we could equally well deal with a constant percentage *loss*. For example, suppose that the value of a car depreciated each year by 15% of its value at the beginning of the year. Then we could write

$$\binom{\text{the average rate of}}{\text{gain in any year}} = \binom{\text{the value at the}}{\text{beginning of that year}} \cdot (-0.15).$$

By continuing in this fashion we could conclude that equation (4) applies if r is a negative, as well as a positive, number.

PROBLEMS

1. In Problem 2(b), **6.3**, you drew graphs of A_1, A_2, A_8 on the same axes from $n = 0$ to $n = 2$ on the assumption that simple interest is paid for a partial period during which a withdrawal is made. Now add graphs of A_{10} and A from $n = 0$ to $n = 1$, both with the same nominal annual rate (20%) as before. Verify from the shapes of the graphs that apparently $\lim_{k \to \infty} [A_k \cdot r] = dA/dn$ for $n = 1$, say.

2. (a) If $y = 2e^{(1/2)x}$, write the formula for dy/dx.
   (b) Make a table of the values of y and dy/dx of part (a) for $x = 0, 1, 2$, and -1.
   (c) Draw a fairly careful, large-scale graph of $y = 2e^{(1/2)x}$ over $-2 \le x \le 3$, using the same scale vertically as horizontally.
   (d) Draw the tangent lines to the curve of (c) at the points corresponding to $x = 0, 1, 2$, and -1.

3. Consider the function given by $f(x) = c \cdot e^{rx}$, $c \ne 0$, $r \ne 0$.
   (a) What is the natural domain?
   (b) What is the range? (Be careful!)
   (c) Is $f(x)$ ever zero?
   (d) Does $f(x)$ have any critical values?
   (e) Are there any points of inflection on the graph of f?

4. A sum of $1000 is on deposit for 10 years at a rate of 7% compounded continuously. Use the derivative to approximate the change in the amount if
   (a) the money is left at interest for one more month;
   (b) the time remains 10 years, but the interest rate for the whole period is 7.1%.

5. For the curve with equation $y = x \cdot e^x$, find maximum and minimum points and points of inflection, and sketch the curve. Where is the flexion an extreme? Maximum or minimum?

★ 6. Same as Problem 5, for the curve with equation $y = x^2 \cdot e^x$.

7. If $f(x) = (e^x + e^{-x})/(e^x - e^{-x})$, find $f'(x)$.

8. For the curve with equation $y = e^x/x$, find maximum and minimum points and points of inflection, if any, and sketch the curve. (Remember to check the natural domain!)

9. Same as Problem 8, for $y = x/e^x$.

10. Same as Problem 8, for $y = \frac{1}{2}(e^x + e^{-x})$. (You should be able to sketch the curve quickly if you first sketch $y = e^x$ and $y = e^{-x}$ on the same set of axes.)

★ 11. Same as Problem 8, for $y = e^x/(e^x + x)$.

C 12. (a) Make rough graphs of $y = e^{x/2}$ and $y = x + 1$ to find approximately where the two curves intersect. Then use the Newton–Raphson method of Section 3.10 to determine the results to two-decimal-place accuracy.

  (b) Same as (a), for the curves $y = e^x$ and $y = x^2$.

  (c) Same as (a), for the curves $y = e^x$ and $y = 2 - x^2$.

★ (d) If you were to use $x_0 = 100$ as an initial approximation in parts (b) and (c), your calculator would show $x_1 = 99.0$, $x_2 = 98.0$, and $x_3 = 97.0$. If you were to use $x_0 = 100$ in part (a), your calculator would show $x_1 = 98.0$, $x_2 = 96.0$, and $x_3 = 94.0$. Explain this strange situation. In each of these cases, would you expect the pattern to continue indefinitely? Or could the process lead to a solution?

6.6 Relative errors and relative rates

For most practical purposes, we need to know the size of a quantity: *How many* bushels of wheat? *How large* a rate of change in the population? However, it is not uncommon to be more interested in the *ratio* of a quantity, Q, to some other quantity, P, than in the size of Q itself. For example, an error of 1 ft in measuring the distance from the earth to the moon is much less significant than an error of 1 ft in measuring the width of a highway. What is often of importance, then, is the value of the ratio:

$$\frac{\text{size of error}}{\text{size of quantity being measured}}.$$

This ratio is called the *relative error*, and it can be expressed as relative error $= 1/25 = 0.04$, say. Sometimes this same relative error is expressed as 4%, and then it would be called the *percentage error*.

Example 1

If the side of a square plate is measured as 8 in. with an error of not more than 0.1 in., what is the approximate maximum possible error in the computed area of the plate?

If we designate the length of the side by x and the area by A, then $A = x^2$. Using the differential approximation, we have $\Delta A \approx dA = 2x\, dx$. Hence, the relative error in the area, $\Delta A / A$, can be expressed as

$$\frac{\Delta A}{A} \approx \frac{dA}{A} = \frac{2x\, dx}{x^2} = 2\frac{dx}{x}. \tag{6}$$

For the given data, then, $\Delta A / A \approx 2 \cdot 0.1 / 8 = 0.025$, or, equivalently, a percentage error of 2.5%.

Note an important aspect of equation (6): Because dx/x is the relative error in the side of the square, equation (6) can be read as follows: "The relative error in the computed area of a square equals two times the relative error in the measurement of a side of the square."

To answer the question in the example, then, it will suffice to know the *relative error* in the measurement of a side – it is not necessary to know the actual error in the side and its length.

We turn now to another "relative" concept – that of the *relative rate of change of a function*, $f(x)$, which is defined as the ratio of the rate of change to the value of the function itself. That is,

$$\text{relative rate of change of } f(x) = \frac{f'(x)}{f(x)}.$$

Or, to use another notation, if $y = f(x)$, the relative rate of change of y is defined as $(dy/dx)/y$. This concept is of particular interest in connection with the exponential function, $y = c \cdot e^{rx}$. As we know, $dy/dx = c \cdot e^{rx} \cdot r = y \cdot r$. Hence,

$$\frac{dy}{dx} \Big/ y = r.$$

In other words, *for the exponential function, the relative rate of change is constant.*

Problem 1

By way of contrast, if it is the *rate of change* of $f(x)$ that is constant, what is an expression for $f(x)$?

Keep in mind the distinction between *rate of change* and *relative rate of change*! In summary, we note three equivalent statements:

If $\quad y = c \cdot e^{rx}$,

then

$$\frac{dy}{dx} = ce^{rx} \cdot r,$$

247

or

$$\frac{dy}{dx} = y \cdot r \quad \text{(``the rate of change is proportional to } y \text{ itself, and the constant of proportionality is } r \text{''}),}$$

or

$$\frac{dy}{dx} \Big/ y = r \quad \text{(``the relative rate of change equals } r \text{, a constant'')}.}$$

With differing phraseology, all say the same thing.

PROBLEMS

(These problems have nothing to do with exponential functions, but they illustrate the concept of relative change.)

2. (a) The diameter of a steel ball is measured as 10 cm, with a maximum possible error of 0.05 cm. What is the approximate possible relative error in the computed volume of the ball?

(b) More generally, what is the relation between the relative error in the diameter of the ball and the relative error in the computed volume of the ball?

(c) What is the relation between the relative error in the diameter of the ball and the relative error in the computed surface area of the ball?

3. (a) A growing pile of sand has the shape of a cone in which the radius of the base always equals the height of the cone. What is the approximate relative change in the volume if the relative change in the height is 0.06?

(b) Same as part (a), if the radius of the base always equals one-half the height.

★ 4. The radius of the base and the height of a cylinder are measured, with certain relative errors, and the volume of the cylinder is then computed. Express the relative error in the computed volume in terms of the relative errors in the radius and the height.

5. Elasticity of demand.

(a) For the sausage problem, as developed in Section 3.9, solve equation (2), **3.9**, $p = 6.4 - 0.03x$, for x, the number of pounds sold daily in steady operation, as a function of p, the selling price per pound. Let us call this the "demand function," $f(p)$.

(b) Express the total revenue, R, as a function of p. For what p's is R increasing? What is the value of p_M, the p corresponding to maximum R? (Note, by the way, that we are not speaking of maximum *profit*.)

(c) Compute the value of $-(dx/x)/(dp/p)$ at $p = p_M$.

The ideas contained in the specific illustration in parts (a)–(c) can be described quite generally: If $x = f(p)$ expresses demand as a function of price, and if price changes by an amount dp, then demand changes approximately by an amount dx. The *relative* change in price is dp/p, and the *relative* change in demand is dx/x. In most cases, f is a decreasing function, so if dp is positive (price goes up), dx is negative (demand goes down), and dx/x is also negative. The **elasticity of demand**, $E(p)$, is defined by $E(p) = -(dx/x)/(dp/p)$, that is, the ratio of relative *increase* in demand to relative increase in price.

(d) Use $R = x \cdot p$ and the formula for the derivative of a product to show that *total revenue is an increasing function of price* if $E(p) < 1$ and *total revenue is a maximum* if $E(p) = 1$.

(e) Show that if demand, x, is a power function of price, p, then $E(p)$ is constant.

6. Assume that the rate of flow, F, of blood varies with the diameter, D, of the blood vessel and the blood pressure, P, as follows: $F = kD^4 \cdot P$, where k is a constant. Find an approximate expression for the relative change in F in terms of the relative changes in D and in P. If the change in F is to be zero (and hence also the relative change is to be zero), what is the relationship between the relative change in D and the relative change in P? (See Problem 11, **3.3**.)

7. What is the limit on the relative error in measuring the edge of a cube if the relative error in the
(a) volume computed from the edge is not to exceed 6%?
(b) surface area computed from the edge is not to exceed 6%?

8. A ball (sphere) is made of metal of known constant density (k g/cc). What is the limit on the relative error in measuring the weight of the ball if the relative error in the surface area computed from the weight is not to exceed 4%? Is it necessary to know the value of k?

6.7 Antiderivatives of the exponential

In the equation $y = ce^{rx}$ we know that the constant r represents the constant relative (or percentage) rate of change and that c represents the initial value of y – the value of y at $x = 0$. As mentioned in the first comment at the end of Section 6.5, we often designate this initial value of y by y_0, so that the basic exponential equation can be written as

$$y = y_0 e^{rx}. \tag{7}$$

We know that if equation (7) holds, then

$$\frac{dy}{dx} = y \cdot r. \tag{8}$$

Is the converse valid? That is, does (8) imply (7)? This is a question about antidifferentiation, and the answer is quickly found to be yes: Suppose that y is any function of x for which equation (8) holds. Let us consider y/e^{rx}

and differentiate this as a quotient:

$$\frac{d}{dx}\left(\frac{y}{e^{rx}}\right) = \frac{e^{rx}\cdot y' - ye^{rx}\cdot r}{e^{2rx}} = \frac{e^{rx}(y'-y\cdot r)}{e^{2rx}} = \frac{y'-y\cdot r}{e^{rx}},$$

or

$$\frac{d}{dx}\left(\frac{y}{e^{rx}}\right) = \frac{0}{e^{rx}} = 0,$$

because equation (8) holds. Hence, $y/e^{rx} =$ a constant, c, or $y = ce^{rx}$. If $y = y_0$ at $x = 0$, we conclude that $c = y_0$, which completes the argument. Thus:

If $\dfrac{dy}{dx} = y\cdot r$ and if $y = y_0$ at $x = 0$, then $y = y_0 e^{rx}$.

Example 1
Suppose that the U.S. water consumption (W billions of gallons per day) varies with time (t 20-year periods after 1800) as follows: dW/dt is always 58.1% of W, and $W = 2$ in 1800. What is the formula for W in terms of t?

Here we have $r = 0.581$ and $W_0 = 2$, so $W = 2\cdot e^{0.581t}$ is the formula.

Problem 1
(a) What is the value of W in 1960? In 1900? (Compare with Table 6-1, **6.1.**)
(b) What is the value of W in 1930? (Compare with the result of Problem 1, **6.1.**)

By what date had the original W increased by a factor of 10? We must solve the equation $2\cdot e^{0.581t} = 20$ for t. We have, then, $e^{0.581t} = 10$.

Problem 2
Use Table B at the end of this book to show that this is equivalent to $e^{0.581t} = e^{2.3}$, so that $t = 3.96$.

Thus, the answer is 1879. (Check against Table 6-1.)

Example 2
If the quantity (Q mg) of a radioactive substance decreases at an instantaneous rate (per day) always equal to 10% of Q, and if $Q = 3$ at $t = 0$, the formula is $Q = 3\cdot e^{-0.1t}$.

Problem 3
Find Q at the end of 15 days.

To find the "half-life" of this radioactive substance (the length of time necessary for it to lose half its weight through decomposition) we must find t corresponding to $Q = 1.5$:

$$3e^{-0.1t} = 1.5, \quad \text{or} \quad e^{-0.1t} = 0.5. \tag{9}$$

(It doesn't matter if we see how long it takes 3 mg to decompose to 1.5 mg, or 1 mg to decompose to $\frac{1}{2}$ mg, or 34 mg to decompose to 17 mg.)

Problem 4

Use Table B to show that equation (9) is equivalent to $e^{-0.1t} = e^{-0.69}$ and, hence, that $t = 6.9$ days.

PROBLEMS

5. The pressure (p lb/in.^{2}) in a space capsule decreased at an instantaneous rate (per hour) always equal to 8% of p.
   (a) If p equaled 75 originally, find p at the end of 10 hr.
   (b) When had p decreased to one-third its original value?

6. The area (A mm^2) of a bacterial colony grows for a limited time at a constant relative rate per hour. If $A = 4.4$ at $t = 0$, and 6.8 at $t = 1$, find the formula for A in terms of t. To what rate, compounded hourly, is this growth equivalent? Use the formula and Table B to compute A for $t = 2, 3, 5$, and 10. (Compare your answers with Table 1-1, **1.2**.)

7. In the interval between 1950 and 1965, the number (N million) of overseas telephone calls originating in the United States varied with the time (t years after 1950) as follows: The relative rate of increase (per year) was constant. At $t = 0$, $N = 2.5$, and at $t = 1$, $N = 2.75$. Find the formula for N in terms of t, and use it and Table B to compute N for $t = 5, 10$, and 15. (Compare your answers with Table 1-5, **1.4**.) Also use the formula and Table B to compute N for 1972, and compare your answer with 45, the actual figure as stated in Section 1.4.

8. A radioactive substance disintegrates at a constant percentage rate per week. If there are 2.5 mg at some time, taken to be $t = 0$, and 0.753 mg 10 weeks later, find a formula for the amount (Q mg) at any t. What is the half-life of this substance?

9. The pressure in a pump cylinder decreases at the constant instantaneous percentage rate of 5% per day. How long does it take the pressure to drop to 0.1 its original value?

10. An estate originally worth $300,000 depreciated so that the instantaneous rate of loss was continually 5% of the current value. What was its value after t years? After 100 years? What percentage was actually lost in a year?

11. Find the area under the curve $y = e^x$ from $x = 0$ to $x = 1$.

12. The area under the curve $y = e^{0.5x}$ from $x = 0$ to $x = 1$ is revolved about the x axis. Find the volume so generated.

13. On the same pair of axes, make sketches of the graphs of $y = e^x$, $y = e^x - 1$, and $y = e^x - e$. Find the area bounded by the graph of $y = e^x - e$, the x axis, and the lines $x = 0$ and $x = 2$. Use Table B to give your answer to three decimal places.

6
Exponential functions

14. Find the coordinates of maximum and minimum points and points of inflection, if any, on the curve $y = e^{(1/2)x}/x$. Sketch the curve.

15. Same as Problem 14, for $y = x/e^{(1/2)x}$.

16. Same as Problem 14, for $y = x^4 e^{-x}$.

17. Sketch the curve $y = 2e^x - 2$. Find the area bounded by this curve, the x axis, and the lines $x = -1$ and $x = 2$, to three decimal places, using Table B.

18. If $f(x) = (e^{2x} + 1)/e^x$, find $f'(x)$ in two ways: by dividing each term of the numerator by the denominator and then differentiating; by using the rule for the derivative of a quotient. Reconcile your results.

19. If $G(x) = (e^{(1/2)x} + e^{-(1/2)x})(e^{(1/2)x} - e^{-(1/2)x})$, find $G'(x)$ in two ways: by multiplying out first, and then differentiating termwise; by using the rule for the derivative of a product. Reconcile your results.

20. Sketch the curve $y = e^x - 1$. Find the area bounded by this curve, the x axis, and the lines $x = -1$ and $x = 1$, to three decimal places, using Table B.

21. Same as Problem 20, for $y = 3e^x - 3$, from $x = -1$ to $x = 2$.

★ 22. When the body has received a dose of a drug, the concentration c of drug in the body fluids declines at an approximately exponential rate according to the relation $c = c_0 e^{-t}$, where c_0 is the initial dose and t the time in suitable units. A dose c_0 is administered repeatedly at constant time intervals t_0.
   (a) Find the concentration immediately after the first, second, and third doses.
   (b) Deduce the concentration immediately after the Nth dose. [Hint: $1 + x + x^2 + \cdots + x^{N-1} = (1 - x^N)/(1 - x)$.]
   (c) Show that when a very large number of doses have been given, the concentration is approximately $c_0/(e^{t_0} - 1)$ immediately before each dose and $c_0 e^{t_0}/(e^{t_0} - 1)$ immediately after.
   (d) Sketch the form of the graph of c against t.

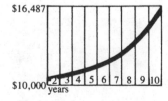

Fig. 6-6

★ 23. Figure 6-6 is a paraphrase of an advertisement designed to make you change your bank.
   (a) Assuming that daily payment of interest is effectively continuous, show that the two statements indicated by an asterisk are *in a sense* correct.
   (b) If you invest $100 with this bank, how much will you have after 10 years?
   (c) If your present bank pays 5% compounded *annually*, how much will you have after 10 years?
   (d) What rate will your present bank need to pay annually to give the same sum after 10 years as in part (b)?
   (e) In what sense are the statements of part (a) *not* correct?
   (f) Is it worth changing your account?

252

6.8 e^u: derivative and antiderivative

The Chain Rule enables us to find dy/dx in case $y = e^u$ and u is a function of x: From equation (5) **6.5**, we know that $dy/du = e^u$, so

$$\frac{dy}{dx} = \frac{dy}{du} \cdot \frac{du}{dx} = e^u \cdot \frac{du}{dx}.$$

Example 1
The graph of the function defined by $y = 4e^{-(1/2)x^2}$ is closely related to what is called the normal probability curve. Let us analyze the graph.

Problem 1
Show that $x = 0$ is the only critical number.

Problem 2
Show that $d^2y/dx^2 = 0$ for $x = \pm 1$.

We have a maximum point at $(0,4)$ and points of inflection at $(\pm 1, 4e^{-1/2})$. The natural domain of the function is the set of all real numbers. The general appearance of the curve is shown in Figure 6-7. In Problem 1 you should have found that

$$\frac{dy}{dx} = 4e^{-(1/2)x^2}(-x),$$

which can be written as

$$\frac{dy}{dx} = y(-x) = -xy.$$

In Chapter 8 we shall "solve the differential equation," $dy/dx = -xy$, to recapture the "primitive," $y = 4e^{-(1/2)x^2}$.

Example 2
To find the area under the curve $y = e^{2x}$ from $x = 0$ to $x = 1$ we must evaluate

$$\int_0^1 e^{2x}\,dx,$$

and to do this we must find an antiderivative of e^{2x}. A bit of experimentation leads to the result:

Because $\dfrac{d}{dx}(e^{2x}) = e^{2x} \cdot 2,$ we realize that $\dfrac{d}{dx}\left(\tfrac{1}{2}e^{2x}\right) = e^{2x}.$

Hence,

$$\int_0^1 e^{2x}\,dx = \tfrac{1}{2}e^{2x}\big|_0^1 = \tfrac{1}{2}(e^2 - 1).$$

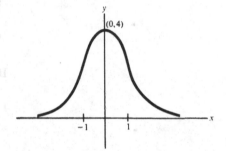

Fig. 6-7

(It is easy to make the mistake of thinking that substitution of the lower limit, 0, gives a zero result, but this is not so.)

PROBLEMS

3. If $y = e^{ax+b}$, where a and b are constants, verify that the relative rate of change of y with respect to x is constant. What is the constant? Rewrite the given expression for y in the form $y = c \cdot e^{rx}$, stating the values of c and r.

4. If $y = e^{x^2}$, is the relative range of change of y with respect to x constant?

5. Let $f(x) = e^x + e^{-x}$ and $g(x) = e^x - e^{-x}$. Show that $f'(x) = g(x)$, $g'(x) = f(x)$, $f''(x) = f(x)$, $g''(x) = g(x)$.

6. If $F(x) = (e^x + e^{-x})^2$, find $F'(x)$ in two ways:
   (a) First square the binomial and then differentiate each term.
   (b) Set $u = e^x + e^{-x}$ and use the Chain Rule directly.
   Reconcile your results.

7. If $y = e^{e^x}$, find dy/dx.

8. Find an antiderivative of each of the following:
   (a) $e^{x^2} \cdot 2x$ (b) $e^{x^2} \cdot x$ (c) $e^{x^2} \cdot 3x$

9. Verify that the area under the curve $y = e^{-x}$ from $x = 0$ to $x = 1$ equals $(1 - 1/e)$ square units.

10. Verify that the volume of the solid generated by revolving the area from Problem 9 about the x axis equals $(\pi/2)(1 - 1/e^2)$ cubic units.

11. Find the volume of the solid generated by revolving the area under the curve $y = e^{(1/3)x}$, between $x = 0$ and $x = 3$, about the x axis. Use Table B to express your answer as $k \cdot \pi$ cubic units, where k is given to three decimal places.

12. Same as Problem 11, for $y = e^{(3/2)x}$, between $x = 0$ and $x = 1$.

13. Same as Problem 11, for $y = \sqrt{x}\, e^{x^2}$, between $x = 0$ and $x = 2$.

14. For the curve $y = x^2 \cdot e^{2-x}$, find the coordinates of maximum and minimum points and points of inflection, if any. What is the natural domain of this function of x? Sketch the curve.

15. Same as Problem 14, for $y = x \cdot e^{1/x}$.

16. If $F(x) = (e^x + e^{-x})/e^x$, find $F'(x)$ in two ways:
   (a) Divide each term of the numerator by the denominator and then differentiate.
   (b) Use the formula for the derivative of a quotient.
   Reconcile your results.

17. Consider the function $f(x) = \sqrt{x}\, e^{-0.01x^2}$
   (a) Find the natural domain.
   (b) Find the coordinates of maximum and minimum points on the graph of f.
   (c) Find the volume obtained by revolving the area under the graph of f, between $x = 0$ and $x = 5$, about the x axis.

★   (d) Find the coordinates of any point of inflection on the graph of f.
18. Consider the curve with equation $y = e^{(1/2)x} - e^{1/2}$.
    (a) Determine the values of y at $x = 0$ and $x = 2$, using Table B.
    (b) Determine where the curve crosses the x axis.
    (c) Determine the general appearance of the curve in the interval $[0, 2]$.
    (d) Find the area bounded by the curve, the x axis, and the lines $x = 0$ and $x = 2$. Use Table B to give the answer to three decimal places.
    (e) Find the volume generated by revolving the area from (d) about the x axis. Express the answer as $k \cdot \pi$ cubic units, where k is a number to three decimal places (Table B).

6.9 Summary

The exponential function appears in so many contexts that it has many names – the law of natural growth, the law of organic growth, and the compound interest law are some of them. This function and its inverse, the logarithmic function, are of central importance in any study of variation. We turn to the logarithmic function in the next chapter.

SAMPLE TEST

These sample test questions include review material from earlier chapters.
1. In how long a time will any sum be trebled (tripled) at 10% interest, compounded continuously. (Use Table B.)
2. If $F(x) = (e^{3x} + e^{-3x})^2$, find $F'(x)$ in two ways: by squaring the binomial and differentiating termwise; by setting $u = e^{3x} + e^{-3x}$. Reconcile your results.
3. Find the volume of the solid generated by revolving the area under the curve $y = e^{-x}$, between $x = 0$ and $x = 2$, about the x axis. Use Table B to express the answer as $k\pi$ cubic units, where k is a number to three decimal places.
4. Find the maximum and minimum values of y on the curve $y = xe^{-(1/2)x^2}$.
5. Sketch a graph of $y = f(x)$ if $f(0) = 1$, $f(3) = -1$, $f(7) = 4$, $f(9) = 3$, $f(12) = 5$, $f'(3) = 0$, $f'(7) = \frac{1}{2}$, $f'(9)$ does not exist, and $f'(x)$ increases throughout $0 < x < 5$, decreases throughout $5 < x < 9$, and increases throughout $9 < x < 12$.
6. An ellipse is an oval that looks like that in Figure 6-8. The area within such an ellipse with semiaxes u and v is given by $A = \pi uv$. What is the volume of an "elliptical cone" – a solid every horizontal section of which x in. below the vertex is an ellipse for which $u = 3x$, $v = e^{(1/10)x^2} - 1$ – from $x = 0$ to $x = 5$? Use Table B.
7. Find all points of inflection on the curve $y = xe^{-(1/2)x^2}$.

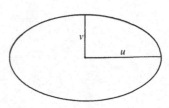

Fig. 6-8

255

Here is another sample test on Chapter 6:

1'. The pressure (p kg/cm^2) in a space capsule *decreases* at the constant instantaneous percentage rate of 11% per week. How long does it take for p to drop to one-third of its initial value? Use Table B.

2'. If $G(x) = (e^{2x} + e^{-2x})/e^x$, find G' in two ways: by dividing each term in the numerator by the denominator and then differentiating; by using the formula for the derivative of a quotient. Reconcile your results.

3'. Sketch the curve $y = e^{2x} - 1$. Find the area bounded by this curve, the x axis, and the lines $x = -1$ and $x = 1$, to three-place accuracy, using Table B.

4'. Find any maximum and minimum values of y on the curve $y = x^2 \cdot e^{1/x}$, testing to show whether your have a maximum or a minimum.

5'. Sketch a graph of $y = f(x)$, $-2 \le x \le 7$, if $f(-2) = 3$, $f(2) = 0$, $f(3) = -1$, $f(7) = 2$, $f'(3) = 0$, $f'(4)$ does not exist, $f''(x) < 0$ throughout $-2 < x < 2$, $f''(x) > 0$ throughout $2 < x < 4$, and $f'(x)$ increases throughout $4 < x < 7$.

6'. A cone is generated by revolving about the x axis the line segment joining the points with coordinates $(0,0)$ and $(5,2)$. Use integration to find the volume of this cone, and check using the formula for the volume of a cone.

7'. The slope of a curve is given by $x \cdot e^{-x^2}$, and the curve passes through the point $(0,1)$.

(a) Find an equation of the curve.

(b) Find the values of x corresponding to maximum flexion and minimum flexion on the curve, testing to show which.

Logarithmic functions

7.1 Introduction

You may have worked with logarithms of numbers as an aid to computation. A review of logarithms in this context is provided in Section 0.13. The availability of calculators and computers has reduced the importance of logarithms for numerical computation, but the *logarithmic functions* are still of central importance in mathematics and its applications, as will be seen in this chapter.

We shall introduce the logarithmic functions as inverses of the exponential functions.

7.2 Inverse functions and the inverse of the exponential

We first encountered the idea of functions that are inverses of each other in Section 1.11. The discussion contained in that section informs us that $f(x) = x^3$, with $D_f = (-\infty, \infty)$ and $R_f = (-\infty, \infty)$, and $g(x) = x^{1/3}$, with $D_g = (-\infty, \infty)$ and $R_g = (-\infty, \infty)$, are inverses of each other, with graphs as shown in Figure 7-1. Likewise, the functions $F(x) = x^2$, with $D_F = [0, \infty)$ and $R_F = [0, \infty)$, and $G(x) = x^{1/2}$, with $D_G = [0, \infty)$ and $R_G = [0, \infty)$, are inverses of each other, with graphs as shown in Figure 7-2.

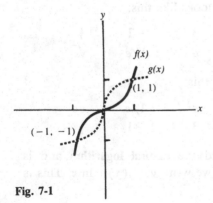

Fig. 7-1

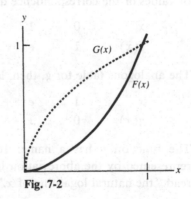

Fig. 7-2

257

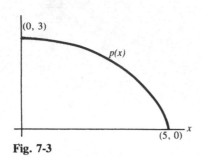

(0, 3)

p(x)

(5, 0)

Fig. 7-3

Problem 1

Why is the domain of F restricted as described? That is, what about an inverse of $H(x) = x^2$, with its natural domain $(-\infty, \infty)$?

Problem 2

Figure 7-3 shows the graph of the function $p(x) = \frac{3}{5}(25 - x^2)^{1/2}$, with $D_p = [0,5]$ and $R_p = [0,3]$. Write the equation for its inverse function $q(x)$, state the domain and range of q, and sketch the graphs of p and q on the same axes.

All this is by way of a review of the idea of inverse functions, as an introduction to the extremely important *inverse* of the *exponential function*. The graph of $y = f(x) = e^x$ is shown in Figure 7-4. $D_f = (-\infty, \infty)$ and $R_f = (0, \infty)$.

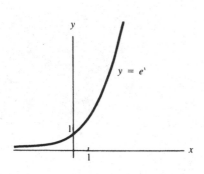

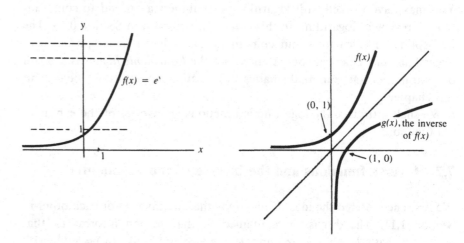

Fig. 7-4

Because $dy/dx = f'(x) = e^x$, and because e^x is always positive, we know that f is a strictly increasing function, so any horizontal line above the x axis meets the curve in only one point. Thus, the inverse of f is a function – call it $g(x)$. $D_g = R_f = (0, \infty)$, and $R_g = D_f = (-\infty, \infty)$. A table of a few pairs of values of the correspondence defining f looks like this:

x	0	1	2	3	-1	$\frac{1}{2}$
$f(x)$	1	e	e^2	e^3	$1/e$	$\sqrt{e}$

The analogous table for g, then, looks like this:

x	1	e	e^2	e^3	$1/e$	$\sqrt{e}$
$g(x)$	0	1	2	3	-1	$\frac{1}{2}$

The function g has a name: It is called the **natural logarithm** and is represented by the abbreviation ln. Thus, we write $y = g(x) = \ln x$. This is read "the natural logarithm of x."

The function $f(x) = e^x$ is not the only exponential that we encounter. We also meet such exponentials as $f_1(x) = 10^x$, $f_2(x) = (1.8)^x$, and, more generally, $F(x) = b^x$ for any positive $b \neq 1$. Each of these functions has an inverse, called a logarithm, and abbreviated log. To distinguish among them, we note the "base":

$$\text{The inverse of} \quad f_1(x) = 10^x \quad \text{is} \quad \log_{10} x.$$
$$\text{The inverse of} \quad f_2(x) = (1.8)^x \quad \text{is} \quad \log_{1.8} x.$$
$$\text{The inverse of} \quad F(x) = b^x \quad \text{is} \quad \log_b x.$$

The expression $\log_b x$ is read "the logarithm of x to the base b."

The natural logarithm, $\ln x$, can also be written $\log_e x$. If there is no ambiguity about the base, it can be omitted, and the function can simply be written $\log x$. Unfortunately, there *is* occasional ambiguity in standard practice: (a) For computational purposes, the base 10 is most convenient, and $\log_{10} x$ – the so-called common logarithm – is often written as $\log x$. (b) For theoretical purposes (the analysis of how functions behave), the base e is most convenient, and $\log_e x$ – the natural logarithm – is also often written as $\log x$. It is to avoid this possible confusion that the notation $\ln x$ for the natural logarithm is adopted.

7.3 Laws of logarithms

Just as the equation $y = x^3$ can be "solved for x" and written $x = y^{1/3}$, so likewise the equation

$$y = \log_b x \tag{1}$$

can be "solved for x" and written

$$x = b^y. \tag{2}$$

Thus, instead of using the language of inverses, we can define *the logarithm of x to the base b* as that number, y, such that $x = b^y$, or equivalently:

The logarithm of x to the base b is the exponent of the power to which b must be raised to equal x.

Problem 1
What is the value of $10^{\log_{10} 100}$? Of $10^{\log_{10} 1000}$? Of $10^{\log_{10} 20}$? Of $e^{\ln 5}$? Of $b^{\log_b x}$? Of $e^{\ln a}$?

The importance of logarithms as an aid to computation is based on two theorems, called the laws of logarithms:

I. $\log_b(x_1 \cdot x_2) = \log_b x_1 + \log_b x_2$: "The log of a product equals the sum of the logs."

II. $\log_b x^n = n \log_b x$: "The log of a power equals the exponent times the log."

These laws can be established using the definition of the logarithm [i.e., the equivalence of equations (1) and (2)] and the known laws of exponents:

$$b^u \cdot b^v = b^{u+v} \quad \text{and} \quad (b^u)^n = b^{un}.$$

Here's the way it goes:

If we set $\log_b(x_1 \cdot x_2) = r$, then $x_1 \cdot x_2 = b^r$.

If we set $\log_b x_1 = s$, then $x_1 = b^s$.

If we set $\log_b x_2 = t$, then $x_2 = b^t$.

Now, $b^r = x_1 \cdot x_2 = b^s \cdot b^t = b^{s+t}$. Hence, $r = s + t$. Substituting the expressions for r, s, and t into this last equation is exactly the statement of (I). Similarly:

If we set $\log_b x^n = u$, then $x^n = b^u$.

If we set $\log_b x = v$, then $x = b^v$.

If we raise both sides of this last equation to the nth power, we get $x^n = (b^v)^n = b^{nv}$. But x^n also equals b^u. Hence, $u = nv$. Substituting the expressions for u and v into this last equation is exactly the statement of (II).

Problem 2

Use (I) and (II) to obtain the two corollary laws of logarithms:

III. $\log_b(x_1/x_2) = \log_b x_1 - \log_b x_2$.

IV. $\log_b(\sqrt[n]{x}) = (1/n)\log_b x$.

Because the operations of addition and subtraction are easier to perform then are multiplication and division, and because multiplying or dividing by n is much easier than raising to the nth power or extracting the nth root, the availability of tables of $\log x$ was a tremendous boon to persons faced with computational problems, and the invention of logarithms was characterized by Laplace as "doubling the life of an astronomer." The great German mathematician, Karl Friedrich Gauss, who did much work in astronomy and physics, is said to have memorized a table of logarithms to save himself the time of having to look up a value each time he needed it. The development of calculators and of computers in the middle of the twentieth century has reduced the importance of logarithms for computation, but not the importance of the logarithmic functions.

The particular simplicity of the base 10 for computation stems from the fact that we use 10 as the "base" of our number system. For example,

$$12.73 = 10(1.273); \quad 127.3 = 10^2(1.273); \quad 0.01273 = 10^{-2}(1.273).$$

Hence,

$$\log_{10} 12.73 = \log_{10} 10 + \log_{10} 1.273 = 1 + \log_{10} 1.273,$$

$$\log_{10} 127.3 = \log_{10} 10^2 + \log_{10} 1.273 = 2 + \log_{10} 1.273, \quad \text{and}$$

$$\log_{10} 0.01273 = \log_{10} 10^{-2} + \log_{10} 1.273 = -2 + \log_{10} 1.273.$$

Thus, if we have a table of common logarithms of numbers in the interval $(1, 10)$, we can quickly write the logarithms of numbers of any size.

When we have occasion to compute with natural logarithms, we use similar relationships:

$$\ln 12.73 = \ln 10 + \ln 1.273,$$

$$\ln 127.3 = \ln 10^2 + \ln 1.273 = 2\ln 10 + \ln 1.273, \quad \text{and}$$

$$\ln 0.01273 = \ln 10^{-2} + \ln 1.273 = -2\ln 10 + \ln 1.273.$$

The table of natural logarithms again needs to cover only the domain $(1, 10)$, but we also need to know the value of $\ln 10$.

Example 1

Suppose that in a standard compound-interest situation the interest rate changes a bit. How will this affect the amount after a certain number of years? To be specific, let us determine the approximate change in the amount of $2000 after 27 years at an annual rate, r, compounded semiannually, that would result from increasing r from 5.9% to 6.1%.

Because we are interested in the change in A corresponding to a change in r, we first express A as a function of r:

$$A = 2000\left(1 + \frac{r}{2}\right)^{54}.$$

Because the change in r is small, we can appropriately use the differential approximation:

$$\Delta A \approx dA = \frac{dA}{dr} \cdot dr.$$

Problem 3

Differentiate to obtain $dA/dr = 54{,}000(1 + r/2)^{53}$.

As long as we are approximating, we might as well simplify matters by using $r = 0.06$, so $dA/dr = 54{,}000(1.03)^{53}$. In this case, $dr = 0.002$, so $dA = 108(1.03)^{53}$. We now must calculate this number:

$$\log_{10} dA = \log_{10} 108 + 53\log_{10}(1.03)$$

$$= 2 + \log_{10} 1.08 + 53\log_{10}(1.03).$$

Problem 4

Use Table C at the end of this book to obtain $\log_{10} dA = 2.7118$.

Working with the body of Table C, we find that the number whose log is 0.7118 is 5.15. Hence, the number whose log is 2.7118 is 515. The answer, then, is that $\Delta A \approx dA = 515$.

c Problem 5

Use a calculator to find the value of $108(1.03)^{53}$ without logs. Also calculate A for $r = 6.1$ and for $r = 5.9$, and subtract to find ΔA without differentiating.

Example 2

In how long a time will any sum be doubled at 8% interest compounded annually?

Starting with $\$P$, we wish A to equal $2P$, with $A = P(1.08)^n$. Thus, we are to solve the equation $2P = P(1.08)^n$ for n. As we have seen in similar cases, this immediately reduces to $(1.08)^n = 2$. Hence, $n \cdot \log_{10} 1.08 = \log_{10} 2$, or

$$n = \frac{\log_{10} 2}{\log_{10} 1.08} = \frac{0.3010}{0.0334} = \frac{30.10}{3.34}.$$

By long division, we obtain $n = 9.01$ years.

Problem 6

For practice, use common logs to perform this division.

Example 3

Same as Example 2, if interest is compounded continuously. This is the same as Problem 8, **6.4**, where we used the small tables of powers of e (Table B) to obtain the answer. We shall now take a slightly different approach.

Problem 7

Show that we must find t satisfying

$$e^{0.08t} = 2. \tag{3}$$

Using natural logs, we have $0.08t = \ln 2$, so $t = (\ln 2)/0.08$.

Problem 8

Use Table B and division to obtain $t = 8.66$.

We can also use common logarithms to solve equation (3): $0.08t \cdot \log_{10} e = \log_{10} 2$, so

$$t = \frac{\log_{10} 2}{(0.08)\log_{10} e} = \frac{\log_{10} 2}{(0.08)\log_{10} 2.718}.$$

Use Table C to obtain the same value, $t = 8.66$.

In this case, natural logs are *much* easier to use!

PROBLEMS

10. The population of the United States was 150 million in 1936 and 200 million in 1964. Assuming a constant relative rate of growth, what will be the population in the year 2000?

11. In how long a time will any sum be quadrupled at 10% interest, compounded semiannually?

12. Use differentiation to determine the approximate change in the amount of \$1500 after 23 years at an annual rate r, compounded semiannually, that would result from decreasing r from 0.072 to 0.069.

13. The population of Erewhon was 3 million in 1900 and 6 million in 1950. Assuming a constant relative rate of growth, what will be its population in 2050? Did you (can you) do this one in your head?

○ 14. Starting with $\log_a x = y$, write this as $x = a^y$, and take logs of both sides to the base b, to obtain, finally

$$\log_b x = \log_a x \cdot \log_b a,$$

for all positive numbers x. Then set $x = b$ to obtain

$$\log_a b \cdot \log_b a = 1.$$

15. The earth's population now produces energy at the rate of 5×10^{19} calories per year. The earth's land mass receives 2.5×10^{23} calories per year from the sun. We have been increasing our production at the rate of 5% per year. (From an editorial in *Science*, October 18, 1974.) If this rate were to continue, in how many years would our production equal the amount received by the land mass from the sun?

16. A formula allegedly used by bankers to get a quick, rough approximation to the length of time (N years) for a sum of money to double when compounded annually at an interest rate $r\%$ is the following: $N = 73/r$. For example, at a rate of 10%, compounded annually, money will double in approximately $73/10 = 7.3$ years; at a rate of 12%, in approximately $73/12 = 6.1$ years; and so forth. Discuss this approximation.

17. (a) If your first child is born when you are 25 years old, then 25 years from now he or she may be at the same stage at your college as you are now. If college costs increase at the rate of 6% per year, compounded annually, what will be the cost of a year at your college for your first child? For your youngest child, born when you are 35 years old?

263

(b) Same as (a), if college costs increase by 7% per year, compounded annually.

7.4 The derivative of the log function

If $y = \ln x$, we can find the rate of change, dy/dx, easily by using implicit differentiation. We use the definition of ln to write $x = e^y$, and knowing that y is a function of x, we differentiate both sides of the equation with respect to x to obtain

$$1 = e^y \frac{dy}{dx}.$$

Hence, $dy/dx = 1/e^y = 1/x$. Thus, we have shown the following:

$$\text{If} \quad y = \ln x, \quad \text{then} \quad \frac{dy}{dx} = \frac{1}{x}. \qquad (4)$$

Example 1
Let us use this formula to analyze the function given by the equation $y = x \cdot \ln x$.

Problem 1
What is the domain of this function? For what number(s) x is $y = 0$?

Problem 2
Show that the only critical number is $x = 1/e$ and that the corresponding value of y is a minimum. What is the minimum value of y?

Problem 3
Show that there are no points of inflection on the graph of this function.

Problem 4
Knowing that $\ln 10 = 2.3026$, compute the values of y corresponding to $x = 0.1, 0.01,$ and 0.001.

The information you have obtained from working these problems enables us to sketch the curve, as shown in Figure 7-5.

The Chain Rule, together with the basic formula, equation (4), leads to a more general statement:

$$\text{If } y = \ln u, \text{ and } u \text{ is a function of } x, \quad \text{then} \quad \frac{dy}{dx} = \frac{1}{u} \cdot \frac{du}{dx}. \qquad (5)$$

Example 2
If $y = \ln(x^3 - 1)$, then $dy/dx = [1/(x^3 - 1)] \cdot 3x^2$.

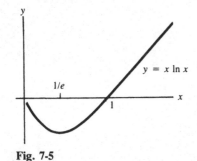

$y = x \ln x$

Fig. 7-5

Example 3

If $y = \ln x^6$, then $dy/dx = (1/x^6) \cdot 6x^5 = 6/x$. (It would have been simpler to write $y = 6 \ln x$ *before* differentiating.)

Example 4

If $y = \ln \sqrt{(x^2 - 9)/(x^2 + 9)}$, it surely pays to simplify before differentiating: $y = \frac{1}{2} \ln(x^2 - 9) - \frac{1}{2} \ln(x^2 + 9)$.

Problem 5

Now differentiate to obtain $dy/dx = 18x/[(x^2 - 9)(x^2 + 9)]$.

Example 5

Finding dy/dx if $y = \sqrt{(x^2 - 9)/(x^2 + 9)}$ seems to have nothing to do with logs, and, indeed, our earlier methods are adequate to answer the question.

Problem 6

Show that $dy/dx = 18x/[(x^2 - 9)^{1/2}(x^2 + 9)^{3/2}]$.

The labor of this differentiation can be reduced by *introducing* logarithms, as follows: Take logs of both sides, before differentiating, to obtain

$$\ln y = \ln \sqrt{\frac{x^2 - 9}{x^2 + 9}}.$$

The purpose of this step is to permit a simplification of the right side, as was done initially in Example 4:

$$\ln y = \frac{1}{2} \ln(x^2 - 9) - \frac{1}{2} \ln(x^2 + 9).$$

Implicit differentiation now gives

$$\frac{1}{y}\frac{dy}{dx} = \frac{1}{2}\frac{2x}{x^2 - 9} - \frac{1}{2}\frac{2x}{x^2 + 9}$$

$$= \frac{18x}{(x^2 - 9)(x^2 + 9)}.$$

Hence, $dy/dx = y \cdot 18x/[(x^2 - 9)(x^2 + 9)]$, which reduces to the result obtained in Problem 6. This technique of *introducing* logs to simplify the work connected with a differentiation is called "logarithmic differentiation."

PROBLEMS

7. Use Table B to find the values of $\ln x$ for $x = \frac{1}{4}, \frac{1}{2}, 1, \frac{3}{2}, 2, \frac{5}{2}$, and 3, and plot a large-scale graph of $y = \ln x$ over the interval $\left[\frac{1}{4}, 3\right]$, using the same scale vertically as horizontally. Draw what look like the tangent lines to this curve at $x = 1$, at $x = \frac{3}{2}$, and at $x = 2$, and check the slopes

of the lines you have drawn against what you know to be the correct values for these slopes.

8. Table B shows $\ln 5$ to be 1.6094. Use differential approximations to find $\ln 5.01$, $\ln 5.02$, $\ln 5.03$, and $\ln 5.04$, and check your results against those given in the table. How large is h before $\ln(5+h)$ found in this way deviates from the tabulated value?

9. After preliminary simplifications, find dy/dx for each of the following cases:

(a) $y = 3\ln x^{1/2}$ (b) $y = 3(\ln x)^{1/2}$ (c) $y = \ln 5x - \ln x^2$

(d) $y = \ln\dfrac{5}{x} + \ln\dfrac{x^3}{8}$ (e) $y = \ln(\ln x)$ (f) $y = \left(x + \ln\dfrac{1}{x}\right)^3$

(g) $y = \ln(x + \sqrt{x^2 - a^2})$, where a is constant

(h) $y = \ln\left(\dfrac{x + \sqrt{x^2 - a^2}}{x + \sqrt{x^2 + a^2}}\right)$, where a is constant

(i) $y = \ln\sqrt{10/x}$

10. Find dy/dx in each of the following cases:

(a) $y = \ln\sqrt{(5+x)/(5-x)}$ (b) $y = \sqrt{(5+x)/(5-x)}$

11. Find dy/dx in each of the following cases:

(a) $y = \ln\sqrt{(25+x^2)/(25-x^2)}$ (b) $y = \sqrt{(25+x^2)/(25-x^2)}$

12. Use logarithmic differentiation to solve Problem 5(e), **6.6**.

★ 13. Consider the curve $y = x^x$, for $x > 0$.

(a) Use logarithmic differentiation to find dy/dx, and thus find the coordinates of any maximum and minimum points on the curve.

(b) Find d^2y/dx^2, and show that the curve is concave up for $x > 0$.

c   (c) Calculate values of y for small positive x's, and hence guess $\lim y$ as $x \to 0$ through positive numbers. Thus extend the function by defining y at $x = 0$ so as to make the extension continuous for all $x \geq 0$.

(d) Sketch the graph of the extension of the function.

c 14. (a) Apply the Newton–Raphson method to approximate the solution of the equation $x + e \cdot \ln x = 0$ after plotting the graph of $y = x + e \cdot \ln x$ to find a good starting point.

(b) Same as (a), for the equation $e^x - 2x - 1 = 0$.

(c) Why would part (b) be awkward (or worse) to solve numerically if you used an initial value near $\ln 2$? Why can no similar difficulty arise in part (a)?

(d) Apply the Newton–Raphson method to find all solutions of $x = 3\ln x$.

7.5 Antiderivatives of $1/x$

In Chapter 5 we found the antiderivatives of x^n to be $x^{n+1}/(n+1) + c$, provided that $n \neq -1$. The answer to the "missing case" – the antiderivatives of $1/x$ – comes from equation (4), which states that the derivative of $\ln x$ is $1/x$. Hence, we now know the following:

For $x > 0$, the antiderivatives of $1/x$ are $\ln x + c$.

We can use this result to obtain the area under the curve $y = 1/x$ from $x = 1$ to $x = 2$ (Figure 7-6):

$$A = \int_1^2 \frac{1}{x}\, dx = \ln x|_1^2 = \ln 2 - \ln 1 = \ln \frac{2}{1} = \ln 2 \text{ (square units).}$$

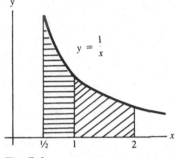

Problem 1
Show that the area under this curve from $x = \frac{1}{2}$ to $x = 1$ also equals $\ln 2$ square units. Likewise for the area under this curve from $x = a$ to $x = 2a$, for any positive number a.

Fig. 7-6

Problem 2
Show that the volume of the solid obtained by revolving the area under $y = 1/x$ from $x = 1$ to $x = 2$ about the x axis equals $\pi/2$ cubic units.

Problem 3
Find the volume of the solid obtained by revolving the area of the first part of Problem 1 about the x axis. The second part of Problem 1.

Problem 4
Find the volume of the solid obtained by revolving the area under $y = 1/\sqrt{x}$ from $x = 5$ to $x = 15$ about the x axis. Similarly, from $x = a$ to $x = 3a$, for any positive number a.

A portion of the curve $y = e^x/(e^x - 1)$ is shown in Figure 7-7. The area (A square units) under this curve between $x = 1$ and $x = 2$ is given by

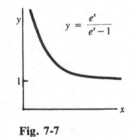

$$A = \int_1^2 \frac{e^x}{e^x - 1}\, dx.$$

We can evaluate this integral as follows: If we set $u = e^x - 1$, then $du/dx = e^x$, and $du = e^x\, dx$. Thus,

Fig. 7-7

$$\frac{e^x}{e^x - 1}\, dx = \frac{1}{u}\, du.$$

We know that an antiderivative of $1/u$ is $\ln u$, and $\ln u$, of course, equals $\ln(e^x - 1)$. Hence,

$$A = \ln(e^x - 1)|_1^2 = \ln(e^2 - 1) - \ln(e - 1) = \ln \frac{e^2 - 1}{e - 1}.$$

Problem 5
Show that this reduces to $A = \ln(e + 1)$. Evaluate this number by use of the tables.

PROBLEMS

6. (a) What is the natural domain of the function given by $y = e^x/(e^x - 1)$?
   (b) Show that the slope of the graph is negative everywhere.
   (c) Sketch the graph.
   (d) Find the area under the curve between $x = 2$ and $x = 4$, evaluating the result by use of the tables.

7. Find the area under $y = e^x/(e^x - 1)^2$ between $x = 2$ and $x = 4$.

8. Without use of tables, find the area under $y = 1/x$ from $x = k$ to $x = ke$, where k is any positive number.

9. If a, b, and k are any positive numbers, show that the area under $y = 1/x$ between $x = a$ and $x = b$ equals the area under this curve between $x = ka$ and $x = kb$.

10. Show that the area under $y = 3/(x - 2)$ between $x = 3$ and $x = 6$ equals $6 \cdot \ln 2$ square units.

11. Evaluate each of the following integrals, simplifying the results as much as possible without using tables:

   (a) $\int_0^5 \dfrac{dx}{3x + 1}$ (b) $\int_0^5 \dfrac{dx}{(3x + 1)^2}$ (c) $\int_0^4 \dfrac{x\,dx}{x^2 + 9}$

   (d) $\int_0^4 \dfrac{x\,dx}{\sqrt{x^2 + 9}}$ (e) $\int_1^3 \dfrac{(x + 1)\,dx}{x^2 + 2x + 3}$ (f) $\int_1^3 \dfrac{(x + 1)\,dx}{(x^2 + 2x + 3)^2}$

   (g) $\int_2^6 \dfrac{e^{(1/2)x}}{e^{(1/2)x} + 1}\,dx$ (h) $\int_2^6 \dfrac{e^{(1/2)x}}{\left(e^{(1/2)x} + 1\right)^2}\,dx$

7.6   Derivatives of b^x and $\log_b x$

We are now in a position to take that final step, referred to near the end of Section 6.5, to find the derivative of $y = c \cdot b^x$. One approach is to find r such that $b = e^r$. But, by the definition of logarithm, $r = \ln b$, so $b = e^{\ln b}$. Hence,

$$y = c \cdot b^x = c \cdot \left(e^{\ln b}\right)^x = c \cdot e^{(\ln b)x},$$

and

$$\frac{dy}{dx} = c \cdot e^{(\ln b)x} \cdot (\ln b) = y \ln b. \tag{6}$$

Problem 1

Chapter 6 began with a consideration of the function $W = 2(1.8)^t$. Follow the method of the preceding paragraph to obtain $dW/dt = W \ln 1.8 \approx (0.5878)W$.

An alternative method is as follows: Take logs of both sides of the equation $W = 2(1.8)^t$ as a preliminary to differentiation:

$$\ln W = \ln 2 + t \ln 1.8.$$

Problem 2

Differentiate implicitly and solve for dW/dt to obtain $dW/dt = W\ln 1.8 = 2(1.8)^t \ln 1.8$.

Problem 3

Use the method of Problem 2 with $y = c \cdot b^x$ to obtain $dy/dx = y \ln b = c \cdot b^x \ln b$.

In practice it is unnecessary to memorize a formula like equation (6) – if the derivative of $y = b^x$ is to be found, it is easy to proceed as in Problem 3.

Problem 4

If $y = 10^x$, find the rate of change of y with respect to x at $x = 2$.

We know the value of dy/dx if $y = \ln x$. What is the derivative if $y = \log_b x$? In Problem 14, **7.3**, we found that $\log_b x = \log_a x \cdot \log_b a$. For $a = e$, this gives $\log_b x = \ln x \cdot \log_b e$. Therefore, we can write $y = \log_b x$ as

$$y = \ln x \cdot \log_b e.$$

The second factor on the right is a constant, of course. Hence, we have immediately

$$\frac{dy}{dx} = \frac{1}{x} \cdot \log_b e.$$

Problem 5

If $y = \log_{10} x$, find dy/dx at $x = 1$. At $x = 5$.

Problem 6

In Section 6.2, we noted, through use of the definition of the derivative, that if $f(x) = c \cdot b^x$, then $f'(x) = c \cdot b^x \cdot \lim_{h \to 0}[(b^h - 1)/h]$. We now know that $f'(x) = c \cdot b^x \cdot \ln b$. Hence, $\lim_{h \to 0}[(b^h - 1)/h] = \ln b$.

C (a) By means of the square-root key of a calculator (in effect, taking h of the form 2^{-n}), approximate $\ln 10$ to three decimal places. Check your result by use of Table B or with a calculator that has a ln function.

C (b) Approximate $\ln 2$, $\ln 5$, and $\ln 8$.

   (c) The merit of this technique is that you can calculate natural logarithms on a calculator having nothing more sophisticated than a square-root function. Discuss the limitations of the technique.

7.7 Log–log and semilog graphs

If the relation between two variables x and y is shown in a table of values – obtained, perhaps, from an experiment, or from data accumulated by a government office – we would like to know if there is an equation in x and

y, or, better still, a formula for y in terms of x, that accords well with the values of the table.

If a plot of the points corresponding to the pairs in the table shows them to lie on a straight line, the answer is easy: $y = ax + b$, where the constants a and b can be determined by using two pairs of values.

As was remarked in Chapter 1, the art of curve fitting in cases other than the linear case can be complicated, but there are two situations of importance that we can now handle.

The power law

$$\text{If} \quad y = ax^n, \quad \text{for some constants } a \text{ and } n, \tag{7}$$

$$\text{then} \quad \log y = \log a + n \log x, \tag{8}$$

where any base of logarithms can be used. If we think of new variables $Y = \log y$ and $X = \log x$, then equation (8) is equivalent to

$$Y = \log a + n \cdot X. \tag{9}$$

But (9) is a linear equation in Y and X. Hence, if we start with pairs of values satisfying the power law (7), and plot $\log x$ horizontally and $\log y$ vertically, the points will lie along a straight line, of slope n.

Problem 1

For the equation $y = 5x^2$, compute the values of y corresponding to $x = 1, 2, 3, 4$, and 5, and look up the natural logs of each value of x and its corresponding y. Plot these logarithmic values, and see that the points lie on a straight line. Do the same thing with common logs for this example.

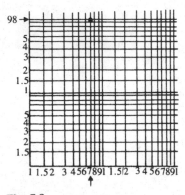

Fig. 7-8

Contrariwise, if we start with pairs of values such that the plot of $\log y$ versus $\log x$ gives points lying on a straight line, we know that we have a relation like equation (9), and, hence, by working back, like equation (7). That is, the relation between x and y is a power law, and the values of a and n can be determined from two pairs of the table. To save looking up logarithms, we can plot the points on "log–log" paper, which has logarithmic scales, rather than uniform scales, both horizontally and vertically (Figure 7-8). When we plot $(7, 98)$ on this paper, we actually are plotting $(\log 7, \log 98)$. Note that the horizontal axis begins with 1, not with 0 – the log of 0 does not exist. If we start with 1, the successive division points on the paper represent

$$1.5, 2, 3, 4, 5, 6, 7, 8, 9, 10, 15, 20, 30, 40, \ldots, 100, 150, 200, \ldots.$$

The distance from 3 to 7, say, should appear the same as the distance from 30 to 70 or the distance from 300 to 700, because we really have the distance from $\log 3$ to $\log 7$, and

$$\log 7 - \log 3 = \log \tfrac{7}{3} = \log \tfrac{70}{30} = \log 70 - \log 30, \text{ etc.}$$

If we start the horizontal axis at 10, the successive division points on the paper represent

$$15, 20, 30, 40, 50, 60, 70, 80, 90, 100, 150, 200, 300, 400, \ldots, 1000, 1500, 2000, \ldots,$$

and if we start with 0.01, they represent

$$0.015, 0.02, 0.03, \ldots, 0.1, 0.15, 0.2, 0.3, \ldots, 1, 1.5, 2, 3, \ldots.$$

The foregoing remarks apply to the vertical axis as well. It is entirely acceptable to start the horizontal axis at 100, say, and the vertical axis at 0.1.

Problem 2

On a piece of log–log paper, plot the points corresponding to this table:

x	1	2	3	4	5	6
y	360	90	40	22.5	14.4	10

You should find that they lie along a straight line, so that the relation between y and x must be of the form $y = ax^n$.

Problem 3

Use two pairs from the table to determine a and n, and check with the other pairs. What is the measured slope of the line you plotted?

The numbers in this example are so simple that you can find a and n almost by inspection. In more complicated cases, you will have to use logs.

A log–log graph, even if not straight, is useful if we need to display a very large domain and a very large range. Because $\log_{10} 100,000 = 5$, large intervals are compressed so as to fit on a graph of reasonable size. Of course, features of the relation between the variables appear different from the way they would appear on a graph with uniform horizontal and vertical scales.

The exponential function

$$\text{If } y = ce^{rx}, \text{ for some constants } c \text{ and } r, \tag{10}$$

$$\text{then} \quad \ln y = \ln c + rx. \tag{11}$$

If we think of new variables $Y = \ln y$ and $X = x$, then equation (11) is equivalent to

$$Y = \ln c + rX.$$

From here on, the discussion proceeds in a fashion analogous to that on the power law earlier. We use "semilog" paper, with a *uniform* scale horizontally and any *logarithmic* scale vertically (Figure 7-9). If the points corresponding to the pairs in a table lie on a straight line, we have an exponential function, as in equation (10), and we determine c and r from two pairs of the table.

There is one essential difference between the case of the power law and the case of the exponential function: If y as a function of x is given as a

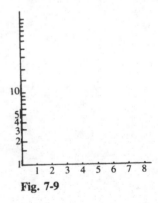

Fig. 7-9

271

power law ($y = ax^n$), then $x = (1/a)^{1/n} \cdot y^{1/n}$ (i.e., x as a function of y) is also a power law. It doesn't matter in our test, then, which variable we plot horizontally. But if y as a function of x is an exponential, x as a function of y is *not* an exponential. This implies that if a semilog plot with x horizontal does not give a straight line, we should try a semilog plot with x vertical before we give up and conclude that we have no exponential function.

Example 1

The diameter (d units) of the pupil of the eye of a dog varied with the intensity of light (l units) as given here:

l	1	2	4	10	50
d	4.6	3.9	3.2	2.3	0.7

We wish to discover the law relating d and l.

Problem 4

Plot on log–log paper, on semilog paper with l plotted horizontally, and on semilog paper with d plotted horizontally.

Only the last gives points lying on a straight line. Hence, the relation is $l = c \cdot e^{r \cdot d}$ for some c and r.

Problem 5

Use two pairs from the table to obtain $r = -1$ and $c = 100$.

Hence, the answer is $l = 100 e^{-d}$.

PROBLEMS

6. The water resistance (R tons) encountered by a ship traveling at various speeds (v knots) varied as in the following table. Discover the law.

v	10	12	16	20	25
R	1581	2494	5120	8944	15,625

7. The number, N, of bacteria remaining in a culture treated with an antibiotic varied with the time (t hr) as in the following table. Discover the law.

t	0	1	2	3	4
N	10,000	6700	4490	3010	2020

8. The distances of the planets from the sun (in terms of the distance of the earth from the sun as a unit) and their periods of revolution (T years) are given below. Discover the law.

	Mercury	Venus	Earth	Mars	Jupiter	Saturn	Uranus	Neptune	Pluto
D	0.387	0.723	1.00	1.52	5.20	9.54	19.2	30.1	41.3
T	0.241	0.615	1.00	1.88	11.9	29.5	84.0	165	265

9. The speed (v ft/sec) of a hailstone falling through the atmosphere varies with its diameter (d mm) as in the following table. Discover the law.

d	0.5	1.0	1.5	2.0	3.0
v	7.07	20	36.74	56.56	103.92

10. The quantity (Q g) of radium left in a medical needle varied with the time (t hr) as in the following table. Discover the law.

t	1	2	3	4	5
Q	0.109	0.098	0.089	0.080	0.073

11. The Pareto distribution. The famous Italian economist Vilfredo Pareto observed that many economic variables exhibit a power-law relationship, $y = ax^n$, at least over a limited domain. Examples are shown in Table 7-1.

Table 7-1

x	y
Annual sales	Number of firms with sales $\geq x$
Wealth	Percentage of population with wealth $\geq x$
Population	Number of towns with population $\geq x$
Salary	Number of government posts with initial salary $\geq x$
Years	Percentage of firms that survive to age $\geq x$

(a) (Steindl) In 1931 the percentages (y) of owners in Sweden with property equal to or greater than x (thousand crowns) were as follows:

x	0	10	20	30	50	100	200	300	500	1000	2000
y	100	34.43	17.61	11.00	5.84	2.41	1.00	0.58	0.29	0.11	0.04

Draw a line that is a reasonable fit of the log–log plot of these data, and find the formula.

(b) (Steindl) For U.S. firms that were born in 1944, the percentages (y) still alive x years later were as follows:

x	0.5	1.5	2.5	3.5	4.5	5.5	6.5	7.5	8.5
y	80.9	56.7	42.2	34.1	29.6	26.5	24.2	22.6	21.2

Draw a line that is a reasonable fit of the log–log plot of these data, and find the formula.

273

(c) If y, the number of people in the United States with annual income $\geq \$x$ is given by $y = (1.9)10^{12}/x^{1.6}$ for $5 \cdot 10^2 < x < 10^7$, find the number of people with incomes over $\$1,000,000$, the lowest income of the 50 people with highest incomes, and the number of people with incomes between $\$50,000$ and $\$100,000$.

12. The relation between distance (x m) from the stack of a smelter and amount (y μg/g) of lead in air-dried soil is found to be:

x	50	100	150	200	250	300
y	13,815	9,000	6,180	4,185	2,630	1,365

Find the formula. (From *Science*, December 20, 1974, pp. 1120–1.)

13. (a) The intensity (I units) of sunlight varies with the depth (x m) below the surface of clear seawater as follows:

x	0	1	2	5	10	20	30
I	100	93.2	86.9	70.5	49.7	24.7	12.2

Find a formula for I in terms of x. How much is I for $x = 100$?

(b) The force of attraction (F units) between two charged particles (unlike charges) varies with their distance apart (x mm) as follows:

x	1	1.5	2	3	5	7	10
F	125	55.56	31.25	13.89	5	2.55	1.25

Find a formula for F in terms of x. How much is F for $x = 100$?

(c) Compare the behavior of the two functions for small x. For large x.

7.8 Summary

In Section 4.16, a summary of results on differentiation was given under the headings of "General results" and "Specific results." We have nothing to add to the general results, but our list of specific results now looks like this:

Specific results
1. Derivative of x^n
2. Derivative of e^x
3. Derivative of $\ln x$
4. Derivative of b^x
5. Derivative of $\log_b x$

In each case, through use of the Chain Rule, we have generalizations:
1. Derivative of u^n
2. Derivative of e^u
3. Derivative of $\ln u$ $\quad\big\}$ where u is a function of x.
4. Derivative of b^u
5. Derivative of $\log_b u$

Moreover, we have results on antidifferentiation:

General results
1. Antiderivatives of the sum of two functions
2. Antiderivatives of a constant times a function

Specific results
1. Antiderivatives of x^n, $n \neq -1$
2. Antiderivatives of x^{-1}
3. Antiderivatives of e^x

Each of these results can be generalized through use of the Chain Rule. It will be worth your while to write out a complete statement for each of the foregoing headings.

PROBLEMS

This set is not restricted to problems on logarithmic functions.

1. The number (p) of pneumococci surviving t sec after treatment with an antiseptic decreased according to the law $p = 10^7 \cdot 2^{-t/15}$. Show that the number of pneumococci halved every 15 sec. Show also that this law can be written in the form $p = 10^7 e^{-rt}$, and find the value of r.

★ 2. Find the maximum, minimum, and inflection points on the curve $y = -4xe^{-(1/2)x^2}$, and sketch the graph. What is the relation of this curve to that of Example 1, **6.8**?

3. Referring to the observations at the beginning of Problem 6, **7.6**, find

   (a) $\lim\limits_{h \to 0} \dfrac{e^h - 1}{h}$ (b) $\lim\limits_{u \to 0} \dfrac{e^{3u} - 1}{u}$ (c) $\lim\limits_{v \to 0} \dfrac{e^v - 1}{2v}$

★ 4. (a) A rectangle of dimensions x and y has area A given by $A = xy$, of course. Take natural logs of both sides of this equation, and then write the differential of each side. Express this result in words, in the language of "relative error."

   (b) Same as part (a), for the volume of a cylinder in terms of radius and height (cf. Problem 4, **6.6**).

5. In any 4-week period, a certain radioactive substance loses 30% of its mass (i.e., of the mass at the beginning of that 4-week period). To what rate per week of continuous compounding is this equivalent?

6. A bank offers 7.5%, compounded continuously, on 4-year certificates of deposit. To what rate, compounded annually, is this equivalent?

7. If the consumer price index goes up by 1% in a month, this is sometimes described as a 12% annual rate of inflation. If the index is actually 1% higher each month than the month before, by how much will prices increase in a year?

8. By counting for 10 sec intervals, it is estimated that the counts (N per hour) from a radioactive source varied with the time (t hr) after the start of an experiment as in Table 7-2.

Table 7-2

t	N
0	320
2	172
4	107
6	76
8	58
10	45
12	37
14	30
16	24

275

(a) Make a semilog plot, and observe that although all the points do *not* lie on a straight line, those corresponding to large t *do* appear to lie on a straight line. We assume, then, that the source is a mixture of more than one radioactive substance, with different half-lives, and that the straight-line portion of the graph corresponds to the remaining substance, after the other has (or the others have) "died out."

(b) Fit an exponential formula to the straight-line portion of the graph, and subtract its ordinates from those given in the table to obtain a table of "residues." Then fit an exponential formula to the graph of residues, to find the formula for N as the sum of two exponential functions.

9. Same as Problem 8, for Table 7-3. (There are three component substances here.)

★ 10. The function given by $f(x) = x^{1/x}$ is continuous for all $x > 0$.
   (a) What do you think to be the value of $\lim_{x \to \infty} f(x)$?
   (b) What is the definition of $f(0)$ in order that the extended function be continuous at $x = 0$?
   (c) Locate maximum and minimum points on the curve $y = x^{1/x}$, and sketch the graph.

★ 11. Which is larger, π^e or e^π?

12. Consider the function $f(x) = (\ln x)/\sqrt{x}$.
   (a) What is the natural domain?
   (b) What are the coordinates of maximum, minimum, and inflection points, if any, on the curve $y = f(x)$?
   (c) Sketch the curve.

★ 13. Same as Problem 12, for $g(x) = \sqrt{x}/\ln x$.

14. Same as Problem 12, for $F(x) = (\ln x)/x^2$.

★ 15. Same as Problem 12, for $G(x) = (\ln x)/x^k$, where k is a positive integer.

★ 16. Same as Problem 12, for $H(x) = \ln(\ln x)$.

17. The area (A mm^2) of a bacterial culture increased with time (t days) as follows:

t	0	2	4	6	8	10	12	14
A	5	6.35	8.10	10.30	13.10	16.60	21.15	26.85

Use log and semilog paper, as needed, and tables to find a formula for A in terms of t.

18. The quantity of energy (Q quads) used per year increased at a constant instantaneous rate of 4% per year, and Q was 70 in 1960. Express Q as a function of t, the number of years after 1960. Use differentiation to express the approximate increase in Q if t increases from 10 to 10.2. Give the answer to three decimal places.

19. The area bounded by $y = \frac{2}{3}e^x$, the x axis, and the lines $x = -1$ and $x = 2$ is revolved about the x axis. Find the volume so generated, expressing

Table 7-3

t	N
0	210
1	127
2	90
3	72
4	61
5	54
6	49
10	37
15	28
20	22
30	13

the answer as $k\pi$ cubic units, where k is a number to three decimal places.

20. The flow (y units) from an orifice varied with the height (x units) of the water surface above the orifice as follows:

x	1	2.25	4	9	16
y	2	6.75	16	54	128

Use log paper and semilog paper, as needed, and tables to find a formula for y in terms of x.

21. The area (A mm^2) of a bacterial culture increased at a constant instantaneous rate of 10% per day, and A was 12 at $t = 0$. Express A as a function of t, the number of days. Use differentiation to find the approximate change in A if t increases from 20 to 20.25. Give the answer to three decimal places.

22. The area bounded by $y = \sqrt{x}\,e^{-x^2}$, the x axis, and the lines $x = 0$ and $x = 1$ is revolved about the x axis. Find the volume so generated, expressing the answer as $k\pi$ cubic units, where k is a number to three decimal places.

23. The relation between body mass (x kg) and stride frequency (y per minute) at the trot–gallop transition is shown in this table:

	mouse	mouse	mouse	rat	cat	dog	dog	horse
x	0.02	0.04	0.05	0.2	1	9	20	1600
y	500	455	443	370	300	225	203	115

On log–log paper plot these points, draw the most reasonable straight line through them, and find an equation for y in terms of x. (Hint: For simplicity, use the pair $x = 1$, $y = 300$.)

24. Find dy/dx if
  (a) $y = \ln\sqrt{(16 - x^2)/(16 + x^2)}$ (b) $y = \sqrt{(16 - x^2)/(16 + x^2)}$

25. The area bounded by $y = e^{2x}$, the x axis, and the lines $x = -1$ and $x = 1$ is revolved about the x axis. Find the volume so generated, expressing the answer as $k\pi$, where k is a number to three decimal places.

26. A certain strain of mouse cancer cells is treated with radiation. The surviving fraction (F) of cells varies with the dose (D, in certain units) as follows:

D	0	5	10	15	20
F	1	0.046	10^{-3}	10^{-4}	10^{-5}

Find an equation relating D and F. [From William D. Bloomer et al., "Astatine-211-Tellurium Radiocolloid Cures Experimental Malignant Ascites," *Science*, Vol. 212 (1981), pp. 340–341.]

27. Approximations of $\ln 3$ and $\ln 2$.
  (a) Find the area under the curve $y = 1/x$, above the x axis, and between the lines $x = 1$ and $x = 3$.

7
Logarithmic functions

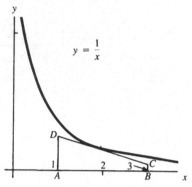

Fig. 7-10

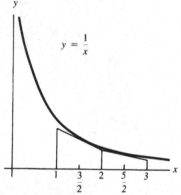

Fig. 7-11

(b) Demonstrate that the tangent to the curve $y = 1/x$ at any point lies beneath the curve.

(c) Find the area of the trapezoid formed by the tangent to the curve at the point where $x = 2$, the x axis, and the lines $x = 1$ and $x = 3$ (i.e., the area of trapezoid $ABCD$ in Figure 7-10).

(d) Use (a), (b), and (c) to demonstrate that $\ln 3 > 1$.

(e) Find the area of the trapezoid formed by the tangent to the curve at the point where $x = \frac{3}{2}$, the x axis, and the lines $x = 1$ and $x = 2$; also find the area of the trapezoid formed by the tangent to the curve at the point where $x = \frac{5}{2}$, the x axis, and the lines $x = 2$ and $x = 3$ (Figure 7-11).

(f) Add the areas in (e) to conclude that $\ln 3 > 2(\frac{1}{3} + \frac{1}{5})$.

(g) Continue the process begun earlier, doubling the number of trapezoids each time, to conclude that

$$\ln 3 > 2\left(\frac{1}{5} + \frac{1}{7} + \frac{1}{9} + \frac{1}{11}\right),$$

$$\ln 3 > 2\left(\frac{1}{9} + \frac{1}{11} + \frac{1}{13} + \frac{1}{15} + \frac{1}{19} + \frac{1}{21} + \frac{1}{23}\right),$$

and guess that

$$\ln 3 = \lim_{n \to \infty} 2\left(\frac{1}{2^{n-1}+1} + \frac{1}{2^{n-1}+3} + \frac{1}{2^{n-1}+5} + \cdots + \frac{1}{3 \cdot 2^{n-1}-1}\right).$$

(h) Use a similar process to conclude that

$$\ln 2 > \frac{2}{3},$$

$$\ln 2 > 2\left(\frac{1}{5} + \frac{1}{7}\right),$$

$$\ln 2 > 2\left(\frac{1}{9} + \frac{1}{11} + \frac{1}{13} + \frac{1}{15}\right),$$

$$\ln 2 > 2\left(\frac{1}{17} + \frac{1}{19} + \cdots + \frac{1}{31}\right),$$

and guess that

$$\ln 2 = \lim_{n \to \infty} 2\left(\frac{1}{2^n+1} + \frac{1}{2^n+3} + \cdots + \frac{1}{2^{n+1}-1}\right).$$

(i) Use a similar process to conclude that

$$\ln 4 > \frac{2}{3} + \frac{2}{5} + \frac{2}{7}$$

and hence that

$$\ln 2 > \frac{1}{3} + \frac{1}{5} + \frac{1}{7}.$$

(j) Continue this process to conclude that

$$\ln 2 > \frac{1}{5} + \frac{1}{7} + \frac{1}{9} + \frac{1}{11} + \frac{1}{13} + \frac{1}{15},$$

$$\ln 2 > \frac{1}{9} + \frac{1}{11} + \frac{1}{13} + \cdots + \frac{1}{31},$$

and guess that

$$\ln 2 = \lim_{n \to \infty} \left(\frac{1}{2^n+1} + \frac{1}{2^n+3} + \cdots + \frac{1}{2^{n+2}-1}\right).$$

(k) Use a calculator and various values of n to approximate $\ln 3$ and $\ln 2$, comparing your results with values found in a table of natural logs. Also compare the accuracy of the process in (h) with that in (j).

28. Carbon 14 decays at a constant instantaneous percentage rate r. If its half-life is 5568 years, what is the value of r?

29. A count of the number (n) of mathematical publications per year between 1860 and 1965 suggests that n varies with t (the number of years since 1860) as follows: $n = 1400e^{0.025t}$. [From K. O. May, "Quantitative Growth of Mathematical Literature," *Science*, Vol. 154 (1966), pp. 1672–3.] There were major deviations from this formula during World Wars I and II, but otherwise the formula applies closely.

(a) According to this formula, how long does it take for n to double?

(b) If this formula continued to apply, how many mathematical titles were published in 1980?

(c) Use this formula, and assume that the mathematical publications up to 1860 totaled 40,000, to find a formula for N, the total number of mathematical titles published up to year t since 1860.

SAMPLE TEST

1. Find dy/dx if

(a) $y = \ln \dfrac{\sqrt{5-x^2}}{5+x}$ (b) $y = \dfrac{\sqrt{5-x^2}}{5+x}$

2. Samples of soil all treated at the same time with the same amount of the insecticide parathion were placed in five test jars, and 2000 fruit flies were introduced into the jars, successively, at 1-day intervals. The number (N) of fruit flies that died within 24 hr varied with the time (t days after the start of the experiment) at which they were placed in the jar as follows:

t	0	1	2	3	4
N	700	285	116	47	19

Express N as a function of t, using log and semilog paper, and the tables, to help in the process.

3. Consider the function determined by $y = (\ln x)/x$.

(a) What is its domain?

(b) Find the extreme value(s) of this function, determining whether maximum or minimum.

4. For the graph of the function in Problem 3, determine points of inflection.

Here is another sample test on Chapter 7.

1′. Find dy/dx if

(a) $y = \ln\sqrt{(10+x^2)/(10-x^2)}$ (b) $y = \sqrt{(10+x^2)/(10-x^2)}$

2'. The heat intensity (I calories/mm^2) at a point varies with the distance (x cm) from the heat source as follows:

x	1	2	3	5	6
I	3600	900	400	144	100

Using log and semilog paper, as needed, find an equation relating I and x.

3'. Consider the function determined by $y = x/\ln x$.
   (a) What is its natural domain?
   (b) Find the extreme value(s) of this function, determining whether maximum or minimum.

4'. Evaluate

   (a) $\int_1^2 \dfrac{e^{2x}}{e^{2x}-1}\,dx$ (b) $\int_1^2 \dfrac{e^{2x}}{\left(e^{2x}-1\right)^2}\,dx$

   Simplify your results as much as possible without the use of tables.

5'. A bacterial population, initially 4,000,000, grows at a rate constantly 20% per day. Express the number, N, after t days. Use differentiating to approximate the increase in N if t changes from 5 days to 5.1 days.

6'. Evaluate $\int_e^{e^2}[(\ln x)/x]\,dx$.

PROJECTS

An individual project can provide an interesting and rewarding experience. The form of the project may vary with the topic studied, but an appropriate norm might be the writing of a paper of five pages or so in which you describe as clearly as you can the nature of the problems treated in what you read, the way that the author(s) attacked the problem, the conclusions reached, and *the role of mathematics in the solution.* The last includes filling in gaps in the mathematical treatment, to the extent that you can. Useful sources are suggested:

1. V. A. Tucker, "The Energetic Cost of Moving About," *American Scientist*, Vol. 63 (1975), pp. 413–19 (including bibliograph y).
2. Josef Steindl, *Random Processes and the Growth of Firms — A Study of the Pareto Law*, Hafner Publishing Co., 1965.
3. David Pilbeam and Stephen Jay Gould, "Size and Scaling in Human Evolution," *Science*, Vol. 186 (1974), pp. 892–901; also, Roger Lewin, "How Did Humans Evolve Big Brains?" *Science*, Vol. 216 (1982), pp. 840–1; and Este Armstrong, "Relative Brain Size and Metabolism in Mammals," *Science*, Vol. 220 (1983), pp. 1302–4.
4. Kosta Tsipis, "Physics and Calculus of Countercity and Counterforce Nuclear Attacks," *Science*, Vol. 187 (1975), pp. 393–7.
5. Fred N. White and James L. Kinney, "Avian Incubation," *Science*, Vol. 186 (1974), pp. 107–15.

6. Norman C. Heglund, C. Richard Taylor, and Thomas A. McMahon, "Scaling Stride Frequency and Gait to Animal Size: Mice to Horses," *Science*, Vol. 186 (1974), pp. 1112–13 (including bibliography).

7. T. T. Liang and E. P. Lichtenstein, "Synergism of Insecticides by Herbicides: Effect of Environmental Factors," *Science*, Vol. 186 (1974), pp. 1128–30.

8. Roy M. Anderson and Robert M. May, "Directly Transmitted Infectious Diseases: Control by Vaccination," *Science*, Vol. 215 (1982), pp. 1053–60.

9. Dennis Epple and Lester Lave, "The Helium Storage Controversy: Modeling Natural Resource Supply," *American Scientist*, Vol. 70 (1982), pp. 286–93 (including bibliography).

10. Bruce Hannon and James Brodrick, "Steel Recycling and Energy Conservation," *Science*, Vol. 216 (1982), pp. 485–91.

11. D. A. Smith, "Human Population Growth: Stability or Explosion," *Mathematics Magazine*, Vol. 50 (1977), pp. 186–97.

An excellent source of material for projects is *Undergraduate Mathematics and Applications* (UMAP). A catalog of UMAP modules can be obtained from Consortium for Mathematics and Its Applications, 271 Lincoln Street, Suite 4, Lexington, MA 02173.

8 Differential equations

8.1 Introduction

We began Chapter 5 with problems of antidifferentiation: For example, if we have a formula for the *slope* of a curve, can we find an equation of the curve, that is, an expression for the *height*, y, in terms of x? In the first example of Chapter 5 we concluded that

$$\text{if} \quad \frac{dy}{dx} = x, \quad \text{for all } x, \tag{1}$$

$$\text{then} \quad y = \frac{x^2}{2} + c, \quad \text{for an arbitrary constant, } c. \tag{2}$$

Similarly, in Section 6.7 we determined that

$$\text{if} \quad \frac{dy}{dx} = 2y, \tag{3}$$

$$\text{then} \quad y = c \cdot e^{2x}, \quad \text{for an arbitrary constant, } c. \tag{4}$$

Equations (1) and (3) are examples of what are called *differential equations*, and equations (2) and (4) are the respective *general solutions* of these differential equations. Other forms of equations (1) are

$$dy = x \cdot dx \tag{1'}$$

and

$$f'(x) = x. \tag{1''}$$

Similarly, other forms of equation (3) are

$$dy = 2y \cdot dx \tag{3'}$$

and

$$g'(x) = 2 \cdot g(x). \tag{3''}$$

Of these equations, only (1′) and (3′) are actually written in *differential* form. It might be more logical to call the other forms "equations involving derivatives."

Note that the general solution of each of the differential equations mentioned here embodies a whole "family" of functions – unlike the situation with the algebraic equation $2x - 6 = 0$, which has the single solution $x = 3$, or that of the algebraic equation $x^2 - x - 2 = 0$, which has the pair of solutions $x = -1$ and $x = 2$. (Some differential equations have no solutions. This can occur with an algebraic equation, too. Consider $0 \cdot x = 5$.)

With equation (3″), if we had the additional information that at $x = 0$, $g(x) = 10$, we would get the *particular* solution $g(x) = 10e^{2x}$.

Thus, a differential equation is an equation involving an independent variable, an unknown function of that variable, and one or more derivatives of that function. A *solution* of the differential equation consists of a *function* such that when it and its derivatives are substituted for the unknown function and its derivatives appearing in the equation, the equation is satisfied for all values of the independent variable in a certain domain.

In view of the definition of a differential equation and its solution, it should not seem surprising that antidifferentiation plays an important role in solving differential equations. Indeed, much of the material in the first part of Chapter 5 can be cast in the language of differential equations.

Example 1

If the velocity (v ft/sec) of a particle moving on a straight line varies with time (t sec) as

$$\frac{dv}{dt} = 10t,$$

then $\quad v = \int 10t\, dt = 5t^2 + c.$

Example 2

If the height (y ft) of a projectile moving under the influence of gravity is such that

$$\frac{d^2y}{dt^2} = -32 \quad \text{for } 0 \le t \le 8,$$

then $\quad \dfrac{dy}{dt} = \int -32\, dt = -32t + k_1,$

and $\quad y = \int (-32t + k_1)\, dt = -16t^2 + k_1 t + k_2, \quad 0 \le t \le 8.$

The equation $dv/dt = 10t$ is called a *first-order* differential equation, because the only derivative that appears is the first derivative, whereas the equation $d^2y/dt^2 = -32$ is called a *second-order* differential equation because it contains a second derivative.

The examples suggest that the general solution of a first-order differential equation contains *one* arbitrary constant, and that of a second-order equation contains *two* arbitrary constants. This statement and its obvious extension to nth-order differential equations are essentially valid, but we shall not try to prove them.

The study of differential equations began with the invention of calculus in the late seventeenth century, and it has continued to this day as a major part of ongoing research in mathematics. The mathematical models of many phenomena in the natural and social sciences contain differential equations, and therefore methods for the solution of such equations are an important ,part of calculus. We shall examine the elementary portions of the subject in this chapter.

8.2   An approximate solution of a differential equation

Of the four differential equations mentioned in Section 8.1,

$$\frac{dy}{dx} = x,$$

$$\frac{dy}{dx} = 2y,$$

$$\frac{dv}{dt} = 10t,$$

$$\frac{d^2y}{dt^2} = -32,$$

all but the second are easily solved by straightforward antidifferentiation, and the second is solved through our knowledge of the basic property of the exponential function.

The important differential equation

$$\frac{dy}{dx} = -xy \tag{5}$$

cannot be treated by either of these approaches. It is a member of a particular class of first-order differential equations to be dealt with in Section 8.3. We wish now to use equation (5) to illustrate an instructive numerical procedure to approximate solutions of differential equations – the very same procedure used at the beginning of Chapter 5 to approximate an antiderivative by means of short line segments.

We choose to approximate the particular solution of $dy/dx = -xy$ for which $y = 4$ at $x = 0$. That is, we seek a curve passing through the point $(0,4)$ such that at any point (x, y), the slope of the tangent to the curve equals $-x \cdot y$.

If $(x_0, y_0) = (0,4)$ is on the curve, the slope there is $-0 \cdot 4 = 0$ – the tangent line is horizontal (Figure 8-1). So, if we use this tangent line as an approximation to the curve for a small interval, say of length $h = 0.1$, we obtain $(0.1,4)$ as a point of our approximation. Call this point (x_1, y_1). Although (x_1, y_1) is not on the curve, we assume that is close enough to warrant repeating the process: If (x_1, y_1) were actually on the curve, the

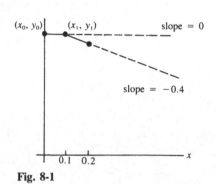

Fig. 8-1

slope of the tangent line there would be $-(0.1)(4) = -0.4$. Then a move to the right along this line by another x increment of $h = 0.1$ produces a drop of 0.04. Thus, the value of what we name y_2 is $4 - 0.04 = 3.96$, and we have $(x_2, y_2) = (0.2, 3.96)$ as the next point of the approximation. We can tabulate the information as follows:

x	0	0.1	0.2	0.3	0.4	0.5
y	4	4	3.96			
$\dfrac{dy}{dx}$	0	-0.4				

Problem 1
Carry on with this table, rounding to two decimal places, through $x = 0.5$.

Proceeding in the manner described yields a sequence of points (x_n, y_n) where each successive point is determined by the equations

$$x_{n+1} = x_n + h = x_0 + (n+1)h, \quad \text{and} \quad y_{n+1} = y_n + (-x_n \cdot y_n)h.$$

There is, in fact, precisely one solution of this differential equation passing through $(0,4)$, and the points obtained numerically by this "linearization" process do not stray far from the solution curve. A smaller increment, h, will give a more accurate approximation and a graph more closely resembling a smooth curve.

PROBLEMS

○ 2. Continue the process begun in Problem 1 as far as $x = 3$. You should arrive at $(x_{30}, y_{30}) = (3, 0.03)$ as the approximating point. Graph these points, and connect successive points with line segments.
3. Sketch on your graph of Problem 2 how you expect the graph to look for $x > 3$.
○ 4. Complete your graph to $x = -3$ by using $h = -0.1$. If it is not immediately clear from the foregoing formulas in the text what is going to happen, calculate a few values.
c 5. With a programmable calculator, prepare similar data based on $h = \pm 0.01$. Compare the results with those of the previous problems.
c 6. Repeat Problems 1 and 2, with $h = \pm 0.1$, and with (x_0, y_0) equal to
  (a) $(0, 10)$ (b) $(1, 3)$ (c) $(0, -4)$
7. If a curve is such that $dy/dx = -xy$, then

$$\frac{d^2y}{dx^2} = -\left(x\frac{dy}{dx} + y\right) = -[x(-xy) + y] = y(x^2 - 1).$$

Use this result to find the abscissas of the points of inflection on the curve. Differentiate once more to find the abscissas of the points where the flexion is an extreme. Maximum or minimum? Note that your

285

answers are independent of "initial conditions" – the answers are the same if $(x_0, y_0) = (0,4)$ or $(0,1)$ or $(0,10)$ or whatever.

8.3 Variables separable

The method for dealing with differential equations of the type known as "variables separable" is illustrated by solving the equation treated approximately in Section 8.2:

$$\text{If}\quad \frac{dy}{dx} = -xy, \tag{6}$$

$$\text{then}\quad dy = -xy\,dx,$$

$$\text{and}\quad \frac{dy}{y} = -x\,dx. \tag{7}$$

The variables are now "separated," the left side being expressed in terms of y and dy, and the right side in terms of x and dx. Our desired solution would express y as a function of x,

$$y = f(x),$$

say. Because $dy = f'(x)\,dx$, equation (7) would read

$$\frac{f'(x)\,dx}{f(x)} = -x\,dx. \tag{7'}$$

The basic principle to which we appeal is that antiderivatives of the left and right sides of equation (7′) differ by no more than a constant. Applying this principle to the differential equation written in form (7) rather than (7′) gives

$$\ln y = -\frac{x^2}{2} + c, \quad \text{if}\quad y > 0, \quad \text{and}$$

$$\ln(-y) = -\frac{x^2}{2} + c, \quad \text{if}\quad y < 0.$$

Problem 1
Show that either of these equations can be recast as

$$y = y_0 \cdot e^{-x^2/2}. \tag{8}$$

What is the value of the constant y_0 in terms of c? By differentiation, verify that equation (8) is actually a solution of equation (6). Sketch the graph of equation (8) for $y_0 = 1$. For $y_0 = 4$. For $y_0 = -4$.

Whenever we can "separate the variables," as was done in this example, we try to find antiderivatives of each side of the equation; if we succeed, the equation is solved.

2. For practice, use the method of separation of variables to solve the characteristic equation of the exponential function: The rate of change of y with respect to x is proportional to y.

3. Find the general solution of each of the following differential equations, and sketch graphs of the solutions for three typical values of the arbitrary constant in each case:

(a) $y\,dx - x\,dy = 0$ (b) $y\,dx + x\,dy = 0$ (c) $\dfrac{dy}{dx} = -\dfrac{x}{y}, \quad y \neq 0$

(d) $\dfrac{dy}{dx} = \dfrac{x}{y}, \quad y \neq 0$ (e) $\dfrac{dy}{dx} = x^3 y^2$

○ 4. Let Y represent national income, S represent stock of capital, I represent total net investment, and C represent total consumption. Each of these quantities is a function of time.

The definitions of the quantities imply that $Y = C + I$ and that $I = dS/dt$. We make the (doubtlessly over-simple) assumptions that $Y = kS$ and $C = lY$ for constants k and l.

(a) Show that $dY/dt = k(1-l)Y$.

(b) Solve the differential equation of (a) to obtain $Y = Y_0 e^{k(1-l)t}$, where Y_0 is income at $t = 0$.

(c) If consumption accounts for 90% of income and if each dollar of capital stock generates 40 cents of income, what is the annual percentage rate of growth of income?

8.4 Comparison of approximate and exact solutions

In the preceding section we found the exact solution of $dy/dx = -xy$, with $y = y_0$ at $x = 0$, to be

$$y = y_0 e^{-x^2/2}. \tag{8}$$

The graph of (8), with $y_0 = 1/\sqrt{2\pi} \approx 0.4$, is called the Normal Curve, of great importance in probability and statistics. The graph you sketched in Problem 4, **8.2**, is an approximation to the Normal Curve, except that all y values were magnified by a factor of about 10 because we started at $(0, 4)$ instead of at $(0, 1/\sqrt{2\pi})$.

The method we used in Section 8.2 is known as *Euler's method*, named for the great eighteenth-century Swiss mathematician. His method is applicable to any differential equation of the form

$$\frac{dy}{dx} = g(x, y), \quad \text{with} \quad y = y_0 \quad \text{at} \quad x = x_0.$$

Note that g is a function of two variables. In the differential equation of

Section 8.2, $g(x, y) = -x \cdot y$. As in the example of Section 8.2, Euler's method gives a sequence of points generated by the formulas

$$x_{n+1} = x_0 + (n+1) \cdot h \quad \text{and}$$

$$y_{n+1} = y_n + g(x_n, y_n) \cdot h, \quad n = 0, 1, 2, \ldots.$$

Euler's method is not accurate enough to be of great practical importance, but simple calculations with the method can provide an idea of the nature of a solution.

Table 8-1

"Exact" solution, using Table B	Euler's method with $h = 0.1$	Euler's method with $h = 0.01$
$x = 0.5$ $y = 4e^{-(0.5)^2/2} = 3.530$	$x_5 = 0.5$ $y_5 = 3.61$ $y_5 - y = 0.08$	$x_{50} = 0.5$ $y_{50} = 3.538$ $y_{50} - y = 0.008$
$x = 2$ $y = 4e^{-2^2/2} = 0.541$	$x_{20} = 2$ $y_{20} = 0.52$ $y_{20} - y = -0.02$	$x_{200} = 2$ $y_{200} = 0.539$ $y_{200} - y = -0.001$

Table 8-1 presents some results of the method applied to the differential equation $dy/dx = -xy$, with $y = 4$ at $x = 0$, as compared with what we now know to be the exact solution, $y = 4e^{-x^2/2}$. Note that the errors with $h = 0.01$ are about one-tenth the errors with $h = 0.1$ – with 10 times as much work, we cut the errors by a factor of 10. This is only a modest improvement, as compared, for example, with the "effort-improvement" ratio in the Newton–Raphson method for solving ordinary equations. And that is why Euler's method has been superseded by similar but more efficient methods.

All these methods face the problem of straying seriously from the exact solution as errors accumulate in moving far from the starting point. The problem we have dealt with, $dy/dx = -xy$, is reasonably well-behaved in this respect. Because initially the curve is concave downward, the tangent lines lie above the curve, and the approximations overestimate the height of the curve (Figure 8-2). But the curve has a point of inflection at $x = 1$ (see Problem 7, **8.2**), and compensation occurs thereafter. Eventually, both the exact solution and its approximation become very small.

For the differential equation $dy/dx = xy$, the Euler method does *not* give good results, as the following problems show.

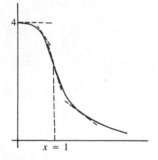

Fig. 8-2

Problem 1

Verify that the solution of $dy/dx = xy$, with $y = 1$ at $x = 0$, is $y = e^{x^2/2}$.

c Problem 2
Program a calculator or computer to approximate the solution of $dy/dx = xy$, with $(x_0, y_0) = (0, 1)$, using Euler's method:

$$x_{n+1} = (n+1)h,$$

$$y_{n+1} = y_n + x_n y_n h.$$

Use $h = 0.1$, and let n go to 39. You should find that the approximation gives $y_{40} \approx 1030$, while $e^{4^2/2} = e^8 = 2981$. Thus, the error is approximately 1951, an error of 65% of the true value.

Use $h = 0.01$, and let n go to 399. You should find that the approximation gives $y_{400} \approx 2633$, for an error of 12% of the true value.

8.5 Population changes

If a species has an unlimited food supply, without predators or competitors, we may assume a birth rate b (per thousand, per year, say) and a death rate d (also per thousand, per year), so that the *net* rate of increase per thousand will be $(b - d)$ per year. In assuming that b and d are constant, we are ignoring the effects of a changing age structure of the population with time.

If y is the population (in thousands), if t is time (in years), and if we set $r = (b - d)/1000$, then the variation described in the preceding paragraph is given by

$$\frac{dy/dt}{y} = r,$$

the "law of natural growth." We know that the solution of this differential equation is

$$y = y_0 e^{rt}. \tag{9}$$

Even with a limited food supply, a population may grow approximately like (9) for some time, until nutrients run short.

PROBLEMS

1. If the population of a country is now 200 million individuals, how long will it take the population to grow to 1 billion if
   (a) the annual birth rate is 45 per thousand and the annual death rate is 20 per thousand?
   (b) the birth rate is as in (a), but the death rate is 15 per thousand?
2. If the annual death rate is 15 per thousand, what must the birth rate be if a population is to double in 100 years?
3. Suppose that a population is now P and that the annual death rate remains fixed at 15 per thousand. If the annual birth rate is 45 per

thousand for the next 20 years, what must the birth rate be for the following 80 years if the population is to be $2P$ one hundred years hence?

4. A colony of bacteria, initially 5×10^4 in number, increases through fission (division of each bacterium into two) every 20 min. Express the population, y, at any time t min after the start in terms of powers of 2. Of powers of e.

8.6   The logistic equation

If the food supply is limited, a population may grow rapidly at first, almost in accord with the law of natural growth; but as food becomes scarce, the rate of growth becomes quite small, and the population stabilizes at some constant level, the *equilibrium population*. The same effect can have a cause other than shortage of food, such as the accumulation of metabolites, as in the example of the bacterial colony in Section 1.2. In such cases the following type of growth is often a satisfactory refinement of the law of natural growth (in which, of course, relative rate of change of population is constant):

The logistic law
The relative rate of change in population is proportional to the amount by which the population falls short of the equilibrium population.

If y represents the population and y_E is the equilibrium population, then the logistic law can be put in the following symbolic form:

$$\frac{dy/dt}{y} = c(y_E - y), \tag{10}$$

where c is the constant of proportionality. Note that during a time interval when y is very small relative to y_E, we have an equation close to that defining the law of natural growth; but when y is very close in size to y_E, then the *relative rate* of increase of y with respect to t is very small, and the same is true of the *rate* of increase.

In equation (10), the differential equation of the logistic law, the variables are separable; so as a step preliminary to solving the equation, we write equation (10) in the form

$$\frac{dy}{y(y_E - y)} = c \, dt. \tag{11}$$

How can we find an antiderivative of the left side? We require an algebraic trick, to be discussed in the next section. But first we shall practice with Euler's method to solve the differential equation approximately.

In Chapter 1 we studied an example of a colony of bacteria the population of which was measured in terms of the area covered by the colony, as shown in Table 1-1, here repeated. It is consistent with biological experience that the logistic law applies to this situation. In Chapter 1 we estimated 40 as the limiting area. Assume, then, that the growth of the colony conforms to the differential equation

$$\frac{dA/dt}{A} = c(40 - A).$$

○ 1. Find $\frac{1}{80}$ to be a reasonable value for c. To do this, you need to estimate dA/dt at some value of t, say $t = 4$, where the variation is fairly regular. Calculating $[A(t+1) - A(t)]/1$ for $t = 2, 3, 4,$ and 5 will be of help. The expression for the approximating $\bar{A}$ in Section 1.7 will also be of help.

○ **C** 2. With $h = 0.5$, apply Euler's method to find approximate values for A:

$$A_{n+1} = A_n + \tfrac{1}{80} A_n (40 - A_n) h, \quad n = 0, 1, 2, \ldots, 19,$$

starting with $t_0 = 0$, $A_0 = 4.4$. Compare your results with those of Table 1-1.

Table 1-1

t (hr)	A (mm^2)
0	4.4
1	6.8
2	10.2
3	14.4
4	19.2
5	24.2
6	28.6
7	32.2
8	34.8
9	36.7
10	38.0

8.7 The method of partial fractions

Suppose that we have the fraction

$$\frac{1}{(x-3)(x+1)},$$

and ask if it can be written as the sum of two fractions whose denominators are the factors $(x-3)$ and $(x+1)$; that is, can we find constants p and q such that

$$\frac{1}{(x-3)(x+1)} = \frac{p}{x-3} + \frac{q}{x+1}?$$

Adding the two fractions on the right, we get

$$\frac{p(x+1) + q(x-3)}{(x-3)(x+1)},$$

so that, if we are to reach our goal, we must have

$$\frac{1}{(x-3)(x+1)} = \frac{(p+q)x + p - 3q}{(x-3)(x+1)}.$$

It should seem reasonable to you that if this is to be a valid equality for all x for which the original fraction is defined (i.e., all x except $x = 3$ and $x = -1$), we must have

$$p + q = 0 \quad \text{and}$$
$$p - 3q = 1.$$

291

Problem 1

Solve this pair of simultaneous equations to obtain the result

$$\frac{1}{(x-3)(x+1)} = \frac{1/4}{x-3} + \frac{-1/4}{x+1}.$$

The method illustrated here is of use in many fields of mathematics. It can be generalized to handle more complicated fractions, but the simple case of the foregoing example is sufficient for dealing with the differential equation of the logistic law.

PROBLEMS

In Problems 2–6 apply the method of partial fractions to break down each of the following fractions into the sum of two fractions:

2. $\dfrac{5}{(x-2)(x+3)}$ 3. $\dfrac{2}{x^2-5x+6}$ 4. $\dfrac{3}{x^2-4x}$ 5. $\dfrac{2x-1}{x^2-2x-3}$

○ 6. $\dfrac{1}{y(y_E-y)}$, where y_E is a constant.

7. Prove that if $r + sx = a + bx$ for all x, then $r = a$ and $s = b$.

★ 8. Generalize Problem 7.

8.8 The logistic equation (*continued*)

Applying your answer to Problem 6, **8.7**, to equation (11), we obtain

$$\left[\frac{1/y_E}{y} + \frac{1/y_E}{y_E-y}\right] dy = c\, dt, \quad \text{or}$$

$$\frac{dy}{y} + \frac{dy}{y_E-y} = cy_E\, dt.$$

Problem 1

Solve this differential equation to obtain, for $0 < y < y_E$,

$$\ln \frac{y}{y_E-y} = cy_E \cdot t + d,$$

and hence

$$\frac{y}{y_E-y} = k \cdot e^{cy_E \cdot t} \tag{12}$$

Problem 2

Verify that if $y = y_0$ at $t = 0$, then $k = y_0/(y_E - y_0)$.

Problem 3

Solve equation (12) for y to obtain

$$y = \frac{k \cdot y_E}{k + e^{-cy_E \cdot t}}. \tag{13}$$

Problem 4

Using the result that if $x < 0$, $\int dx/x = \ln(-x) + c$, carry through the steps leading to the same equation (13) for $y < 0$. For $y > y_E$.

Thus, equation (13) applies for all $y \neq 0$; it is the general solution of the logistic differential equation. Equation (13) and its S-shaped graph (Figure 8-3) are of far-reaching importance in applications of calculus to problems of organic growth.

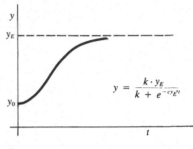

$$y = \frac{k \cdot y_E}{k + e^{-cy_E t}}$$

Fig. 8-3

PROBLEMS

★ 5. Beginning with equation (13), differentiate twice to locate the point of inflection on the graph of the logistic equation. You should find that the height (y coordinate) of the point of inflection equals $\frac{1}{2}y_E$.

★ 6. Begin with the differential equation of the logistic equation, $dy/dt = cy(y_E - y)$, and differentiate once with respect to t to obtain an alternative form of the second derivative you found in Problem 5.

7. For a certain population, y individuals, the relationship with time (t days) is given by equation (13), with $y_E = 200$, with $c = 0.001$, and with k determined by the fact that at $t = 0$, $y = 50$. Make a table of values for t and y for all integral (whole-number) values of t, $0 \leq t \leq 20$, and plot a graph of this function carefully on a full-size sheet of graph paper.

8. In the basic differential equation (11), for values of y very much smaller than y_E, we may assume that $y_E - y \approx y_E$, so that equation (11) becomes (approximately) $dy/dt = cy_E \cdot y$. Solve this equation for y in terms of t, using the same constants as in Problem 7, make a table of values of t and y for integral values of t, $0 \leq t \leq 10$, and plot the graph of this function on the same axes that you used for Problem 7.

9. Use Euler's method to find approximate solutions of the logistic differential equation (10), with the data given in Problem 7. Use $h = 1$, and plot the results on the same axes you used in Problems 7 and 8.

C 10. Show that the logistic differential equation $dA/dt = \frac{1}{80}A(40 - A)$ for the bacterial colony reintroduced in the problems of Section 8.6, with $A = 4.4$ at $t = 0$, has for its solution (to five-decimal-place accuracy) $A = 4.94382/(0.12360 + e^{-t/2})$. Use this formula to calculate A for $t = 0, 1, 2, \ldots, 10$, and compare these numbers with (a) the values of A as listed in Table 1-1; (b) the values of A obtained by Euler's method in Problem 2, **8.6**.

You should find the largest discrepancy between the exact solution and the Euler approximate solution to be 1.18 (at $t = 4.5$) where $A = 21.59$. That is a fairly large error accumulation, considering that the curve has an inflection point at $t = 4.18$. (The result of Problem 5 gives the location of the inflection point.)

In examining your results, refer to the discussion in Section 8.4 of error in Euler's method. The case before us shows that the method, although it produces entirely satisfactory results for the coarse data of the example, is not so good when greater precision is required. Of course, the use of smaller h reduces the errors, but at the cost of considerable effort. For example, with $h = 0.1$, the error at $t = 4.5$ is 0.23, an improvement by a factor of 5 in the error noted earlier with $h = 0.5$.

8.9 Linear differential equations with constant coefficients

We shall introduce an important class of differential equations by considering a model of predator–prey relationships. Different conditions call for different models; we shall assume separate generations of predators (rather than overlapping generations), with a cohort of predators replaced by their offspring in a unit of time. Then the model goes as follows: In the absence of predators, the population (X) of prey would follow the logistic differential equation

$$\frac{dX}{dt} = aX(b - X).$$

If, in unit time, each predator kills a number of prey proportional to the abundance of prey (i.e., number killed in unit time by one predator equals cX), and if the population of predators is Y, then the total number of prey killed in unit time is cXY. Hence, a differential equation for the rate of change of prey population in the *presence* of predators is

$$\frac{dX}{dt} = aX(b - X) - cXY. \tag{14}$$

Now, if the number of offspring produced by each predator in unit time is proportional to the number of prey killed by that predator (i.e., $k \cdot cX$), then Y predators give rise to $kcXY$ offspring, and the increment of predator population in unit time is $kcXY - Y$, or $(kcX - 1)Y$, so the rate of change of the predator population is given by

$$\frac{dY}{dt} = (kcX - 1)Y. \tag{15}$$

Equations (14) and (15) are difficult to analyze. But if we assume that there are certain equilibrium levels X_E and Y_E, of X and Y, if we let x and y represent deviations from these equilibrium levels (i.e., $x = X - X_E$, $y = Y - Y_E$), and if we assume that x and y are small, so that terms involving x^2 and xy can be ignored relative to terms involving x and y, then equations (14) and (15) are equivalent to

$$\frac{dx}{dt} = -px - qy \tag{16}$$

8.9
Linear differential equations
with constant coefficients

and

$$\frac{dy}{dt} = rx, \tag{17}$$

where p, q, and r are positive constants. [See Problem 13 at the end of this section for the details of how to get from (14) and (15) to (16) and (17).]

Problem 1

Eliminate x between equations (16) and (17) by solving equation (17) for x, differentiating with respect to t, and making substitutions in the first equation to obtain

$$\frac{d^2y}{dt^2} + p\frac{dy}{dt} + qry = 0. \tag{18}$$

Equation (18) is a "second-order linear differential equation with constant coefficients and with right member equal to zero":

second order, because the highest derivative is the second;

linear, because wherever y or one of its derivatives appears, it is raised to the first power;

with constant coefficients, because the coefficients of d^2y/dt^2, dy/dt, and y are all constants;

with right member equal to zero to distinguish equation (18) from another important class of differential equations:

$$A\frac{d^2y}{dt^2} + B\frac{dy}{dt} + Cy = f(t).$$

Later we shall show how to use Euler's method to obtain approximate solutions of the pair of simultaneous differential equations (14) and (15); in this section and the next we shall discuss exact solutions of linear differential equations with constant coefficients.

Problem 2

(a) Show that if $y = F(t)$ is a solution of equation (18), then $y = k \cdot F(t)$, for any constant k, is also a solution.

(b) Show that if $y = F(t)$ and $y = G(t)$ are solutions of equation (18), then $y = F(t) + G(t)$ is also a solution.

The results of Problem 2 are valid for a linear differential equation with constant coefficients and right member equal to zero, whatever its order, and are basic to solving such equations. We illustrate a method of solution by consideration of a particular example:

$$\frac{d^2y}{dt^2} + 5\frac{dy}{dt} + 6y = 0. \tag{19}$$

Experience with the exponential function and the way in which its successive derivatives "replicate" the function itself suggests that the function

$$y = e^{mt} \tag{20}$$

may be a solution of equation (19) for some value or values of the constant m.

Problem 3

Find dy/dt and d^2y/dt^2 from equation (20) and substitute into equation (19) to show that if equation (20) is a solution of equation (19), then

$$m^2 + 5m + 6 = 0. \tag{21}$$

The roots of equation (21) are -2 and -3. Hence, we expect $y = e^{-2t}$ and $y = e^{-3t}$ to be solutions of equation (19). By Problem 2, we know that if these functions *are* solutions, then so are $y = c_1 e^{-2t}$, $y = c_2 e^{-3t}$, and $y = c_1 e^{-2t} + c_2 e^{-3t}$.

Problem 4

Verify that all the functions mentioned in the two preceding sentences are solutions of equation (19).

Problem 5

Prove the converse of the result of Problem 3 for the general second-order linear differential equation with constant coefficients and with right member equal to zero; that is, show that if r is a root of

$$am^2 + bm + c = 0,$$

then $y = e^{rt}$ is a solution of

$$a\frac{d^2y}{dt^2} + b\frac{dy}{dt} + cy = 0.$$

The function $y = c_1 e^{-2t} + c_2 e^{-3t}$, involving two arbitrary constants, is called the *general solution* of the second-order equation (19). In any given case we may have initial conditions that provide information to determine c_1 and c_2.

Problem 6

Find the solution of

$$\frac{d^2y}{dx^2} + 5\frac{dy}{dx} + 4y = 0$$

for which $y = 5$ and $dy/dx = 7$ at $x = 0$. Sketch on the same axes the graphs of $y = 9e^{-x}$, $y = -4e^{-4x}$, and $y = 9e^{-x} - 4e^{-4x}$, $0 \le x \le 4$.

Equation (21) is called the *characteristic equation* of the differential equation (19). In both examples so far considered, the roots of the characteristic equation have been real and unequal. We shall learn in Chapter 10 that if the roots of the characteristic equation are imaginary, the solutions of the differential equation can be expressed in terms of trigonometric functions. Let us now investigate what happens if the roots of the characteristic equation of a second-order equation are equal: The characteristic equation of

$$\frac{d^2y}{dx^2} - 6\frac{dy}{dx} + 9y = 0 \qquad (22)$$

is

$$m^2 - 6m + 9 = (m-3)^2 = 0.$$

Following the pattern of our previous work, we know that

$$y = c_1 e^{3x} + c_2 e^{3x} \qquad (23)$$

is a solution of equation (22). But equation (23) does not really involve *two* arbitrary constants. It can be written as

$$y = Ce^{3x}, \quad \text{where} \quad C = c_1 + c_2.$$

Thus, we suspect that we do not yet have the general solution of equation (22). Experimentation leads to the result that $y = xe^{3x}$ is a solution of the differential equation.

Problem 7
Verify that $y = xe^{3x}$ is a solution of equation (22).

Hence, the general solution of equation (22) is

$$y = c_1 e^{3x} + c_2 x e^{3x} = (c_1 + c_2 x)e^{3x}.$$

The method that has been illustrated to solve second-order linear differential equations with constant coefficients and with right member zero, and the modification for the case of a repeated root, are applicable to higher-order differential equations of this type as well.

Example 1
The characteristic equation of

$$\frac{d^3y}{dt^3} - \frac{d^2y}{dt^2} - 4\frac{dy}{dt} + 4y = 0$$

is

$$m^3 - m^2 - 4m + 4 = (m-1)(m-2)(m+2) = 0.$$

Hence, the general solution of this third-order differential equation is

$$y = c_1 e^t + c_2 e^{2t} + c_3 e^{-2t}.$$

Example 2

The characteristic equation of

$$\frac{d^3y}{dx^3} - 5\frac{d^2y}{dx^2} + 8\frac{dy}{dx} - 4y = 0 \tag{24}$$

is

$$m^3 - 5m^2 + 8m - 4 = (m-1)(m-2)^2 = 0.$$

Hence, the general solution of this third-order differential equation is

$$y = c_1 e^x + (c_2 + c_3 x)e^{2x}. \tag{25}$$

Problem 8

Verify that equation (25) is a solution of equation (24).

PROBLEMS

9. Find the general solution of each of the following:

   (a) $\dfrac{d^2y}{dx^2} + 7\dfrac{dy}{dx} + 12y = 0$ (b) $\dfrac{d^3y}{dt^3} + \dfrac{d^2y}{dt^2} - 6\dfrac{dy}{dt} = 0$

   (c) $\dfrac{d^2y}{dx^2} - 4y = 0$ (d) $\dfrac{d^2y}{dx^2} - 4\dfrac{dy}{dx} = 0$

   (e) $\dfrac{d^3y}{dt^3} + 2\dfrac{d^2y}{dt^2} - \dfrac{dy}{dt} - 2y = 0$ (f) $\dfrac{d^3y}{dt^3} - 3\dfrac{dy}{dt} - 2y = 0$

10. What does the solution to Problem 9(b) become if the initial conditions are $y = 0$ and $dy/dt = 3$ at $t = 0$, and if y is to remain finite as $t \rightarrow \infty$?

★ 11. If $b^2 - 4ac = 0$, both roots of $am^2 + bm + c = 0$ are $\bar{m} = -b/2a$. Show that if $b^2 - 4ac = 0$, then $y = (p + qx)e^{\bar{m}x}$ satisfies $a(d^2y/dx^2) + b(dy/dx) + cy = 0$ for all constants p and q.

★ 12. Conjecture the general solution of

$$\frac{d^3y}{dx^3} + 6\frac{d^2y}{dx^2} + 12\frac{dy}{dx} + 8y = 0,$$

and verify that your conjectured solution does satisfy the differential equation.

13. Recall the model of predator–prey relationship set out in this section.

   (a) With $X = x + X_E$, $Y = y + Y_E$, why does $dX/dt = dx/dt$ and $dY/dt = dy/dt$?

   (b) When $X = X_E$, why does $dX/dt = 0$?

   (c) When $X = X_E$, $Y = Y_E$, why does $dX/dt = aX_E(b - X_E) - cX_EY_E$?

★   (d) Use the results of (a), (b), and (c), and ignore terms involving x^2 and xy, to obtain equations (16) and (17) from equations (14) and (15).

8.10
Linear differential equations
with constant coefficients
(*continued*)

8.10 Linear differential equations with constant coefficients (*continued*)

We turn now to a type of differential equation mentioned but not discussed in the previous section: the linear equation with constant coefficients and with right member *not* equal to zero, like

$$A\frac{d^2y}{dt^2} + B\frac{dy}{dt} + Cy = f(t),\qquad(26)$$

where $f(t)$ is not identically zero.

Problem 1
The basic results on the equations studied in Section 8.9 are contained in Problem 2 of that section. Show that analogous results are *not* valid for equation (26).

However, there *is* a result bearing some similarity to those of Problem 2, **8.9**:

Theorem
If $y = F(t)$ is a solution of equation (26) and if $y = G(t)$ is a solution of the "reduced equation"

$$A\frac{d^2y}{dt^2} + B\frac{dy}{dt} + Cy = 0,\qquad(27)$$

then $y = F(t) + G(t)$ is also a solution of equation (26).

Problem 2
Prove this theorem.

A solution of equation (26) is called *a particular integral*; the general solution of equation (27) is called the *complementary function*. Because the complementary function involves two arbitrary constants, we hold the key to the problem:

> The general solution of equation (26) equals a particular integral plus the complementary function.

We know how to find the complementary function, but we do not yet have a method for finding a particular integral. Good guessing and good luck are helpful, but some general principles will appear in the following examples.

Example 1

$$\frac{d^2y}{dt^2} - 5\frac{dy}{dt} + 6y = 10e^{4t}.\qquad(28)$$

The same "replicating" property of the exponential function we exploited in Section 8.9 suggests that

$$y = ke^{4t} \tag{29}$$

may be a particular integral of equation (28) *for some value of k*.

Problem 3
Use equation (29) to substitute for y and its derivatives in equation (28).

You should have found that equation (29) is a particular integral of equation (28) if and only if $k = 5$.

Problem 4
Find the complementary function and thus show that the general solution of equation (28) is $y = 5e^{4t} + c_1 e^{2t} + c_2 e^{3t}$.

Example 2

$$\frac{d^2 y}{dt^2} - 5\frac{dy}{dt} + 6y = 10 e^{2t}. \tag{30}$$

Problem 5
Try to find a particular integral of equation (30) by the method of Example 1, and explain what goes wrong.

Because e^{2t} is part of the complementary function of equation (30), substitution of $y = ke^{2t}$ and its derivatives into that equation will make the left side equal to zero, not $10e^{2t}$. What to do? In rough analogy to what we found in Section 8.9 for the case of two equal roots of the characteristic equation, we try

$$y = kte^{2t} \tag{31}$$

Problem 6
Use equation (31) and its derivatives to find that $(-10te^{2t})$ is a particular integral of equation (30).

The general solution of equation (30), then, is $y = (c_1 - 10t)e^{2t} + c_2 e^{3t}$.

Example 3

$$\frac{d^2 y}{dt^2} - 5\frac{dy}{dt} + 6y = 12. \tag{32}$$

Because the right side of equation (32) is a constant, we try $y = k$ as a possible particular integral. We immediately find that $k = 2$, so the general solution of equation (32) is $y = 2 + c_1 e^{2t} + c_2 e^{3t}$.

8.10

Linear differential equations
with constant coefficients
(*continued*)

Example 4

$$\frac{d^2y}{dt^2} - 5\frac{dy}{dt} + 6y = 11 - 6t. \tag{33}$$

In this case we try $y = k + lt$ as a particular integral.

Problem 7

Show that this trial function leads to

$$(6k - 5l) + 6lt = 11 - 6t. \tag{34}$$

As with the method of partial fractions (Section 8.7, especially Problem 7), we conclude that equation (34) will be valid for all values of t if and only if $6k - 5l = 11$ and $6l = -6$. This means that $l = -1$ and $k = 1$. Hence, the general solution of equation (33) is $y = 1 - t + c_1 e^{2t} + c_2 e^{3t}$.

Problem 8

Try $y = p + qt + rt^2$ as a particular integral of equation (33), and determine values of the constants p, q, and r.

The preceding examples indicate how to find a particular integral if the right member is an exponential function or a polynomial in the independent variable. If the right member is the *sum* of an exponential and a polynomial function, say $4e^{-2t} + 19 + 2t + 12t^2$, then a particular integral is the *sum* of particular integrals corresponding to right members that are $4e^{-2t}$ and $19 + 2t + 12t^2$.

Several variations on these techniques will arise in the following problems and in Section 10.10. Indeed, there is a summary comment on this matter at the end of the text of Section 10.10.

PROBLEMS

9. (a) Find the general solution of

$$\frac{d^2y}{dx^2} + 7\frac{dy}{dx} + 12y = 4e^{-2x}.$$

(b) Find the general solution of

$$\frac{d^2y}{dx^2} + 7\frac{dy}{dx} + 12y = 19 + 2x + 12x^2.$$

(c) Hence, write the general solution of

$$\frac{d^2y}{dx^2} + 7\frac{dy}{dx} + 12y = 4e^{-2x} + 19 + 2x + 12x^2.$$

10. Find the general solution of

$$\frac{d^2y}{dt^2} + 7\frac{dy}{dt} + 12y = 5e^{-3t}.$$

11. Find the general solution of

$$\frac{d^2y}{dx^2} - 4y = e^{-2x}.$$

12. Modify Problem 4, **8.3**, by assuming that consumption has an "autonomous" (constant) component as well as a component proportional to income (i.e., that $C = p + lY$).
   (a) Show that $dY/dt - k(1-l)Y = -kp$.
   (b) Find the general solution of this differential equation.
   (c) Show that the solution for which $Y = Y_0$ at $t = 0$ can be written

$$Y = \left(Y_0 - \frac{p}{1-l}\right)e^{k(1-l)t} + \frac{p}{1-l}.$$

Thus, national income is the sum of two terms, one growing with time and the other constant.

SAMPLE TEST ON SECTIONS 8.1–8.10

1. Use separation of variables and the method of partial fractions to show that the solution of $dy/dx = y^2 - 5y + 6$, for which $y = 4$ at $x = 0$ can be written in the form $y = (2e^x - 6)/(e^x - 2)$. What is the value of $\lim_{x \to \infty} y$? (Justify your answer.)
2. Find the general solution of $d^2y/dx^2 - 9y = 0$.
3. Find the general solution of $d^2y/dx^2 - 9(dy/dx) = 0$.
4. Find the general solution of $d^2y/dx^2 - 9y = 5e^{2x}$.
5. Find the solution of $dy/dx = -y/x$ for which $y = 3$ at $x = 3$.
★ 6. Find the general solution of $d^2y/dx^2 - 9(dy/dx) = 18e^{9x}$.

★ 8.11 Approximating the solutions of a pair of simultaneous differential equations

In Section 8.9, a certain model of the relationship between a prey population, X, and a predator population, Y, was described by the pair of differential equations

$$\frac{dX}{dt} = aX(b - X) - cXY \qquad (14)$$

and

$$\frac{dY}{dt} = (kcX - 1)Y \qquad (15)$$

for suitable constants a, b, c, and k. Because these equations are difficult to analyze – for example, they are not linear – we reduced them to a related, but not equivalent, pair of equations of relatively simple form:

$$\frac{dy}{dt} = rx \qquad (17)$$

and

$$\frac{d^2y}{dt^2} + p\frac{dy}{dt} + qry = 0, \tag{18}$$

where p, q, and r are positive constants, and x and y are the differences between X and Y and their presumed equilibrium, or steady-state, values, X_E and Y_E.

Equations (17) and (18) are useful in analyzing the behavior of X and Y only when x and y are small (i.e., when X and Y are close to equilibrium). Indeed, the main use of these equations is to determine whether X and Y approach equilibrium. Moreover, the characteristic equation of (18) not uncommonly has imaginary roots, a situation that we can handle only after we have developed trigonometric functions in Chapter 10. But we can solve (14) and (15) *approximately*, by Euler's method, as we shall now illustrate.

Suppose that x and y are two quantities whose behavior is described by the pair of differential equations

$$\frac{dx}{dt} = F(t, x, y) \quad \text{and} \quad \frac{dy}{dt} = G(t, x, y),$$

where F and G are specific functions of t, x, and y. Starting with initial conditions – at $t = t_0$, $x = x_0$ and $y = y_0$ – we have numerical values for the derivatives:

$$\frac{dx}{dt} = F(t_0, x_0, y_0) \quad \text{and} \quad \frac{dy}{dt} = G(t_0, x_0, y_0).$$

At $t_1 = t_0 + h$, let $x = x_1$ and $y = y_1$. Then

$$\Delta x = x_1 - x_0 \approx h \cdot \frac{dx}{dt} = h \cdot F(t_0, x_0, y_0)$$

and

$$\Delta y = y_1 - y_0 \approx h \cdot \frac{dy}{dt} = h \cdot G(t_0, x_0, y_0).$$

(Note the similarity of this process to the approximate increments of Section 3.3.) Thus,

$$x_1 \approx x_0 + h \cdot F(t_0, x_0, y_0) \quad \text{and} \quad y_1 \approx y_0 + h \cdot G(t_0, x_0, y_0).$$

Using these approximate values for x_1 and y_1 at t_1, we repeat the process at $t_2 = t_1 + h = t_0 + 2h$:

$$x_2 \approx x_1 + h \cdot F(t_1, x_1, y_1) \quad \text{and} \quad y_2 \approx y_2 + h \cdot G(t_1, x_1, y_1).$$

Continuing, we obtain the approximate solution by the following:

The Euler method for the system of two first-order differential equations, $dx/dt = F(t, x, y)$, $dy/dt = G(t, x, y)$, at $t = t_0$, $x = x_0$, and $y = y_0$:

$$t_{n+1} = t_0 + h(n+1),$$

$$x_{n+1} = x_n + h \cdot F(t_n, x_n, y_n),$$

$$y_{n+1} = y_n + h \cdot G(t_n, x_n, y_n), \quad n = 0, 1, 2, \ldots.$$

303

Example 1

Let us apply this Euler iterative method to the system (14) and (15), with $a = 0.01$, $b = 150$, $c = 0.05$, $k = 0.2$ – constants that have been chosen to correspond to neat equilibrium values of $X_E = 100$ and $Y_E = 10$. Then the Euler equations are

$$X_{n+1} = X_n + h\left[0.01X_n(150 - X_n) - 0.05X_nY_n\right]$$

and

$$Y_{n+1} = Y_n + h\left[(0.01X_n - 1)Y_n\right].$$

Note that t does not appear explicitly in this case.

We must choose initial values, X_0 and Y_0, which can be done with knowledge of the natural circumstances. If the ecological system actually leads to a balance of the populations, we should attain values close to the equilibria. For simplicity's sake, let us start with $X_0 = 102$ and $Y_0 = 9.8$ – numbers very close to the equilibria. Later we shall consider initial values much farther from the equilibrium values.

We choose $h = 1$ to stand for the duration of one generation in the prey population. Then, to two decimal places,

$$X_1 = 102 + \left[(0.01)(102)(150 - 102) - (0.05)(102)(9.8)\right] = 100.98,$$

and

$$Y_1 = 9.8 + \left[(0.01)(102) - 1\right][9.8] = 10.00.$$

Both X_1 and Y_1 are closer to X_E and Y_E than X_0 and Y_0 are. Let us take another step:

$$X_2 = 100.98 + \left[(0.01)(100.98)(150 - 100.98) - (0.05)(100.98)(10.00)\right]$$
$$= 99.99,$$

and

$$Y_2 = 10.00 + \left[(0.01)(100.98) - 1\right][10.00] = 10.10.$$

This time, X has moved toward X_E, but Y has moved away from Y_E. Further steps display similar behavior – occasional improvements, occasional setbacks – but over a substantial sequence of steps, the values of X_n and Y_n will approach their equilibria. We shall return to this example in the problems at the end of this section.

Example 2

Here is a quite different example in which the Euler method can also provide useful information:

$$\frac{d^2z}{dt^2} - t\frac{dz}{dt} - z = 0, \tag{35}$$

with $z = 0$ and $dz/dt = 1$ at $t = 0$. This is a second-order linear differential equation, with right member equal to zero, but the coefficient of dz/dt is *not* a constant. Tricks are needed to solve equations of this sort – if, indeed, they can be solved at all. In the problems at the end of this section we shall learn that the desired particular solution of this equation can be written as

$$z = e^{t^2/2} \int_0^t e^{-s^2/2} \, ds. \tag{36}$$

The integral appearing in this solution cannot be evaluated by antidifferentiation, as we shall note in Chapter 9. Thus, the existence of this formula for the solution does not obviate the need for numerical approximation, so we may as well do the approximation at the outset.

To apply Euler's method to equation (35), we reduce it to a pair of first-order equations through the substitution $x = z$ and $y = dz/dt$.

Then

$$\frac{dx}{dt} = \frac{dz}{dt} = y, \quad \text{and} \quad \frac{dy}{dt} = \frac{d^2z}{dt^2} = t\frac{dz}{dt} + z = ty + x.$$

Thus, equation (35) is equivalent to the system

$$\frac{dx}{dt} = y \quad \text{and} \quad \frac{dy}{dt} = ty + x,$$

with $x_0 = 0$ and $y_0 = 1$ at $t_0 = 0$. The Euler method gives us for this system

$$t_{n+1} = h(n+1),$$
$$x_{n+1} = x_n + h \cdot y_n, \quad \text{and}$$
$$y_{n+1} = y_n + h(t_n y_n + x_n).$$

Choosing $h = 0.1$, we obtain

$$x_1 = 0 + (0.1)(1) = 0.1, \quad y_1 = 1 + (0.1)(0 \cdot 1 + 0) = 1,$$
$$x_2 = 0.1 + (0.1)(1) = 0.2, \quad y_2 = 1 + (0.1)[(0.1)(1) + (0.1)] = 1.02,$$
$$x_3 = 0.2 + (0.1)(1.02) = 0.302,$$
$$y_3 = 1.02 + (0.1)[(0.2)(1.02) + (0.2)] = 1.0604,$$

and so on.

Because $x = z$, the values of x_n constitute the approximate solution. We can compare the values of x for $t = 0.1, 0.2, 0.3, \ldots$ as thus obtained with the values provided by equation (36), approximating the integral in that equation by using statistical tables or one of the techniques developed in Chapter 9.

The accuracy of the Euler method for systems of differential equations is again of order h: Halving h will tend to halve the error, reducing h by a factor of 10 will tend to reduce the error by a factor of 10, and so forth. As with a single equation, there exist other methods of greater efficiency, but we shall not develop them here.

PROBLEMS

1. Extend the calculations of Example 1 to $n = 10$. Observe the oscillations about $X_E = 100$, $Y_E = 10$ and their tendency to diminish in magnitude as n increases.

C 2. (a) Prepare a program to calculate X_n and Y_n of Example 1, and use it to extend the sequence of Problem 1 to $n = 20$. To $n = 50$.

(b) Without knowledge of X_E and Y_E in Example 1, a person might choose more "distant" X_0 and Y_0. With understanding of the conditions of the population problem, a choice as close as $X_0 = 110$, $Y_0 = 20$ is likely. Use these initial conditions to see whether you tend toward X_E and Y_E. (Although the first few steps are discouraging, after $n = 10$ things look better.)

(c) Try other starting values, even such wild ones as $X_0 = 15$, $Y_0 = 1$. Can you account for the consequences on ecological grounds?

○ 3. Extend the calculations of x_n and y_n of Example 2 to $t_n = 1.0$ (i.e., $n = 10$).

○ 4. (a) Same as Problem 3, with $h = 0.01$.

(b) For $t = 1$, equation (36) gives $z = e^{1/2} \int_0^1 e^{-s^2/2} ds$. By use of statistical tables, we find this z to equal 1.41068614, to eight decimal places. Calculate the error, $z - x_{10}$, in Problem 3, and the error $z - x_{100}$ in Problem 4(a).

5. The aim of this problem is to find an exact solution of equation (35):

$$\frac{d^2z}{dt^2} - t\frac{dz}{dt} - z = 0.$$

(a) Verify that the left side of the differential equation is

$$\frac{d}{dt}\left(\frac{dz}{dt} - tz\right)$$

and hence that $dz/dt - tz = c_1$, where c_1 is an arbitrary constant.

(b) Multiply both sides of this last equation by $e^{-t^2/2}$, and thus obtain

$$\frac{d}{dt}\left(e^{-t^2/2} \cdot z\right) = c_1 e^{-t^2/2}.$$

(In differential equations texts, the multiplier – in this case, $e^{-t^2/2}$ – is called an "integrating factor".)

(c) Because

$$e^{-t^2/2} = \frac{d}{dt}\left(\int_0^t e^{-s^2/2} ds\right),$$

we can rewrite the equation of part (b) as

$$\frac{d}{dt}\left(e^{-t^2/2} \cdot z\right) = c_1 \frac{d}{dt}\left(\int_0^t e^{-s^2/2} ds\right).$$

Hence, $e^{-t^2/2} \cdot z = c_1 \int_0^t e^{-s^2/2} ds + c_2$. Solve for z to obtain the general solution.

(d) Show that the initial conditions, $z = 0$ and $dz/dt = 1$ at $t = 0$, imply that $c_1 = 1$ and $c_2 = 0$, and hence that the particular solution of equation (35) is $z = e^{t^2/2}\int_0^t e^{-s^2/2}\,ds$.

The remaining problems have exact solutions involving trigonometric functions. Save your results on these problems to compare with results in Chapter 10 where such functions are discussed.

○ C 6. Apply Euler's method with $h = 0.1$ to the system of equations

$$\frac{dx}{dt} = x - y \quad \text{and}$$

$$\frac{dy}{dt} = x + y, \quad \text{with}$$

$$x = 1 \quad \text{and} \quad y = 0 \quad \text{at} \quad t = 0,$$

to obtain approximate solutions for $t = 0.1, 0.2, \ldots, 1.0$.

○ C 7. Same as Problem 6, with $h = 0.01$.

○ 8. (a) Reduce the differential equation $d^2z/dt^2 + 2(dz/dt) + 2z = 0$, with initial conditions $z = 0$ and $dz/dt = 1$ at $t = 0$, to the system $dx/dt = y$ and $dy/dt = -2(x + y)$, with $x = 0$ and $y = 1$ at $t = 0$.

  C   (b) Apply Euler's method with $h = 0.1$ to this system to obtain approximate solutions for $t = 0.1, 0.2, \ldots, 1.0$.

○ C 9. Same as Problem 8(b), with $h = 0.01$.

9 Further integration

9.1 Introduction

There are five topics to be treated in this chapter:
1. Review of the role of the Chain Rule in antidifferentiation
2. More applications of integration, in the spirit of the discussion of Chapter 5
3. New sorts of "elements" – besides rectangles, disks, and slabs – to sum for the "whole"
4. The idea of *mean* (or *average*) *value* and its applications
5. Quadrature: numerical (approximate) integration

9.2 Review of the use of the Chain Rule in integration (antidifferentiation)

We illustrate our first topic with a familiar example:

$$\int_2^3 \frac{dx}{x} = \ln x \big|_2^3 = \ln 3 - \ln 2 = \ln \tfrac{3}{2}.$$

It is also true that $\int_2^3 du/u = \ln u \big|_2^3 = \ln 3 - \ln 2 = \ln \tfrac{3}{2}$.

The symbolism, of course, implies that the limits of integration refer to the variable appearing in the integrand – x in the first case and u in the second. The value of the integral, then, depends on the function appearing in the integrand and on the limits of integration – it does not depend on the particular letter or letters used. To emphasize this basic fact, we can write $\int_2^3 d(*)/* = \ln * \big|_2^3$. Mathematicians with a primitive sense of humor sometimes write

$$\int_2^3 \frac{d(\text{cabin})}{\text{cabin}} = \log \text{cabin} \big|_2^3.$$

We have already made use of these observations extensively in Sections 5.9, 6.8, and 7.5, and we take this opportunity for some review.

Example 1

$$I = \int_0^2 \frac{9x+3}{3x^2+2x+9} dx$$

can be evaluated as follows: If we set $u = 3x^2 + 2x + 9$, then $du = (6x+2)\,dx$. Hence, we write

$$I = \int_0^2 \frac{3(3x+1)\,dx}{3x^2+2x+9} = 3\int_0^2 \frac{(3x+1)\,dx}{3x^2+2x+9} = \tfrac{3}{2}\int_0^2 \frac{2(3x+1)\,dx}{3x^2+2x+9}$$

$$= \tfrac{3}{2}\int_{x=0}^{x=2} \frac{du}{u} = \tfrac{3}{2}\ln u\Big|_{x=0}^{x=2} = \tfrac{3}{2}\ln(3x^2+2x+9)\Big|_0^2$$

$$= \tfrac{3}{2}(\ln 25 - \ln 9) = \tfrac{3}{2}\ln\tfrac{25}{9} = \tfrac{3}{2}\ln\left(\tfrac{5}{3}\right)^2 = 3\ln\tfrac{5}{3}.$$

Example 2

$$J = \int_0^2 \frac{9x+3}{(3x^2+2x+9)^{1/2}} dx$$

can be evaluated through the same substitution as was used in Example 1:

$$J = \tfrac{3}{2}\int_{x=0}^{x=2} \frac{du}{u^{1/2}} = \tfrac{3}{2}\frac{u^{1/2}}{1/2}\Big|_{x=0}^{x=2} = 3(3x^2+2x+9)^{1/2}\Big|_0^2$$

$$= 3(\sqrt{25} - \sqrt{9}) = 3(5-3) = 6.$$

Don't fall into the trap of antidifferentiating the function in this case as $\ln(3x^2+2x+9)^{1/2}$ just because $(3x^2+2x+9)^{1/2}$ appears in the denominator!

Example 3
The integral

$$\int_0^1 \frac{e^{2x} - e^{-2x}}{e^{2x} + e^{-2x}} dx$$

looks forbidding, but it is susceptible to the same approach we have used in the preceding examples: set $e^{2x} + e^{-2x} = u$.

Problem 1
Work out this example to show that the given integral equals $\tfrac{1}{2}\ln(e^{2x} + e^{-2x})|_0^1 = \tfrac{1}{2}\ln[(e^4+1)/2e^2]$. Check by verifying that the derivative of $\tfrac{1}{2}\ln(e^{2x} + e^{-2x})$ equals $(e^{2x} - e^{-2x})/(e^{2x} + e^{-2x})$.

PROBLEMS

2. Evaluate each of the following:

    (a) $\int_0^2 (x^2+2)x\,dx$ (b) $\int_0^2 3(x^2+2)x\,dx$ (two ways)

(c) $\int_0^2 \dfrac{\sqrt{x^2+2}\cdot x\,dx}{2}$ (d) $\int_0^2 \dfrac{4x\,dx}{x^2+2}$ (e) $\int_0^2 \dfrac{-x\,dx}{2(x^2+2)^3}$

(f) $\int_0^2 \dfrac{5x\,dx}{\sqrt{x^2+2}}$ (g) $\int_2^{11} \sqrt[3]{x^2+4}\,\dfrac{8x}{7}\,dx$ (h) $\int_0^2 9x^2\sqrt{x^3+1}\,dx$

(i) $\int_0^1 \dfrac{t+1}{t^2+2t+3}\,dt$ (j) $\int_0^1 \dfrac{t+1}{\sqrt{t^2+2t+3}}\,dt$ (k) $\int_0^1 \dfrac{t+1}{(t^2+2t+3)^2}\,dt$

3. Evaluate $\int_0^1 (e^x + e^{-x})(e^x - e^{-x})\,dx$ in three ways:
   (a) Multiply out the two factors of the integrand, and find the antiderivative of each term.
   (b) Set $e^x + e^{-x} = u$.
   (c) Set $e^x - e^{-x} = u$.

4. Evaluate $\int_1^2 (e^{3x}\,dx)/(e^{3x}-1)$. Show that the answer can be written in the form $\frac{1}{3}\ln(e^3+1)$.

5. Evaluate $\int_1^3 (e^{2x}\,dx)/(e^{2x}-1)$. Show that the answer can be written in the form $\frac{1}{2}\ln(e^4+e^2+1)$.

6. Evaluate $\int_{-2}^2 e^{-x^2}\cdot x\,dx$.

7. Evaluate $\int_1^2 (e^{3x}\,dx)/(e^{3x}-1)^2$. Show that the answer can be written in the form $e^3/3(e^6-1)$.

8. Evaluate $\int_1^3 (e^{2x}\,dx)/(e^{2x}-1)^2$. Show that the answer can be written in the form $e^2(e^2+1)/2(e^6-1)$.

9.3   Force of attraction

Newton's law of universal gravitation states that any two particles attract each other with a force proportional to the product of their masses and inversely proportional to the square of the distance between them. According to this law, the force of attraction (F dynes) between two particles, P_1 and P_2, of masses m_1 g and m_2 g, respectively, at a distance r cm apart (Figure 9-1) is given by $F = Gm_1m_2/r^2$, where G is the constant of proportionality – the so-called gravitational constant.

There are many quantities besides the force of gravitational attraction that vary with distance in an "inverse square" fashion: The intensity of light (or heat) at a distance from a source (like the sun) is inversely proportional to the square of that distance; the force of repulsion between two electrically charged particles of the same sign is inversely proportional to the square of the distance between the particles, and so forth. The power law, $Q = kr^n$, with $n = -2$, is one of the most important types of variation for many applications.

If we have two extensive bodies, composed of many particles, the total force of gravitational attraction between the bodies is the sum of the forces between their pairs of particles (Figure 9-2). Force, like velocity, is specified

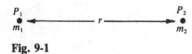

Fig. 9-1

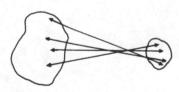

Fig. 9-2

by *direction* as well as *magnitude*, and the problem of summing forces of differing directions as well as different magnitudes can be quite complicated. We shall deal only with a problem involving a *single* direction; it will suffice to illustrate the use of integration in such a problem. Suppose a particle of mass m g is located at point P, in line with a bar AB, of length l cm, and that the distance PA is a cm (Figure 9-3). Suppose, moreover, that the bar has a uniform density, k g/cm. What is the total force of attraction between the particle and the bar?

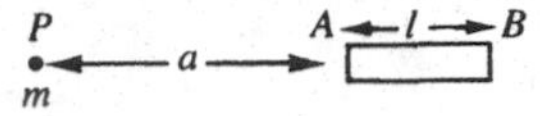

Fig. 9-3

The crux of the problem lies in the varying distances between the particle at P and the particles of the bar AB. A particle at A is at distance a from P, and a particle at B is at distance $a + l$ from P. Hence, according to the law of universal gravitation, the forces differ. All the varying forces, though, are directed along the straight line PAB, so that we can get the magnitude of the total force by summing the magnitudes of the constituent parts. If we consider a small piece of the bar, of length Δx cm, at distance x from P, as in Figure 9-4, the mass of that piece is $k \cdot \Delta x$ g. The force of attraction, ΔF dynes, between the particle at P and this small piece of the bar is given approximately by

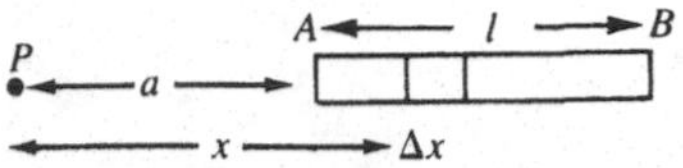

Fig. 9-4

$$\Delta F = \frac{G \cdot m \cdot k \cdot \Delta x}{x^2}.$$

Problem 1
Explain why this equation is only *approximately* (not exactly) valid.

The total force of attraction is the *limit of the sum* of such "elementary" forces, as $\Delta x \to 0$, and the limit of this sum is the integral

$$F = \int_a^{a+l} \frac{Gmk\,dx}{x^2},$$

where the limits of integration correspond to the values of x at the ends of the bar.

Problem 2
Perform the integration to verify that

$$F = \frac{Gmkl}{a(a+l)} \text{ dynes.}$$

Now let us change the problem by making the density of the bar variable: Suppose that the bar's density, ρ g/cm, increases uniformly from ca g/cm at A to $c(a+l)$ g/cm at B.

Problem 3
Find a formula for ρ in terms of x. (Answer: $\rho = cx$.)

311

Problem 4

Set up the integral for the total force of attraction between the particle at P and the bar in this case, evaluate the integral, and thus verify that

$$F = Gmc \ln \frac{a+l}{a} \text{ dynes.}$$

Problem 5

(a) Suppose that the bar's density, ρ g/cm, increases uniformly from 0 at A to k at B. Find the formula for ρ in terms of x.

(b) Set up the integral for the total force of attraction between the particle at P and the bar, and verify that it can be written in the form

$$F = \frac{Gmk}{l} \int_a^{a+l} \frac{dx}{x} - \frac{Gmka}{l} \int_a^{a+l} \frac{dx}{x^2}.$$

(c) Evaluate these integrals, and thus verify that

$$F = Gmk \left[\frac{1}{l} \ln \frac{a+l}{a} - \frac{1}{a+l} \right] \text{ dynes.}$$

9.4 Loads

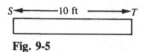

S ◄——— 10 ft ———► T

Fig. 9-5

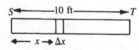

S ◄——— 10 ft ——► T
◄— x —► Δx

Fig. 9-6

If the load on the beam ST in Figure 9-5 were uniformly distributed – 25 lb/ft, say – then the total load on the beam would be found by simple arithmetic: 25 lb/ft $\times$ 10 ft = 250 lb. However, if the loading is variable, we may need integration. Let us assume that the loading (y lb/ft) varies with the distance (x ft) from S as follows: $y = 2 + 5x - 0.3x^2$. Then the load on the small piece of the beam marked Δx in Figure 9-6 is approximately $y \cdot \Delta x = (2 + 5x - 0.3x^2)\Delta x$, and the total load ($L$ lb) on the whole beam is the *limit of the sum* of such "elementary" loads. That is, L is given by

$$L = \int_0^{10} (2 + 5x - 0.3x^2) \, dx.$$

Problem 1

Evaluate this integral to verify that $L = 170$ lb.

Problem 2

Suppose that the loading on this beam is given by $y = 2 + 5x$.

(a) Find the total load, L lb, in this case.

(b) At $x = 0$, $y = 2$; at $x = 10$, $y = 52$. What do you think to be the average value of y? Hence, what total L?

(c) Interpret your results in (a) and (b) in terms of the area under $y = 2 + 5x$ from $x = 0$ to $x = 10$.

(d) Can the simple "averaging method" be used for the original loading function, $y = 2 + 5x - 0.3x^2$? Explain.

Now we consider a two-dimensional loading problem. Suppose that the loading (y lb/ft^2) on a rectangular floor, 20 ft by 50 ft, varies thus with the distance (x ft) from one of the 20-ft sides: $y = 8 + 1.4x + 0.06x^2$. Then we proceed to find the total load (L lb) as follows: All points on a line segment, UV, parallel to the 20-ft sides have the same value of x, and hence the same loading, y (Figure 9-7). On the shaded strip of width Δx, the load is approximately (y lb/ft^2)·($20 \cdot \Delta x$ ft^2) = $20y\Delta x$ lb. The total load on the whole floor is the *limit of the sum* of such "elementary" loads; that is,

$$L = 20 \int_0^{50} (8 + 1.4x + 0.06x^2)\, dx.$$

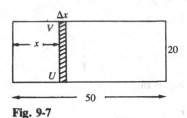

Fig. 9-7

(a)

Problem 3
Evaluate the integral to verify that $L = 46.5$ tons.

Here is a two-dimensional problem that uses language somewhat different from that of the preceding example, but a similar argument: The graph of $x^2/100 + y^2/36 = 1$ is a closed oval curve, called an *ellipse*, as shown in Figure 9-8(a). Suppose that a metal plate has the shape of the region bounded by the coordinate axes and the arc of this ellipse in the first quadrant, with x and y measured in inches, and suppose that the density (ρ lb/in.^{2}) of the plate varies thus with the distance (x in.) from the y axis: $\rho = \frac{1}{2}x$. We find the total weight, W, of the plate as follows: The area, ΔA, of a strip of width Δx at distance x from the y axis is given approximately by $\Delta A = y \cdot \Delta x$, where y is the height of the curve corresponding to the abscissa x, as shown in Figure 9-8(b). The weight of the strip is given approximately by $\Delta W = \rho \cdot y \cdot \Delta x$. Because all points in the strip have approximately the same abscissa, x, we know that $\Delta W = \frac{1}{2}x \cdot y \cdot \Delta x$, approximately.

(b)

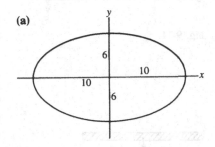

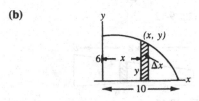

Fig. 9-8

Problem 4
There are *two* approximations involved in this equation. What are they?

Problem 5
Solve the given equation of the ellipse for y in terms of x to obtain $y = \frac{3}{5}\sqrt{100 - x^2}$.

Hence, $\Delta W = \frac{3}{10}\sqrt{100 - x^2}\, x\, \Delta x$. The entire weight of the plate is the *limit of the sum* of the "elementary" weights of the strips. That is, $W = \int_0^{10} \frac{3}{10}\sqrt{100 - x^2}\, x\, dx$.

Problem 6
Evaluate this integral to obtain $W = 100$ lb.

313

9
Further integration

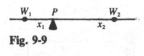

Fig. 9-9

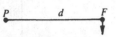

Fig. 9-10

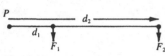

Fig. 9-11

Fig. 9-12

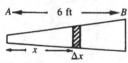

Fig. 9-13

9.5 Moment of a force

As a child learns early from experience on a seesaw, one can balance a heavier weight by sitting farther from the fulcrum on which the board rests. In fact, in the situation in Figure 9-9, if $W_1 = 2W_2$, then x_2 must equal $2x_1$ for balancing. In general, for equilibrium, $W_1 \cdot x_1 = W_2 \cdot x_2$. The measure of the tendency of a force to produce rotation about a point, P, is given by the magnitude of the force times its "lever arm" – the distance from P to the line of action of the force (Figure 9-10). This measure is called the *moment of the force about* P:

$$M = F \cdot d.$$

Moments are additive: If we have two forces, F_1 and F_2, at distances d_1 and d_2 from P, as in Figure 9-11, then the total moment about P is given by

$$M = F_1 \cdot d_1 + F_2 \cdot d_2.$$

Suppose, though, that we have not one or two or ten forces, but a continuous distribution of forces, as is in fact the case even in our initial example if we do not ignore the weight of the seesaw board itself. If the density of the board is constant, it can be shown that the moment of its weight about P is the same as though its entire weight were concentrated at its midpoint (Figure 9-12). If the density varies, we must use integration: Suppose that the horizontal beam has a weight (w lb/ft) that varies thus with the distance (x ft) from A: $w = 5 + 2x$.

To find M_A, the moment of its weight about A, we proceed as suggested in Figure 9-13: At distance x from A, a small piece of length Δx has a weight, ΔW, given approximately by $\Delta W = (w\ \text{lb/ft}) \cdot (\Delta x\ \text{ft}) = w\Delta x$ lb. The moment of this piece about A is given approximately by

$$\Delta M_A = (w \cdot \Delta x) \cdot x$$
$$= (5 + 2x) \cdot \Delta x \cdot x\ \text{(lb-ft)}.$$

Problem 1
There are *two* approximations in this last equation. What are they?

The desired M_A is the *limit of the sum* of the "elementary" moments, ΔM_A, so

$$M_A = \int_0^6 (5 + 2x) x\, dx.$$

Problem 2
Evaluate this integral to verify that $M_A = 234$ lb-ft.

Problem 3
The density of this beam varies uniformly from 5 lb/ft at A to 17 lb/ft at B. What, then, do you think to be its "average density"? Hence, what do you

think to be its total weight? Check by integration to verify that its total weight is, in fact, 66 lb. Can one consider its total weight as being concentrated at the midpoint of the beam to compute M_A?

Problem 4

Suppose the density of a 6-ft beam to be constant: $w = 5$ lb/ft. Use integration to verify that the moment of the weight of the beam about one end of the beam is the same as though the entire weight were concentrated at its midpoint.

Problem 5

For the case of the 6-ft beam in which $w = 5 + 2x$, where x is the distance from A, find M_B, the moment of the weight of the beam about B. (Hint: What is the weight of a piece of length Δx at distance x from A? What is its lever arm?)

Problem 6

The answer to Problem 5 is $M_B = 162$ lb-ft, which is less than M_A, as found in Problem 2. Can you give an argument that one should expect M_B to be less than M_A?

○ **Problem 7**

The loading (l lb/ft) on a beam, AB, 10 ft long varies thus with the distance (x ft) from A: $l = 2 + 3x - 0.12x^2$.

(a) Find the total load, L lb, on the beam.

(b) Find the moment of the load about A.

(c) Find the moment of the load about B.

9.6 Consumers' and producers' surpluses

Suppose that the demand curve for a product is as shown in Figure 9-14, with the price ($\$p$) per item at which there is a market demand for q items given by $p = f(q)$, $0 \le q \le a$. That is, corresponding to a demand for q items is a price $f(q)$. If the producer increases his production from q to $q + \Delta q$, his incremental revenue, ΔR, is given by

$$\Delta R \approx f(q) \cdot \Delta q.$$

His *total revenue*, R, is the limit of the sum of such incremental revenues, that is,

$$R = \int_0^a f(q) \, dq.$$

In other words, *the total revenue is the area under the demand curve*. This is valid on the assumption that the producer has *perfect discrimination*, and

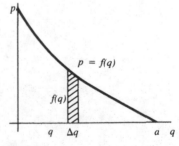

Fig. 9-14

can practice it; that is, he can charge exactly "what the traffic will bear," obtaining from some eager, well-heeled customers the relatively high price that they are willing to pay for his product, and scaling down the price to sell also to those who can't or won't meet the premium prices.

Actually, it is more likely that the producer sells q_0 items, say, all at the fixed price $p_0 = f(q_0)$, obtaining total revenue, R_0, given by

$$R_0 = q_0 \cdot p_0 = q_0 \cdot f(q_0),$$

which appears graphically as the area of the shaded rectangle in Figure 9-15. If he were able to practice perfect discrimination, his revenue in selling q_0 items would be

$$R = \int_0^{q_0} f(q) \, dq.$$

The difference, $R - R_0$, is called *consumers' surplus* and is shown graphically as the horizontally shaded area in the figure. It represents the total financial advantage to those customers who were willing to pay more than $\$p_0$ for the item.

Problem 1

If the demand function is given by $p = 99 - 2q + 0.01q^2$, find the consumers' surplus if the constant price corresponds to a demand of $q_0 = 30$. If $q_0 = 60$.

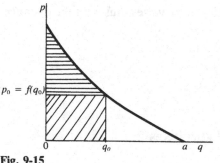
Fig. 9-15

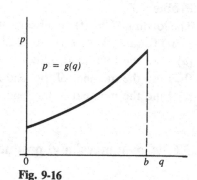
Fig. 9-16

The foregoing discussion has an analogue, leading to the concept of producers' surplus: If the supply curve for a product is as shown in Figure 9-16, with the price ($\$p$) per item at which the producer is willing to supply q items given by $p = g(q)$, $0 < q \le b$, then the total revenue obtained by selling b items in accord with this pricing function is

$$R = \int_0^b g(q) \, dq.$$

However, if the producer sells $\bar{q}$ items, all at the fixed price of $\bar{p} = g(\bar{q})$ each, his total revenue $\bar{R}$ is given by

$$\bar{R} = \bar{q} \cdot \bar{p} = \bar{q} \cdot g(\bar{q}).$$

$\bar{R}$ appears graphically as the area of the shaded rectangle in Figure 9-17, R as the area under the curve between $q = 0$ and $q = \bar{q}$, and the *producers' surplus*, $\bar{R} - R$, as the doubly shaded area in the figure.

Problem 2
If the supply function is given by $p = 15 + q + 0.07q^2$, find the producers' surplus if the supply is $\bar{q} = 10$. If $\bar{q} = 30$.

For steady operation, the quantity supplied should equal the quantity demanded. The quantity (and price) corresponding to such an equilibrium situation can be found by solving simultaneously the formulas for the supply and demand functions.

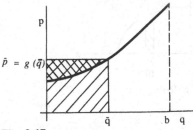

Fig. 9-17

Problem 3
For the demand function of Problem 1 and the supply function of Problem 2, find the equilibrium, q, and its corresponding p.

Problem 4
For the constants p and q found in Problem 3, find the consumers' surplus and the producers' surplus.

9.7 Horizontal rectangular strips and circular strips

Thus far, in setting up integrals, we have had occasion to select "elements" like those in Figure 9-18: short pieces of bars, thin vertical rectangles, cylindrical disks, and slabs of rectangular and other shapes. In every case, the criteria are as follows: (1) Can we express the desired quantity (area, volume, mass, moment, load, etc.) as the limit of a sum of the quantity for such elements? (2) Can we express the quantity for the element itself in the form $f(x) \cdot \Delta x$, for some function, f? If both criteria are met, then we know that the desired quantity equals

$$\int_a^b f(x)\, dx,$$

where the limits are such as to encompass the entire figure. We can then try to evaluate the integral by finding an antiderivative of $f(x)$.

The theory of integration does not restrict us to elements of the sort pictured earlier, and we shall now consider some examples that necessitate the use of other types of elements.

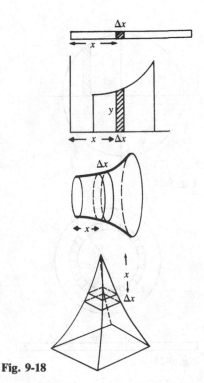

Fig. 9-18

Example 1
Suppose that the quarter-elliptical plate encountered at the end of Section 9.4 has a density (ρ lb/in.^{2}) that varies thus with the distance (y in.) from the x axis: $\rho = \frac{1}{3}y$. If we set out to find the total weight, W, of the plate by

9
Further integration

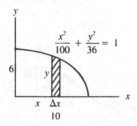

Fig. 9-19

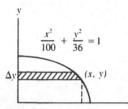

Fig. 9-20

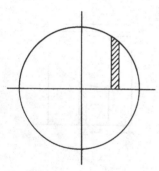

Fig. 9-21

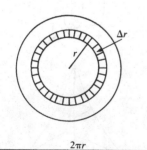

Fig. 9-22

318

taking vertical strips as before (Figure 9-19), we have no difficulty with the area, ΔA, of the strip: $\Delta A = y\Delta x$, approximately, or $\Delta A = \frac{3}{5}\sqrt{100-x^2}\,\Delta x$, approximately.

But there are many different values of y in the strip (indefinitely many, in fact), so it is *not* correct to say that for this strip $\rho = \frac{1}{3}y = \frac{1}{5}\sqrt{100-x^2}$. (This value of ρ in terms of x is valid only at the top of the strip, where the equation of the curve is $y = \frac{3}{5}\sqrt{100-x^2}$.)

What we must do is to take a *horizontal* strip (Figure 9-20), within which all values of y are essentially the same, and hence, also, all values of the density are essentially constant.

Problem 1
Solve the given equation of the ellipse for x in terms of y.

The area, ΔA, of the horizontal strip shown in Figure 9-20 is given approximately by

$$\Delta A = x \cdot \Delta y = \frac{5}{3}\sqrt{36-y^2}\,\Delta y.$$

Hence, the weight, ΔW, of the strip is given approximately by

$$\Delta W = \rho \cdot \Delta A = \frac{1}{3}y \cdot \frac{5}{3}\sqrt{36-y^2}\,\Delta y.$$

Then, the total weight, W, of the plate is the *limit of the sum* of the weights of the "elementary" strips, so

$$W = \frac{5}{9}\int_0^6 \sqrt{36-y^2}\,y\,dy.$$

Problem 2
Evaluate this integral to obtain $W = 40$ lb.

Example 2
Suppose that the loading (w lb/ft^2) on a circular floor of radius 6 ft varies with the distance (r ft) from the center of the floor as follows: $w = 2 + \frac{1}{2}r$. To find the total load (L lb) on the floor, we *cannot* use vertical rectangular strips, as in Figure 9-21, because r varies within such a strip, and hence w does, also. Horizontal strips would be no better. To obtain an element within which r (and w) are essentially constant, we must take a *circular strip*, as in Figure 9-22. We can approximate the area of such a circular strip by thinking of its being made of paper, snipping it with scissors, and laying it out as a long thin rectangle, whose length is $2\pi r$, the circumference of the original circular strip, and whose width is Δr. Thus, $\Delta A = 2\pi r \cdot \Delta r$, approximately.

Digression The *exact* value of ΔA is the difference between the area within a circle of radius $r + \Delta r$ and the area within a circle of radius r.

Problem 3

Assuming the formula for the area of a circle, verify that $\Delta A = 2\pi r \cdot \Delta r + \pi(\Delta r)^2$, exactly.

We argue that if Δr is small, $(\Delta r)^2$ is *much* smaller and can safely be ignored in an approximation. Note the analogy with Example 1, **3.2**.

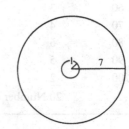

Returning to our circular floor, we say that the load, ΔL, on a circular strip is given by

$$\Delta L = (w \text{ lb/ft}^2) \cdot (\Delta A \text{ ft}^2) = w \cdot \Delta A \text{ lb}$$
$$= (2 + \tfrac{1}{2}r) \cdot 2\pi r \cdot \Delta r \text{ lb}.$$

The total load, L, is the *limit of the sum* of the "elementary" loads, or

$$L = \int_0^6 (2 + \tfrac{1}{2}r)2\pi r \, dr.$$

Problem 4

Evaluate the integral and thus verify that $L = 144\pi$ lb.

In this section we have seen how horizontal rectangular strips and circular strips can be used in setting up integrals. There are still other useful types of "elements" that could be taken up if time were available.

PROBLEMS

5. Assuming knowledge of the formula for the circumference of a circle, use integration with circular strips to find the area of a circle of radius R.
6. A washer consists of the region between two concentric circles of radii 1 mm and 7 mm, as in Figure 9-23. The density (ρ g/mm^2) of the material forming the washer varies thus with the distance (r mm) from the center: $\rho = 0.08\sqrt{15 + r^2}$. Find the total weight of the washer.
★ 7. The bottom of a tin can is in the form of a circle of radius 4 cm, with a hole of radius 1 cm at the center. The bottom is covered uniformly with fine dust particles, 10^5 of them per square centimeter. The dust is drawn off through the hole by "suction," each particle traveling in a straight line toward the center. Find the aggregate (i.e., total) distance traveled by all particles in reaching the edge of the hole.
★ 8. The ellipse with equation $x^2/a^2 + y^2/b^2 = 1$ has semiaxes a and b. According to a statement made at a number of points in this book, the area within this ellipse equals πab square units. Obtain this result in the following way:
   (a) Using vertical rectangular strips, set up the integral expressing the area, A_C, within the first quadrant of the circle $x^2 + y^2 = a^2$ (Figure 9-24).

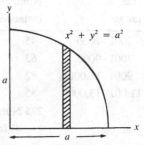

Fig. 9-24

Fig. 9-23

319

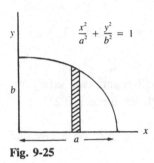

$$\frac{x^2}{a^2} + \frac{y^2}{b^2} = 1$$

Fig. 9-25

(b) Using vertical rectangular strips, set up the integral expressing the area, A_E, within the first quadrant of the ellipse $x^2/a^2 + y^2/b^2 = 1$ (Figure 9-25).

(c) From (a) and (b), express the relationship between A_E and A_C.

(d) Knowing the formula for the area within a circle, verify the formula for the area within an ellipse.

9.8 The idea of an average

Simple arithmetical averages are entirely familiar: If scores on three 1-hr tests are 70, 80, and 90, then the *average score* is $(70 + 80 + 90)/3 = 80$. Exactly the same idea is involved in averaging the homework grades in Table 9-1, although there is more arithmetic: The average grade is

$$\frac{60 + 60 + 60 + 70 + 70 + \cdots + 100}{20} = 79.5.$$

The arithmetic could be condensed by writing

$$\text{average grade} = \frac{60 \cdot 3 + 70 \cdot 4 + 80 \cdot 6 + 90 \cdot 5 + 100 \cdot 2}{3 + 4 + 6 + 5 + 2}.$$

In this form, we can think of the coefficients 3, 4, 6, 5, and 2 as being *weights* indicating the relative "importance" of the individual grades of 60, 70, 80, 90, and 100.

A generalization is seen in defining an average income for the data in Table 9-2. It is customary to choose the mid-income of each range to represent that range, leading, in this case, to the equation

$$\text{average income} = \frac{6000 \cdot 45 + 8000 \cdot 63 + 10000 \cdot 82 + 12000 \cdot 85}{45 + 63 + 82 + 85}.$$

In doing this, we are assuming that incomes are uniformly distributed within each range.

Table 9-1.

Grade	Number of instances
60	3
70	4
80	6
90	5
100	2
	20 Number of grades

Table 9-2

Income range	Number of instances
5001–7000	45
7001–9000	63
9001–11,000	82
11,001–13,000	85
	275 Number of families

9.9 Average velocity

Another familiar notion about averages is embodied in the query "What was your average velocity in driving from Boston to Washington?" The velocity (v mph) of the car is a function of the time (t hr),

$$v = f(t),$$

and it is likely that v is continuously changing. There are, then, infinitely many values of v to average, and the ideas of the preceding section do not immediately apply.

Problem 1

Before reading further, try to formulate a definition of *average velocity*.

There are two approaches we can use, leading (fortunately) to the same result:

1. If we cover M miles in T hr, we define our average velocity ($\bar{v}$ mph) by $\bar{v} = M/T$. In other words, if a journey at a variable velocity takes T hr, the same journey, at the constant velocity $\bar{v}$, will also take T hr. Because $v = ds/dt$, where s is distance, we know that s is an antiderivative of v; and if M miles are covered in T hr, it must be that

$$M = \int_0^T v\, dt.$$

Hence,

$$\bar{v} = \frac{M}{T} = \frac{\int_0^T v\, dt}{T} = \frac{\int_0^T f(t)\, dt}{T}.$$

2. Another approach is to think of the graph of $v = f(t)$ over $0 \le t \le T$, as in Figure 9-26. If we divide the interval $[0, T]$ into n subintervals each of length Δt, we can choose some value of v within each subinterval as representing the v's of that subinterval – in analogy with the income example of the preceding section, we might choose, in each subinterval, the value of v halfway between the minimum v and the maximum v of that subinterval, but *any* choice will do. Let us designate by $v_1 = f(t_1)$, $v_2 = f(t_2), \ldots, v_n = f(t_n)$ the values of v in the first through nth subintervals. The arithmetical average (also called the *arithmetic mean*) of these n values of v,

$$\frac{v_1 + v_2 + \cdots + v_n}{n} = \frac{f(t_1) + f(t_2) + \cdots + f(t_n)}{n},$$

is an approximation to what we think of as the average of the infinitely many v's. If n is larger (Δt is smaller), we expect the approximation to be better, and we say that

$$\bar{v} = \lim_{n \to \infty} \frac{v_1 + v_2 + \cdots + v_n}{n} = \lim_{n \to \infty} \frac{f(t_1) + f(t_2) + \cdots + f(t_n)}{n}.$$

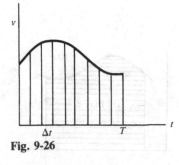

Fig. 9-26

But, as so often happens, we can't immediately tell what the limit is: The numerator of the fraction gets larger with increasing n (more terms), and the denominator obviously does, too. The trick in this case is to multiply numerator and denominator by Δt:

$$\frac{f(t_1)+f(t_2)+\cdots+f(t_n)}{n} = \frac{[f(t_1)+f(t_2)+\cdots+f(t_n)]\Delta t}{n\Delta t}.$$

Problem 2

(a) What is a symbol for the limit of the numerator of the right side, as $\Delta t \to 0$?

(b) What is the value of the denominator of the right side? Hence, its limit?

This second approach, then, gives the same result as obtained with the first approach:

> If $v = f(t)$, then $\bar{v}$, the average of v over the interval $0 \leq t \leq T$, is given by
>
> $$\bar{v} = \frac{\int_0^T f(t)\, dt}{T}.$$

9.10 The average of a function defined on an interval

The same arguments used to define average velocity, when v is a function of t, can be used to define the average value of y, when y is any function of x:

> If $y = f(x)$, then $\bar{y}$, the average of y over the interval $a \leq x \leq b$, is defined by
>
> $$\bar{y} = \frac{\int_a^b f(x)\, dx}{b - a}.$$

A standard name for $\bar{y}$ is "mean value," or, in graphical language, "mean ordinate" (synonym, average height of the curve). The defining relation for $\bar{y}$ can also be written as

$$\int_a^b f(x)\, dx = \bar{y} \cdot (b - a). \tag{1}$$

Because $\int_a^b f(x)\, dx$ can be interpreted as the area under the curve $y = f(x)$ from $x = a$ to $x = b$, we see from equation (1) that $\bar{y}$ is the height of that rectangle with base $(b - a)$ such that the area of the rectangle equals the area under the curve: In Figure 9-27, the area of rectangle $ABCD$ equals the area under the curve $y = f(x)$ from $x = a$ to $x = b$.

This interpretation can be used to obtain a useful approximation to the average value of y, if $y = f(x)$, over the interval $a \leq x \leq b$: We plot the graph of $y = f(x)$ and then estimate the position of a horizontal line, CD,

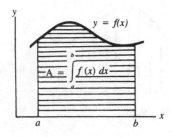

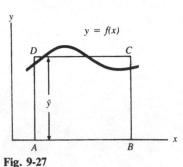

Fig. 9-27

such that the area(s) under the curve above CD equal(s) the area(s) under CD above the curve. Then the height of CD equals $\bar{y}$.

By way of emphasis, we note how the remarks on mean ordinate apply to average velocity: Because $\bar{v} = (\int_0^T v\, dt)/T$, we see that $\bar{v}$ is the mean ordinate of a "speed–time" graph. But the area under a speed–time graph equals distance traveled. Hence, $\bar{v} = M/T$, as before.

Example 1

Suppose that the density (ρ g/cm) of a bar of length 8 cm (Figure 9-28) varies thus with x, the distance from one end of the bar: $\rho = 3 + 2x$. To find the average density, $\bar{\rho}$, we make use of the general formula given at the beginning of this section, and we write

$$\bar{\rho} = \frac{\int_0^8 (3 + 2x)\, dx}{8}.$$

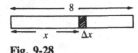

Fig. 9-28

Problem 1

Evaluate the integral to obtain $\bar{\rho} = \frac{88}{8} = 11$.

The integral represents the mass of the bar, so that if we note the units of the component parts, we have 88 g/8 cm = 11 g/cm. In words, the average density equals the total mass divided by the length; or, the average density is such that a bar of the same length, with a constant density equal to this average, has a mass equal to that of the given bar.

Note that, in this case, at $x = 0$, $\rho = 3$; at $x = 8$, $\rho = 19$. Because ρ increases at a uniform rate, we expect that $\bar{\rho}$ will be the arithmetic average of the initial and final values of ρ, and, indeed, this is the case:

$$\bar{\rho} = \tfrac{1}{2}(3 + 19) = 11.$$

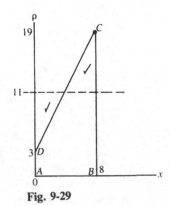

Fig. 9-29

We can visualize this result in Figure 9-29, where $ABCD$ is a trapezoid whose area represents the mass of the bar. The area of a trapezoid equals its altitude times half the sum of its bases. In this case,

$$A = 8 \cdot \tfrac{1}{2}(3 + 19) = 8 \cdot 11 = 88,$$

and the mean ordinate, $\bar{\rho}$, is that height such that the two triangular areas with check marks are equal in area.

Example 2

In Problem 7, **9.5**, the loading (l lb/ft) on a 10-ft beam varied thus with the distance (x ft) from one end: $l = 2 + 3x - 0.12x^2$. To find the average loading, $\bar{l}$, we again make use of the general formula given at the beginning of this section, and we write

$$\bar{l} = \frac{\int_0^{10} (2 + 3x - 0.12x^2)\, dx}{10}.$$

323

Problem 2

Evaluate the integral to obtain $\bar{l} = \frac{130}{10} = 13$ lb/ft.

The integral represents the total load on the bar, so we can say that the average loading equals the total load divided by the length, or the average loading is such that a beam of the same length, with a constant loading equal to this average, has a total load equal to that of the given beam.

Problem 3

Using a large scale, sketch the graph of $l = 2 + 3x - 0.12x^2$ from $x = 0$ to $x = 10$, and estimate $\bar{l}$ by "balancing areas."

PROBLEMS

4. Find the mean ordinate of the curve $y = kx^2$ over the interval
   (a) $[0,1]$ (b) $[1,4]$ (c) $[p,q]$ (Simplify.)
5. Find the mean ordinate of the curve $y = 2x - x^2$ over the interval $[0,3]$. Interpret the result in a sketch of the graph of the curve.
6. Find the mean ordinate of the curve $y = e^x$ over the interval $[0, \ln 5]$.
7. If the loading (y lb/ft) on a beam 10 ft long varies with the distance (x ft) from one end as $y = 2 + 5x - 0.3x^2$, find the average loading.
8. If the velocity (v ft/sec) varies with elapsed time (t sec) as $v = v_0 - 32t$, where v_0 is a constant, find the average velocity in the interval
   (a) $[0,1]$ (b) $[1,2]$ (c) $[2,3]$
   ★ (d) $[t_0, t_1]$. Verify that $\bar{v}$ is the arithmetic average of the velocities at the ends of the time interval.
9. An advertisement for the Porsche 924 Turbo provides the following data:

Gear	1st	2nd	3rd	4th	5th
t (sec)	2.3	7	12	23.2	49
v (mph)	30	52	78	100	120

   (a) Assuming that the car is stationary at $t = 0$, find the average acceleration in the first 2.3 sec. In the first 7 sec. In each case, express the answer in terms of ft/sec^2, and thus as a certain fraction of g, the acceleration due to gravity, which is about 32 ft/sec^2.
   (b) Sketch a large graph of v in terms of t, assuming that the speed of the car is constant for 0.2 sec at each change of gears (i.e., assuming that $v = 30$ from $t = 2.3$ to 2.5, etc.).
   (c) Use your graph to estimate the average speed of the car in each 7-sec interval.
   (d) Use your answers to (c) to estimate the distance traveled in the 49-sec interval.
   (e) Estimate the average speed in the 49-sec interval.
   (f) Find an approximate formula for v in terms of t.

We illustrate by a couple of examples some slightly more complicated ideas about averages.

Example 1

In Problem 7, **9.5**, a beam, AB, 10 ft long, had a loading (l lb/ft) that varied with the distance (x ft) from A as follows: $l = 2 + 3x - 0.12x^2$. Computation in that problem showed that the total load was $L = 130$ lb, that the moment of the load about A was $M_A = 800$ lb-ft, and that the moment of the load about B was $M_B = 500$ lb-ft. In Example 2, **9.10**, we found the average loading, $\bar{l}$ lb/ft. We now ask: With respect to A, what is the *average lever arm*?

Problem 1

Before reading further, try to formulate a sensible definition of the *average lever arm with respect to A*.

The definition is analogous to those we have encountered for average velocity, average density, and the like: *The average lever arm with respect to A is a distance, $\bar{x}$, such that if the entire load, L, were concentrated at that distance from A, the moment of the load would equal M_A.* In other words, $L \cdot \bar{x} = M_A$, or

$$\bar{x} = \frac{M_A}{L} = \frac{\int_0^{10} x(2 + 3x - 0.12x^2)\, dx}{\int_0^{10} (2 + 3x - 0.12x^2)\, dx}.$$

Because $M_A = 800$ and $L = 130$, $\bar{x} = \frac{80}{13} = 6+$ ft. Figure 9-30 shows a point, C, at distance $\bar{x}$ from A.

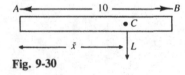

Fig. 9-30

Problem 2

Compute $\bar{y}$, the average lever arm with respect to B.

You should have found that $\bar{y} = \frac{50}{13}$. But the same point, C, is at distance $\frac{50}{13}$ from B! In fact, the average lever arm of the load with respect to any point P in the line AB can be shown to equal PC.

Problem 3

For the practice in integration, help to verify the preceding sentence by computing the average lever arm with respect to P, if P is (a) 1 ft to the left of A, (b) 2 ft to the right of B, (c) 3 ft to the right of A. [In this case you will have to take the difference between the (clockwise) moment of the load on the portion PB of the beam and the (counterclockwise) moment of the load on the portion PA of the beam.]

The point C is called the *center of gravity* of the load.

Problem 4

Let AB be a beam of length k ft with a loading (y lb/ft) at any point given by $y = f(x)$, where x (ft) is the distance of the point from A. If P is the point on AB such that $AP = p$, $p < k$, let M_P be the (clockwise) moment about P of the load on PB minus the (counterclockwise) moment about P of the load on AP.

(a) Show that $M_P = \int_0^k (x - p) f(x) \, dx$. (Hint: Use the result of Problem 13, **5.7**.)

(b) Let $\bar{x}$ be the average lever arm of the load with respect to A. Let C be the point on AB such that $AC = \bar{x}$ (i.e., let C be the center of gravity of the load). Show that $M_C = 0$.

Example 2

In Problem 7, **9.7**, the total distance traveled by all dust particles in reaching the edge of the hole (Figure 9-31) is given by

$$D = \int_1^4 (r - 1) 10^5 \cdot 2\pi r \, dr$$

$$= 2.7\pi \cdot 10^6 \text{ cm}.$$

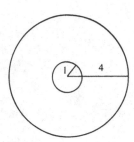

Fig. 9-31

Observe that the number of dust particles on the bottom of the can is *finite*, albeit extremely large, so that finding the aggregate distance traveled by them in reaching the edge of the hole is merely an arithmetical problem – an extraordinarily long arithmetical problem! Assuming a continuous distribution of particles, so that calculus can be used, provides a simpler solution. This is characteristic of our use of calculus in many physical situations: The atomic theory implies that a piece of matter is composed of a finite number of particles, but we find it easier to assume that matter is continuous when we have to compute densities, moments, and so forth.

A natural question is "What is the *average distance* traveled by a particle?"

Problem 5

Before reading further, try to formulate a sensible definition of average distance in this case.

A definition of *average distance* consistent with our previous discussion is the following: The average distance, $\bar{x}$ cm, is that distance such that if every particle traveled that distance, the total distance traveled by all particles would equal the actual total distance traveled. Or, equivalently, the average distance equals the total distance divided by the number of particles. Thus, if N = number of particles, $N \cdot \bar{x} = D$, or $\bar{x} = D/N$.

Problem 6

Compute N and thus determine $\bar{x}$ to equal $\frac{9}{5}$ cm.

Problem 7

Note that $\bar{x}$ as found in Problem 6 is greater than $\frac{3}{2}$, the distance from the "midring" to the edge of the hole (Figure 9-32). Explain why this result is reasonable.

There is a notation that can be used in all sorts of problems of finding averages that helps to unify the subject and to make it easier to remember how to proceed. We present the notation now in the context of examples and problems that we have encountered.

In Section 9.9, the *average velocity*, $\bar{v}$, in the interval $0 \leq t \leq T$ was seen to be

$$\bar{v} = \frac{\int_0^T v \, dt}{T},$$

where $v = f(t)$ expresses the way that v varies with t. We can write $T = \int_0^T dt$, obtaining

$$\bar{v} = \frac{\int_0^T v \, dt}{\int_0^T dt}.$$

To find $\bar{v}$, we must go back and insert the formula for v in terms of t; but the symbolic formula is a convenient one to remember.

In Section 9.10, we expressed the *mean ordinate*, $\bar{y}$, as

$$\frac{\int_a^b f(x) \, dx}{b - a},$$

where $y = f(x)$ is the formula for y in terms of x. Because $b - a = \int_a^b dx$, we can write

$$\bar{y} = \frac{\int_a^b y \, dx}{\int_a^b dx}.$$

In Section 9.10, Example 1, we obtained the *average density*, $\bar{\rho}$, as

$$\bar{\rho} = \frac{\int_0^8 (3 + 2x) \, dx}{8},$$

where $\rho = 3 + 2x$ is the formula for the density in terms of x. In general, we can write

$$\bar{\rho} = \frac{\int_a^b \rho \, dx}{\int_a^b dx}.$$

9.11
Further averages

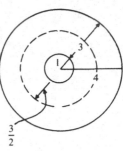

Fig. 9-32

327

Similarly, the average loading, $\bar{l}$, is given by

$$\bar{l} = \frac{\int_a^b l\, dx}{\int_a^b dx}.$$

Now we come to a couple of more interesting cases. In Example 1 in this section, we determined $\bar{x}$, the average lever arm of the load with respect to A, by

$$\bar{x} = \frac{M_A}{L} = \frac{\int_0^{10} x(2 + 3x - 0.12x^2)\, dx}{\int_0^{10} (2 + 3x - 0.12x^2)\, dx}.$$

The expression $2 + 3x - 0.12x^2$ equals l, the loading (in lb/ft). Observe that $l = dL/dx$, so that

$$(2 + 3x - 0.12x^2)\, dx = l\, dx = \frac{dL}{dx}\, dx = dL.$$

Hence, a useful symbolic form for $\bar{x}$ is the following:

$$\bar{x} = \frac{\int_a^b x \cdot dL}{\int_a^b dL}.$$

Finally, in Example 2 in this section, we saw that $\bar{x}$, the average distance traveled by the dust particles, is given by

$$\bar{x} = \frac{D}{N} = \frac{\int_1^4 (r - 1) 10^5 \cdot 2\pi r\, dr}{\int_1^4 10^5 \cdot 2\pi r\, dr}.$$

If N is the total number of particles and A is the area of the bottom of the can, then $N = 10^5 \cdot A$, and $dN = 10^5\, dA = 10^5 \cdot 2\pi r\, dr$. Hence,

$$\bar{x} = \frac{\int_1^4 (r - 1)\, dN}{\int_1^4 dN}.$$

For the sake of symmetry, we might introduce the quantity $\bar{r}$ as the average distance to the *center* of the hole, whereas $\bar{x}$ is the average distance to the *edge* of the hole. Then $\bar{x} = \bar{r} - 1$, and we have

$$\bar{r} - 1 = \frac{\int_1^4 (r - 1)\, dN}{\int_1^4 dN}$$

9.12 Summary

The problems of this chapter have followed a pattern:

(a) The quantity sought is approximated as the sum of "elementary" quantities, which are expressed in terms of one variable and an increment of that variable.

(b) The quantity sought is then expressed exactly as the limit of the sum referred to in (a) (i.e., as an integral).

(c) The integral is evaluated by the Fundamental Theorem (i.e., by antidifferentiation).

This pattern is characteristic of the way that integration is used in scientific applications. Practice is needed, as in most mathematics, to learn how to apply steps (a), (b), and (c) to an unfamiliar problem.

An excellent example of this pattern is provided by the derivation of a formula needed in the article "Physostigmine: Improvement of Long-Term Memory Processes in Normal Humans," by K. L. Davis and associates [*Science*, Vol. 201 (July 21, 1978), pp. 272–4]. The drug was administered by slow intravenous infusion (1 mg in 1 hr, at a constant rate) to human subjects who had learned a list of words shortly before the start of the infusion. About 2 hr later, subjects were tested for the number of words recalled. The results were compared with the number of words recalled after a saline injection used as a control on another day. Subjects did not know which substance they were receiving, of course. The drug appeared to have a significant effect in improving memory.

One of the critical issues was the amount of the drug present in the blood plasma. If the drug had remained unchanged in the plasma after infusion, there would have been no problem: With time (t hr) measured from the beginning of the infusion, the graph of the rate of infusion (r mg/hr) would appear as in Figure 9-33, and the amount (u mg) in the plasma at any time would be given by

$$u = \begin{cases} t, & 0 \le t \le 1, \\ 1, & t > 1. \end{cases}$$

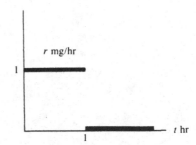

Fig. 9-33

But the situation is more complicated than this, for physostigmine decomposes at a constant percentage rate, with half-life of $\frac{1}{2}$ hr, so the amount of the drug actually present in the blood plasma at any time T is less than the total amount administered up to time T. The problem, then, is to express the *actual* amount (U mg) of the drug present in the blood plasma in terms of the time (T hr) since the start of infusion. We proceed to solve this problem.

Because the drug decomposes at a constant percentage rate, we know that if we administer y_0 mg at some time, the amount (y mg) at time x hr later is given by $y = y_0 \cdot e^{-kx}$, where k is the percentage rate of decomposition. We determine k from the given information about half-life: At $x = \frac{1}{2}$, $y = \frac{1}{2}y_0$. Thus, $\frac{1}{2}y_0 = y_0 e^{-(1/2)k}$.

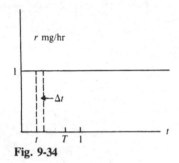

Fig. 9-34

Problem 1

Solve this equation for k to obtain $k = 2 \ln 2$.

Hence,

$$y = y_0 \cdot e^{(-2\ln 2)x}. \tag{2}$$

To determine the amount of the drug present at time T, we must add up the amounts resulting from the infusion administered for all times $0 \leq t \leq T$. In an interval of length Δt beginning at time t, the amount administered is $1 \cdot \Delta t$ (Figure 9-34). This is the y_0 of equation (2). The time interval from t to T is, of course, $T - t$; this is the x of equation (2). Hence, the amount of the drug present in the plasma at time T resulting from the infusion in the interval $[t, t + \Delta t]$ is given approximately by $y = \Delta t \cdot e^{(-2\ln 2)(T-t)}$, or $y = \Delta t \cdot e^{(2\ln 2)(t-T)}$.

The *total* amount of the drug present in the plasma at time T is obtained by adding up the amounts resulting from each of the short intervals prior to T, or, more precisely, by taking the *limit* of such sums, as $\Delta t \to 0$ – in other words, an *integral*:

$$U = \int_0^T e^{(2\ln 2)(t-T)} \, dt.$$

We are integrating with respect to t. The letter T represents any time ≤ 1.

Problem 2

Making the substitution $v = (2\ln 2)(t - T)$, perform the antidifferentiation to obtain

$$U = \frac{1}{2\ln 2} e^{(2\ln 2)(t-T)} \Big|_0^T.$$

Hence, $U = [1/(2\ln 2)][1 - e^{(-2\ln 2)T}]$. Because $e^{\ln 2} = 2$, we can write $e^{(-2\ln 2)T} = (e^{\ln 2})^{-2T} = 2^{-2T} = 1/4^T$.

Thus, $U = [1/(2\ln 2)](1 - 1/4^T)$.

This formula is valid for all T in the interval $0 \leq T \leq 1$.

Problem 3

Check that $U = 0$ at $T = 0$. Make the calculation to show that at $T = 1$, $U = 3/(8\ln 2) \approx 0.54$.

Because no drug is administered after $T = 1$, the amount present in the plasma for any $T > 1$ is simply the amount left from the decomposition of the $3/(8\ln 2)$ mg present at $T = 1$. In other words, in equation (2), $y_0 = 3/(8\ln 2)$, and $x = T - 1$. Thus, for $T > 1$, $U = [3/(8\ln 2)]e^{(-2\ln 2)(T-1)}$.

Problem 4

Simplify this last equation to $U = [3/(2\ln 2)](1/4^T)$.

We can finally put our results together as follows:

$$U = \begin{cases} \dfrac{1}{2\ln 2}\left(1 - \dfrac{1}{4^T}\right), & 0 \le T \le 1, \\[3ex] \dfrac{3}{2\ln 2}\,\dfrac{1}{4^T}, & T > 1. \end{cases}$$

There appears to be a mistake in the *Science* article, for the authors present a different formula for U – perhaps just a typographical error. It may be useful to realize that statements are not valid just because they appear in print. *Caveat lector!*

Problem 5

Make the calculations to show that 30 min after the start of the infusion, the amount of the drug present in the plasma is about 0.36 mg, and that the amount 80 min after the start of the experiment is about 0.34 mg.

PROBLEMS

6. Find the weight of a flat plate bounded by $y = e^{x^2}$, $x = 1$, $x = 2$, and the x axis, if linear dimensions are measured in inches and the surface density of the plate is given by $\rho = 3x$ oz/in.^{2}

7. Same as Problem 6, if the plate is bounded by $y = x^{1/2}$, $x = 0$, $x = 4$, and the x axis, with $\rho = (1 + x)$ oz/in.^{2}

8. Same as Problem 6, if the plate is bounded by $y = \ln x$, $x = 1$, $x = e$, and the x axis, with $\rho = 1/x$ oz/in.^{2}

9. Find the volume obtained by revolving about the x axis the area under $y = xe^{x^3}$ from $x = -1$ to $x = 1$.

10. Same as Problem 9, for $y = xe^{-x^3}$.

★ 11. If the plate of Problem 6 has a constant surface density of k oz/in.^{2} and is in a horizontal position, find the moment of its weight about the y axis.

★ 12. In his *History of Geometrical Methods*, J. L. Coolidge states that Isaac Barrow, some years before the development of the calculus by Isaac Newton, obtained the formula for the area of a circle in the following way, reminiscent of the method by which the area under a parabola was obtained in Section 5.5: Divide the radius, R, into n equal parts, each of length h. Draw circles, centered at the center of the given circle, through each of the division points. Assuming that the circumference of a circle of radius x is known to be $2\pi x$, the areas of the various rings are approximately $h \cdot 2\pi h$, $h \cdot 2\pi(2h)$, $h2\pi(3h),\ldots,h2\pi(nh)$. The area of the given circle is the sum of the areas of these rings. Knowing that $1 + 2 + 3 + \cdots + n = n(n+1)/2$, finish the argument.

In connection with the definition of the integral,

$$\int_a^b f(x)\,dx = \lim_{\Delta x \to 0}\left[f(x_1) + f(x_2) + \cdots + f(x_n)\right]\Delta x,$$

we have noted several times that the quantity in brackets increases without bound, while $\lim_{\Delta x \to 0}\Delta x = 0$, of course. We may say, for short, that we have an expression $\infty \cdot 0$ to "evaluate."

The mathematician C. L. Dodgson (Lewis Carroll) wrote as follows in making his annual report as bursar of Christ Church, Oxford:

> The consumption of Madeira (B) has been, during the past year, zero. After careful calculation I estimate that, if this rate of consumption be steadily maintained, our present stock will last us an indefinite number of years. And although there may be something monotonous and dreary in the prospect of such vast cycles spent in drinking second-class Madeira, we may yet cheer ourselves with the thought of how economically it may be done.

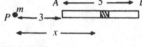

Fig. 9-35

13. As in Figure 9-35, a particle of mass m is at point P, in line with a bar of length 5 cm, PA equaling 3 cm. The density (ρ g/cm) of the bar increases uniformly from 0 at A to 10 at the other end, B.
    (a) Find a formula for the density at any point in the bar in terms of x, the distance of the point from P.
    (b) Find the force of attraction between the bar and the particle at P.

14. Same as Problem 13, if $PA = 1$, $AB = 5$, and ρ increases uniformly from 2 at A to 17 at B.

15. A plate has the shape of the region bounded by the portion of the ellipse $x^2/25 + y^2/16 = 1$ in the first quadrant and the x and y axes. The loading (l oz/in.^{2}) on the plate at any point (x, y) is given by $l = 6x$. Find the total load on the plate.

16. Same as Problem 15, for the ellipse $x^2/9 + y^2/16 = 1$, if $l = 5y$.

17. The density (ρ lb/ft^2) of a circular plate varies thus with the distance (r ft) from its center: $\rho = 2 + 3r$. If the radius of the plate is 5 ft, find the weight of the plate.

18. A floor has the shape of a circle of radius 5 m, with a hole of radius 2 m at the center. The loading (l kg/m^2) varies thus with the distance (r m) from the center of the hole: $l = 5 - r$. Find the total load (L kg) on the floor.

19. The velocity (v ft/sec) of a point moving on a straight line varies thus with time (t sec): $v = 3 - 2t + 0.6t^2$. Find the average velocity between $t = 5$ and $t = 10$.

20. Same as Problem 19, with $v = 2 + 5t - 0.3t^2$, between $t = 2$ and $t = 10$.

★ 21. The density (ρ oz/in.) of a bar AB of length 3 in. varies thus with distance (x in.) from the end A: $\rho = 2 + 4x^2$. Find the distance of the center of gravity from the end B.

★ 22. The density (ρ oz/in.) of a bar AB of length 6 in. varies thus with the distance (x in.) from the end A: $\rho = 5 + \frac{1}{2}x^2$. Find the distance of the center of gravity from the end A.

23. Evaluate

(a) $\int_1^2 \dfrac{e^{3x}}{\sqrt{e^{3x}-1}}\,dx$ (b) $\int_0^1 \dfrac{dx}{3x+e}$

(c) $\int_0^{\ln 2} e^{3x}\,dx$ (d) $\int_{-1}^0 \dfrac{x^2\,dx}{\sqrt{x^3+5}}$

24. Find the area bounded by $y = (4x-2)^3$, $x = 0$, and $x = 1$.

SAMPLE TEST

1. (a) Evaluate

$$\int_0^1 \frac{e^{(1/2)x}+2}{e^{(1/2)x}+x}\,dx,$$

and show that the result can be written in the form $\ln(e + 2e^{1/2}+1)$.

(b) Evaluate

$$\int_0^1 \frac{e^{(1/2)x}+2}{\left(e^{(1/2)x}+x\right)^2}\,dx,$$

and show that the result can be written in the form $2\sqrt{e}\,/(\sqrt{e}+1)$.

2. As in Figure 9-36, a particle of mass m is in line with a bar of length 4 cm, the particle being 2 cm from the end of the bar. The density (ρ g/cm) of the bar increases uniformly from 1 at A to 9 at the other end, B.

(a) Find a formula for the density at any point of the bar in terms of x, the distance of the point from the particle of mass m.

(b) If the force of attraction between two particles of masses m_1 and m_2, at distance r apart, is given by $F = Gm_1m_2/r^2$, where G is a constant, find the force of attraction between the bar and the particle of mass m.

3. A metal plate is in the shape of the region bounded by the portion of the ellipse $x^2/25 + y^2/16 = 1$ in the first quadrant and the x and y axes (Figure 9-37). The density (d oz/in.^{2}) of the plate at any point varies thus with the y coordinate of the point: $d = 6y$. Find the total weight (W oz) of the plate.

4. A floor is in the shape of a circle of radius 5 ft, with a hole of radius 1 ft at the center. The loading (l lb/ft^2) varies thus with the distance (r ft) from the center of the hole: $l = 1 + 3r$. Find the total load (L lb) on the floor.

5. The velocity (v ft/sec) of a point moving on a straight line varies thus with the time (t sec): $v = 1 + 4t - 0.6t^2$. Find the average velocity between $t = 1$ and $t = 5$.

Fig. 9-36

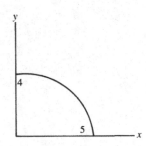

Fig. 9-37

★ 6. The density (ρ oz/in.) of a bar AB of length 5 in. varies thus with the distance (x in.) from the end A: $\rho = 2 + 12x^2$. Find the distance of the center of gravity from the end A.

9.13 Quadrature

As we know from Chapter 5, the integral of a continuous function f over an interval $[a, b]$, $a < b$, can be thought of as the area of the region bounded by the graph of f, the x axis, and the lines $x = a$ and $x = b$, provided the graph lies on or above the axis. More generally, the integral $\int_a^b f(x)\,dx$ is the area of that part of the region lying above the axis minus the area of that part lying below it, as indicated in Figure 9-38. And as we reasoned in Chapter 5, these areas can be approximated by adding together the areas of narrow rectangles, the vertical length of each being the numerical value of the function at some point in the base of the rectangle. In the figure, these lengths are $\pm f(x)$, where x is taken to be the left end point of the subintervals.

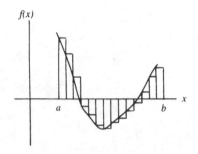

Fig. 9-38

We also discovered in Chapter 5 that

(a) these approximations, if they are to be accurate, involve laborious computations, and

(b) it is preferable to find the integral exactly (if possible) through the Fundamental Theorem: If F is an antiderivative of f, then $\int_a^b f(x)\,dx = F(b) - F(a)$.

Unfortunately, it may be difficult or impossible to find an antiderivative. Even with the elementary functions we have met, antidifferentiation may require an ingenious trick, like a by-no-means-obvious substitution. During the century and a half following the invention of the calculus, mathematicians devoted much effort to discovering such tricks. The results are impressive, but often complicated, and the end computations are frequently troublesome. Worse still, although the theory ensures the *existence* of the antiderivatives of continuous functions, the antiderivatives of many such functions cannot be expressed explicitly in terms of elementary functions: $\int e^{-x^2/2}\,dx$ is one such example. Then a numerical approach becomes mandatory, and even when antidifferentiation is possible, numerical solutions may be more useful.

Numerical integration is called *quadrature*, reflecting the ancient technique of building complex areas out of small squares. Its essence is in the rectangular method used in Chapter 5. This section and the next will be given to two refinements of that method, refinements that yield more precise approximations for similar computation.

To be specific about finding an approximation to $\int_a^b f(x)\,dx$, with $a < b$, let n be a positive integer, set $x_0 = a$ and $x_n = b$, and choose $x_1, x_2, \ldots, x_{n-1}$

as $n-1$ equally spaced numbers between a and b, as in Figure 9-39. Thus,

$$x_1 - x_0 = x_2 - x_1 = \cdots = x_{n-1} - x_{n-2} = x_n - x_{n-1}.$$

Designate by h this common length of a subinterval on the x axis. In this partition, we have n subintervals. In these terms, the first rectangle in the approximation contributes $h \cdot f(x_0)$; the second, $h \cdot f(x_1)$; the third, $h \cdot f(x_2)$; the nth, $h \cdot f(x_{n-1})$. Note that some of these contributions are negative numbers for the function shown in Figure 9-40. Adding up these n contributions gives

$$L_n = h\big[f(x_0) + f(x_1) + \cdots + f(x_{n-1})\big],$$

which is the **left-end-point rule** for approximating $\int_a^b f(x)\,dx$.

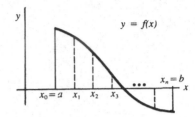

Fig. 9-39

As is stated later in Problem 7, this is the same result as would be obtained by applying Euler's method (Chapter 8) to the differential equation $dy/dx = f(x)$.

Example 1
If we use the left-end-point rule to approximate $\int_0^1 x^3\,dx$, with $n=2$, we have $x_0 = 0$, $x_1 = \frac{1}{2}$, and $h = \frac{1}{2}$, whence

$$L_2 = \tfrac{1}{2}\Big[0^3 + \big(\tfrac{1}{2}\big)^3\Big] = \tfrac{1}{16}.$$

By antidifferentiation, $\int_0^1 x^3\,dx = \frac{1}{4}x^4\big|_0^1 = \frac{1}{4}$, so the error in this case is $\int_0^1 x^3\,dx - L_2 = \frac{1}{4} - \frac{1}{16} = \frac{3}{16}$. Pretty bad!

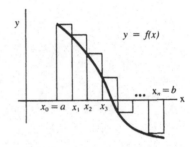

Fig. 9-40

Problem 1
For the same integral, show that the error using L_4 is $\frac{7}{64}$. Still not good.

With these small values of n, it is not surprising that the errors are large. Increased accuracy comes with larger n: How much improvement will there be for the additional work of using a large n? As stated later in Problem 6, the result is as follows:

Bound on error with L_n
If the numerical value of $f'(x)$ does not exceed some constant, B', for all x in $[a, b]$, then the difference between L_n and $\int_a^b f(x)\,dx$ is at most $[B'(b-a)/2]h$.

Example 2
In approximating $\int_0^1 x^3\,dx$, we have $f(x) = x^3$, so $f'(x) = 3x^2$. Over the interval $[0,1]$, the maximum value of $3x^2$ is 3, so we can choose $B' = 3$. Thus, the error in approximating the integral by L_2 does not exceed $[3(1-0)/2]\cdot\frac{1}{2} = \frac{3}{4}$, and the error with L_4 does not exceed $[3(1-0)/2]\cdot\frac{1}{4} = \frac{3}{8}$. These results are consistent with those of the foregoing Example 1 and Problem 1.

By doubling n (and hence halving h), the error bound is halved. As numerical processes go, that is not very good. The process would be much more efficient if the error bound involved h^2, or better yet a higher power of h, for then doubling n would reduce the error bound at least fourfold. In the next section we shall develop another quadrature method and a modification of that method. Both guarantee efficiencies of higher order, and both take only slightly more complicated computation than the left-end-point rule.

PROBLEMS

2. Write a formula for a right-end-point rule. What would you expect a bound on the error to be?

3. Approximate $\int_0^1 x^4 \, dx$ by L_2. By L_4. What are the actual errors in these approximations? What are the bounds on error provided by our formula?

4. Same as Problem 3, for $\int_0^1 [1/(1+x)] \, dx$.

5. (a) Approximate $\int_0^2 e^{-x^2/2} \, dx$ by L_2. By L_4. By L_8. (This integral cannot be obtained by antidifferentiation. Its value to four decimal places, obtained by more refined numerical methods, is 1.1963.)

   ★ (b) If $f(x) = e^{-x^2/2}$, what is the maximum value of $|f'(x)|$ in $[0,2]$? Hence, what are the bounds on the errors in (a) provided by our formula?

★ 6. The purpose of this problem is to establish our formula for the bound on the error for the left-end-point rule. We assume that f is continuous on $[a, b]$ and that $|f'(x)| \leq B'$ for all x in (a, b).

   (a) Show that $\int_{x_0}^{x_1} f(x) \, dx - f(x_0) \cdot h = \int_{x_0}^{x_1} [f(x) - f(x_0)] \, dx$.

   (b) Use the Mean-Value Theorem to show that $|f(x) - f(x_0)| \leq B' \cdot |x - x_0| = B'(x - x_0)$ for all x in $[x_0, x_1]$.

   (c) For any continuous function f, it is true that $|\int_a^b f(x) \, dx| \leq \int_a^b |f(x)| \, dx$; a sketch of a graph lying partly above and partly below the x axis makes the relation seem reasonable. Use the relation to show that $|\int_{x_0}^{x_1} f(x) \, dx - f(x_0) \cdot h| \leq B' \int_{x_0}^{x_1} (x - x_0) \, dx = B' h^2/2$.

   (d) By repeated use of the inequality $|u + v| \leq |u| + |v|$, show that the bound on the error, which is $|\int_a^b f(x) \, dx - L_n|$, is less than or equal to $[B'(b - a)/2]h$.

7. Show that Euler's method (Section 8.4) applied to the differential equation $dy/dx = f(x)$, with $y_0 = 0$ at $x = x_0$, produces $y_n = L_n$.

9.14 More on quadrature: the trapezoidal rule and its adjustment

The left-end-point rule uses a horizontal line segment, like PH in Figure 9-41(a), as a substitute for $f(x)$ over any subinterval of the partition. For a

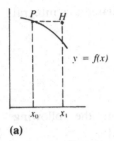

(a)

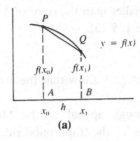

(a)

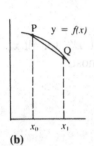

(b)

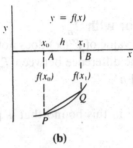

(b)

Fig. 9-41   **Fig. 9-42**

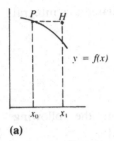

(a)

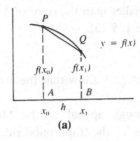

(a)

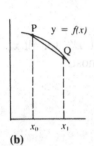

(b)

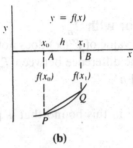

(b)

Fig. 9-41   **Fig. 9-42**

better approximation, we take the line segment determined by the points on the graph at the end points of the subinterval – PQ in Figure 9-41(b). If P and Q are both above the x axis, as in Figure 9-42(a), then the integral over this subinterval will be approximated by the area of the trapezoid $ABQP$, which equals $\frac{1}{2}[f(x_0)+f(x_1)]\cdot h$. If P and Q are both below the x axis, as in Figure 9-42(b), then the integral over this subinterval will be approximated by the *negative* of the area of the trapezoid $ABQP$, which again equals $\frac{1}{2}[f(x_0)+f(x_1)]\cdot h$. [Note that both $f(x_0)$ and $f(x_1)$ are negative numbers in this case.] If $f(x_0)$ is positive and $f(x_1)$ is negative, then the integral over this subinterval will be approximated by the area of triangle PAC *minus* the area of triangle QBC (Figure 9-43). Once again, this equals $\frac{1}{2}[f(x_0)+f(x_1)]\cdot h$. (See Problem 21 at the end of this section.) Similarly, as shown in Figure 9-44, we approximate the integral over the second subinterval by $\frac{1}{2}[f(x_1)+f(x_2)]\cdot h$, and so forth. We add all n parts to obtain

$$T_n = \tfrac{1}{2}h\big[\,f(x_0)+f(x_1)+f(x_1)+f(x_2)+\cdots+f(x_{n-1})+f(x_n)\,\big]$$

or

$$T_n = \frac{h}{2}\big[\,f(x_0)+2f(x_1)+2f(x_2)+\cdots+2f(x_{n-1})+f(x_n)\,\big],$$

which is the **trapezoidal rule** for approximating $\int_a^b f(x)\,dx$.

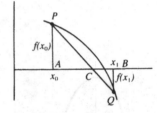

Fig. 9-43

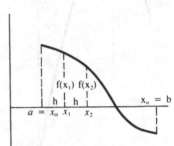

Fig. 9-44

Example 1
For $\int_0^1 x^3\,dx$, with $n=2$, we have $T_2 = \frac{1}{2}\cdot\frac{1}{2}[0^3+2(\frac{1}{2})^3+1^3]=\frac{5}{16}$. Because $\int_0^1 x^3\,dx=\frac{1}{4}$, the error in this approximation is $\frac{1}{16}$, still fairly large, but

substantially smaller than the error of $\frac{1}{4}$ when we approximated this integral by L_2 (Example 1, **9.13**).

Problem 1
For the same integral, show that the error using T_4 is $\frac{1}{64}$.

Basically through appeal to the Mean-Value Theorem, the following bound on errors for the trapezoidal rule can be established:

Bound on error with T_n
If the numerical value of $f''(x)$ does not exceed some constant, B'', for all x in $[a, b]$, then the difference between T_n and $\int_a^b f(x)\,dx$ is at most $B'' \cdot [(b-a)/12] \cdot h^2$.

It is the factor h^2 in this bound that is particularly helpful.

Example 2
For $f(x) = x^3$, $f''(x) = 6x$. Hence, for any x in $[0,1]$, we know that $|f''(x)| \le 6$. Thus, a bound on the error in T_2 is $[6(1-0)/12](\frac{1}{2})^2 = \frac{1}{8}$, and a bound on the error in T_4 is $[6(1-0)/12](\frac{1}{4})^2 = \frac{1}{32}$. These results are consistent with those obtained in Example 1 and Problem 1.

Example 3
As a more complicated example (but still solvable exactly by antidifferentiation), consider bounds on the error for the trapezoidal rule applied to $\int_{-1}^2 (x^2 - x^4)\,dx$.

Here $f(x) = x^2 - x^4$; $f'(x) = 2x - 4x^3$; $f''(x) = 2 - 12x^2$. We see from Figure 9-45 that the largest numerical value of $f''(x)$ over $-1 \le x \le 2$ is 46, so $B'' = 46$. Hence, a bound on the error for the trapezoidal rule used to approximate $\int_1^2 (x^2 - x^4)\,dx$ is

$$46 \frac{2-(-1)}{12} h^2 = \frac{23}{2} h^2.$$

If we seek to ensure accuracy to, say, three decimal places, we need to have

$$\frac{23}{2} h^2 < 5 \cdot 10^{-4}.$$

Because $h = (b-a)/n = 3/n$, we need

$$\frac{23}{2}\left(\frac{3}{n}\right)^2 < 5 \cdot 10^{-4},$$

or

$$n^2 > \frac{(23)(9)}{2 \cdot 5 \cdot 10^{-4}} = (20.7)10^4, \quad \text{or} \quad n > \sqrt{20.7} \cdot 10^2 \approx 454.97.$$

Thus, a partition into 455 subintervals will assure the desired accuracy. That

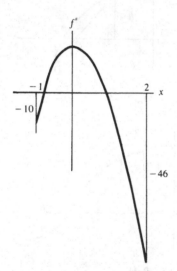

Fig. 9-45

would take a lot of work, but it beats approximating by the left-end-point rule, as the following problem shows.

Problem 2

Show that the largest numerical value of $f'(x) = 2x - 4x^3$ over $-1 \leq x \leq 2$ is 28 and that to obtain three-decimal-place accuracy with the left-end-point rule takes a partition into 252,000 subintervals (!).

Even with patience to apply the trapezoidal rule with $n = 455$, there are likely to be blunders. A computer or programmable calculator can get the approximate value easily, but writing the program and running it take some time. (One calculator took 11 min to run this problem. Incidentally, the result was -3.600108440, comfortably within the desired tolerance of 0.0005 of the exact value, which is -3.6.)

Another commonly used quadrature method, called Simpson's rule, uses parabolic arcs instead of line segments to approximate the integrand. It generally gives a better result than T_n. For instance, in Example 3, a partition into 16 subintervals with Simpson's rule would assure the desired accuracy – lots better than the 455 subintervals of the trapezoidal rule. Simpson's rule is a popular method of quadrature, and it is a wired-in feature of some calculators.

Surprisingly, a simple adjustment of the trapezoidal rule generally is even better than Simpson's rule: We simply add $\{[f'(a) - f'(b)]/12\}h^2$ to T_n:

$$AT_n = \frac{h}{2}\left[f(x_0) + 2f(x_1) + 2f(x_2) + \cdots + 2f(x_{n-1}) + f(x_n)\right]$$
$$+ \frac{f'(a) - f'(b)}{12}h^2$$

is the **adjusted trapezoidal rule** for approximating $\int_a^b f(x)\,dx$.

Error bound with AT_n

If the numerical value of $f^{(iv)}(x)$ does not exceed some constant, $B^{(iv)}$, for all x in $[a, b]$, then the difference between AT_n and $\int_a^b f(x)\,dx$ is at most $B^{(iv)} \cdot [(b - a)/720]h^4$.

Example 4

For $\int_0^1 x^3\,dx$, we saw in Example 1 that $T_2 = \frac{5}{16}$. Hence,

$$AT_2 = \frac{5}{16} + \frac{3 \cdot 0^2 - 3 \cdot 1^2}{12}\left(\frac{1}{2}\right)^2 = \frac{5}{16} - \frac{1}{16} = \frac{1}{4},$$

which is the *exact* value of the integral. Because, if $f(x) = x^3$, $f^{(iv)}(x) = 0$ for all x, the error bound with AT_n predicts that we should obtain the exact value in this case. Indeed, the adjusted trapezoidal rule gives exact results for all polynomials of degree 3 or less. (See Problem 22 at the end of this section.)

Example 5

Let us apply AT_n to the problem of Example 3, that is, to $\int_{-1}^{2}(x^2 - x^4)\,dx$. In this case, $f^{(iv)}(x) = -24$, so $B^{(iv)}$ can be taken as 24. Thus, a bound on the error with AT_n is $[(24)(3)/720]h^4 = \frac{1}{10}h^4 = 3^4/(10n^4)$. Hence, to ensure three-decimal-place accuracy, we need $3^4/(10n^4) < 5 \cdot 10^{-4}$, or $n^4 > (3^4/50)10^4$, or $n > 30\sqrt[4]{1/50}$. Two strokes of a square-root key tell us that n must be greater than 11.3. Thus, AT_{12} will provide three-decimal-place accuracy in this case, contrasting with the T_{455} we found in Example 3. Computation shows AT_{12} to be -3.5996, within the desired tolerance of 0.0005 of the exact value, -3.6.

Because of its simplicity and its efficiency, the adjusted trapezoidal rule is an excellent quadrature method. Simpson's rule can be adjusted, too, but it requires calculations involving f'''. When increased accuracy is required, it is usually easier to rerun the computation with a larger n.

A note of caution on the error bounds: They take account of imprecision in the approximation of the integral, not of round-off errors in the calculation. Consequently, the bounds need to be accorded some tolerance when the partition is fine (n is large) and each subinterval's contribution to the approximation is correspondingly small.

PROBLEMS

3. Approximate $\int_{-2}^{3} x^4\,dx$ by
   (a) T_2 (b) AT_2 (c) T_4 (d) AT_4
4. (a)–(d) Calculate the error bound for each of the approximations in Problem 3.
5. (a)–(d) Calculate the actual error in each of the approximations in Problem 3.
6. (a)–(d) Same as Problem 3, for $\int_0^1 [1/(1+x)]\,dx$.
7. (a)–(d) Similar to Problem 4, relative to Problem 6.
8. (a)–(d) Similar to Problem 5, relative to Problem 6.
9. (a)–(d) Same as Problem 3, for $\int_1^2 (1/x)\,dx$.
10. (a)–(d) Similar to Problem 4, relative to Problem 9.
11. (a)–(d) Similar to Problem 5, relative to Problem 9.
12. Show geometrically that if the graph of $f(x)$ is concave upward, then T_n overestimates $\int_a^b f(x)\,dx$ and that if the graph of $f(x)$ is concave downward, the reverse is true.
13. (a) How large must n be in order to ensure three-decimal-place accuracy in approximating $\int_{-2}^{2}(2x^4 - 0.6x^5)\,dx$ by T_n?
    (b) Same as (a), for seven-decimal-place accuracy.
14. (a),(b) Same as Problem 13, using AT_n.

15. Suppose you use AT_n to approximate $\int_1^x (1/t)\,dt$ for x in increments of 0.01, $1 \leq x \leq 10$, in order to reproduce Table B of four-place natural logs at the back of the book. How large must n be to cover all cases?

For Problems 16 and 18 you will need a computer or a calculator that you have programmed for the trapezoidal rule. The program should allow a change in the function being integrated from one run to the next. The adjustment to the trapezoidal rule is probably most easily done by hand.

C 16. Determine AT_n for the following cases. Find or guess the exact value of the integral when you can.

(a) $\int_0^1 e^x\,dx$, with $n = 10$; $n = 20$

(b) $\int_1^e \dfrac{1}{x}\,dx$, with $n = 10$; $n = 20$

(c) $\int_0^1 \dfrac{4}{1+x^2}\,dx$, with $n = 10$; $n = 20$

(d) $\int_0^5 \dfrac{4}{25}\sqrt{25 - x^2}\,dx$, with $n = 50$; $n = 200$

(e) $\int_1^{10} \dfrac{1}{x}\,dx$, with $n = 100$; $n = 200$

17. Which is the more accurate estimate of $\ln 10$: the second number you obtained in Problem 16(e) or the entry 2.3026 in Table B?

C 18. The function given by $f(x) = (1/\sqrt{2\pi})e^{-x^2/2}$ is the standard normal frequency function of probability and statistics. Justification of the constant $1/\sqrt{2\pi}$ should appear through this problem.

(a) Sketch a graph of $f(x)$.

(b) To eight decimal places, $\int_0^1 f(x)\,dx = 0.34134475$. Get as close to this number as your calculator will allow through use of AT_n.

(c) Use AT_{50} to approximate $\int_1^2 f(x)\,dx$; $\int_2^3 f(x)\,dx$; $\int_3^4 f(x)\,dx$.

(d) Use your results in (b) and (c) to approximate $\int_{-1}^1 f(x)\,dx$; $\int_{-2}^2 f(x)\,dx$; $\int_{-3}^3 f(x)\,dx$; $\int_{-4}^4 f(x)\,dx$.

(e) Guess the approximate value of $\int_{-10}^{10} f(x)\,dx$. Of $\int_{-100}^{100} f(x)\,dx$. Why would it make little sense to use your program here?

(f) Give a reasonable definition of $\int_{-\infty}^{\infty} f(x)\,dx$. What should its value be?

★ 19. For $f(x) = (1/\sqrt{2\pi})e^{-x^2/2}$, find the extremes of f'' on the interval $[0,1]$ and of $f^{(iv)}$ on the same interval. Show that the bound, B'', for f'' on $[0,1]$ can be taken to be 0.4 and that the bound, $B^{(iv)}$, for $f^{(iv)}$ on $[0,1]$ can be taken to be 1.2. Thus, find error bounds in approximating $\int_0^1 f(x)\,dx$ by the trapezoidal rule with $n = 100$. With $n = 1000$. Do the same for the adjusted trapezoidal rule.

★ 20. An improvement on the Euler method for approximating the solution of the differential equation $dy/dx = g(x, y)$, with $y = y_0$ at $x = x_0$, comes

341

about by applying the trapezoidal rule to approximate $\int g(x, y)\,dx$. The result, called the *Heun method*, is as follows:

$$x_{n+1} = x_0 + (n+1)h,$$

$$y_{n+1} = y_n + \frac{g(x_n, y_n) + g(x_n + h, y_n + hg(x_n, y_n))}{2}h.$$

If the solution, y, has a continuous third derivative, it can be shown that the Heun method's order of accuracy is h^2, as contrasted with h for the Euler method.

In Section 8.11, Problem 5, we found that the solution of $d^2z/dt^2 - t(dz/dt) - z = 0$ for which $z = 0$ and $dz/dt = 1$ at $t = 0$ can be written as $z = e^{t^2/2}\int_0^t e^{-s^2/2}\,ds$.

We also found that the differential equation can be recast as $dz/dt - tz = c_1$, or, because of the initial conditions, as $dz/dt - tz = 1$.

In our present notation, this equation becomes

$$\frac{dy}{dx} = xy + 1, \tag{3}$$

with $y = 0$ at $x = 0$, the exact solution of which is $y = e^{x^2/2}\int_0^x e^{-s^2/2}\,ds$.

(a) Show that the Heun method for the approximate solution of equation (3) takes the form

$$x_{n+1} = (n+1)h,$$

$$y_{n+1} = y_n + \frac{x_n y_n + 1 + (x_n + h)[y_n + h(x_n y_n + 1)] + 1}{2}h.$$

c   (b) Apply the Heun method with $h = 0.1$ to approximate y at $x = 0.1, 0.2, \ldots, 1.0$.

c   (c) Same as (b), with $h = 0.01$.

(d) Compare your answers in (b) and (c) with the Euler method for this problem [your answers to Problems 3 and 4(a), **8.11**] and the errors [your answers to Problem 4(b), **8.11**].

21. For the situation depicted in Figure 9-43, show that the area of triangle *PAC minus* the area of triangle *QBC* equals $\frac{1}{2}[f(x_0) + f(x_1)] \cdot h$, where $h = x_1 - x_0$. (Hint: The triangles are similar.)

22. Verify that if $f(x) = c_0 + c_1 x + c_2 x^2 + c_3 x^3$, where the c's are constants, AT_1 gives the exact value of $\int_a^b f(x)\,dx$. Hence, AT_n for $n > 1$ also gives the exact value.

Trigonometric functions

10

10.1 Introduction

This chapter focuses on three basic functions: the sine, cosine, and tangent functions of trigonometry. They differ in a fundamental respect from functions treated thus far in being *periodic* – they repeat themselves after a fixed period.

In exact language, a nonconstant function f is called *periodic* if there is a positive number p such that for any number x with both x and $x + p$ in the domain of f, $f(x + p) = f(x)$. Part of the graph of a periodic function is shown in Figure 10-1.

For the periodic function g pictured in Figure 10-2, we have $g(x) = g(x+1) = g(x+2) = g(x+3) = g(x+4) = \cdots$ for all x. The *smallest* number p such that $g(x + p) = g(x)$ is called the *period* of g. Thus, in this case, the period is 1, not 2 or 3 or anything else.

The importance of the functions of trigonometry lies less in the business of measuring triangles (which accounts for the origin of the word *trigonometry*) than in describing phenomena that are basically periodic. A great many aspects of our lives are of this nature, for instance, the angle of rotation of the earth on its axis, which we see in the sun's apparent motion through the sky each day. Another example: The mean (average) monthly temperature at your home town undoubtedly shows considerable variation, but all the same it comes close to repeating itself annually. Except for minor variations, the height of an ocean wave, your body temperature, the voltage in an

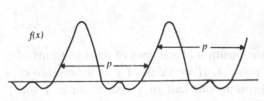

Fig. 10-1

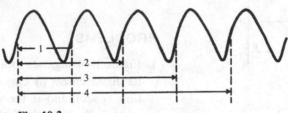

Fig. 10-2

343

alternating-current circuit, and the position of a planet in its orbit are all periodic functions of time.

According to the study cited in Section 1.13, Problem 4, the approval rating of a U.S. president during his first term has varied in much the same way for all presidents in recent times. Similarly, numerous economic and business data show a generally cyclical pattern. A dramatic example of this was observed in 1925 by N. D. Kondratieff, a Soviet economist, who inferred a 50-year cycle in world economic levels. A greatly simplified Kondratieff wave, with economic levels taken relative to 100 for 1967, is shown in Figure 10-3.

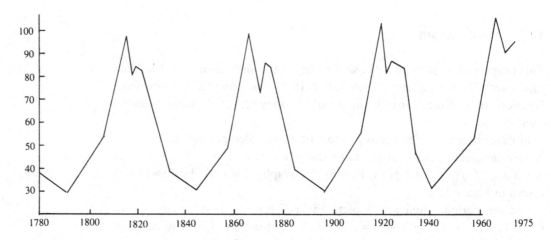

Fig. 10-3

The functions we are about to study are by no means so complicated as this one. But because they are periodic, unlike rational, exponential, and logarithmic functions, they are useful for the analysis of cyclic phenomena. We come a bit closer to these functions in the following simple example.

Imagine a wheel rotating on its axle, with an attached flashlight, arranged to remain horizontal regardless of the wheel's position. The flashlight is aimed at a nearby wall, and as the wheel rotates steadily, the spot of light on the wall rises and falls. After one revolution of the wheel, the spot of light is back exactly where it started, and it repeats its position periodically in every subsequent revolution.

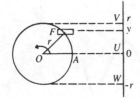

Fig. 10-4

PROBLEMS

○ 1. Figure 10-4 suggests locating the position of the spot of light in terms of its displacement (y cm) from point A. If we think of y as a function of time (t sec), and if the wheel rotates one-half of a revolution in 1 sec, what is the period of the light spot's motion? What is its maximum

displacement? Its minimum displacement? At what point(s) is the light
spot moving fastest? Slowest?

○ 2. In Figure 10-4, consider the central angle, *AOF*. Describe how this angle
changes as the wheel rotates.

3. Extrapolate the Kondratieff wave of Figure 10-3 for another 50 years.
Does your extension conform with your sense of what has happened to
the world's economy since 1975? What does further extension portend?

4. Each of the sketches in Figure 10-5 is intended to represent the graph of
a function. Making reasonable assumptions about these intentions, state
which of the functions are periodic. For those that are periodic, what are
their periods?

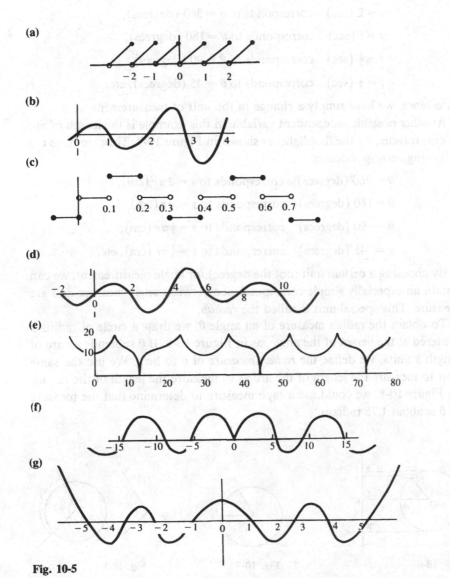

(a)

(b)

(c)

(d)

(e)

(f)

(g)

Fig. 10-5

345

10.2   Angle measure

In many examples of periodic functions, time is the independent variable. However, nothing in the definition of a periodic function requires that time be the independent variable, and it is often useful to choose an independent variable with a geometrical interpretation. For instance, in the case of the flashlight on the rotating wheel (Figure 10-6), the displacement (y cm) could be considered in relation to the central angle, θ (degrees). (The symbol θ is the Greek letter theta.) Indeed, the change from *time* to *angle* is extremely simple; if the wheel rotates one-half revolution in 1 sec (as proposed in Problem 1 in the preceding section), then

$$t = 2 \text{ (sec)}\quad \text{corresponds to } \theta = 360 \text{ (degrees)},$$
$$t = 1 \text{ (sec)}\quad \text{corresponds to } \theta = 180 \text{ (degrees)},$$
$$t = \tfrac{1}{2} \text{ (sec)}\quad \text{corresponds to } \theta = 90 \text{ (degrees)},$$
$$t = \tfrac{1}{4} \text{ (sec)}\quad \text{corresponds to } \theta = 45 \text{ (degrees), etc.}$$

In essence, we have simply a change in the unit of measurement.

Another possible independent variable in this example is the length of arc (s cm) traveled by the flashlight, as shown in Figure 10-7. Then we have the following correspondence:

$$\theta = 360 \text{ (degrees)}\quad \text{corresponds to } s = 2\pi r \text{ (cm)},$$
$$\theta = 180 \text{ (degrees)}\quad \text{corresponds to } s = \pi r \text{ (cm)},$$
$$\theta = 90 \text{ (degrees)}\quad \text{corresponds to } s = \tfrac{1}{2}\pi r \text{ (cm)},$$
$$\theta = 45 \text{ (degrees)}\quad \text{corresponds to } s = \tfrac{1}{4}\pi r \text{ (cm), etc.}$$

By choosing a certain unit (not the degree) for angle measurement, we can obtain an especially simple correspondence between angle measure and arc measure. This special unit is called the *radian*.

To obtain the radian measure of an angle θ, we draw a circle of radius 1 centered at the vertex of the angle, as in Figure 10-8. If θ subtends an arc of length s units, we define the *radian measure* of θ to be s. We use the same unit to measure the length of the arc as to measure the length of the radius. In Figure 10-8, we could use a tape measure to determine that the measure of θ is about 1.75 radians.

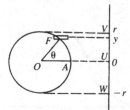

Fig. 10-6

Fig. 10-7

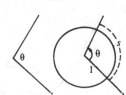

Fig. 10-8

Because the circumference of a circle of radius 1 is 2π, we have the following correspondences:

an angle with degree measure	has radian measure
360	2π
180	π
90	$\pi/2$
60	$\pi/3$
45	$\pi/4$
30	$\pi/6$
$\vdots$	$\vdots$

It is often useful to consider what might be called "directed angles," or "angles in trigonometric position," as in Figure 10-9. We think of the origin of a coordinate system at the vertex of the angle, and the positive x axis extending along what we choose as the *initial* side of the angle. Then, if it requires a *counterclockwise* motion to rotate the first (initial) side of the angle into the second (terminal) side, we say that the measure of θ is *positive*. On the other hand, if it requires a *clockwise* motion to rotate the initial side into the terminal side, we say that the measure of the angle is *negative*. In Figure 10-10, the angle α (Greek letter alpha) has a measure of about -1.75 radians. In Figure 10-11, the angle β (Greek letter beta) has a measure of about $+4.5$ radians.

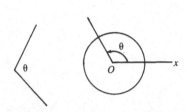

Fig. 10-9

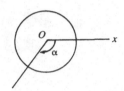

Fig. 10-10

Because of our interest in periodic functions, we do *not* restrict our attention to angles between -2π and 2π radians. We may have more than one revolution. In Figure 10-12, the angle θ_1 has a measure of about $(2\pi + \frac{1}{2})$ radians, or about 6.8 radians, and the angle θ_2 has a measure of $(-4\pi - \pi/2)$ radians, or about -14.137 radians. The radian measure of an angle in "trigonometric position" can be any real number.

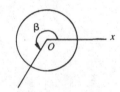

Fig. 10-11

Problem 1
Verify that an angle of 1 radian has degree measure of about 57.3, and that an angle of 1 degree has radian measure of about 0.017.

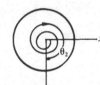

Fig. 10-12

We can use the definition of radian to obtain useful formulas for arc length and area in circles of arbitrary radius: Suppose that we have a central angle of θ radians, $0 \le \theta \le 2\pi$, in a circle of radius r. Let s be the length of the intercepted arc, and A the area of the sector, as in Figure 10-13. Then s is the same fraction of the entire circumference that θ is of the entire angle around the center:

$$\frac{s}{2\pi r} = \frac{\theta}{2\pi},$$

so

$$s = \theta \cdot r.$$

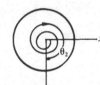

Fig. 10-13

347

Likewise, A is the same fraction of the entire area of the circle that θ is of the entire angle around the center:

$$\frac{A}{\pi r^2} = \frac{\theta}{2\pi},$$

so

$$A = \frac{\theta \cdot r^2}{2}.$$

Radian measure is so much the standard in calculus that it is customary not to label it as such. In statements like "the measure of θ is $\pi/2$," or the less precise "$\theta = \pi/2$," it is understood that the unit is *radian*. Calculators are equipped to handle both radians and degrees. Make sure that yours is set correctly, or you will get weird results.

PROBLEMS

2. Find the radian equivalent of each of the following:
   (a) $15°$ (b) $22\frac{1}{2}°$ (c) $120°$ (d) $135°$ (e) $240°$
   (f) $270°$ (g) $315°$ (h) $-17°$ (i) $900°$ (j) $d°$
3. Find the degree equivalent of each of the following:
   (a) $-\frac{\pi}{2}$ (b) $\frac{3\pi}{4}$ (c) $-\frac{3\pi}{2}$ (d) -6π (e) 0.035
   (f) 1.047 (g) -3.14159 (h) $\frac{\pi}{90}$ (i) 10 (j) r
4. Find the length of the arc subtended on a circle of radius 4 by a central angle of
   (a) $60°$ (b) $25°$ (c) $\frac{2\pi}{3}$ (d) $\frac{5\pi}{6}$
5. Find the area of the sector of a circle of radius 4 determined by a central angle of
   (a) $60°$ (b) $25°$ (c) $\frac{2\pi}{3}$ (d) $\frac{5\pi}{6}$
6. Find the measure of the central angle that on a circle of radius 4 subtends an arc of length
   (a) π (b) $\frac{9\pi}{2}$ (c) 7.2
7. Find the radius of a circle on which an arc of length 2 is subtended by a central angle of
   (a) 4 (b) π (c) $25°$
8. (a) When we view the full moon from the earth, its diameter fills about $\frac{1}{2}°$ of our total field of vision. Assuming that for this small angle it is adequate to think of the moon's diameter as being the arc s in Figure 10-13, and that r, the distance to the moon, is about 235,000 miles, find the diameter of the moon.
   (b) The sun appears to us as approximately the same size as the moon, as illustrated during a total eclipse, when the moon blocks out the sun,

with little or no excess. If the distance to the sun is about 93,000,000 miles, about how many times the moon's diameter is the sun's?

9. The A.U. (astronomical unit) is a unit of distance: the mean distance between the earth and the sun. The *parsec* is another unit of distance used by astronomers. The parsec is defined as the distance at which 1 A.U. subtends an angle of 1 second of arc $(\frac{1}{3600}°)$ at the eye of the observer. Find the length of the parsec in the following units:

(a) A.U.'s.

(b) Miles, if 1 A.U. equals approximately 93,000,000 miles.

(c) Light-years (where 1 light-year equals the distance traveled by light in 1 year), if light travels at about 186,000 miles per second.

10.3   The sine and cosine functions

We define these functions in terms of relationships on a circle. Let u be any real number, represented in Figure 10-14 as the (radian) measure of an angle in trigonometric position. Draw a circle of radius r centered at the origin. Then A, the point where the circle intersects the initial side of the angle, has coordinates $(r, 0)$. Let the coordinates of P, the point where the circle meets the terminal side of the angle, be called (x, y).

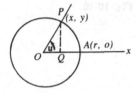

Fig. 10-14

As u increases, the number y behaves exactly like the displacement of the light spot in the flashlight example. The number y varies periodically between a minimum of $-r$ and a maximum of r. The same is true of x, although it is not the same as y, of course. Indeed, when $y = \pm r$, $x = 0$; and when $y = 0$, $x = \pm r$.

We define the *sine* and *cosine* functions by the equations

$$\sin u = \frac{y}{r},$$

$$\cos u = \frac{x}{r}.$$

Note that, depending on u, x and y may be positive, negative, or zero, but r, the length of the radius, is always *positive*.

The numbers $\sin u$ and $\cos u$ depend only on u, not on the size of the circle we choose. For example, if we choose a circle of radius r', as in Figure 10-15, but keep the same number u as before, then triangle $OQ'P'$ of Figure 10-15 is similar to triangle OQP of Figure 10-14, so $y'/r' = y/r$.

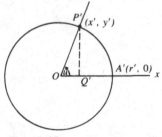

Fig. 10-15

The domain of the sine or the cosine is the set of all real numbers. From the figure used in their definition, it is clear that the range of each is the interval $[-1,1]$ and that each is periodic of period 2π. That is, for all u,

$$-1 \le \sin u \le 1, \quad \text{and} \quad \sin(u + 2\pi) = \sin u,$$
$$-1 \le \cos u \le 1, \quad \text{and} \quad \cos(u + 2\pi) = \cos u. \tag{1}$$

It is not hard to verify that for all u,

$$\sin(-u) = -\sin u,$$
$$\cos(-u) = \cos u,$$
$$\sin(u + \pi) = -\sin u,$$
$$\cos(u + \pi) = -\cos u.$$

(2)

Problem 1
Draw figures to support the assertions in (2).

From (2), we obtain

$$\sin(\pi - u) = -\sin(-u) = \sin u,$$
$$\cos(\pi - u) = -\cos(-u) = -\cos u.$$

(3)

Applying the Pythagorean Theorem to any version of the basic figure, such as that shown in Figure 10-16, we have

$$x^2 + y^2 = r^2, \quad \text{or} \quad \frac{x^2}{r^2} + \frac{y^2}{r^2} = 1.$$

Now, $x/r = \cos u$, so $x^2/r^2 = (\cos u)^2$. It is customary to write $\cos^2 u$ for $(\cos u)^2$, to save the need for parentheses. Similarly, $y^2/r^2 = \sin^2 u$. Thus,

$$\text{for all } u, \quad \sin^2 u + \cos^2 u = 1.$$

(4)

Equations (1), (2), (3), and (4) are called trigonometric "identities," meaning that they are valid *for all u*. Two other trigonometric identities, called the *addition formulas*, are as follows:

For all u and v,

$$\sin(u + v) = \sin u \cos v + \cos u \sin v, \quad \text{and}$$
$$\cos(u + v) = \cos u \cos v - \sin u \sin v.$$

(5)

An indication of how the addition formulas can be proved will be found in Problem 9, **10.4**. Many other identities follow directly from these. We shall encounter some of them in problems.

Graphing the sine function may at first seem difficult, because we cannot readily calculate most values of $\sin u$. A calculator or the sine table at the back of the book will give us the value of the function for many values of the independent variable, but even without them we can accomplish a good deal.

First, because the sine function has period 2π, it is necessary to graph it only over an interval of that length; it repeats indefinitely on either side. Next, because $\sin(u + \pi) = -\sin u$, we need only the values over the interval $[0, \pi]$. Then, because $\sin(\pi - u) = \sin u$, we need only the values over the

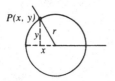

$P(x, y)$

Fig. 10-16

interval $[0, \pi/2]$. This is the reason that the sine table gives values of $\sin u$ only for $0 \le u \le \pi/2$.

Now, from the defining circle, $\sin 0 = 0$ and $\sin \pi/2 = 1$.

Problem 2

Use geometry to obtain

$$\sin \frac{\pi}{6} = \sin 30° = \frac{1}{2},$$

$$\sin \frac{\pi}{3} = \sin 60° = \frac{\sqrt{3}}{2} \approx 0.866,$$

$$\sin \frac{\pi}{4} = \sin 45° = \frac{1}{\sqrt{2}} \approx 0.707.$$

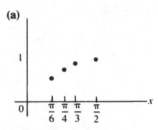
(a)

We are now ready to graph the sine function. To keep to our usual conventions, we shall use x, rather than u, as independent variable, and y as dependent variable. Thus, we graph $y = \sin x$. Plotting the values already found gives the five points of Figure 10-17(a). Remembering the defining circle (or the flashlight example), we expect the graph to be continuous and smooth, as in Figure 10-17(b). Because $\sin(\pi - x) = \sin x$, we extend the graph as in Figure 10-17(c). Because $\sin(x + \pi) = -\sin x$, we extend it further, as in Figure 10-17(d). We now have the function graphed over an entire period, and we can extend it for numbers x greater than 2π or less than zero, as in Figure 10-17(e).

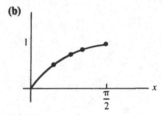
(b)

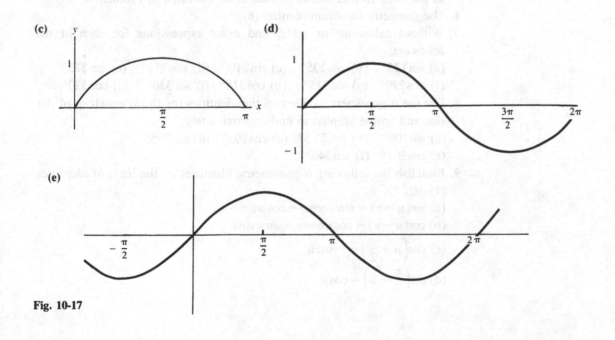
(c) (d) (e)

Fig. 10-17

351

To graph $y = \cos x$, we can repeat the process used for graphing the sine function, but it is easier to apply another identity: By use of the first addition formula, and the values $\sin \pi/2 = 1$, $\cos \pi/2 = 0$, we obtain

$$\sin\left(u + \frac{\pi}{2}\right) = \cos u, \quad \text{for all } u. \tag{6}$$

Problem 3
Verify the identity (6).

Thus, $y = \cos x = \sin(x + \pi/2)$, so the graph of $y = \cos x$ can be obtained from that of $y = \sin x$ by moving the y axis $\pi/2$ units to the right and relabeling points on the x axis. Details are left to Problems 4 and 5.

PROBLEMS

4. Complete the following table:

u	0	$\frac{\pi}{6}$	$\frac{\pi}{4}$	$\frac{\pi}{3}$	$\frac{\pi}{2}$	$\frac{2\pi}{3}$	$\frac{3\pi}{4}$	$\frac{5\pi}{6}$	π
$u + \frac{\pi}{2}$		$\frac{\pi}{2}$	$\frac{2\pi}{3}$	$\frac{3\pi}{4}$					
$\cos u = \sin\left(u + \frac{\pi}{2}\right)$	1	$\frac{\sqrt{3}}{2}$	$\frac{1}{\sqrt{2}}$						

5. Graph $y = \cos x$. For practice, use a calculator or the table at the back of the book to find values besides those obtained in Problem 4.
6. Use geometry to obtain identity (6).
7. Without calculator or table, find exact expressions for each of the following:
   (a) $\sin 135°$ (b) $\cos 135°$ (c) $\sin 210°$ (d) $\cos 210°$ (e) $\sin 270°$
   (f) $\cos 270°$ (g) $\sin 315°$ (h) $\cos 315°$ (i) $\sin 330°$ (j) $\cos 330°$
8. Use the trigonometric table and the identities (or the symmetries of the sine and cosine graphs) to find approximately:
   (a) $\sin 100°$ (b) $\cos 27.5°$ (c) $\cos 205°$ (d) $\sin 205°$
   (e) $\cos 301°$ (f) $\sin 340°$
○ 9. Establish the following trigonometric identities on the basis of identities (1)–(6).
   (a) $\sin(u - v) = \sin u \cos v - \cos u \sin v$
   (b) $\cos(u - v) = \cos u \cos v + \sin u \sin v$
   (c) $\cos\left(u + \frac{\pi}{2}\right) = -\sin u$
   (d) $\sin\left(\frac{\pi}{2} - u\right) = \cos u$

(e) $\cos\left(\dfrac{\pi}{2} - u\right) = \sin u$

(f) $\sin 2u = 2\sin u\cos u$

(g) $\cos 2u = \cos^2 u - \sin^2 u = 1 - 2\sin^2 u = 2\cos^2 u - 1$

(h) $\sin\dfrac{u}{2} = \pm\sqrt{(1 - \cos u)/2}$

(i) $\cos\dfrac{u}{2} = \pm\sqrt{(1 + \cos u)/2}$

10. Check identities (h) and (i) of Problem 9 for $u = \pi,\ 3\pi/2$, and $7\pi/4$. Concoct rules by which the ambiguity of the signs can be resolved in general.

11. On the basis of identities (d) and (e) in Problem 9, explain how tables of $\sin u$ and $\cos u$ for $0 \le u \le \pi/4$ are adequate for all needs.

12. (a) Suppose that the wheel, of radius r cm, with the horizontally aimed flashlight revolves twice per second. Graph the displacement of the light spot as a function of time, t, through three revolutions. Let $t = 0$ correspond to the flashlight's being closest to the wall.

(b) Consider the same situation as in (a), except that the flashlight is aimed straight downward, so that the light spot appears on the floor beneath the wheel. On the same axes used in (a), graph the displacement of the light spot from the point directly below the center of the wheel.

(c) Express each of the functions graphed in (a) and (b) by a formula.

13. Sketch graphs of

(a) $y = 2\sin x$ (b) $y = \sin 2x$ (c) $y = -\tfrac{1}{2}\sin x$

(d) $y = \sin\tfrac{1}{2}x$ (e) $y = 3\cos(\tfrac{1}{2}\pi x)$ (f) $y = 10\cos(3\pi x)$

(g) $y = \sin x + 1$ (h) $y = \sin(x + 1)$

14. Find all x in $[0, 2\pi]$ such that

(a) $\sin x = 1$ (b) $\cos 2x = -1$ (c) $\cos 7x = \dfrac{1}{\sqrt{2}}$

(d) $\sin^2 x = \dfrac{1}{2}$ (e) $\dfrac{1}{\cos 2x} = \sqrt{2}$ (f) $\sin x = 0.940$

15. Find all values of

(a) $\sin x$, if $\cos x = \tfrac{1}{2}$ (b) $\sin x$, if $\cos x = \tfrac{3}{5}$

(c) $\cos x$, if $\sin x = 0$ (d) $\cos x$, if $\sin x = -\tfrac{3}{5}$

(e) $\sin x$, if $\cos x = \dfrac{\sqrt{3}}{2}$ (f) $\cos x$, if $\sin x = -\dfrac{\sqrt{3}}{2}$

(g) $\cos x$, if $\sin x = \tfrac{2}{3}$ (h) $\sin x$, if $\cos x = \tfrac{4}{5}$

(i) $\sin x$, if $\cos x = 1$ (j) $\sin x$, if $\cos x = -\dfrac{1}{\sqrt{2}}$

(k) $\cos x$, if $\sin x = -\tfrac{5}{13}$ (l) $\cos x$, if $\sin x = 1$

16. Where possible without using calculator or table, state the number(s) x in $[0, 2\pi]$ satisfying each part of Problem 15.

10.4   The tangent function, and application of the basic functions to triangles

The *tangent* is defined by the equation

$$\tan u = \frac{\sin u}{\cos u}.$$

Its domain consists of all real numbers except those at which $\cos u = 0$. Because the numbers in $[0, 2\pi]$ for which $\cos u = 0$ are $u = \pi/2$ and $u = 3\pi/2$, we have $\cos u = 0$ for $u = \pi/2$, $\pi/2 \pm 2\pi$, $\pi/2 \pm 4\pi$, $\pi/2 \pm 6\pi, \ldots$ and also for $u = 3\pi/2$, $3\pi/2 \pm 2\pi$, $3\pi/2 \pm 4\pi$, $3\pi/2 \pm 6\pi, \ldots$. Putting these together into a single sequence, we conclude that $\cos u = 0$ for $u = \pm \pi/2$, $\pm 3\pi/2$, $\pm 5\pi/2$, $\pm 7\pi/2, \ldots$. Therefore:

> The tangent function is defined everywhere except at odd multiples of $\pi/2$.

As u approaches any one of these exceptions, $\sin u$ approaches ± 1, and $\cos u$ approaches zero. Hence, the numerical value of $\tan u$ becomes unboundedly large. For example, as $u \to \pi/2$ from the left (through numbers slightly less than $\pi/2$), both $\sin u$ and $\cos u$ are positive. Hence, $\tan u$ is positive and very large. As u approaches $\pi/2$ from the right, $\sin u$ is positive and $\cos u$ is negative. So $\tan u$ is negative and numerically very large. We can describe the situation symbolically as follows:

$$\lim_{u \to \pi/2^-} (\tan u) = \infty; \qquad \lim_{u \to \pi/2^+} (\tan u) = -\infty.$$

C Problem 1

Find $\tan u$ at $u = \pi/2 \pm 0.01$. At $u = \pi/2 \pm 0.001$.

Certainly $\tan(u + 2\pi) = \tan u$ for all u. But in fact, $\tan(u + \pi) = \sin(u + \pi)/\cos(u + \pi) = (-\sin u)/(-\cos u) = \tan u$. That makes π a candidate for the period of the tangent function. Could any smaller number be the period? No, because in the open interval $(-\pi/2, \pi/2)$, which is of length π, there is one and only one number where the sine, and hence the tangent, is zero.

Thus, to graph $y = \tan x$, it is necessary to work hard on only one interval, say $-\pi/2 < x < \pi/2$. It is easy to determine these values:

$$\tan 0 = 0,$$

$$\tan \frac{\pi}{6} = \frac{1}{\sqrt{3}} \approx 0.577,$$

$$\tan \frac{\pi}{4} = 1,$$

$$\tan \frac{\pi}{3} = \sqrt{3} \approx 1.732.$$

Problem 2

Verify these four values, and use a calculator or the trigonometric table to obtain values of the tangent function for three or four more numbers in $(0, \pi/2)$.

The information at hand enables us to sketch Figure 10-18(a). Because $\tan(-u) = \sin(-u)/\cos(-u) = (-\sin u)/(\cos u) = -\tan u$ for all u for which the function is defined, and because the function has period π, the graph of the tangent function looks like Figure 10-18(b).

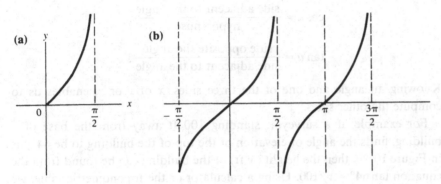

Fig. 10-18

Two identities involving the tangent function come from one of our earlier identities:

$$\sin^2 u + \cos^2 u = 1.$$

If we divide through by $\cos^2 u$, we obtain

$$\tan^2 u + 1 = \frac{1}{\cos^2 u}, \tag{7}$$

provided that $\cos u \neq 0$, whereas if we divide through by $\sin^2 u$, we obtain

$$1 + \frac{\cos^2 u}{\sin^2 u} = \frac{1}{\sin^2 u},$$

or

$$1 + \frac{1}{\tan^2 u} = \frac{1}{\sin^2 u}, \tag{8}$$

provided that $\sin u \neq 0$.

Tradition gives standing to three other trigonometric functions: the cosecant, the secant, and the cotangent, defined by

$$\csc u = \frac{1}{\sin u}, \quad \sec u = \frac{1}{\cos u}, \quad \cot u = \frac{1}{\tan u}.$$

Except for making the writing of such expressions as (7) and (8) somewhat simpler, these functions serve no purpose in our development of trigonometry. We shall make no use of them.

10
Trigonometric functions

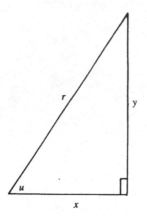

Fig. 10-19

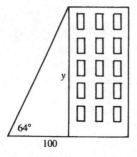

Fig. 10-20

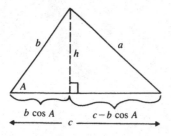

Fig. 10-21

This section concludes with some work with triangles. In right triangles, where one of the angles has measure $\pi/2$, each of the other angles must be less than $\pi/2$. The basic figure, like Figure 10-14 or Figure 10-16, looks like Figure 10-19 in this case, and the equations

$$\sin u = \frac{y}{r}, \quad \cos u = \frac{x}{r}, \quad \tan u = \frac{\sin u}{\cos u} = \frac{y}{x}$$

become, in traditional language,

$$\sin u = \frac{\text{side opposite the angle}}{\text{hypotenuse}},$$

$$\cos u = \frac{\text{side adjacent to the angle}}{\text{hypotenuse}},$$

$$\tan u = \frac{\text{side opposite the angle}}{\text{side adjacent to the angle}}.$$

Knowing an angle and one of the three sides, x or y or r, enables us to compute the other sides.

For example, if a surveyor, standing 100 ft away from the base of a building, finds the angle of elevation of the top of the building to be 64°, as in Figure 10-20, then the height (y ft) of the building can be found from the equation $\tan 64° = y/100$. Using a calculator or the trigonometric table, we find y to be 205 (ft).

More intricate formulas are needed for "solving" triangles in which no angle is $\pi/2$. To derive the most important of these, we label the lengths of the sides of a triangle by a, b, and c, and the angle (more precisely, the measure of the angle) opposite a by A. In Figure 10-21, A is less than $\pi/2$. As you will be asked to show, if A is greater than $\pi/2$, the geometric construction changes, but not the algebra. Let h be the length of the altitude onto side c. Then $h = b \sin A$, and the part of c to the left of the foot of the altitude has length $b \cos A$. In the figure we see a right triangle with hypotenuse a and legs $b \sin A$ and $c - b \cos A$. By the Pythagorean Theorem, we have

$$a^2 = (b \sin A)^2 + (c - b \cos A)^2$$
$$= b^2 \sin^2 A + c^2 - 2bc \cos A + b^2 \cos^2 A$$
$$= b^2 (\sin^2 A + \cos^2 A) + c^2 - 2bc \cos A,$$

or

$$a^2 = b^2 + c^2 - 2bc \cos A. \tag{9}$$

Equation (9) is known as the Law of Cosines: "In any triangle, the square of a side equals the sum of the squares of the other two sides, minus twice their product times the cosine of their included angle."

Note that if $A = \pi/2$, then $\cos A = 0$, and the Law of Cosines reduces to $a^2 = b^2 + c^2$ – the Pythagorean Theorem. Thus, the Law of Cosines is a generalization of the Pythagorean Theorem. Moreover, if for some triangle, $a^2 = b^2 + c^2$, then the Law of Cosines tells us that $-2bc \cdot \cos A = 0$. Because neither b nor c is zero, we conclude that $\cos A = 0$, and hence that $A = \pi/2$. This means that the triangle is a right triangle. Thus, we have derived the *converse* of the Pythagorean Theorem: "If the square of one side of a triangle equals the sum of the squares of the other two sides, then the triangle is a right triangle."

PROBLEMS

3. Complete the proof of the Law of Cosines by dealing with the case in which A is greater than $\pi/2$. (Draw a figure to show that the foot of the altitude lies outside the triangle, but that there is a right triangle with sides a, $b \sin A$, and $c - b \cos A$, so that the proof goes through in this case as well.)

4. Use the trigonometric table to find
   (a) $\tan 140°$ (b) $\tan 220°$ (c) $\tan 320°$ (d) $\dfrac{1}{\tan 40°}$
   (e) $\tan 170°$ (f) $\tan 370°$

5. Find all x such that
   (a) $\tan x = 1$ (b) $4x \tan 7x = 0$
   (c) $\cos^2 x \tan^2 x = \frac{1}{2}$ (d) $3 \tan^2(12.7\pi) - \dfrac{3}{\cos^2(12.7\pi)} = x$

6. Let a, b, and c be the sides of a triangle, with A the angle opposite side a. Fill in the missing elements of Table 10-1.

7. When the sun is 60° above the horizon, a tree casts a shadow 46 ft long. How tall is the tree?

○ 8. *The distance formula.*
   (a) Let $P(x_1, y_1)$ and $Q(x_2, y_2)$ be two points with $x_2 \neq x_1$ and $y_2 \neq y_1$. Let R be the point with coordinates (x_2, y_1). Draw PR and RQ, and use the Pythagorean Theorem to show that the distance between P and Q is given by $\overline{PQ} = \sqrt{(x_2 - x_1)^2 + (y_2 - y_1)^2}$.
   (b) Show that the formula for $\overline{PQ}$ is true also if $x_2 = x_1$ or $y_2 = y_1$.

★ 9. This problem outlines a proof of the addition formulas for the sine and cosine [equations (5)].
   (a) Angles u and v are represented in Figure 10-22. The circle is of radius 1. Let P be the point with coordinates $(\cos u, \sin u)$ and Q the point with coordinates $(\cos v, \sin v)$. The angle POQ is either $u - v$ or $v - u$, depending on whether $u > v$ or $v > u$. It doesn't matter which, because we shall need $\cos POQ$ and $\cos(u - v) = \cos(v - u)$.
   Calculate $\overline{PQ}^2$ by means of the distance formula (Problem 8). If the points O, P, Q form a triangle and angle $POQ < \pi$, the Law of Cosines will give another expression for $\overline{PQ}^2$. Equate the two to obtain $\cos(u - v) = \cos u \cos v + \sin u \sin v$.

Table 10-1

a	b	c	A
	2	2	$\frac{\pi}{3}$
3	3		$\frac{\pi}{4}$
5	3	4	
1		2	$\frac{\pi}{6}$

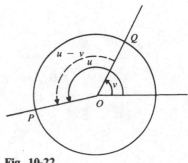

Fig. 10-22

357

(b) Show that this last equation is also correct if angle $POQ > \pi$, or if O, P, Q do not form a triangle.

(c) By writing $u + v = u - (-v)$, obtain the addition formula for the cosine.

(d) Using the result of Problem 9(e), **10.3**, write $\sin(u + v) = \cos[\pi/2 - (u + v)] = \cos[(\pi/2 - u) - v]$, and use the result of part (a) of this problem to obtain the addition formula for the sine.

10. A surveyor on one bank of a river stands directly opposite a post on the opposite bank. Another post is located 30 m downstream from the first, and the angle between his lines of sight to the posts is 20°. How wide is the river?

11. A barge is being towed by a tugboat, with the towline angled upward at an angle $A = 10°$ to the horizontal. The effective force pulling the barge equals the total force pulling on the towline times $\cos A$. What percentage of the total force is effective?

12. A triangle has sides of lengths 3 and 4, with an angle of $\pi/3$ between them. How long is the third side?

13. Same as Problem 12, if the angle is $2\pi/3$.

14. Find all angles of a triangle with sides
    (a) 2, 3, 4   (b) 2, 2, 3   (c) 3, 4, 5

15. The sides of a triangle have lengths 3, 5, and 7. What is the size of the largest angle?

16. One side of a triangle is of length 7, and the opposite angle has measure 60°. If a second side is of length 3, what is the length of the third side?

★ 17. Each of three successive sides of a hexagon inscribed in a circle is of length 3 units; each of the remaining three sides is of length 5 units. Find the area of the hexagon and the radius of the circle.

10.5   Differentiation of the trigonometric functions

The graph of $y = \sin x$ over one period is shown in Figure 10-17(d). Its smoothness suggests that the sine function is differentiable everywhere.

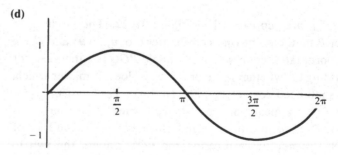

Fig. 10-17

Estimating slopes by eye, we guess that

at $y =$	0	$\dfrac{\pi}{3}$	$\dfrac{\pi}{2}$	$\dfrac{2\pi}{3}$	π	$\dfrac{4\pi}{3}$	$\dfrac{3\pi}{2}$	$\dfrac{5\pi}{3}$	2π
$\dfrac{dy}{dx} = \dfrac{d}{dx}(\sin x) =$	1	$\dfrac{1}{2}$	0	$-\dfrac{1}{2}$	-1	$-\dfrac{1}{2}$	0	$\dfrac{1}{2}$	1

A plot of these points leads to the sketch of the first derived curve shown in Figure 10-23, which looks like the cosine curve. Thus, we guess that if $y = \sin x$, $dy/dx = \cos x$. We proceed to show that this is correct by reverting to the definition of derivative:

$$\text{For any } x, \quad f(x) = \sin x.$$
$$\text{At } x + h, \quad f(x+h) = \sin(x+h).$$

The change in the function $= f(x+h) - f(x) = \sin(x+h) - \sin x.$

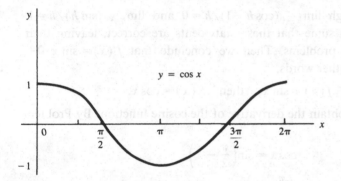

$$y = \cos x$$

Fig. 10-23

The average rate of change of the function $= \dfrac{f(x+h) - f(x)}{h}$

$$= \dfrac{\sin(x+h) - \sin x}{h}.$$

Then, $f'(x) = \lim\limits_{h \to 0} \dfrac{\sin(x+h) - \sin x}{h}.$

As is inevitable with a difference quotient for a differentiable function, we are faced with a fraction in which both the numerator and the denominator approach zero. We must employ some gambit to determine the limit.

Problem 1
Expand $\sin(x+h)$ by the addition formula, and combine terms to obtain $f'(x) = \lim_{h \to 0}[\sin x(\cos h - 1) + \cos x \sin h]/h.$

Note that $\sin x$ and $\cos x$ are unchanged as h approaches zero. Hence,

359

through use of the Limit Theorems, we can write

$$f'(x) = \sin x \cdot \lim_{h \to 0} \frac{\cos h - 1}{h} + \cos x \cdot \lim_{h \to 0} \frac{\sin h}{h}.$$

Problem 2

Experiment with your calculator, or read values from the trigonometric table, for small values of h, to guess values of $\lim_{h \to 0}(\cos h - 1)/h$ and $\lim_{h \to 0}(\sin h)/h$. (The trigonometric functions are evaluated as though the real number h is the radian measure of an angle.)

For example:

$$\frac{\cos(0.035) - 1}{0.035} \approx \frac{0.999 - 1}{0.035} = -\frac{0.001}{0.035} = -0.0029,$$

$$\frac{\cos(0.017) - 1}{0.017} \approx \frac{1 - 1}{0.017} = 0, \quad \frac{\sin(0.035)}{0.035} \approx \frac{0.035}{0.035} = 1.$$

It appears as though $\lim_{h \to 0}(\cos h - 1)/h = 0$ and $\lim_{h \to 0}(\sin h)/h = 1$. For now we shall assume that these statements are correct, leaving their justification to the problems. Then we conclude that $f'(x) = \sin x \cdot 0 + \cos x \cdot 1 = \cos x$. In other words,

$$\text{if} \quad f(x) = \sin x, \quad \text{then} \quad f'(x) = \cos x.$$

We can now easily obtain the derivative of the cosine function: By Problem 9(d), **10.3**,

$$\cos x = \sin\left(\frac{\pi}{2} - x\right).$$

Hence, if $g(x) = \cos x = \sin\left(\frac{\pi}{2} - x\right)$, we use the Chain Rule to conclude that

$$g'(x) = \cos\left(\frac{\pi}{2} - x\right)(-1).$$

But, by Problem 9(e), **10.3**,

$$\cos\left(\frac{\pi}{2} - x\right) = \sin x.$$

Hence,

$$\text{if} \quad g(x) = \cos x, \quad \text{then} \quad g'(x) = -\sin x.$$

We can differentiate the tangent function by writing $\tan x = (\sin x)/(\cos x)$ and using the formula for the derivative of a quotient.

Problem 3

Do this to obtain the result that

$$\text{if} \quad F(x) = \tan x, \quad \text{then} \quad F'(x) = \frac{1}{\cos^2 x}.$$

Many books write this as $F'(x) = \sec^2 x$.

Example 1

10.5
Differentiation

We apply our knowledge of differentiation to find maximum and minimum points and points of inflection on the curve $y = \sin x + \cos x$ in the interval $[0, 2\pi]$.

Here, $dy/dx = \cos x - \sin x$, and $d^2y/dx^2 = -\sin x - \cos x$. Critical numbers are the roots of

$$\cos x - \sin x = 0,$$

which is equivalent to

$$\sin x = \cos x \quad \text{or} \quad \tan x = 1.$$

The roots of this equation in $[0, 2\pi]$ are $x = \pi/4, 5\pi/4$.

At $x = \pi/4$, $y = 1/\sqrt{2} + 1/\sqrt{2} = 2/\sqrt{2} = \sqrt{2}$, and $d^2y/dx^2 = -1/\sqrt{2} - 1/\sqrt{2}$, which is a negative number. So $(\pi/4, \sqrt{2})$ is a relative maximum point on the curve.

At $x = 5\pi/4$, $y = -1/\sqrt{2} - 1/\sqrt{2} = -2/\sqrt{2} = -\sqrt{2}$, and $d^2y/dx^2 = 1/\sqrt{2} + 1/\sqrt{2}$, which is a positive number. So $(5\pi/4, -\sqrt{2})$ is a relative minimum point on the curve.

To find points of inflection, we set $d^2y/dx^2 = 0$:

$$\sin x = -\cos x, \quad \text{or} \quad \tan x = -1.$$

The roots of this equation in $[0, 2\pi]$ are $x = 3\pi/4, 7\pi/4$.

At $x = 3\pi/4$, $y = 1/\sqrt{2} - 1/\sqrt{2} = 0$, and at $x = 7\pi/4$, $y = -1/\sqrt{2} + 1/\sqrt{2} = 0$. So $(3\pi/4, 0)$ and $(7\pi/4, 0)$ are points of inflection on the curve. At the left end of the interval, $x = 0$, $y = \sin 0 + \cos 0 = 1$; and at the right end of the interval, $x = 2\pi$, $y = \sin 2\pi + \cos 2\pi = 1$. Because the turning-point maximum, $(\pi/4, \sqrt{2})$, is higher than either of these end points, and the turning-point minimum, $(5\pi/4, -\sqrt{2})$, is lower than either of these end points, we conclude that the turning-point extremes are also the absolute extremes.

Problem 4

Sketch $y = \sin x$ and $y = \cos x$ on the same axes over $[0, 2\pi]$, and use your sketches to sketch $y = \sin x + \cos x$, illustrating the results just obtained in Example 1.

Example 2

Let us practice with the Chain Rule by finding the critical numbers in $[0, 2\pi]$ for the function $f(x) = \sin^2(x/2)$.

It may be clearer if we write the function as

$$f(x) = \left(\sin \frac{x}{2} \right)^2.$$

Then

$$f'(x) = 2 \left(\sin \frac{x}{2} \right) \cdot \left(\text{derivative of } \sin \frac{x}{2} \right)$$

$$= 2 \left(\sin \frac{x}{2} \right) \cdot \cos \frac{x}{2} \cdot \frac{1}{2}.$$

361

Thus,

$$f'(x) = \sin\frac{x}{2} \cdot \cos\frac{x}{2}.$$

Then $f'(x) = 0$ if and only if $\sin(x/2) = 0$ or $\cos(x/2) = 0$. Now $\sin(x/2) = 0$ provided $x/2 = 0$ or π, that is, provided $x = 0$ or 2π. And $\cos(x/2) = 0$ provided $x/2 = \pi/2$ or $3\pi/2$, that is, provided $x = \pi$ or 3π. But 3π is not within the interval $[0, 2\pi]$, so the critical numbers in $[0, 2\pi]$ for this function are 0, π, and 2π.

It would have been simpler in this case had we used the "double-angle formula" [Problem 9(f), **10.3**] to write $f'(x)$ in a different form:

$$f'(x) = \sin\frac{x}{2} \cdot \cos\frac{x}{2} = \frac{1}{2}\sin x.$$

Now, very quickly, $\sin x = 0$ for $x = 0, \pi, 2\pi$.

PROBLEMS

5. Differentiate:

  (a) $\sin 3x$ (b) $x \cdot \cos(x^2 + 1)$ (c) $\dfrac{1}{\tan x}$

  (d) $\dfrac{x^3}{6} + \cos\left(\dfrac{x^3}{6}\right)$ (e) $2\cos^2\left(\dfrac{x}{2}\right)$ (f) $\sin^3 x + \cos 4x$

  (g) $\cos^2(3x) + \sin^2(3x)$ (h) $\dfrac{1}{\cos(3x+1)}$ (i) $\tan\dfrac{x}{3} - \sin^2(x^3)$

  (j) $\dfrac{\sin(5x^{3/2})}{x+3}$ (k) $\tan(\sin x)$ (l) $\dfrac{1}{\sin x}$

  (m) $\dfrac{\cos 4x}{\sin 4x}$ (n) $\dfrac{1}{\sin(x^2)}$ (o) $\sqrt{1 - \sin^2 x}$

6. The light in a lighthouse 5 miles off shore turns at 2 revolutions per minute. If the shore is straight, at what rate is the lighted spot on the beach moving at a point 10 miles from the nearest point on shore?

7. An alternating current (i amperes) is determined as a function of time (t sec) by the formula $i = 100\sin(\pi t/30 - 1)$. For what values of t does i attain its maximum value?

8. Find the maximum and minimum points and the points of inflection in the interval $[0, \pi]$ on the curve $y = \cos^2(2x)$. Sketch the curve. What is the period of $\cos^2(2x)$?

9. Here is the argument leading to $\lim_{h \to 0}(\sin h)/h = 1$, a result required to obtain the derivative of the sine function: In Figure 10-24, AOB is a sector of radius 1, in which the arc length, AB, is h. Hence, the central angle is also h. In this figure, h is any real number in the interval $(0, \pi/2)$. BC and DA are perpendicular to OA. From the figure: area of triangle BOC < area of sector BOA < area of triangle DOA.

  (a) Show that the area of triangle $BOC = \frac{1}{2}\sin h \cos h$.

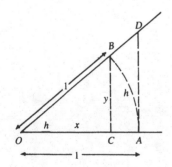

Fig. 10-24

362

(b) Show that the area of sector $BOA = \frac{1}{2}h$.

(c) Show that the area of triangle $DOA = \frac{1}{2}\tan h = \frac{1}{2}(\sin h)/(\cos h)$.
Hence, $\frac{1}{2}\sin h\cos h < \frac{1}{2}h < \frac{1}{2}(\sin h)/(\cos h)$.

(d) Multiply through by the positive number $2/(\sin h)$ to obtain

$$\cos h < \frac{h}{\sin h} < \frac{1}{\cos h}.$$

(e) Why is it legitimate to conclude that $1/(\cos h) > (\sin h)/h > \cos h$?
Because $\cos h = x$, $\cos h$ approaches 1 as $h \rightarrow 0$.

Because $(\sin h)/h$ is pinched between two quantities, each of which approaches 1, $(\sin h)/h$ also approaches 1.

The argument can be carried through for negative h approaching zero, leading to the same result.

10. The other result we need in finding the derivative of the sine function is $\lim_{h \rightarrow 0}(1 - \cos h)/h = 0$.
Write $(1 - \cos h)/h$ as

$$\frac{1 - \cos h}{h} \cdot \frac{1 + \cos h}{1 + \cos h},$$

and reduce this to

$$\frac{1}{1 + \cos h} \cdot \frac{\sin h}{h} \cdot \sin h.$$

How does this give us the desired result?

11. Obtain the formula for the derivative of the cosine function in the same way that the derivative of the sine function is obtained in the text – by going back to the definition of derivative.

★ 12. (a) Find $\lim_{h \rightarrow 0}(\sin 2h)/h$. (Hint: Let $2h = q$.)

c (b) Set your calculator to its "degree" mode, and find $(\sin h)/h$ for $h = 0.1$, 0.01, and 0.001. What is the apparent limit as h approaches zero? Can you account for this?

10.6 Antidifferentiation and integration of trigonometric functions

From the differentiation formulas of the three basic trigonometric functions we immediately obtain three antidifferentiation formulas. We write them using the symbolism of the indefinite integral:

$$\int \cos x \, dx = \sin x + c,$$

$$\int \sin x \, dx = -\cos x + c,$$

$$\int \frac{1}{\cos^2 x} \, dx = \tan x + c.$$

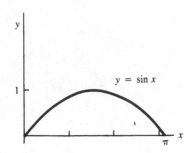

Fig. 10-25

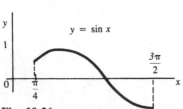

Fig. 10-26

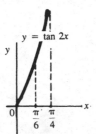

Fig. 10-27

We can apply these formulas to solve integration problems similar to those encountered in Chapter 5 and subsequent chapters.

Example 1
The area under an arch of the curve $y = \sin x$, as in Figure 10-25, is given by

$$A = \int_0^\pi \sin x \, dx = -\cos x|_0^\pi = -(\cos \pi - \cos 0) = -(-1-1) = 2 \text{ units.}$$

Example 2
To find the area bounded by the curve $y = \sin x$, the x-axis, and the lines $x = \pi/4$ and $x = 3\pi/2$, as in Figure 10-26, we must observe that part of the curve is below the x axis. Hence, the desired area is

$$\int_{\pi/4}^\pi \sin x \, dx - \int_\pi^{3\pi/2} \sin x \, dx = -\cos x\Big|_{\pi/4}^\pi - (-\cos x)|_\pi^{3\pi/2}$$

$$= -\left(\cos \pi - \cos \frac{\pi}{4}\right) + \left(\cos \frac{3\pi}{2} - \cos \pi\right)$$

$$= -\left(-1 - \frac{1}{\sqrt{2}}\right) + 0 - (-1) = 2 + \frac{1}{\sqrt{2}} \approx 2.707 \text{ units.}$$

Example 3
To find the area under the curve $y = \tan 2x$ between $x = 0$ and $x = \pi/6$, as in Figure 10-27, we must evaluate the integral $\mathscr{I} = \int_0^{\pi/6} \tan 2x \, dx$. If we write $\tan 2x = (\sin 2x)/(\cos 2x)$, we see a substitution that will work. Let us set $w = \cos 2x$, so that $dw = -\sin 2x \cdot 2 \, dx$. Then

$$\mathscr{I} = \int_0^{\pi/6} \frac{\sin 2x \, dx}{\cos 2x} = -\frac{1}{2} \int_0^{\pi/6} \frac{-\sin 2x \cdot 2 \, dx}{\cos 2x} = -\frac{1}{2} \int_{x=0}^{\pi/6} \frac{dw}{w}$$

$$= -\frac{1}{2} \ln w\Big|_{x=0}^{x=\pi/6} = -\frac{1}{2} \ln(\cos 2x)\Big|_0^{\pi/6}.$$

Problem 1
Complete the evaluation to obtain $\mathscr{I} = \frac{1}{2} \ln 2 \approx 0.3466$.

Example 4
To find the area bounded by the curve $y = \cos^3 x$, the x axis, and the lines $x = 0$ and $x = \pi/2$, we first observe that the curve $y = \cos^3 x$ does not dip below the x axis in this interval, because $\cos x$ is nonnegative over $[0, \pi/2]$. Hence, we must evaluate $J = \int_0^{\pi/2} \cos^3 x \, dx$. It is *not* correct to say that this integral equals $(\cos^4 x)/4|_0^{\pi/2}$. Why?

The following trick works: Write $\cos^3 x = \cos^2 x \cdot \cos x = (1 - \sin^2 x)\cos x$. Then $J = \int_0^{\pi/2}(1 - \sin^2 x)\cos x \, dx = \int_0^{\pi/2}\cos x \, dx - \int_0^{\pi/2}\sin^2 x \cos x \, dx$. The first of these integrals is one of our three basic forms. The second can be handled with the substitution $w = \sin x$.

Problem 2
Show that $J = [\sin x - (\sin^3 x)/3]|_0^{\pi/2} = \frac{2}{3}$.

PROBLEMS

3. Find the following antiderivatives:

(a) $\int \dfrac{4}{\cos^2(2x)}\,dx$ (b) $\int x \cdot \sin(x^2 - 4)\,dx$

(c) $\int \dfrac{\sin u}{\cos^2 u}\,du$ (d) $\int \sin^3(2t)\,dt$

(e) $\int \left(\sin^3 x \cos x + \dfrac{x}{\cos^2(x^2)} \right) dx$

(f) $\int (\cos^2 x - 1)\cos x \sin x\,dx$

(g) $\int \dfrac{1}{\tan v}\,dv$ (h) $\int \cos 3x \cdot \tan 3x\,dx$

(i) $\int \dfrac{\sin u}{\tan u}\,du$ (j) $\int \dfrac{\tan 3x}{\cos 3x}\,dx$

4. Evaluate

(a) $\displaystyle\int_0^{2\pi} \cos x\,dx$ (b) $\displaystyle\int_0^{\pi/4} \dfrac{7\,dx}{\cos^2 x}$

(c) $\displaystyle\int_0^{\pi} \sin^{3/2} x \cos x\,dx$ (d) $\displaystyle\int_{-\pi/24}^{0} \dfrac{\sin 4y}{\cos^3(4y)}\,dy$

(e) $\displaystyle\int_0^{\pi/4} [9 + 7y\cos(y^2)]\,dy$ (f) $\displaystyle\int_{-\pi/4}^{\pi/4} \tan^2 x\,dx$

(g) $\displaystyle\int_0^{\pi/4} \dfrac{\tan^4 x}{\cos^2 x}\,dx$

5. Find $\int \sin x \cos x\,dx$ in three ways:
   (a) by setting $u = \sin x$,
   (b) by setting $v = \cos x$,
   (c) by using the identity $\sin 2x = 2\sin x \cos x$.
   Reconcile your results.

6. Find maximum and minimum points and points of inflection on the graph of $y = \cos^3 x$, and sketch the curve on the same axes on which you sketch $y = \cos x$.

7. Find maximum and minimum points and points of inflection on the graph of $y = \tan^2 x$, and sketch the curve on the same axes on which you sketch $y = \tan x$.

10.7 Inverse trigonometric functions

If f is a periodic function, its inverse cannot be a function, for any horizontal line in the range of f meets the graph of f more than once – in

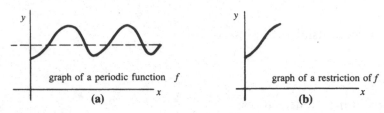

Fig. 10-28

fact, in infinitely many points, as seen in Figure 10-28(a). However
restriction of a periodic function to an interval on which the function
strictly increasing, or is strictly decreasing, will have an inverse that i:
function, as in Figure 10-28(b). This is what we do with the basic trigoi
metric functions. It turns out that we can express antiderivatives of so
commonly appearing functions in terms of these inverses.

Let us start with the sine. It is strictly increasing over $[-\pi/2, \pi/2]$, o
$[3\pi/2, 5\pi/2]$, over $[-5\pi/2, -3\pi/2]$, and over infinitely many more int
vals, as we see in Figure 10-29. For simplicity's sake, we restrict the dom
to the interval including the number 0. That is, let F be the function gi\
by

$$F: y = \sin x, \ -\pi/2 \le x \le \pi/2.$$

The range of F is the interval $-1 \le y \le 1$, as for the unrestricted s
function. The inverse of F is expressed by interchanging x and y in
equation for F, and solving for y in terms of x. But we have no symbol
this, and we have to invent a new expression:

$$F^{-1}: y = \arcsin x.$$

The right side is read "the arc length whose sine is x," or, for short, "the :
sine of x."

The range of F^{-1} is $-1 \le x \le 1$, and the domain of F^{-1} is $-\pi/2 \le y$
$\pi/2$.

The graph of F^{-1} is constructed by reflecting the graph of F in the l
$y = x$, as shown in Figure 10-30. It is, of course, true that $\sin(\arcsin x)$ =

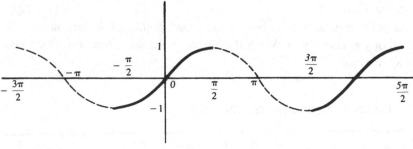

Fig. 10-29

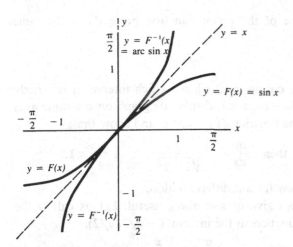

Fig. 10-30

for all numbers x for which this equation is meaningful ($-1 \leq x \leq 1$): "The sine of (the arc length whose sine is x) is x." It is also true that

$$\arcsin(\sin x) = x, \quad \text{provided that } -\pi/2 \leq x \leq \pi/2.$$

This equation is *not* true for all numbers x for which the equation is meaningful. For example,

$$\text{if } x = \pi, \quad \sin x = \sin \pi = 0,$$
$$\text{and} \quad \arcsin(\sin x) = \arcsin(0) = 0.$$
$$\text{But} \quad 0 \neq \pi.$$

A calculator will not protect you from making this error. (However, if you insult the calculator by asking it for arcsin u, with u numerically greater than 1, it will insult you back.)

The graph of the inverse of the sine function suggests that the derivative exists over the interval $-1 < x < 1$. (Because the tangent lines to the graph of the sine function are *horizontal* at $x = -\pi/2$ and at $x = \pi/2$, we expect the tangent lines to the graph of the inverse sine function to be *vertical* at $x = -1$ and at $x = 1$.) Thus, assuming the existence of the derivative, we write $y = \arcsin x$ in the form $x = \sin y$, and differentiate implicitly, using the Chain Rule:

$$1 = \cos y \frac{dy}{dx}, \quad \text{so} \quad \frac{dy}{dx} = \frac{1}{\cos y}.$$

Using the identity $\sin^2 y + \cos^2 y = 1$, we write $\cos y = \pm\sqrt{1 - \sin^2 y}$. But because, for our restricted function, y lies in $[-\pi/2, \pi/2]$, we know that $\cos y \geq 0$. Hence, $\cos y = \sqrt{1 - \sin^2 y} = \sqrt{1 - x^2}$. We have our result:

$$\text{If} \quad y = \arcsin x, \quad \frac{dy}{dx} = \frac{1}{\sqrt{1 - x^2}}, \quad -1 < x < 1.$$

As predicted, the derivative does not exist at $x = \pm 1$.

367

Discussion of the inverse of the cosine function proceeds in the same fashion and is left to you:

Problem 1

Restrict the domain of the cosine to $[0, \pi]$, on which interval it is strictly decreasing. Define $\arccos x$, $-1 \le x \le 1$, display its graph on the same axes as you sketch a graph of the (restricted) $y = \cos x$, and show that:

$$\text{If} \quad y = \arccos x, \quad \text{then} \quad \frac{dy}{dx} = -\frac{1}{\sqrt{1 - x^2}}, \quad -1 < x < 1.$$

This provides us nothing new for antidifferentiation.

The tangent function *does* give us something useful. Let us call G the restriction of the tangent function to the interval $(-\pi/2, \pi/2)$:

$$G: y = \tan x, \quad -\frac{\pi}{2} < x < \frac{\pi}{2}.$$

The range of G is the set of all real numbers, $-\infty < y < \infty$. We define G^{-1} by

$$G^{-1}: y = \arctan x,$$

which we read as "the arc length whose tangent is x," or "the arc tangent of x."

The domain of G^{-1} is $-\infty < x < \infty$, and the range of G^{-1} is $-\pi/2 < y < \pi/2$, as we see in Figure 10-31.

Assuming the existence of the derivative, we proceed as we did with the arcsine function: If $y = \arctan x$, then $x = \tan y$. Hence,

$$1 = \frac{1}{\cos^2 y} \frac{dy}{dx}, \quad \text{or} \quad \frac{dy}{dx} = \cos^2 y.$$

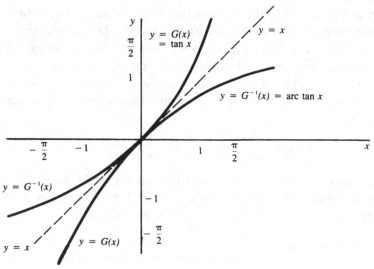

Fig. 10-31

For all y in $(-\pi/2, \pi/2)$,

$$\cos^2 y = \frac{1}{1+\tan^2 y}.$$

Hence, $\cos^2 y = 1/(1+x^2)$, and we have our result:

> If $y = \arctan x$, $\dfrac{dy}{dx} = \dfrac{1}{1+x^2}$, for all real numbers x.

In terms of antidifferentiation, this result takes the form

$$\int \frac{1}{1+x^2}\, dx = \arctan x + c.$$

The earlier result takes the form

$$\int \frac{1}{\sqrt{1-x^2}}\, dx = \arcsin x + c.$$

Here are some applications:

Example 1
To find an expression for $\int (1/\sqrt{4-x^2})\, dx$, we make the substitution $x = 2u$, so that $\sqrt{4-x^2}$ becomes $\sqrt{4-4u^2} = 2\sqrt{1-u^2}$. Now, $dx = 2\, du$, so we have $\int (1/\sqrt{4-x^2})\, dx = \int (1/2\sqrt{1-u^2}) \cdot 2\, du = \arcsin u + c = \arcsin x/2 + c$.

Example 2
Evaluating

$$\int_{2/\sqrt{3}}^{\sqrt{2}} \frac{dx}{x\sqrt{x^2-1}}$$

does not seem to be related to what we have been developing. But the trial substitution $u = \sqrt{x^2-1}$ proves fruitful: We have $u^2 = x^2 - 1$, so $x^2 = 1 + u^2$, and $2x\, dx = 2u\, du$. Thus,

$$\frac{dx}{x\sqrt{x^2-1}} = \frac{u\, du/x}{x\cdot u} = \frac{du}{x^2} = \frac{du}{1+u^2}.$$

Hence,

$$\int_{2/\sqrt{3}}^{\sqrt{2}} \frac{dx}{x\sqrt{x^2-1}} = \int_{x=2/\sqrt{3}}^{\sqrt{2}} \frac{du}{1+u^2} = \arctan u \Big|_{x=2/\sqrt{3}}^{\sqrt{2}}$$

$$= \arctan\sqrt{x^2-1}\,\Big|_{2/\sqrt{3}}^{\sqrt{2}} = \arctan 1 - \arctan \frac{1}{\sqrt{3}}$$

$$= \frac{\pi}{4} - \frac{\pi}{6} = \frac{\pi}{12}.$$

PROBLEMS

2. Evaluate without a calculator or tables:
   (a) arcsin 1   (b) arctan 0   (c) $\arccos(-\frac{\sqrt{2}}{2})$

369

(d) $\arctan(-1)$ (e) $\arcsin \dfrac{\sqrt{2}}{2}$ (f) $\arccos 0$

3. Find in degrees, using a calculator or the trigonometric tables:
   (a) $\arccos(-0.1736)$ (b) $\arctan(5.671)$ (c) $\arcsin(-0.7660)$
   (d) $\arccos(0.9848)$ (e) $\arctan(-0.8391)$ (f) $\arcsin(0.1736)$

4. Evaluate:
   (a) $\cos(\arcsin 0)$ (b) $\tan\left(\arccos \dfrac{\sqrt{2}}{2}\right)$ (c) $\sin[\arctan(-1)]$

   (d) $\sin\left(\arccos \dfrac{\sqrt{3}}{2}\right)$ (e) $\cos\left(\arcsin \dfrac{1}{5}\right)$ (f) $\cos(\arctan 7)$

5. Find dy/dx if
   (a) $y = 4 \arcsin \dfrac{x}{2}$ (b) $y = x^2 \arctan\left(\dfrac{x+1}{3}\right)$

   (c) $y = 3(\arccos 2x)^3$ (d) $y = \dfrac{2}{\arctan x}$

   (e) $y = \dfrac{\arcsin x}{\arctan 5x}$ (f) $y = (1-4x^2)^{3/2} \arccos 2x$

6. Do the following antidifferentiations:
   (a) $\displaystyle\int \dfrac{-4\,dx}{\sqrt{1-x^2}}$ (b) $\displaystyle\int \dfrac{dz}{3+3z^2}$

   (c) $\displaystyle\int \dfrac{dt}{\sqrt{3-27x^2}}$ (d) $\displaystyle\int \dfrac{2\,dx}{x\sqrt{4x^2-1}}$

7. Evaluate:
   (a) $\displaystyle\int_{-1/2}^{0} \dfrac{3\,du}{\sqrt{1-u^2}}$ (b) $\displaystyle\int_{-1/2}^{1/2} \dfrac{3\,dx}{8x^2+2}$

   (c) $\displaystyle\int_{0}^{\sqrt{2}/2} \dfrac{dt}{\frac{3}{2}-6x^2}$ (d) $\displaystyle\int_{-1}^{0} \dfrac{dx}{x^2+2x+2}$

8. Show that if a is a constant, $\int dx/\sqrt{a^2-x^2} = \arcsin(x/a)+c$.

9. Show that if a is a constant, $\int dx/(a^2+x^2) = (1/a)\arctan(x/a)+c$.

10. (a) An aircraft is flying at speed v (mph), climbing at an angle θ (degrees) to the horizontal. Express the rate (R mph) at which it is gaining altitude in terms of v and θ.

    (b) The speed (v mph) of an aircraft at full throttle is roughly proportional to the square of the cosine of its climbing angle ($\theta°$). What θ gives maximum rate of climb (R mph)? If the aircraft's top speed in level flight is 1500 mph, what is the maximum value of the rate of climb, expressed in feet per second?

★ 11. Show that because $(d/dx)(\arcsin x + \arccos x) = 0$, $\arcsin x = \pi/2 - \arccos x$.

○ 12. (a) Divide 1 by $1+x^2$ to obtain $1/(1+x^2) = 1 - x^2 + x^4 - x^6 + \cdots$. The right side is a geometric series with 1 as first term and $(-x^2)$ as common ratio. For $x^2 < 1$, the "sum to infinity" of this infinite series does equal $1/(1+x^2)$, as the division suggests. (See the last paragraphs of Section 0.12.) It is correct, but beyond us to justify at this stage, to integrate "term by term" to obtain

$$\int_0^1 \frac{1}{1+x^2}\, dx = \int_0^1 1\, dx - \int_0^1 x^2\, dx + \int_0^1 x^4\, dx - \int_0^1 x^6\, dx + \cdots .$$

(b) Perform the integrations to obtain $\arctan x\big|_0^1 = \cdots$.

(c) Evaluate the first 10 terms of the right side of your result in (b) to obtain an approximation to $\pi/4$. The convergence is slow, so this is not a useful way to approximate π.

10.8 Further integration involving trigonometric functions

Virtually all the antidifferentiation problems faced so far have called for little more than direct reversal of differentiation formulas, or changes of variable through reasonably transparent substitutions, to permit reversal of differentiation formulas.

With the introduction of the trigonometric functions and their inverses, our collection of "elementary" functions is essentially complete. We have, then, no further means to perform antidifferentiations other than through tricks, like the one used to find $\int dx / x\sqrt{x^2-1}$, or through approximation. The lore of antidifferentiation techniques is large, many based on the use of trigonometric functions, a consequence of the wealth of relationships – the identities – involving those functions. We undertake a tiny sampling of these techniques in the following examples.

Example 1
$\int \sin^2 x\, dx$. The identity $\cos 2u = 1 - 2\sin^2 u$ permits us to write $\sin^2 x = \frac{1}{2}(1 - \cos 2x)$. Thus,

$$\int \sin^2 x\, dx = \int \tfrac{1}{2}(1 - \cos 2x)\, dx = \tfrac{1}{2}\int dx - \tfrac{1}{2}\int \cos 2x\, dx$$

$$= \tfrac{1}{2}x - \tfrac{1}{4}\sin 2x + c.$$

Repeated application of a "double-angle" formula can be used to find expressions for $\int \sin^4 x\, dx$, $\int \sin^8 x\, dx$, and so forth.

Example 2
$\int \cos^6 x\, dx$. The identity $\cos 2u = 2\cos^2 u - 1$ permits us to write $\cos^2 x = \frac{1}{2}(1 + \cos 2x)$. Thus,

$$\cos^6 x = \left[\tfrac{1}{2}(1 + \cos 2x)\right]^3 = \tfrac{1}{8}\left[1 + 3\cos 2x + 3\cos^2(2x) + \cos^3(2x)\right].$$

Now, the antiderivatives of 1 and of $3\cos 2x$ are easy to obtain. To antidifferentiate $3\cos^2(2x)$, repeat application of a double-angle formula. To antidifferentiate $\cos^3(2x)$, apply the method described in Example 4, **10.6**.

371

Problem 1

Complete the antidifferentiation sketched above to obtain

$$\int \cos^6 x \, dx = \tfrac{5}{16}x + \tfrac{1}{4}\sin 2x + \tfrac{3}{64}\sin 4x - \tfrac{1}{48}\sin^3 2x + c.$$

Example 3

$\int_0^5 \sqrt{25 - x^2} \, dx$. We use a substitution based on the identity $1 - \cos^2 u = \sin^2 u$:
Set $x = 5\cos u$. Then $dx = -5\sin u \, du$.

At $x = 0$, $\cos u = 0$, so $u = \pi/2$; at $x = 5$, $\cos u = 1$, so $u = 0$. Now,

$$\sqrt{25 - x^2} = \sqrt{25 - 25\cos^2 u} = 5\sqrt{1 - \cos^2 u} = 5\sqrt{\sin^2 u} = 5\sin u,$$

a positive sign being justified in this last step because $\sin u \geq 0$ for u in the interval $[0, \pi/2]$. Thus,

$$\int_0^5 \sqrt{25 - x^2} \, dx = \int_{x=0}^5 (5\sin u)(-5\sin u) \, du = -25 \int_{x=0}^5 \sin^2 u \, du.$$

Using the result of Example 1, we have

$$\int_0^5 \sqrt{25 - x^2} \, dx = -25 \left(\tfrac{1}{2}u - \tfrac{1}{4}\sin 2u \right)\big|_{x=0}^5.$$

Now, because $x = 5\cos u$, $u = \arccos(x/5)$, and

$$\sin 2u = 2\sin u \cos u = 2 \cdot \frac{\sqrt{25 - x^2}}{5} \cdot \frac{x}{5} = \frac{2x}{25}\sqrt{25 - x^2},$$

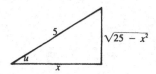

Fig. 10-32

from Figure 10-32. Hence,

$$\int_0^5 \sqrt{25 - x^2} \, dx = -25 \left[\tfrac{1}{2}\arccos\frac{x}{5} - \frac{x}{50}\sqrt{25 - x^2} \right]\Big|_0^5$$

$$= -25\left[\tfrac{1}{2}\arccos 1 - 0 - \left(\tfrac{1}{2}\arccos 0 - 0 \right) \right]$$

$$= -25\left[0 - \frac{1}{2}\frac{\pi}{2} \right] = \frac{25\pi}{4}.$$

Note that if $P(x, y)$ is any point on the arc in the first quadrant of the circle of radius 5, as shown in Figure 10-33, the Pythagorean Theorem gives us

$$x^2 + y^2 = 25, \quad \text{or} \quad y = \sqrt{25 - x^2}.$$

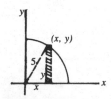

Fig. 10-33

Hence, $\int_0^5 \sqrt{25 - x^2} \, dx$ equals the area within a quarter circle, and our result in Example 3 confirms that the area within a circle of radius r is πr^2.

The methods illustrated in the three foregoing examples are typical of the material of a large topic in calculus called "techniques of integration." Application of those techniques leads to the compilation of "tables of integrals." Faced with a complicated $F(x)$, we can search in a table of integrals for $\int F(x) \, dx$.

$$\int \frac{dx}{\sqrt{a^2 - x^2}} = \arcsin \frac{x}{a} + c \quad \text{and}$$

$$\int \frac{dx}{a^2 + x^2} = \frac{1}{a} \arctan \frac{x}{a} + c,$$

are typical of entries in a table of integrals. So is a generalization of Example 3:

$$\int \sqrt{a^2 - x^2}\, dx = \frac{1}{2} x \sqrt{a^2 - x^2} - \frac{a^2}{2} \arccos \frac{x}{a} + c.$$

It takes practice to recognize which entry in a table of integrals corresponds to a given $\int F(x)\, dx$, for expressions may not obviously correspond, and a change of variable may be required.

Unfortunately, the antiderivatives of many continuous functions cannot be expressed in terms of elementary functions, no matter how ingenious our attempts. A significant example is the function

$$f(x) = \begin{cases} \dfrac{\sin x}{x}, & \text{for } x \neq 0, \\ 1, & \text{at } x = 0. \end{cases}$$

For all nonzero x, $(\sin x)/x$ is continuous. Moreover, as we discovered in obtaining the derivative of the sine function,

$$\lim_{x \to 0} \frac{\sin x}{x} = 1 = f(0),$$

so this function, f, is continuous everywhere. Hence, its antiderivatives exist, but it can be proved that no ingenious substitutions or other trickery will enable us to express $\int f(x)\, dx$ as combinations of elementary functions.

It would be natural, then, to approximate integrals of this function, such as $\int_0^1 f(x)\, dx$, by quadrature – say the Trapezoidal Rule. In fact, because the derivatives of this $f(x)$ exist everywhere (as we shall verify in the problems), we could use the Adjusted Trapezoidal Rule.

There is an even better way, based on *infinite series*, one of the most powerful concepts in calculus. We have occasionally referred to infinite series, but we have not developed the topic systematically, and we shall not do so now. We shall merely provide a few illustrations.

It can be shown that for all x,

$$\sin x = x - \frac{x^3}{3!} + \frac{x^5}{5!} - \frac{x^7}{7!} + \cdots,$$

where $n!$, called "n factorial," is defined by

$$n! = 1 \cdot 2 \cdot 3 \cdots n.$$

(Thus, $3! = 1 \cdot 2 \cdot 3 = 6$; $5! = 1 \cdot 2 \cdot 3 \cdot 4 \cdot 5 = 120$, etc.)

It is correct, although beyond our resources to justify, that

$$f(x) = \begin{cases} \dfrac{\sin x}{x}, & \text{for } x \neq 0, \\ 1, & \text{at } x = 0, \end{cases}$$

$$= 1 - \frac{x^2}{3!} + \frac{x^4}{5!} - \frac{x^6}{7!} + \cdots,$$

the result of dividing each term of the series for $\sin x$ by x. It is also correct, but again beyond our resources to justify at this stage, to integrate this last series "term by term," as we did in Problem 12(b), **10.7**. That is,

$$\int_0^1 f(x)\, dx = \int_0^1 1\, dx - \int_0^1 \frac{x^2}{3!}\, dx + \int_0^1 \frac{x^4}{5!}\, dx - \int_0^1 \frac{x^6}{7!}\, dx + \cdots,$$

or

$$\int_0^1 f(x)\, dx = x\Big|_0^1 - \frac{x^3}{3\cdot 3!}\Big|_0^1 + \frac{x^5}{5\cdot 5!}\Big|_0^1 - \frac{x^7}{7\cdot 7!}\Big|_0^1 + \cdots$$

$$= 1 - \frac{1}{3\cdot 3!} + \frac{1}{5\cdot 5!} - \frac{1}{7\cdot 7!} + \cdots.$$

The sum of 1 term, $S_1 = 1$.

The sum of 2 terms, $S_2 = 1 - \dfrac{1}{3\cdot 3!} = 0.9444\ldots$.

The sum of 3 terms, $S_3 = 1 - \dfrac{1}{3\cdot 3!} + \dfrac{1}{5\cdot 5!} = 0.946111\cdots$.

The sum of 4 terms, $S_4 = 1 - \dfrac{1}{3\cdot 3!} + \dfrac{1}{5\cdot 5!} - \dfrac{1}{7\cdot 7!} = 0.9460828\cdots$.

This convergence is rapid: S_4 is correct as far as the sixth decimal place.

PROBLEMS

2. Find $\int \sin^4 x\, dx$.

3. Find $\int \sin^5 x\, dx$.

4. (a) Use the substitution $x = a\cos u$, in which a is a constant, to obtain

$$\int \sqrt{a^2 - x^2}\, dx = \tfrac{1}{2} x\sqrt{a^2 - x^2} - \frac{a^2}{2}\arccos\frac{x}{a} + c.$$

(b) Differentiate the right side of the preceding equation to obtain $\sqrt{a^2 - x^2}$.

5. Find:

(a) $\int \cos^2(3x)\, dx$ (b) $\int \sqrt{1 - 4x^2}\, dx$

(c) $\int \dfrac{dx}{x\sqrt{2x^2 - 18}}$ (see Example 2, **10.7**)

(d) $\int \dfrac{x^3\, dx}{\sqrt{x^2 + 9}}$ (set $x = 3\tan u$)

6. Find:

(a) $\int \dfrac{x+2}{x^2+4x+8}\,dx$ (b) $\int \dfrac{x+2}{\sqrt{x^2+4x+8}}\,dx$

(c) $\int \dfrac{x+2}{\tan(x^2+2x+8)}\,dx$ (d) $\int \dfrac{1}{x^2+4x+8}\,dx$

(e) $\int \dfrac{2x\,dx}{x^2+4x+8}$ (f) $\int \dfrac{x+1}{x^2+4x+8}\,dx$

7. Evaluate:

(a) $\displaystyle\int_{-1}^{1} \dfrac{dx}{\sqrt{4-x^2}}$ (b) $\displaystyle\int_{0}^{3}\sqrt{9-x^2}\,x^2\,dx$

8. Some antidifferentiations can be successfully attacked by a small amount of trial. For example, to seek antiderivatives of $dy/dx = \ln x$, we might try $y = x\ln x$. Then $dy/dx = 1+\ln x$, which means that we do not have the right y, but we are not far off. Fix things up to obtain $\int \ln x\,dx = x\ln x - x + c$.

9. Apply the idea of Problem 8 to find $\int xe^x\,dx$.

10. Apply the idea of Problem 8 to find $\int e^x \sin x\,dx$ and $\int e^x \cos x\,dx$.

C 11. Calculate S_5 and S_6 of the infinite series for $\int_0^1 [(\sin x)/x]\,dx$. As for accuracy, S_6 is correct to 10 decimal places!

12. Differentiate term by term the series for

$$ f(x) = \begin{cases} \dfrac{\sin x}{x}, & \text{for } x \neq 0, \\ 1, & \text{at } x = 0, \end{cases} $$

and set $x = 0$. The result suggests that $f'(0) = 0$. Find $f''(0)$, $f'''(0)$, and $f^{(iv)}(0)$ in similar fashion.

C 13. With a programmable calculator, apply the Adjusted Trapezoidal Rule to $\int_0^1 [(\sin x)/x]\,dx$, with $n = 50$; $n = 100$; $n = 500$. Compare with your results in Problem 11.

★ 10.9 Other periodic functions

Most periodic functions do not exhibit the simplicity of the basic trigonometric functions. A good example is the function sketched in Figure 10-1, reproduced here with scales on the axes to make its period 2π and its

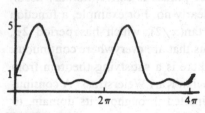

Fig. 10-1

10
Trigonometric functions

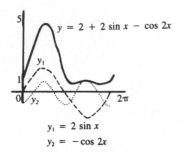

$y = 2 + 2 \sin x - \cos 2x$

$y_1 = 2 \sin x$
$y_2 = -\cos 2x$

Fig. 10-34

maximum 5. As a matter of fact, Figure 10-1 is the graph of $f(x) = 2 + 2\sin x - \cos 2x$; at any x, the height of f is the sum of the heights of

$$y_0 = 2,$$
$$y_1 = 2\sin x, \quad \text{and}$$
$$y_2 = -\cos 2x,$$

as seen in Figure 10-34.

That the period of $\cos 2x$ equals π is clear from the graph. Likewise, $\sin 2x$ has period π. More generally, $\sin 3x$ and $\cos 3x$ have period $2\pi/3$; $\sin 4x$ and $\cos 4x$ have period $2\pi/4; \cdots$ for any positive integer n, $\sin nx$ and $\cos nx$ have period $2\pi/n$.

Problem 1
Show that if $g(x)$ is a function of period p, then $G(x) = g(nx)$ has period p/n.

In Problem 1, n does not have to be an integer. For example, $\sin(\frac{1}{2}x)$ has period $2\pi/\frac{1}{2} = 4\pi$; $\cos(\frac{2}{3}x)$ has period $2\pi/\frac{2}{3} = 3\pi$; $\sin(\pi x)$ has period $2\pi/\pi = 2$; and so forth.

Problem 2
Using 1 in. as a unit on both the x and y axes, sketch graphs of the following equations on the same axes:

$$y = 2\sin x; \quad y = \sin \tfrac{1}{2}x; \quad y = \tfrac{1}{2}\sin 3x; \quad y = \sin \pi x; \quad y = \sin 6x.$$

The addition of constant multiples of $\sin nx$ and of $\cos nx$ to a function of period 2π gives a new function with oscillations within a subinterval of length only $2\pi/n$. We can generate in this way very complicated periodic functions.

The big questions are "How general are these functions?" and "Can any function of period 2π be expressed as a 'trigonometric polynomial,'

$$a_0 + a_1 \cos x + a_2 \cos 2x + \cdots + a_n \cos nx$$
$$+ b_1 \sin x + b_2 \sin 2x + \cdots + b_n \sin nx,$$

for some positive integer n and real numbers $a_0, a_1, a_2, \ldots, a_n, b_1, b_2, \ldots, b_n$?" (This is not actually a polynomial, but the phrase is conventionally used.)

The answer to the second question is clearly no. For example, a function with some points of discontinuity, like $\tan(x/2)$, which has period 2π, cannot be expressed as the sum of functions that are everywhere continuous.

But for *continuous* periodic functions, there is a satisfying theorem from the nineteenth-century German mathematician Karl Weierstrass: A continuous function of period 2π can be approximated throughout its domain, to any degree of accuracy, by some trigonometric polynomial.

376

Simplification is what this discussion is all about. Practical application of the Weierstrass theorem takes the form of decomposing a complicated periodic function, at least approximately, into its *simple* trigonometric components. It could be revealing, for example, to approximate the nearly periodic Kondratieff wave (Section 10.1) by a trigonometric polynomial. With those components having large coefficients we would try to associate historical, cyclic factors of shorter period than the roughly 50 years of the wave. These factors would be the ones most likely to account for the long cyclic nature Kondratieff conjectured for the world's economy.

That is a bigger problem than we are prepared to tackle. Instead, we shall illustrate a method of approximating the components of the less complicated function

$$f(x) = 2 + 2\sin x - \cos 2x.$$

But first a word about the period. The period, 2π, has arisen naturally because it is the period of our building blocks, $\sin x$ and $\cos x$. However, that does not preclude treatment of functions of quite different periods. For example, if the Kondratieff wave is called $K(t)$ for $t \geq 1780$, and if we attach a period of 50 years to the wave, we would write components of $K(t)$ in terms of a new variable

$$x = \frac{2\pi}{50}(t - 1780).$$

Problem 3
Verify that $\sin[(2\pi/50)(t - 1780)]$ and $\cos[(2\pi/50)(t - 1780)]$ have period 50, and that the first is 0, the second 1, at $t = 1780$.

The method we use for approximating components of a function of period 2π calls for numbers, x, evenly distributed through $[0, 2\pi]$, numbers for which corresponding values of the function are at least approximately known. The data thus consist of N points

$$(x_0, \bar{y}_0), (x_1, \bar{y}_1), (x_2, \bar{y}_2), \ldots, (x_{N-1}, \bar{y}_{N-1}),$$

where

$$x_0 = 0, \; x_1 = \frac{2\pi}{N}, \; x_2 = \frac{2\pi}{N} \cdot 2, \ldots, x_{N-1} = \frac{2\pi}{N}(N-1),$$

and the $\bar{y}$'s are the corresponding observed values of the function. It takes quite a bit of manipulation to derive the formulas for the coefficients of an approximating trigonometric polynomial

$$y = a_0 + a_1\cos x + a_2\cos 2x + \cdots + a_n\cos nx$$

$$+ b_1\sin x + b_2\sin 2x + \cdots + b_n\sin nx.$$

The formulas turn out to be

$$a_0 = \frac{1}{N}(\bar{y}_0 + \bar{y}_1 + \bar{y}_2 + \cdots + \bar{y}_{N-1}),$$

$$a_1 = \frac{2}{N}(\bar{y}_0 + \bar{y}_1\cos x_1 + \bar{y}_2\cos x_2 + \cdots + \bar{y}_{N-1}\cdot\cos x_{N-1}),$$

$$a_2 = \frac{2}{N}(\bar{y}_0 + \bar{y}_1\cos 2x_1 + \bar{y}_2\cos 2x_2 + \cdots + \bar{y}_{N-1}\cos 2x_{N-1}),$$

$$\vdots$$

The formula for a_3 is the same as that for a_2 except that $3x_1, 3x_2, \ldots, 3x_{N-1}$ replace $2x_1, 2x_2, \ldots, 2x_{N-1}$. The b's, starting with b_1, are the same as the corresponding a's, except that cos is replaced by sin.

If there are more data points than the number of coefficients to be determined, we have a desirable smoothing effect. The formulas create that trigonometric polynomial that best fits the data in the sense that the sum of the squares of the errors is minimal.

We proceed to experiment with the function $f(x) = 2 + 2\sin x - \cos 2x$ by pretending that it is known to us only through the seven points recorded in the first two columns of Table 10-2. As you note, these are not very precise readings of the function. Let us see what happens. The first of the formulas yields

$$a_0 = \tfrac{1}{7}(0.9 + 3.9 + 4.8 + 2.4 + 0.5 + 0.9 + 0.7) \approx 2.014.$$

With the help of a calculator, we obtain from the formulas $a_1 \approx 0.011$, $b_1 \approx 2.002$, $a_2 \approx -1.010$, $b_2 \approx 0.022$. It is by no means inconceivable that one would guess from this information alone that the underlying function is

$$2 + 0\cdot\cos x + (-1)\cos 2x + 2\sin x + 0\sin 2x = 2 + 2\sin x - \cos 2x,$$

which indeed it is. That is not too bad, considering that our data were

Table 10-2

x	$\bar{y}$	$f(x)$ (to 3 decimal places)
$x_0 = 0$	$\bar{y}_0 = 0.9$	1.0
$x_1 = \dfrac{2\pi}{7}$	$\bar{y}_1 = 3.9$	3.786
$x_2 = 2\cdot\dfrac{2\pi}{7}$	$\bar{y}_2 = 4.8$	4.851
$x_3 = 3\cdot\dfrac{2\pi}{7}$	$\bar{y}_3 = 2.4$	2.244
$x_4 = 4\cdot\dfrac{2\pi}{7}$	$\bar{y}_4 = 0.5$	0.509
$x_5 = 5\cdot\dfrac{2\pi}{7}$	$\bar{y}_5 = 0.9$	0.951
$x_6 = 6\cdot\dfrac{2\pi}{7}$	$\bar{y}_6 = 0.7$	0.659

neither numerous nor especially accurate. A larger sample would have
assured greater reliability in the guess.

PROBLEMS

4. In each of the following cases, find a simple formula for a function, f, of
period p, satisfying the given conditions:

(a) $f(0) = 1$; $p = \dfrac{\pi}{2}$ (b) $f(0) = 0$; $p = 1$ (c) $f\left(\dfrac{\pi}{2}\right) = 4$; $p = 2$

(d) $f(1) = 0$; $p = \pi$ (e) $f(2) = -1$; $p = \frac{1}{2}$ (f) $f(5) = 2$; $p = 100$

5. Suppose that the following data have been collected for what is suspected
to be a function of period 2:

t	1	$\frac{9}{7}$	$\frac{11}{7}$	$\frac{13}{7}$	$\frac{15}{7}$	$\frac{17}{7}$	$\frac{19}{7}$
$\bar{y}$	0.9	3.9	4.8	2.4	0.5	0.9	0.7

(a) Find a change of variable of the form $x = at + b$ that puts these data
in the form $(x, \bar{y})$, with x in $[0, 2\pi]$. (The data will then be those we
have just used in the text.)

(b) Convert the function $f(x) = 2 + 2\sin x - \cos 2x$ to a function $g(t)$, of
period 2, that approximates the data of this problem.

6. On a single set of axes, sketch reasonably large and accurate graphs of
the following functions on $[0, 2\pi]$:

(a) $y_1 = 2\cos x$ (b) $y_2 = -\frac{1}{2}\sin 2x$ (c) $y_3 = 2\cos 2x$

(d) $y_4 = \sin 3x$ (e) $y = y_1 + y_2 + y_3 + y_4$

C 7. The following data are drawn, with small changes, from the function
$y = 2\cos x - \frac{1}{2}\sin 2x + 2\cos 2x + \sin 3x$ of Problem 6. See how accurately
you can reconstruct this function from the data:

x	0	$\frac{2\pi}{9}$	$2 \cdot \frac{2\pi}{9}$	$3 \cdot \frac{2\pi}{9}$	$4 \cdot \frac{2\pi}{9}$	$5 \cdot \frac{2\pi}{9}$	$6 \cdot \frac{2\pi}{9}$	$7 \cdot \frac{2\pi}{9}$	$8 \cdot \frac{2\pi}{9}$
$\bar{y}$	3.9	2.3	-2.5	-1.5	0.8	-1.6	-2.4	-0.4	1.5

[The coefficients turn out to be (to three places) $a_0 = 0.011$, $a_1 = 2.002$,
$b_1 = 0.010$, $a_2 = 1.916$, $b_2 = -0.501$, $a_3 = -0.011$, $b_3 = 1.012$.]

C 8. Repeat Problem 7 for the following data drawn from the same function:

x	0	$\frac{2\pi}{5}$	$2 \cdot \frac{2\pi}{5}$	$3 \cdot \frac{2\pi}{5}$	$4 \cdot \frac{2\pi}{5}$
$\bar{y}$	3.9	-1.9	0.5	-2.4	-0.1

What goes wrong?

C 9. Find $a_0, a_1, b_1, a_2, b_2, a_3, b_3$ for the following data:

x	0	$\frac{2\pi}{9}$	$2 \cdot \frac{2\pi}{9}$	$3 \cdot \frac{2\pi}{9}$	$4 \cdot \frac{2\pi}{9}$	$5 \cdot \frac{2\pi}{9}$	$6 \cdot \frac{2\pi}{9}$	$7 \cdot \frac{2\pi}{9}$	$8 \cdot \frac{2\pi}{9}$
$\bar{y}$	-0.4	1.9	1.6	-1.3	2.1	2.9	0.3	3.4	3.2

Guess the underlying function.

379

10.10 A return to differential equations

As a final topic, we apply trigonometric functions to the solution of linear differential equations with constant coefficients. In Chapter 8 we deferred treating the case in which the roots of the characteristic equation are imaginary numbers. We shall take up this case now and shall find that the solution of such a differential equation involves sines and cosines.

As before, we begin with a second-order equation having right member zero:

$$A\frac{d^2y}{dt^2} + B\frac{dy}{dt} + C = 0. \tag{10}$$

We observed in Section 8.9 that if the reasonable guess

$$y = e^{mt} \tag{11}$$

is to be a solution, then m must satisfy the characteristic equation

$$Am^2 + Bm + C = 0. \tag{12}$$

We verified, conversely, that if m *is* a solution of equation (12) and is also a real number, then (11) is a solution of the differential equation (10).

If m is an *imaginary* number, say

$$m = a + bi,$$

where a and b are real, and $i = \sqrt{-1}$, we face the problem of interpreting the expression e^{mt}. To do this, we need the definition

$$e^{bi} = \cos b + i\sin b, \quad \text{where } b \text{ is a real number.}$$

It turns out that *with this definition*, *all the usual laws of exponents apply to imaginary numbers*. (It is not hard to show this, but it is somewhat lengthy and would take us too far afield.) We now have

$$e^{mt} = e^{(a+bi)t} = e^{at+bti} = e^{at}\cdot e^{bti} = e^{at}(\cos bt + i\sin bt).$$

This is not yet much of a help. But let us review the algebra of solving the characteristic equation (12). By the quadratic formula, the roots are

$$\frac{-B \pm \sqrt{B^2 - 4AC}}{2A}.$$

If the roots are imaginary numbers, $B^2 - 4AC < 0$, so the roots can be expressed as

$$\frac{-B \pm \sqrt{4AC - B^2}\,i}{2A} = \frac{-B}{2A} \pm \frac{\sqrt{4AC - B^2}}{2A}i,$$

where $4AC - B^2$ is positive, so $\sqrt{4AC - B^2}$ is real. Thus, if we set $a = -B/2A$ and $b = \sqrt{4AC - B^2}/2A$, we can express the two roots of the characteristic equation as $m_1 = a + bi$ and $m_2 = a - bi$.

We therefore have trial solutions of the form

$$y_1 = e^{m_1 t} = e^{(a+bi)t} = e^{at}(\cos bt + i \sin bt),$$

$$y_2 = e^{m_2 t} = e^{(a-bi)t} = e^{at}[\cos(-b)t + i \sin(-b)t] = e^{at}(\cos bt - i \sin bt).$$

With $i = \sqrt{-1}$ appearing in y_1 and y_2, these potential solutions still have an undesired imaginary form. But because our differential equation (10) is linear with right member zero, if y_1 and y_2 are solutions of the differential equation, $\frac{1}{2}y_1 + \frac{1}{2}y_2$ should also be a solution. Now

$$\tfrac{1}{2}y_1 + \tfrac{1}{2}y_2 = e^{at}\cos bt.$$

Likewise, $-\frac{1}{2}iy_1 + \frac{1}{2}iy_2 = e^{at}\sin bt$ might be a solution. We must confirm that each of these functions is a solution of (10); so far we have seen merely that they appear to be good candidates.

Problem 1

Verify that if $y = e^{at}\cos bt$, then $dy/dt = ae^{at}\cos bt - be^{at}\sin bt$, and $d^2y/dt^2 = (a^2 - b^2)e^{at}\cos bt - 2abe^{at}\sin bt$.

Substitution of these expressions into the left side of (10) gives

$$A\frac{d^2y}{dt^2} + B\frac{dy}{dt} + Cy = A(a^2 - b^2)e^{at}\cos bt - 2Aabe^{at}\sin bt$$

$$+ Bae^{at}\cos bt - Bbe^{at}\sin bt + Ce^{at}\cos bt$$

$$= e^{at}\cos bt\left[A(a^2 - b^2) + Ba + C\right] - be^{at}\sin bt\left[2Aa + B\right].$$

Problem 2

Remembering that $a = -B/2A$ and that $b = \sqrt{4AC - B^2}/2A$, verify that $A(a^2 - b^2) + Ba + C = 0$ and that $2Aa + B = 0$.

Thus, when we set $y = e^{at}\cos bt$, we find that

$$A\frac{d^2y}{dt^2} + B\frac{dy}{dt} + Cy = e^{at}\cos bt \cdot 0 - be^{at}\sin bt \cdot 0 = 0;$$

that is, $y = e^{at}\cos bt$ *is*, indeed, a solution of (10).

Problem 3

Verify that $y = e^{at}\sin bt$ also satisfies the differential equation.

We now know the general solution of (10):

$$y = c_1 e^{at}\cos bt + c_2 e^{at}\sin bt,$$

for arbitrary (real) constants c_1 and c_2.

The analysis of second-order linear differential equations with constant coefficients and right member zero is now complete:

	Roots of the characteristic equation	General solution
1.	m_1 and m_2, real and distinct	$y = c_1 e^{m_1 t} + c_2 e^{m_2 t}$
2.	m, a double root	$y = e^{mt}(c_1 + c_2 t)$
3.	$a + bi$ and $a - bi$	$y = c_1 e^{at} \cos bt + c_2 e^{at} \sin bt$

Example 1

The differential equation $d^2 y/dt^2 + y = 0$ has characteristic equation $m^2 + 1 = 0$, which has roots $\pm i$, or $0 + i$ and $0 - i$. Hence, $y_1 = e^{0 \cdot t} \cos t = \cos t$ and $y_2 = e^{0 \cdot t} \sin t = \sin t$ are solutions, and the general solution is

$$y = c_1 \cos t + c_2 \sin t.$$

(Note the contrast with the differential equation $d^2 y/dt^2 - y = 0$, which has solutions $y_1 = e^t$ and $y_2 = e^{-t}$, and therefore the general solution $y = c_1 e^t + c_2 e^{-t}$.)

Problem 4

Show that the solution of $d^2 y/dt^2 + y = 0$, for which $y = 1$ at $t = 0$ and $y = -1$ at $t = \pi/4$, is $y = \cos t - (1 + \sqrt{2})\sin t$.

Example 2

In Section 8.9 we considered a complicated system of differential equations describing the relationship of a population X of prey and a population Y of predators. We found that it is possible to simplify this system to one involving linear differential equations with constant coefficients by introducing new variables $x = X - X_E$ and $y = Y - Y_E$, the deviations of the populations from their presumed equilibria, and assuming these deviations to be small.

For the data used in Section 8.11 to approximate solutions of the original system, the reduced system is

$$\frac{d^2 y}{dt^2} + \frac{dy}{dt} + \frac{y}{2} = 0 \quad \text{and}$$

$$\frac{dy}{dt} = \frac{x}{10}.$$

The characteristic equation of the first of these equations is

$$m^2 + m + \tfrac{1}{2} = 0,$$

the roots of which are

$$m = -\tfrac{1}{2} \pm \tfrac{1}{2} i.$$

Therefore, the general solution is

$$y = c_1 e^{-t/2} \cos \frac{t}{2} + c_2 e^{-t/2} \sin \frac{t}{2}, \quad \text{or} \quad y = e^{-t/2}\left(c_1 \cos \frac{t}{2} + c_2 \sin \frac{t}{2} \right).$$

For any given constants, c_1 and c_2, the expression $c_1\cos(t/2)+c_2\sin(t/2)$ oscillates above and below zero by a maximum amount determined by c_1 and c_2. But as t increases, $e^{-t/2}$ approaches zero, so the quantity $e^{-t/2}$ "damps" these oscillations, forcing y toward zero as t increases.

Problem 5

Differentiate the expression found for y, and substitute the derivative in $dy/dt = x/10$ to obtain

$$x = 10e^{-t/2}\left(\frac{c_2-c_1}{2}\cos\frac{t}{2} - \frac{c_1+c_2}{2}\sin\frac{t}{2} \right).$$

Hence, x behaves in the same way that y does. This information is consistent with what was obtained in the much more elaborate numerical experiment of Section 8.11.

When the right member of the differential equation is not zero, we have the same situation as encountered in Chapter 8, namely,

$$\text{for the equation} \quad A\frac{d^2y}{dt^2} + B\frac{dy}{dt} + Cy = f(t),$$

the general solution = (the complementary function) + (a particular integral).

Finding a particular integral is the tricky part. We make the same kinds of guesses that we did in Chapter 8.

Example 3

As a particular integral of

$$\frac{d^2y}{dt^2} - 2\frac{dy}{dt} + 5y = e^t,$$

we see if $y = ke^t$ will work for some constant, k.

Problem 6

Find the first and second derivatives of ke^t, and substitute in the differential equation to show that k must equal $\frac{1}{4}$.

Conversely, $y = \frac{1}{4}e^t$ *does* satisfy the differential equation, so we have a particular integral. The characteristic equation is $m^2 - 2m + 5 = 0$, with roots

$$m = \frac{2 \pm \sqrt{4-20}}{2} = 1 \pm 2i.$$

Hence, the complementary function is

$$c_1e^t\cos 2t + c_2e^t\sin 2t,$$

and the general solution is

$$y = c_1e^t\cos 2t + c_2e^t\sin 2t + \tfrac{1}{4}e^t.$$

Example 4

For the differential equation

$$\frac{d^2y}{dt^2} + 4y = \cos 2t,$$

we encounter the same sort of difficulty met in Example 2, **8.10**. Because the characteristic equation,

$$m^2 + 4 = 0,$$

has roots $m = \pm 2i$, the complementary function is

$$c_1 \cos 2t + c_2 \sin 2t;$$

hence, the most obvious attempt at a particular integral,

$$y = k \cos 2t, \quad \text{for some } k,$$

won't work – substituting this in the left side of the differential equation gives zero for all k. Instead, we try $y = kt \cos 2t$.

Problem 7

Find the first and second derivatives of $kt \cos 2t$, and substitute in the left side of the differential equation to obtain $-4k \sin 2t$.

For no value of k can this expression equal $\cos 2t$ for all t, so this attempt has not given us a particular integral, either. But it does suggest a fruitful approach.

Problem 8

Try $y = kt \sin 2t$, and show that this is a particular integral if and only if $k = \frac{1}{4}$.

Hence, the general solution of our differential equation is

$$y = c_1 \cos 2t + \left(c_2 + \tfrac{1}{4}t \right) \sin 2t.$$

It can be frustrating to probe an equation with nonzero right member to search for a particular integral by trial. However, experienced workers will often proceed by trial despite the existence of systematic methods, because these methods are sometimes too cumbersome to apply.

Nevertheless, it is useful to know that in principle all equations of this kind can be solved. It takes us only a little beyond our experience here and in Chapter 8 to observe that every constant-coefficient linear differential equation, of any positive integral order, with right member zero, has all its solutions expressible in terms of two kinds of functions,

$$t^n e^{at} \cos bt \quad \text{and} \quad t^n e^{at} \sin bt,$$

with $n = 0, 1, 2, \ldots$, and a and b real numbers.

From these functions, a method called *variation of parameters* will provide a particular integral for an equation whose right member is a continuous function, although more often than not that integral will not be expressible in terms of elementary functions.

The universal solvability of constant-coefficient linear differential equations is a major reason for the importance they hold.

PROBLEMS

9. Find the general solution of each of the following:

(a) $\dfrac{d^2y}{dx^2} + 16y = 0$ (b) $\dfrac{d^2y}{du^2} - 5\dfrac{dy}{du} + \dfrac{13}{2}y = 0$

(c) $4\dfrac{d^2y}{dt^2} + 4\dfrac{dy}{dt} + 3y = 0$ (d) $\dfrac{d^3y}{dt^3} - 3\dfrac{d^2y}{dt^2} + 4\dfrac{dy}{dt} - 2y = 0$

10. Find y if $d^2y/dt^2 + dy/dt + y = 0$, and $y = 1$ and $dy/dt = -\frac{1}{2}$ at $t = 0$.

11. Find the general solution of each of the following:

(a) $\dfrac{d^2y}{dt^2} + 2\dfrac{dy}{dt} + 2y = e^{-t}$. By inspection of your answer, what is the value of $\lim_{t \to \infty} y$?

(b) $\dfrac{d^2y}{dt^2} + 4\dfrac{dy}{dt} + 5y = 2 + t$ (c) $\dfrac{d^2y}{dt^2} + 4\dfrac{dy}{dt} + 5y = 2 + t + 10t^2$

(d) $\dfrac{d^2y}{dt^2} + y = 2\sin t$

(e) $\dfrac{d^2y}{dt^2} + 2\dfrac{dy}{dt} + y = e^t\cos 2t$. What function, then, approximates y for large t?

12. Physical principles, mainly Newton's second law of motion, lead to the following differential equation for the motion of a pendulum:

$$L\dfrac{d^2\theta}{dt^2} = -g\sin\theta,$$

where L (ft) is the length of the pendulum, g is the gravitational constant (approximately 32 ft/sec^2), and θ is the angle shown in Figure 10-35.

(a) Justify approximating $\sin\theta$ by θ when θ is small.
(b) Find the general solution of the resulting differential equation.
(c) If the pendulum bob is gently released from an initial $\theta_0 = \pi/24 \approx 7.5°$, and $L = 128$, find a formula for θ at any time (t sec).
(d) What is the period of the motion (i.e., how long does it take the bob to return to its initial position)? Would the period be changed if θ_0 were changed somewhat?
(e) Is the approximation $\sin\theta \approx \theta$ appropriate in this case?
(f) Find the length of a "second's pendulum" – a pendulum that has a period of 1 sec for small oscillations.

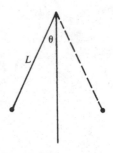

Fig. 10-35

385

13. Suppose that the reduced predator–prey equations of Example 2 had turned out to be

$$\frac{d^2y}{dt^2} - \frac{dy}{dt} + \frac{y}{2} = 0 \quad \text{and}$$

$$\frac{dy}{dt} = \frac{x}{10}.$$

Solve these equations. Do you expect the population to approach equilibrium over a long period?

★ 14. Let y be a solution of the differential equation $A(d^2y/dt^2) + B(dy/dt) + Cy = 0$, $B^2 - 4AC < 0$ and A and B are of the same sign. Discuss $\lim_{t \to \infty} y(t)$.

★ 15. Problem 6, **8.11**, calls for obtaining approximate solutions of the system $dx/dt = x - y$, $dy/dt = x + y$, with $x = 1$ and $y = 0$ at $t = 0$.

    (a) Obtain from this system a second-order differential equation in y. Solve that equation and then solve for x to obtain $y = e^t \sin t$, $x = e^t \cos t$.

C   (b) Prepare a table of values of x and y for $t = 0.1, 0.2, \ldots, 1.0$, and compare its entries with the results you obtained in Problems 6 and 7, **8.11**.

★ 16. Problem 8, **8.11**, involves the equation $d^2z/dt^2 + 2(dz/dt) + 2z = 0$, with $z = 0$ and $dz/dt = 1$ at $t = 0$.

    (a) Solve this equation to obtain

$$z = e^{-t} \sin t, \, dz/dt = e^{-t} \cos t - e^{-t} \sin t.$$

C   (b) Prepare a table of values of z and dz/dt for $t = 0.1, 0.2, \ldots, 1.0$, and compare its entries with the results you obtained in Problems 8 and 9, **8.11** (where z was called x and dz/dt was called y).

10.11 Summary

Periodic functions are essential in the mathematical description and treatment of cyclic behavior, whether it be that of the ocean's tides, the populations of two competing species, or the seasonal changes of a labor market. Two periodic functions – the sine and cosine – are the key elements of this chapter, because from their properties the nature of other periodic functions can be uncovered.

In this chapter we have defined radian measure, defined the sine, cosine, and tangent functions, and determined the fundamental properties of them and their restricted inverses. We have made use of these functions in geometric problems involving circles and triangles, and in antidifferentiations we could not perform earlier. We have applied the sine and cosine to complete the solution of second-order constant-coefficient linear differential equations with right member zero. And with those of you who worked through Section 10.9, we have explored the powerful methods by which the sine and cosine can be used to account for general periodic behavior.

PROBLEMS

1. Which of the following phenomena do you believe to be, at least in some rough approximate sense, periodic? Discuss, naming the variables involved.
   (a) the motion of a tuning fork
   (b) the value of the American dollar in British pounds
   (c) a human heartbeat
   (d) the angular velocity of a record turntable
   (e) the mass of a piece of carbon 14
   (f) the percentage of Americans who consider Lincoln our greatest president
   (g) the annual number of California earthquakes exceeding 4 on the Richter scale
   (h) a woodpecker's pecks
   (i) sunspot activity
   (j) the Dow-Jones industrial average
   (k) the lemming population in Norway

2. A *grad* is $\frac{1}{100}$th of a right angle. How many degrees are there in a grad? How many radians?

3. Establish the following identities, determining any limitations on s and t for their validity:

   (a) $\dfrac{1+\sin s}{1+\sin s - \cos^2 s} = \dfrac{1}{\sin s}$
   (b) $\dfrac{1-\tan^2 s}{1+\tan^2 s} = \cos 2s$

   (c) $\left(\dfrac{1}{\cos s} + \dfrac{1}{\sin s}\right)^2 \tan s = \dfrac{1+\sin 2s}{\cos^3 s \sin s}$

   (d) $\tan(s+t) = \dfrac{\tan s + \tan t}{1 - \tan s \tan t}$
   (e) $\sin 3s = 3\sin s - 4\sin^3 s$

4. By means of the identities

$$\cos(s+t) = \cos s \cos t - \sin s \sin t$$
$$\cos(s-t) = \cos s \cos t + \sin s \sin t$$

   obtain the identities

$$\cos s \cos t = \tfrac{1}{2}\cos(s-t) + \tfrac{1}{2}\cos(s+t)$$
$$\sin s \sin t = \tfrac{1}{2}\cos(s-t) - \tfrac{1}{2}\cos(s+t).$$

5. Use the results of Problem 4 to show that for any positive integers m and n,

$$\int_0^{2\pi} \sin mx \sin nx \, dx = \int_0^{2\pi} \cos mx \cos nx \, dx = \begin{cases} 0 & \text{if } m \neq n, \\ \pi, & \text{if } m = n. \end{cases}$$

387

10
Trigonometric functions

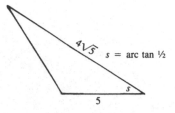

$4\sqrt{5}$ $s = \text{arc tan } \frac{1}{2}$

s

5

Fig. 10-36

6. A radar operator locates an airplane 50 miles away at 30° east of due north and another at the same elevation 40 miles away at 10° west of due north. How far apart are the planes?

7. Deduce the remaining parts (side and angles) of the triangle in Figure 10-36.

8. Find all s in $[0, 2\pi]$ such that

   (a) $\dfrac{1}{\cos s} = -2$ (b) $\tan^2 s = 3$

   (c) $\sin x(1 + \sin x) = 0$ (d) $2\sin^2 x - 3\sin x + 1 = 0$

9. In Section 10.10, the formula $e^{a+bi} = e^a(\cos b + i\sin b)$ was used.

   (a) Use it again to find e^{a+bi} when $a + bi = 1, \pi i, 1 + \pi i, (\pi/2)i, (\pi/4)i$.

   (b) Assuming that $(e^{(\pi/4)i})^2 = e^{2\cdot(\pi/4)i}$, find a square root of i. Check your result.

   (c) Find a fourth root of i.

10. A central angle of 2 subtends an arc of length 10 on a circle. Find the radius of the circle. Find the area of the sector determined by the angle.

11. (a) Express $1/(1 + x^2)$ as an infinite geometric series.

    (b) Assuming that this series can be integrated term by term, show that

$$\frac{\pi}{4} = \arctan 1 = 1 - \frac{1}{3} + \frac{1}{5} - \frac{1}{7} + \frac{1}{9} - \frac{1}{11} + \cdots .$$

(This is the basis for Problem 31, **2.9**.)

12. (a) By differentiating term by term the infinite series for $\sin x$, introduced in Section 10.8, show formally that

$$\cos x = 1 - \frac{x^2}{2!} + \frac{x^4}{4!} - \frac{x^6}{6!} + \frac{x^8}{8!} - \cdots .$$

(b) Calculate the sum of the first three terms of this series with $x = \pi/2 \approx 1.57$. How close is this value to the correct answer? Include the next term and answer the same question.

c   (c) Use the cosine series to approximate $\sqrt{2}$ to seven decimal places.

13. Find the general solution of each of the following differential equations:

    (a) $\dfrac{d^2y}{dt^2} + 2\dfrac{dy}{dt} + 5y = 0$ (b) $\dfrac{d^2y}{dt^2} + \dfrac{dy}{dt} - 2y = 0$

    (c) $\dfrac{d^2y}{dt^2} - 2\dfrac{dy}{dt} - 3y = \sin t$

14. Find the maxima and minima of the function $f(x) = \sin x - \frac{1}{4}\cos 2x$. Sketch the graph of f.

15. Find dy/dx if

    (a) $y = \tan 2x$ (b) $y = \cos^2(3x)$ (c) $y = \sqrt{\arcsin x}$

    (d) $y = \sin^3(3x)\tan^2(2x)$

16. Verify that $y = \cos^2 t$ is a solution of the differential equation $d^2y/dt^2 + dy/dt = -2(\cos 2t + \sin 2t)$, and obtain the general solution.

17. Find antiderivatives of the following functions:

(a) $f(x) = \cos(3x+2)$ (b) $g(x) = \left(\dfrac{\tan x}{\cos x}\right)^2$ (c) $h(x) = \dfrac{2x}{1+x^4}$

(d) $F(x) = \dfrac{1}{x \ln x}$ (e) $G(x) = x \sin x$ (f) $H(x) = \dfrac{x^2}{x^2+4}$

★ 18. In Problems 9 and 10, **10.5**, we demonstrated that

$$\lim_{h \to 0} \frac{\sin h}{h} = 1 \quad \text{and} \quad \lim_{h \to 0} \frac{1-\cos h}{h} = 0.$$

In Section 10.8 we stated that

$$\sin x = x - \frac{x^3}{3!} + \frac{x^5}{5!} - \frac{x^7}{7!} + \cdots.$$

In Problem 12(a) of this section, we differentiated this series term by term to guess that

$$\cos x = 1 - \frac{x^2}{2!} + \frac{x^4}{4!} - \frac{x^6}{6!} + \cdots.$$

(a) Use the series for $\sin x$ to give a plausible argument that $\lim_{h \to 0}(\sin h)/h = 1$.

(b) Use the series for $\cos x$ to give a plausible argument that $\lim_{h \to 0}(1-\cos h)/h = 0$. What does the same argument suggest about $\lim_{h \to 0}(1-\cos h)/h^2$?

(c) Multiply $(1-\cos h)/h^2$ by $(1+\cos h)/(1+\cos h)$, and simplify to demonstrate that $\lim_{h \to 0}(1-\cos h)/h^2 = \frac{1}{2}$.

(d) Use Figure 10-24 to show that for $0 < h < \pi/2$, $\sin h < h < \tan h$.

(e) If $f(x)$ is defined by

$$f(x) = \begin{cases} (\sin x)/x, & \text{for } x \neq 0, \\ 1, & \text{at } x = 0, \end{cases}$$

use the *definition* of the derivative to show that $f'(0) = -\lim_{h \to 0}(h - \sin h)/h^2$.

(f) Now, for $0 < h < \pi/2$,

$$0 < \frac{h - \sin h}{h^2} < \frac{\tan h - \sin h}{h^2},$$

by (d). Find $\lim_{h \to 0}(\tan h - \sin h)/h^2$. Hence, $f'(0) = ?$

(g) For the function $G(x) = (\sin x)/x$, $x \neq 0$, find $G'(x)$ and $\lim_{x \to 0} G'(x)$. Is $f'(x)$ continuous at all x?

SAMPLE TEST

1. A light is spotted directly off a straight coastline at the same time an observation post 10 miles away reports that its sighting forms a 60° angle with the coastline. How far is the light off shore?

2. Determine $\int_0^1 (1/\sqrt{4-x^2})\, dx$.

3. For what numbers x is $\sin 2x = \sin x$?

4. Find $(d/dx)\arctan(\sin x)$.

5. What is the general solution of the differential equation $d^2y/dt^2 + 9y = 0$?

6. Find the solution y of the equation $d^2y/dt^2 + 9(dy/dt) = t$ such that $y = 1$ and $dy/dt = -\frac{26}{9}$ when $t = 0$.

★ 7. Find

$$\lim_{x \to 0} \frac{\sin x^2}{x^2} \quad \text{and} \quad \lim_{h \to 0} \frac{\cos 2h - 1}{h}.$$

Answers to selected problems

Answers to most odd-numbered problems and sample test problems are given here. There are also answers to certain even-numbered problems of particular interest or novelty.

Chapter 0

Section 0.1
1. (a) 20 (b) 6 (c) 2 (d) 300 (e) 0 (f) 40 (g) 1 (h) 9 (i) 6 (j) 9 (k) 4 (l) 35   2. (a) $a(p+q)$ (b) $c^2 + 4d^2$ (c) $b^2 - a^2$ (d) y^2 3. (a) $x^3 + 3x^2h + 3xh^2 + h^3$ (b) $x^4 + 4x^3h + 6x^2h^2 + 4xh^3 + h^4$ (c) $x^5 + 5x^4h + 10x^3h^2 + 10x^2h^3 + 5xh^4 + h^5$

Section 0.2
1. (a) $-3, -2, 0$ (b) $-3, -5$, not possible (c) 3, 4, not possible (d) 0, not possible, not possible   3. (a) $(a+b)q$ (b) $x(v-u)$ (c) $x^2 + y^2$ (d) 0 (e) $ef - cd$ (f) $b - a$ 4. (a) $x^2 - y^2$ (b) $x^3 - y^3$ (c) $x^4 - y^4$

Section 0.3
1. (a) $\frac{2}{5}$ (b) $\frac{25}{6}$ (c) $\frac{14}{3}$ (d) $\frac{2}{15}$ (e) 16 (f) 1 (g) $6h/(6+h)$ (h) $x/[5(x+2)]$ (i) $3y^3/(y+3)$ (j) $3/[y(y+3)]$ 3. (a) $-1/[4(4+h)]$ (b) $-3(2x+h)/[x^2(x+h)^2]$ (c) $-(3x^2 + 3xh + h^2)/[x^3(x+h)^3]$

Section 0.4
1. $x + 1/x$ 3. $(x^3 - x)/2$ 5. $(x-1)/(4x^2 - 1)$ 7. xy
9. $[xy/(x+y)]^2$ 11. $b^2/2a^2$ 13. a^{13}/b^7 15. $(w+1)^2/w^6$

Section 0.5
1. (a) $2\sqrt[4]{3}$ (b) $(\frac{2}{3})^{2/3}$ (c) $\sqrt{a}(a-b)$ or $\sqrt{a}(b-a)$, whichever is nonnegative, provided $a \ge 0$ (d) $2(a^3 - b^3)^{1/3}$ (e) $\sqrt{5}/(a+b)^{3/2}$ 3. (a) yes (b) no, yes when $p \ge 0$ (c) no, yes when a or $b = 0$ (d) no, yes when $pq = 0$ (e) same

as (d) (f) yes (g) no, no (h) no, no **5.** (a) $18y/(81-y^4)$
(b) $18y/[(9-y^2)\sqrt{81-y^4}]$

Section 0.7
1. 10 **3.** $\frac{13}{3}$ **5.** 3 **7.** 4, -1 **9.** 7, -2 **11.** -5 **13.** $\frac{4}{3}$, $-\frac{1}{2}$

Section 0.8
1. $x > 3$ or $x < -1$ **3.** $-6 < x < -1$ **5.** $x < -4$ **7.** $x > \frac{27}{4}$ or $x < 3$
9. $x > 3$ or $x < -3$ **11.** $-2 < x < 2$

Section 0.9
1. $x = \frac{7}{2}$ **3.** $x = 3, y = 1$ **5.** x can be any number; then $y = \frac{4}{3}x - 4$
6. no solution **8.** $x = \frac{3}{13}, y = \frac{-17}{13}$ **10.** $x = \pm 2, y = 1$

Section 0.10
1. $\frac{1}{2}$, 3 **3.** $(5 \pm \sqrt{7}\,i)/2$ **5.** 0, 4 **7.** 2 (double root) **9.** $(7 \pm \sqrt{57})/2$
11. 7, -4 **13.** $(3 \pm \sqrt{5})/2$

Section 0.11
1. 5, $\frac{5}{3}$, $\frac{-3}{2}$ **2.** 5, $(1+\sqrt{91})/6 \approx 1.8$, $(1-\sqrt{91})/6 \approx -1.4$ **3.** To two
decimal places, which is greater precision than graphical methods easily
afford, the roots are 5.01, 1.65, -1.49 **4.** 2 is the only real root **5.** 3, -1,
-4 **7.** 1, 2, -7 **9.** $-6, (-3 \pm \sqrt{17})/4$ **11.** 3, -2, $-2 \pm \sqrt{2}$
13. (a) 12, 226, -58, -54 (b) approximately -8.76, -2.22×10^7,
2.70×10^{15}

Section 0.12
1. (a) 12, 42 (b) 64, 126 **3.** (a) -28, -78 (b) -64, -42 **5.** (a) 11, 36
(b) 243, 364 **7.** $\frac{2}{11}$ **9.** (a) $\frac{4}{11}$ (b) $\frac{7}{9}$ (c) $\frac{7}{37}$ (d) $\frac{74}{99}$

Section 0.13
1. 0.646 **3.** 2.51 **5.** (a) 5 (b) -2 (c) 4 (d) $\frac{1}{2}$ (e) 4 (f) 3 (g) 9 (h) 1
7. (a) $\log_{10}(x+2)$ (b) $\log_p \frac{1}{2}$ (c) $\frac{2}{3}$ (d) $\log_a(x+1)+6$

Section 0.14
1. (a) 4° per hour (b) 8° (c) 8 hours **3.** ≈ 1.2

Chapter 1

Section 1.5
5. 35.4, 3.8 **7.** (a) 2.1737 (b) 2.1772 (c) 10.2183

Section 1.6

6. (a) $y-4=2(x+1)$ (b) $3x+4y=19$ (c) $y=7$ (d) $y=cx$ (e) $y+3=2(x+2)$ (f) $3x+4y+1=0$ (g) $y=x+2$ (h) $3y=2x$ (i) $by=a(x-1)$
8. (a) $E=-\frac{1}{10}P+9$ **10.** (b) $A=P(1.1)^n$ **13.** $C=\frac{5}{216}v^3$ **15.** $s=2600-16t^2$ **17.** $d=(1.23)\sqrt{h}$ **19.** $D=24{,}000/\sqrt{p}$, 8000, 4800, \$5.76
21. (b) 5 times as strong, 3.125 times as strong **22.** $v=100-\frac{2}{9}t$, $p=100-\frac{8}{9}t$, $c=100-\frac{61}{45}t$, approximately **25.** 6.67×10^{-20}

Section 1.7

3. 9.1, 22.9 **5.** (b) $k\approx-0.01$

Section 1.9

1. yes, no **3.** (a) $-2\le v\le\frac{7}{4}$ (b) no function (c) $\frac{1}{10}\le v<1$ (d) $\frac{1}{2}\le v\le100$ or $-100<v<-\frac{1}{2}$ (e) v is a function for $0\le u\le2$, $u\ne1$; $v\le-3$ or $v\ge3$ (f) $-1<v\le1$ (g) $0\le v<25$ (h) $0\le v\le2^n$ **5.** (a) maximum $Q=15$ attained at $x=\frac{7}{2}$, no minimum (b) minimum $P=20$ attained at $x=-3$, no maximum

Section 1.10

1. $1, 0, 6, a^3+2a^2-4a+1, 3h+5h^2+h^3$ **3.** (a) $4, 5, 4\sqrt{2}-5$ (b) 0.6063, 0.6006 (c) $(6+h)/\left[\sqrt{16+(3+h)^2}+5\right]$ **5.** (a) x (b) x (c) x (d) $1/(1-x)$ (e) $1/(1-x)$ (f) $1/(1-x)$ **7.** (a) $[\frac{1}{3},\infty)$ (b) $(-\infty,\infty)$

Section 1.11

1. (b) $[1,5]$ (d) no (e) $g(x)=x^2+1$, $0\le x\le2$; or $g(x)=x^2+1$, $-2\le x\le0$ **3.** (b) $[0,25]$ (d) no (e) $f(x)=10x-x^2$, $0\le x\le5$; or $f(x)=10x-x^2$, $5\le x\le10$ **7.** (a) $[1,6]$, $[500,18{,}000]$ (b) $p=\sqrt{18{,}000}/\sqrt{N}$, $[500,18{,}000]$, $[1,6]$ (c) all $x\ne0$, $(0,\infty)$ (d) $y=\pm\sqrt{18{,}000}/\sqrt{x}$, not a function, $y=18{,}000/x^2$, $x>0$

Section 1.12

1. (a) $-5<x<5$ (b) $-5\le x<0$ or $0<x\le5$ (c) $3\le x<5$ or $-5<x\le-3$ **3.** Figure A-1 **4.** $|xy|=|x|\cdot|y|, |x/y|=|x|/|y|$

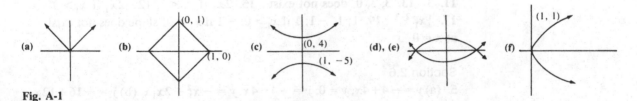

Fig. A-1

Section 1.13

6. (a) 1.33333206, $\frac{4}{3}$ (b) 0.79999924, $\frac{4}{5}$ **8.** (b) 6 (c) $12(2-\sqrt{3})^{1/2}\approx6.2117$ (d) 6.2653, 6.2787, 6.2821 (e) 3.1410

Sample test
1. $y = 3600/x^2$

2.
x	2	4	6	8	10	12
y	900	225	100	56.25	36	25,

Figure A-2 **3.** 30.5 **4.** It should be larger **5.** $y - 36 = -16(x - 10)$
6. $x \neq 0$ **7.** [9, 3600] **8.** not a function **9.** $y = 60/\sqrt{x}$, $9 \leq x \leq 3600$
10. Figure A-3

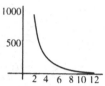

Fig. A-2

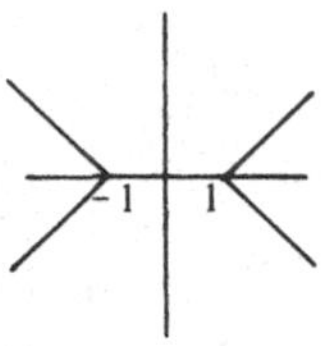

Fig. A-3

Chapter 2

Section 2.1
3. (b) $t = 2$ (c) 64 (d) (i) 16, 16 (ii) 16, -16 (iii) $26\frac{2}{3}$, 16 (iv) 32, 0
5. 54.4, 55.84, 55.984 **7.** (a) $24\frac{2}{3}$ (b) 44, 52, $48\frac{1}{3}$ (c) 8, 44, -52, -8

Section 2.2
5. (a) 4 (b) -2 (c) -3 (d) 0 **7.** (a) 11 (b) 0 (c) 6 (d) $-2h + h^2$ (e) 0
(f) 0 (g) $-2 + 2x$ **9.** (a) 1000 (b) does not exist (c) 1331 (d) does not
exist **11.** (a) 2 (b) 0 (c) 4 (d) 6 (e) 0 (f) $2t$ (g) $2t$ (h) $12 - 2t$
13. (a) 0.25 (b) 0 (c) -0.0625 **15.** average velocity $= 66\frac{2}{3}$ mph

Section 2.3
6. (a) 50 (b) does not exist (c) 3 (d) 0 (e) does not exist (f) does not exist
7. a^n if $a \neq 0$

Section 2.5
5. 6, 0, -6, $3x_1$; 3 **7.** 16, 0, 8, $2x_1 + 3x_1^2$ **9.** 6, 2, 0, -2, $6 - 2x_1$; 1
11. 3 **13.** 3, 3, 0, does not exist **15.** $2x_1$ if $x_1 \leq 3$, $12 - 2x_1$ if $x_1 > 3$
17. $\frac{1}{3}x_1^{-2/3}$ **19.** 1; 1; -1; 1 if $x_1 > 0$, -1 if $x_1 < 0$; slope does not exist
at $x = 0$

Section 2.6
5. (a) $y = -4 + 4x$, $y = 0$, $y = -4 - 4x$, $y = -x_1^2 + 2x_1 x$ (b) $y = -16 + 12x$,
$y = 0$, $y = 16 + 12x$, $y = -2x_1^3 + 3x_1^2 x$ (c) $y = -48 + 32x$, $y = 0$, $y = -48$
$-32x$, $y = -3x_1^4 + 4x_1^3 x$ **7.** $y = 1 + 4x$, $y = 5$, $y = 1 - 4x$, $y = 5 - x_1^2 + 2x_1 x$
9. (a) $y = 1 - \dfrac{x}{4}$, $y = 2 - x$, $y = -2 - x$, $y = 2/x_1 - x/x_1^2$ (b) $y = (3 - x)/4$,
$y = 3 - 2x$, $y = 3 + 2x$, $y = 3/x_1^2 - (2/x_1^3)x$ (c) $y = \frac{1}{2} - \frac{3}{16}x$, $y = 4 - 3x$,

$y = -4 - 3x$, $y = 4/x_1^3 - (3/x_1^4)x$ **11.** $y = -21 - 4x$, $y = -11$, $y = -3 + 8x$, $y = (4x_1 + 8)x - (2x_1^2 + 3)$

Section 2.7

9. (a) 7 (b) x^2 (c) $-2/x^2$ (d) $1/x^3$ (e) 0 (f) $2 - 2t$ (g) $4x^3 + 12x$
(h) $2x - 2/x^3$ (i) $(6t + 1)/2$ (j) $-4/9x^3$ (k) $b + 2ct$ (l) $-4x + 18\sqrt{2}\,x^2$
(m) $4t + 1$ (n) $4x^3 - 2x$ (o) $5 - 1/t^2$ (p) $-3/2x^2 + 4/x^3$

11. (a)

x	-1	0	1	2	3	4
$f(x)$	$-4\frac{1}{3}$	1	$2\frac{1}{3}$	$1\frac{2}{3}$	1	$2\frac{1}{3}$
$f'(x)$	8	3	0	-1	0	3

(b) Figure A-4 (c) $y = 1 + 3x$, $x + y = \frac{11}{3}$, $y = 1$

13. (a)
$$f'(x) = \begin{cases} 1, & x > -2 \\ -1, & x < -2 \end{cases}$$

(b)
$$f'(x) = \begin{cases} 3x^2, & x \geq 0 \\ -3x^2, & x < 0 \end{cases}$$

(c) $f'(x) = 0$, $x \neq 0$

(d)
$$f'(x) = \begin{cases} 2x + 1, & x < -2, x > 1 \\ -2x - 1, & -2 < x < 1 \end{cases}$$

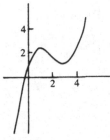

Fig. A-4

15. (a) $101 \cdot 3^{100}$, 0, does not exist

17. (a) and (b)
$$f'(x) = g'(x) = \begin{cases} 2x, & 0 \leq x < 2 \\ 8 - 2x, & 2 < x \leq 5 \end{cases}$$

(c)
$$F'(x) = \begin{cases} 2x, & 0 \leq x < 2 \\ 9 - 2x, & 2 < x \leq 5 \end{cases}$$

Section 2.8

1. (b) $\frac{1}{4}$ **3.** (d) In the first case, if x is the limit, then $x > 0$ and $x = (6 + x)^{1/2}$. Thus, $x^2 - x - 6 = 0$ and $x = 3$.

Section 2.9

Sample test

1. (a) $-2/x^4$ (b) $-2 + 6t$ (c) $2x - 2/x^3$ **2.** $f(0) = 4$, $f'(0) = 0$, $f(1) = 2$, $f'(1) = -3$, $f(2) = 0$, $f'(2) = 0$, $y - 2 = -3(x - 1)$, $y = 0$ **3.** (a) $v(t) = 20 - 2t$
(b) $t = 10$, $s(10) = 100$ (c) $t = 20$ (d) 200 ft (e) 20 ft/sec   **4.** $|(5x + 2) - 17|$
$= |5(x - 3)| < 0.01$, or ε, if and only if $|x - 3| < 0.01/5$, or $\varepsilon/5$ **5.** $(3, 10)$
6. Figure A-5   **7.** (a) $(0, 4)$

(b)
$$g'(x) = \begin{cases} 2x, & 0 < x \leq 2 \\ 8 - 2x, & 2 < x < 4 \end{cases}$$

8. $D_f = (-\infty, 2) \cup (2, \infty)$; $R_f = (0, \infty)$; no, because two values of x correspond to each positive y **9.** (a) $3x - 5 - 1/x^3$ (b) $3t^2 - 4t + 1$ (c) $18u$

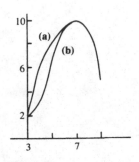

Fig. A-5

Answers to
selected problems

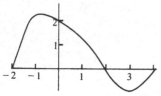

Fig. A-6

11. $(2,13)$, $y - 13 = -6(x - 2)$ **13.** Figure A-6   **15.** (a) $x = -1$ (c) -5
17. 9, $9 + 6h + h^2$, $6h + h^2$, $6 + h$, 6 **21.** $0.05/6$, $\varepsilon/6$ **23.** $y - 2x + 4$,
$y = 9$ **27.** (a) 80 (b) $v(t) = 64 - 32t$ (c) $t = 2$ (d) $t = 5$ (e) 208 (f) 96
29. Figure A-7   **32.** The limits are (a) 4 (b) $\frac{9}{4}$ (c) $\frac{25}{16}$ **33.** $D_F = [-2, \infty)$,
$R_F = [0, \infty)$ The function given by $y = (x^2 - 8)^{1/3}$ is the inverse of F
35. $D'_f = [0, 2) \cup (2, 5) \cup (5, 7]$ Figure A-8   **37.** $f(0) = 0$, $f(1) = 1$, $f(2) = 0$,
$f'(0) = 4$, $f'(1) = -1$, $f'(2) = 0$, $y - 1 = -(x - 1)$, $y = 0$ **39.** (a) $f'(x) = \frac{10}{3} x^2$
$-4 - 3/x^2$ (b) $f'(t) = 4t^3 - 2 - 2/t^3$ (c) $f'(u) = -2/u^6$ **41.** $y(-1) = 0$,
$y(0) = -2$, $y(1) = -4$, $y'(-1) = 0$, $y'(0) = -3$, $y'(1) = 0$, $y + 2 = -3x$,
$y + 4 = 0$ **43.** (a) $f'(x) = 4x^2 - 10x + 1/x^3$ (b) $f'(t) = 3 - 1/t^2$ (c) $f'(u) = 2u - 18/u^3$ **45.** (a) $y(2) = 3$, $y(6) = 23$ (b) $y - 3 = 5(x - 2)$ (c) $(4, 9)$
(d) $y - 9 = 5(x - 4)$ **47.** $D_G = (-\infty, \infty)$, $R_G = (-\infty, 4]$, $y = (64 - x^3)^{1/2}$
is the inverse of G for $x \leq 4$

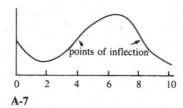

A-7

Fig. A-8

Chapter 3

Section 3.1
4. $z = 1$, $[0, \infty)$, $(0, \infty)$, yes   **6.** $z = \frac{4}{3}$, yes

Section 3.2
3. (a) strictly decreasing on $[-1, 1]$ (b) strictly increasing on $[-1, 2]$
(c) strictly increasing on $(-\infty, \infty)$ (d) strictly decreasing on $[0, 3]$
(e) strictly increasing on $(-\infty, \infty)$ (f) strictly increasing on $[4, \infty)$
(g) strictly increasing on $[2, \infty)$ **4.** $f(x) = x^3$ on $[-1, 1]$ is strictly
increasing, but $f'(0) = 0$

Section 3.3
7. (a) 1.5, 126.5   **9.** $C = (0.005)x^3$, 2.7   **11.** (a) 1.08 (b) 0.15, 5%   **13.** 300
units   **15.** 384

Section 3.4
7. 0.76, 0.60, 0.40   **9.** 6, 5.04, 5.01   **10.** (a) 25, 26 (b) $E/x = 400/x + 24$
$+ 0.01x$ (c) 200 (d) $0 < x < 200$, $x > 200$ (e) 28, 28

Section 3.5
3. 10.5, 11.525   **5.** 140; by completing the square, we obtain $E = 324 - 0.01(x - 140)^2$, which show that $E(140) = 324$ is the maximum.

7. $(-4, 25)$ is the highest point. There is no lowest point; 9 is the y intercept; 1 and -9 are the x intercepts. **9.** (a) $(0.25, 6.125)$, —, 6, $(1 \pm 7)/4$ (c) $(0.25, 1.125)$, —, 1, $(1 \pm 3)/4$ (e) $(0.25, -1.875)$, —, -2, — (g) —, $(2, 0)$, 8, 2 (i) —, $(2, -4)$, 4, $2 \pm \sqrt{2}$

Section 3.6

3. yes; check $f(x) = x^3$ and $c = 0$, for example **5.** The only rel. maximum occurs at $x = -1$; the only rel. minimum occurs at $x = 3$.

Section 3.7

1. Figure A-9 **3.** 128, 320 **4.** (a) 6 is the maximum; 2 is the minimum (b) 35 is the maximum; -19 is the minimum; -1 is a rel. maximum at $x = 0$; $-\frac{7}{3}$ is a rel. minimum at $x = -2$ (c) 33 is the maximum; 5 is the minimum (d) -19 is the minimum; 1 is the maximum, which is attained at both 0 and 3; -3 is a rel. minimum attained at 2 (e) $(0, 0)$ is the minimum point; $(1, 1)$ the maximum point; $(2, \frac{1}{2})$ is a rel. minimum point; $(-\frac{1}{2}, \frac{1}{4})$ is a rel. maximum point (f) 8 is the maximum; -5 is the minimum; a rel. maximum of 1 occurs at 1; a rel. minimum of 0 occurs at 0 **5.** (a) The minimum is 0. (b) The minimum is 0; it occurs at $\pm\sqrt{3}$; $(0, 3)$ is a rel. maximum point. (c) The minimum is -4, at 0, $\pm\sqrt{3}$; there are rel. maxima of value $-\frac{10}{3}$ at ± 1.

Fig. A-9

Section 3.8

12. a square **14.** One side is twice as long as the other. **15.** (a) 10×20 (b) 5×10 **17.** 5×10 **19.** (a) radius $= 5$, height $= 5$, 75π in.^{2} (b) radius $= 5$, height $= 5$, 125π in.^{3} **21.** $h = \frac{40}{3}$, $r = h/\sqrt{2}$ **23.** 5 weeks from now **25.** $8/\sqrt[3]{\pi}$ **27.** 25 trees, 625 bushels **30.** (a) $b = 30$, $h = 40$ (b) $b = 25$, $h = 57.6$ (c) $b = 20\sqrt{3}$, $y = 30$ (d) $b = 10\sqrt{10}$, $h = 36$

Section 3.9

6. (a) $y = 120$, $x = 30 \cdot 10^7$, $P = 5 \cdot 10^9$ **9.** (a) 40, 502.5 (b) 1.602, 2.1 **11.** (c) 1000, minimum (d) 14.5 (e) 14.5

Section 3.10

6. (a) If $x_0 = 4$, $x_1 = 4.125$, $x_2 = 4.123106$, $x_3 = 4.12310563$ (b) If $x_0 = 5$, $x_1 = 4.933$, $x_2 = 4.932424$, $x_3 = 4.93242415$ (c) If $x_0 = 3$, $x_1 = 3.037$, $x_2 = 3.036370$, $x_3 = 3.03637028$ **8.** To eight decimal places, $\sqrt[3]{7} = 1.91293118$, and with $x_0 = 2$, $x_1 = 1.91666667$, $x_1 - \sqrt[3]{7} = 0.00373548$, $x_2 = 1.91293846$, $x_2 - \sqrt[3]{7} = 0.00000728$, $x_3 = 1.91293118$, $x_3 - \sqrt[3]{7} = 0$

Section 3.11

1. (a) $V = (9 - 2x)(24 - 2x)x$, $0 < x < \frac{9}{2}$ (b) $x = 2$ (c) $A = (9 - 2x)(24 - 2x)$ has no extremes for $0 < x < \frac{9}{2}$. It approaches 216 as x approaches 0, and it approaches 0 as x approaches $\frac{9}{2}$.   **5.** 600 tons, \$3,500   **7.** $F = 1000/x^2$, $+0.6$ dynes   **9.** $f'(x)$ does not exist at $x = \pm 3$; relative minima occur at $x = -5$ and 3; relative maxima at 0 and 5. The absolute minimum $= -2$; the absolute maximum $= 9$ **11.** (a) $R = (7 - 0.02x)x$ (b) $P = -100 + 3x - 0.01x^2$ (c) 150 (d) 125 (e) $4 - 0.02x$, 3.8, 2   **13.** (a) $1000/v$ (b) $160,000/v - 10v^2$ (c) 20 (d) 12,000   **15.** f is stationary at $x = -1, 4$, decreasing for $-1 \leq x \leq 4$, and increasing elsewhere   **17.** $I = 4000/x^2$, -19.2 calories/min/ft   **19.** (a) $r = 3$, $h = 6$ (b) $r = \frac{5}{2}$, $h = \frac{216}{25}$ (c) $r = \sqrt[3]{9}$, $h = 6\sqrt[3]{9}$

Sample test
1. f is stationary at $x = -1$; it is increasing elsewhere   **2.** 800 units less
3. 8 in. $\times$ 16 in.   **4.** \$44.6, \$43, \$41

5. (a)
$$f(x) = \begin{cases} 4x + 16, & -4 \leq x \leq 3 \\ -x + 1, & -3 < x \leq 0 \\ x + 1, & 0 < x \leq 3 \\ -4x + 16, & 3 < x \leq 4 \end{cases}$$

(b) $-3, 0, 3$ (c) 0 at -3 and 3; 1 at 0 (d) 0 at -3 and 3   **6.** Each side length $= 10$. Because the base costs twice what the sides do for each square inch, the problem is equivalent to that of a closed box with uniform cost
7. (a) $x = 30 = y$ (b) $x = (l + w)/4 = y$

Chapter 4

Section 4.1
7. (a) s is at a maximum of 9 at $t = -1$, a minimum of -23 at $t = 3$ (b) v is at a minimum of -12 at $t = 1$ (c) s decreases on $[-1, 3]$ and increases elsewhere (d) v decreases on $(-\infty, 1]$, increases on $[1, \infty)$ (e) Figure A-10
9. (a) The absolute maximum of s is 108 in., at $t = 6$; the absolute minimum is 0, at $t = 0$ and 9 (b) The absolute maximum of v is 27 in./sec, at $t = 3$; the absolute minimum is -81 in./sec, at $t = 9$ (c) $-36, 18$ (d) $[0, 6], [6, 9]$ (e) $[0, 3], [3, 9]$ **11.** (a) $y(-2) = -3$, $y(-1) = 8$, $y(2) = -19$, $y(4) = 33$; absolute maximum $= 33$, absolute minimum $= -19$ (b) $y'(-2) = 24$, $y'(\frac{1}{2}) = -\frac{27}{2}$, $y'(4) = 60$; absolute maximum $= 60$, absolute minimum $= -\frac{27}{2}$ (c) $-30, 42$

Section 4.2
9. (a) $s''(t) = 6(t - 1)$; because $s''(-1) < 0$, s is at a maximum at $t = -1$; because $s''(3) > 0$, s is at a minimum at $t = 3$ (b) $v''(t) = 6$, which means that v is at a minimum at $t = 1$ (c) $(1, -7)$, $[-2, 1)$, $(1, 4]$ **11.** The maximum of s is 27, at $t = 3$; $(2, 16)$ is a point of inflection [if the natural domain is considered, $(0, 4)$ is also a point of inflection].   **13.** $y(0) = 1$ is minimal. The maxima occur at the end points; $(1, 12)$ and $(3, 28)$ are points

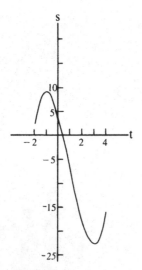

Fig. A-10

of inflection. **16.** (c) $\bar{A}$ is only an approximate model for the growth of the bacterial colony. The problem points to a respect in which it fails to reflect reality. **17.** $(0,0)$ is the minimum point for f_1, the maximum point for f_2, and the horizontal point of inflection for f_3.

Section 4.3
3. $t=9$ gives maximum velocity; $t=6$ gives maximum acceleration
5. 2400, decreasing at 120 ft/min, 2700 **7.** 8, decreasing at 60 ft/min, 83

Section 4.4
3. w **7.** $F'(x)=12(x^2-2x+5)^{11}(2x-2)$, $x=1$ corresponds to minimum, $F(1)=4^{12}$; there is no maximum **9.** (a) $C'=4-0.002n$, $n'=-2x$
(b) $2x-0.004x^3$, -176 (c) 2.2 **11.** (a) $x \neq 1$, $y \neq 0$ (b) $-12/(x-1)^2$
(c) $x \neq 1$, $(-\infty,0)$ (d) $-\frac{4}{3}$, $-\frac{3}{4}$ **13.** (a) $-\frac{1}{3}$ (b) 0 (c) ≈ -0.09 (by the Newton–Raphson method) (d) ± 1 (e) $-\sqrt[3]{4}$, 2 (f) 0 (g) 0 (h) 0 (i) 3
(j) $-\frac{9}{2}$ (k) no critical numbers

Section 4.5
1. (a) $(-\infty,\infty)$ (b) $x \neq 0$ (c) $[-2,2)$ (d) $t \neq 1$ (e) $w \neq \pm 3$ (f) $[0,30]$
(g) $h \neq 0$ (h) $(-\infty,\infty)$ (i) $x \neq 0$ (j) $x \neq 3$ (k) only at $x=0$ (l) $h \neq 0$
3. $\lim_{x \to p}(f+g)(x)=\lim_{x \to p}f(x)+\lim_{x \to p}g(x)=f(p)+g(p)=$
$(f+g)(p)$; $f \cdot g$ and kf are continuous; f/g is continuous provided $g(p) \neq 0$
5. In $0<|u-p|<\delta \Rightarrow |f(u)-L|<\varepsilon$, replace u with $p+h$.

Section 4.8
5. $dr/dt=-1/(600\pi)$ in./hr **7.** $dr/dt=1/(5\pi)$ in./hr **9.** (a) 20
cm³/min, $+\frac{10}{3}$ cm³ (b) $dA/dt=8\pi r(dr/dt)$, $dV/dt=4\pi r^2(dr/dt)=$
$(r/2)(dA/dt)=(a \cdot A)/2$

Section 4.10
3. (a) $x^{-1/2}+2x^{-2/3}-3x^{1/2}$ (b) $-x^{-3/2}-2x^{-4/3}+3x^{-5/2}+3x^{7/4}$
(c) $2x/(3x^2)^{2/3}+1/(2\sqrt{5}\,x^{3/2})$ (d) $8t(t^2-3)^3$ (e) $-[1/x^3+2/(x\sqrt{x})]$
(f) $(1/\sqrt{x})(\sqrt{2}/2+1)$ (g) $-(t^{-3/2}+2t^{-4/3})$ (h) $5(5-x^{-1/2})$ (i) $\frac{1}{4}-1/v^2$

(j)
$$-\frac{x}{2\sqrt{1-x^2}\sqrt{1+\sqrt{1-x^2}}}$$

5. (a) $dy/dx=-x/4y$, 0, $-\frac{3}{8}$, vertical tangent (b) $dy/dx=$ $-x/(2\sqrt{25-x^2})$ (c) $3x+8y=25$ (d) Figure A-11

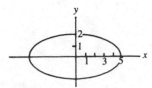

Fig. A-11

Section 4.11
5. $45/\sqrt{5}$ ft/sec **7.** 420 yards/min **9.** 32 ft/sec

Section 4.12

1. (a) dV/dt (b) dS/dx (c) dy (d) dy **3.** (a) $dy = 3u^2\,du$ (b) $dy = 3(x^2 + 1)^2 2x\,dx$ (c) yes, given in (a) that $u = x^2 + 1$

Section 4.13

5. (a) $(16 - 2x^2)/(16 - x^2)^{1/2}$, $[-4, 4]$, $(-4, 4)$ (b) $-(x^2 + 1)(1 - x)^3(7x^2 - 4x + 3)$, $(-\infty, \infty)$, $(-\infty, \infty)$ (c) $-\frac{4}{3}(x - 2)^{-2}$, $x \neq 2$, $x \neq 2$ (d) $(15 - 2x)/[2(4 - x)^{3/2}]$, $(-\infty, 4)$, $(-\infty, 4)$ (e) $(1 - 3x^2)/(2\sqrt{x})$, $[0, \infty)$, $(0, \infty)$
(f) $-(3x^2 + 1)/(x\sqrt{x})$, $(0, \infty)$, $(0, \infty)$ (g) $(3x^2 - 26x + 27)/(x^2 - 9)^2$, $x \neq \pm 3$, $x \neq \pm 3$ (h) $9/(9 - x^2)^{3/2}$, $(-3, 3)$, $(-3, 3)$ (i) $-11x/[(25 - x^2)^{1/2}(36 - x^2)^{3/2}]$, $[-5, 5]$, $(-5, 5)$ (j) not a function, because there is no domain   **7.** minimum of -1 at $x = 0$; $(1/\sqrt{3}, -1/2)$ and $(-1/\sqrt{3}, -1/2)$ are points of inflection   **9.** valid for $x \neq 1$ **11.** volume increases at $\frac{22.5}{3}\pi$ in.^{3}/min   **13.** (a) $G'' = fg'' + 2f'g' + f''g$, $G''' = fg''' + 3f'g'' + 3f''g' + f'''g$, $G'''' = fg'''' + 4f'g''' + 6f''g'' + 4f'''g' + f''''g$ (b) $G^{(n)} = f^{(0)}g^{(n)} + nf^{(1)}g^{(n-1)} + [n(n-1)/2]f^{(2)}g^{(n-2)} + \cdots + \binom{n}{k}f^{(k)}g^{(n-k)} + \cdots + f^{(n)}g^{(0)}$, where

$$\binom{n}{k} = \frac{n!}{(n-k)!k!}$$

Section 4.14

3. $n > 2500$ **5.** $P = -200 + 2n - 0.001n^2$, $n = 1000$, $P(1000) = 800$
9. TUC $= 200/n + 5 - 0.001n$ **11.** $MC = 5 + 0.0004n$; $MR = 8 - 0.002n$; C is always increasing; R is decreasing for $n > 4000$; $P = -200 + 3n - 0.0012n^2$; $P(1250) = 1675$ is maximal; TUC $= 200/n + 5 + 0.0002n$; $(\text{TUC})' = -200/n^2 + 0.0002$; < 0 (economies of scale) if $0 < n < 1000$; > 0 (diseconomies of scale) if $n > 1000$

Section 4.15

5. a square   **7.** $h = 2r$ gives maximum volume for given area or minimum area for given volume   **9.** $r = 10(\frac{2}{3})^{1/2}$, $h = 20/\sqrt{3}$ **11.** In either case the radii must be equal.   **13.** $x = a/\sqrt{2}$, $y = b/\sqrt{2}$

Section 4.16

1. (a) 285 (b) 125 (c) 117.15 (d) $p(a^2 + b^2)^{1/2}$ if $qa/(p^2 - q^2)^{1/2} \geq b$ and $p > q$; $ap^2/(p^2 - q^2)^{1/2} + q[b - qa/(p^2 - q^2)^{1/2}]$ if $qa/(p^2 - q^2)^{1/2} < b$ and $p > q$, $p(a^2 + b^2)^{1/2}$ if $p \leq q$ **3.** (a) \$18,750 (b) no; yes, to \$37,500
(c) 8 knots   **5.** (a) $50\sqrt{x} - x$, $x = 625$, $p = 2$, profit $= 625$
(b) $2500 (1/p - 1/p^2)$, $p = 2$, $x = 625$, profit $= 625$ **7.** (a) $-6/(5\pi)$
(b) $dx/dt = -30/\pi x^2$ **9.** 3.4 ft/sec   **11.** (a) $\frac{50}{3}$ and $\frac{100}{3}$ (b) For any specific length, maximum volume goes with maximum cross-sectional area.
(c) radius $= 100/(3\pi)$, length $= 100/3$ **13.** maximum at $(0, 1)$ and $(0, -1)$; minimum at $(1, 0)$ and $(-1, 0)$ **14.** height $= 4r$, radius $= \sqrt{2}\,r$, where r is the radius of the sphere   **15.** (b) $18/(16\pi)$, $18/(9\pi)$, $18/(4\pi)$,

$18/\pi$ in./sec (c) $72/\pi$ **17.** $10\times10\sqrt{3}$ in. **19.** $R=4-48x+12x^2$,
minimum at $x=2$ **21.** $6x^2+8xy$, $x=10$ and $y=15$, $x=5\sqrt{5}$ and $y=12$
23. (a) The domain consists of all x except those between 0 and $-a$.
(b) $(0,0)$ and $(-a,0)$ are minimum points; at each, the tangent is vertical
(c) concave down throughout the domain (d) Figure A-12 **25.** (a) $f'(x)=$
$-(x+3)(x+1)/(x+2)^2$, $f''(x)=-2/(x+2)^3$ (b) $x=-1$ corresponds
to a maximum; $x=-3$ corresponds to a minimum; there are no points of
inflection (c) maximum $=\frac{3}{2}$, minimum $=1.5$ **27.** (a) $(3,-13.5)$ is minimum
point (b) $(0,0)$, $(2,-8)$ (c) Figure A-13 (d) 18, -6 **29.** $(0,0)$ is maximum
point; $(0.4,-0.326)$ is minimum point; $(-0.2,-0.410)$ is point of
inflection; dy/dx is not defined at $x=0$; Figure A-14 **31.** -0.01 ft/min
33. $2\times2\times4$ m, $4\times4\times2$ m **35.** $h=15$, $r=10\sqrt{2}$ in. **37.** $x=6$ miles
38. One piece should be the shortest possible that can be formed into a
square (better yet is to defy instructions and make only the circle).
41. (a) $[-5,5]$ (b) 0 (c) $|x|/(25-x^2)^{1/2}$, $(-5,5)$ (d) $(-5,5)$ (e) $f''(0)$ does
not exist

(f)
$$f''(x)=\begin{cases} -25(25-x^2)^{-3/2}, & -5<x<0 \\ 25(25-x^2)^{-3/2}, & 0<x<5 \end{cases}$$

(g) all

Sample test

1. (a) all real numbers except -2 (b) $f'(x)=(x+5)(x-1)/(x+2)^2$,
$f''(x)=18/(x+2)^3$ (c) 1 (min), -5 (max), no points of inflection (d) 2,
-10 **2.** $-4\pi/3$ in./min **3.** $\frac{3}{2}$ ft/sec **4.** 0; decreasing, $y''(2)=-48$;
24 **5.** $\frac{5}{2}\sqrt{5}\times\frac{5}{2}\sqrt{5}\times8$ **6.** $y''=2(a+b^2)/(x+b)^3$, which cannot change
sign at any point in the domain; $x=-b-(a+b^2)^{1/2}$ corresponds to the
only maximum, and $x=-6+(a+b^2)^{1/2}$ corresponds to the only
minimum, provided $a+b^2>0$ **7.** 6.5 ft/sec **8.** (a) $10\times10\times7.5$
(b) $\frac{15}{2}\times\frac{15}{2}\times\frac{40}{3}$ **9.** (a) height $=\frac{3}{4}$ side length (b) height $=$ (volume)$/(7.5)^2$

Chapter 5

Section 5.3

9. (a) $(4x+1)^4+c$ (b) $\frac{2}{3}(25+x^2)^{3/2}+c$ (c) $\frac{1}{2}(25+x^2)^2+c$ or $25x^2+$
$x^4/2+k$; $k=\frac{625}{2}+c$ (d) $2(10-2x+x^2)^{1/2}+c$ (e) $\frac{2}{3}[x^3+(9-x^2)^{3/2}]+c$
(f) $-(5+3x-x^3)^{-1}+c$ (g) $\frac{1}{3}(\sqrt{x}-\sqrt{2})^3+c$ or $\frac{1}{3}x^{3/2}-\sqrt{2}\,x+2\sqrt{x}+k$;
$k=c-\frac{2}{3}\sqrt{2}$ (h) $\frac{2}{3}(3x^{1/3}+1)^{3/2}+c$ (i) $\frac{2}{3}(x^2+4x-3)^{3/2}+2(x+2)^2+c$
(j) $2(16-x^2)^{1/2}+c$ **11.** $y=\frac{1}{3}x^{3/2}+x^{1/2}+x-20$ **13.** minimum of c
at $x=\pm3$; there are no maxima or points of inflection **15.** 3744 gallons
16. (a) 100 (b) $C=100x-0.1x^2+400$ (c) $R=120x-0.2x^2$ (d) $P=20x-$
$0.1x^2-400$ (e) 600

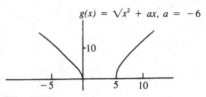

Fig. A-12

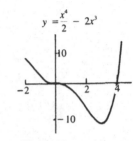

Fig. A-13

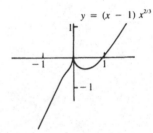

Fig. A-14

401

Section 5.4

7. 2 sec, 112 ft/sec   **9.** $(1, \frac{11}{6})$ is a horizontal point of inflection; there are
no maxima or minima   **11.** $-16t^2 + 192t + 7500$, $t = 2$, $t = 10$, 128 ft/sec,
$t = 6$, 8076   **13.** (a) 180 ft (b) 50 ft/sec   **15.** (a) $y = -8t^2 + 48t + 56$
(b) 128 ft (c) $t = 1, 5$ (d) $t = 7$, 64 ft/sec

Section 5.5

9. (a) 0.385 (b) 0.33835, 0.33383   **11.** (a) 0.75999 (b) 0.61510, 0.60150
13. (a) 0.71051 (b) 0.67146, 0.66716   **15.** (a) 10.52 (b) 11.2532, 11.32533

Section 5.6

1. πx^2 **5.** 1.20951, 1.12705   **7.** (a) 65.99 (b) 65.99 (c) 66.005

Section 5.7

7. (a) 12 (b) 18 (c) 18   **9.** (a) 16 (b) 12 (c) 14   **11.** $\frac{196}{3}$ **15.** (a) 70.25
(b) 61.25 (c) 66.125   **17.** (a) 66.08990 (b) 65.86468 (c) 65.95498

Section 5.8

13. 500   **15.** $\frac{1}{12}$ **17.** $32\pi/5$ **18.** 500   **19.** 1782   **20.** 250   **21.** $\frac{19}{3}$
23. $9\pi/2$

Section 5.9

1. $\frac{49}{3}$ **3.** 12   **5.** $\frac{3}{4}(\frac{1}{25} - \frac{1}{129})$ **7.** $\frac{26}{9}$ **9.** $\frac{3}{4}(4 - 9^{1/3})$

Section 5.10

1. no; $\int f'(x)\,dx$ is undetermined to the extent of a constant

Section 5.11

1. (a) $x^3/3 + 1/x + c$ (b) $2\sqrt{x}(x/3 + 1) + c$ (c) $\frac{2}{5}x^{5/2} - \frac{2}{3}x^{3/2} + 3x^{1/3} + c$
(d) $\frac{1}{4}(x + 1)^4 + c$ or $x^4/4 + x^3 + \frac{3}{2}x^2 + x + k$, where $k = c + \frac{1}{4}$ **3.** $y = 2x^3$
$-1/x - 2x + 4$ **5.** $1.24 \cdot 10^6$ **7.** 32, Figure A-15   **9.** $y = x^3 - x$, Figure
A-16   **11.** (a) $(1, 5)$ (min), $(3, 9)$ (max), $(2, 7)$ (inflection) (c) 8 (e) 40.5
15. $\frac{109}{6}$; 0, $\frac{16}{3}$ **17.** $500/\sqrt{3}$ **19.** $-2x^{-1/2} - x + 2x^3 - 3$ **21.** 500π
23. 55   **25.** (a) $54 - 32t - 16t^2$ (b) $t = 1$ (c) 46.64 mph   **27.** 216 cc
29. $y = \frac{1}{3}(x^2 - 16)^{3/2} + 4$ **31.** $\frac{64}{3}$ **33.** $\frac{32}{3}$, Figure A-17
35. (a) $(k/6)(b - a)^3$ **37.** (a) $(5 - \sqrt{19})/3$, $(5 + \sqrt{19})/3$ (b) $(\frac{5}{3}, \frac{56}{27})$, $-\frac{19}{3}$
(e) $\frac{253}{12}$

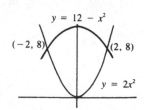

Fig. A-15

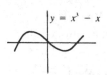

Fig. A-16

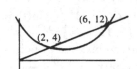

Fig. A-17

Sample test
1. 20   **2.** 3525   **3.** (a) $y = -16t^2 + 128t + 144$ (b) $t = 9$ (c) 4, 400
(d) 2, 6 (e) -1, 160   **4.** $y = (25 - x^2)^{1/2} + 5$ **5.** 1000   **6.** 8; 0 and 4

Sample test
1′. $y = 12t^3 - 3t^2 + 200t + y(0)$ **2′.** $800\sqrt{2}$ ft, $-3/(20\sqrt{2})$ ft/sec
3′. Figure A-18, $\frac{1}{20}$ **4′.** $17 - (10 - t^2)^{1/2}$ **5′.** $32\pi/5$ **6′.** $(\pm 6, 0)$, 576

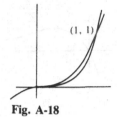

Fig. A-18

Chapter 6

Section 6.1
11. (a) e^{-2x} (b) $e^{-0.1x}$ (c) $e^{2\sqrt{x}}$ (d) $3e^{1/x}$ **12.** (a) $x = \pm 2$ (b) $x = \pm 1$
(c) no solution (d) $x = 0, 3$ (e) $x = 0$ **15.** (a) $c = 1, b = \frac{1}{2}$ (b) $c = 3, b = 4$
(c) $c = \frac{1}{3}, b = 4$ (d) $c = -1, b = 2$ (e) $c = 10, b = \frac{1}{2}$ (f) $c = -3, b = 2$
(g) $c = 5, b = \sqrt{2}$ (h) $c = 1, b = 1/\sqrt{2}$ (i) $c = 1, b = 9$

Section 6.3
4. $(1.321)(0.690) - 1 = -0.08851 = -8.851\%$

Section 6.4
7. (a) 22.1% (b) 21.0% (c) 20.5%   **9.** 6.18%   **11.** 1.98 years, 3.96 years
13. 1

Section 6.5
3. (a) $(-\infty, \infty)$ (b) $(0, \infty)$ if $c > 0$, $(-\infty, 0)$ if $c < 0$ (c) no (d) no (e) no
5. $(-1, -1/e)$ is the minimum point; $(-2, -2/e^2)$ is the point of
inflection; Figure A-19; the flexion is minimal at $x = -3$ **6.** minimum
point $= (0, 0)$; maximum point $= (-2, 4e^2)$; points of inflection at $x = -2 \pm$
$\sqrt{2}$ **7.** $-4/(e^x - e^{-x})^2$ **9.** $(1, 1/e)$ is the maximum point; $(2, 2/e^2)$ is
the point of inflection; the curve approaches the x axis for large positive
x **11.** minimum point $= (1, e/(e + 1))$; y approaches the line $x = 1$ from
below as $x \to \pm \infty$, $+\infty$ as x approaches the solution of $e^x + x = 0$
(≈ -0.57) from the right, $-\infty$ as x approaches this value from the left;
the only inflection point is $\approx (2.27, 0.81)$

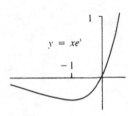

Fig. A-19

Section 6.6
3. (a) 0.18 (b) $\frac{0.18}{4}$ **4.** twice the relative error of the radius plus the relative
error of the height   **5.** (b) $R = pf(p) = (640p - 100p^2)/3$, $p_M = 3.2$
(c) -1 **7.** (a) 2% (b) 3%

Section 6.7
3. 0.669 mg   **5.** (a) 33.7 lb/in.² (b) 13.7 hr   **7.** $N = 2.5e^{0.1t}$ **9.** 46 days
11. $e - 1$ **13.** $(e - 1)^2 \approx 2.952$ **15.** maximum point $= (2, 2/e)$; inflection

403

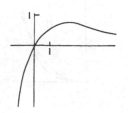

Fig. A-20

point $= (4, 4/e^2)$; Figure A-20   **17.** $2(e^2 - 3 + 1/e) \approx 9.514$ **19.** $e^x + e^{-x}$
22. (b) $c_0(1 - e^{-Nt_0})/(1 - e^{-t_0})$ **23.** (b) \$164.87 (c) \$162.89 (d) 5.13%

Section 6.8
3. $a, c = e^b, r = a$ **7.** $\exp(e^x + x)$ **11.** $(3\pi/2)(e - 1) \approx 8.097$
13. $(\pi/4)(e^8 - 1)$ **15.** minimum point $= (1, e)$, concave upward for $x > 0$,
concave downward for $x < 0$; $y \to 0$ as $x \to 0$ from the left, $+\infty$ as $x \to 0$
from the right, and the line $y = x$ as $x \to +\infty$ **17.** (a) $[0, \infty)$ (b) minimum
point $= (0, 0)$; maximum point $= (5, \sqrt{5}\, e^{-0.25})$ (c) $50\pi(1 - 1/\sqrt{e})$
(d) inflection point $\approx (10.29, 1.11)$

Section 6.9

Sample test
1. about 11 years   **2.** $6(e^{6x} - e^{-6x})$ **3.** $(\pi/2)(1 - e^{-4}) \approx \pi(0.491)$
4. $e^{-1/2}, -e^{-1/2}$ **5.** Figure A-21   **6.** $15\pi(e^{2.5} - 3.5)$ **7.** $(0, 0)$,
$(\sqrt{3}, \sqrt{3}\, e^{-3/2}), (-\sqrt{3}, -\sqrt{3}\, e^{-3/2})$

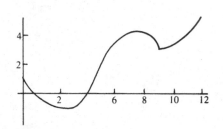

Fig. A-21

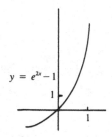

Fig. A-22

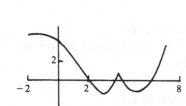

Fig. A-23

Sample test
1′. about 10 weeks   **2′.** $e^x - 3e^{-3x}$ **3′.** Figure A-22, $\frac{1}{2}(e + e^{-1})^2$
4′. minimum of $e^2/4$ **5′.** Figure A-23   **6′.** $20\pi/3$ **7′.** (a) $-\frac{1}{2}e^{-x^2} + \frac{3}{2}$
(b) maximum flexion at $x = \pm(\frac{3}{2})^{1/2}$, minimum flexion at $x = 0$

Chapter 7

Section 7.3
11. 14.21 years   **13.** 24 million   **15.** 170 years   **17.** (a) $4.3C, 7.7C$, where
C = present yearly cost (b) $5.4C, 10.7C$

Section 7.4
9. (a) $3/(2x)$ (b) $\frac{3}{2}[1/(x\sqrt{\ln x})]$ (c) $-1/x$ (d) $2/x$ (e) $1/(x\ln x)$
(f) $3(x - \ln x)^2(1 - 1/x)$ (g) $(x^2 - a^2)^{-1/2}$ (h) $(x^2 - a^2)^{-1/2} - (x^2 + a^2)^{-1/2}$ (i) $-1/(2x)$ **11.** (a) $50x/(625 - x^4)$

(b) $[(25 + x^2)/(25 - x^2)]^{1/2}[50x/(625 - x^4)]$ **13.** (a) $x^x(1 + \ln x)$,
minimum point $= (1/e, e^{-1/e})$ (b) $x^x[(1 + \ln x)^2 + 1/x]$
(d) Figure A-24   **14.** (a) 0.756945 (b) 1.256431, 0 (d) 1.857184, 4.536404

Answers to selected problems

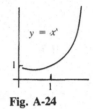

Fig. A-24

Section 7.5
3. $\pi, \pi/2a$ **7.** $e^2/(e^4 - 1)$ **11.** (a) $\frac{4}{3}\ln 2$ (b) $\frac{5}{16}$ (c) $\ln\frac{5}{3}$ (d) 2 (e) $(\ln 3)/2$
(f) $\frac{1}{18}$ (g) $2\ln(e^2 - e + 1)$ (h) $2e(e - 1)/(e^3 + 1)$

Section 7.6
4. 230.26   **5.** 0.434, 0.087

Section 7.7
6. $R = 5v^{2.5}$ **7.** $N = 10,000e^{-0.4t}$ **8.** $T = D^{1.5}$ **9.** $v = 20d^{1.5}$ **10.** $Q = 0.12e^{-0.1t}$ **11.** (a) $y \approx 695x^{-1.23}$ (b) $y \approx 73x^{-0.6}$ (c) 480, \$4.1 million, 39,000   **12.** $y \approx 20,000e^{-0.008x}$ **13.** (a) $I = 100e^{-0.07x}$ (b) $F = 125/x^2$
(c) For large x, both approach 0; for small x, only one is bounded.

Section 7.8
1. $r = (\ln 2)/15$ **3.** (a) 1 (b) 3 (c) $\frac{1}{2}$ **5.** -0.089 **7.** $(1.01)^{12} - 1 = 12.68\%$
9. $N \approx 60e^{-0.05t} + 90e^{-0.5t} + 60e^{-1.5t}$ **11.** e^π **13.** (a) $x > 0, x \neq 1$
(b) $g'(x) = [1/(2\sqrt{x}\,)][\ln x - 2)/(\ln x)^2]$, $g''(x) =$
$[8 - (\ln x)^2]/[4x^{3/2}(\ln x)^3]$; $(e^2, e/2) =$ minimum point; $(e^{2\sqrt{2}}, e^{\sqrt{2}}/(2\sqrt{2}\,))$
and $(e^{-2\sqrt{2}}, -e^{-\sqrt{2}}/(2\sqrt{2}\,))$ are points of inflection (c) The graph is concave
upward on $(0, e^{-2\sqrt{2}})$ and $(1, e^{2\sqrt{2}})$, concave downward elsewhere; it
approaches zero as $x \to 0$, ∞ as $x \to \infty$ and as $x \to 1$ from the right, $-\infty$
as $x \to 1$ from the left.   **15.** (a) $(0, \infty)$ (b) $(e^{1/k}, 1/ke) =$ maximum point,

$$\left(e^{(2k+1)/k(k+1)}, \frac{2k+1}{k(k+1)} e^{-(2k+1)/(k+1)} \right) = \text{point of inflection}$$

(c) Figure A-25   **17.** $A \approx 5e^{0.12t}$ **19.** $k = 12.103$ **21.** $A = 12e^{0.1t}$,
$dA = 2.217$ **23.** $y \approx 300x^{-0.13}$ **24.** (a) $-32x/(256 - x^4)$
(b) $-32x/(16 - x^2)^{1/2}(16 + x^2)^{3/2}$ **29.** (a) 27.7 years (b) 28,000
(c) $40,000 + 56,000e^{0.025t}$

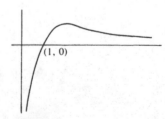

Fig. A-25

Sample test
1. (a) $-5(x + 1)/(5 + x^2)(5 + x)$ (b) $-5(x + 1)/(5 - x^2)^{1/2}(5 + x^2)^2$
2. $N \approx 700e^{-0.9x}$ **3.** (a) $(0, \infty)$ (b) $(e, 1/e) =$ maximum point
4. $(e^{3/2}, \frac{3}{2}e^{-3/2}) =$ point of inflection

Sample test
1'. (a) $20x/(100 - x^4)$ (b) $[20x/(100 - x^4)][(10 + x^2)/(10 - x^2)]^{1/2}$
2'. $I = 3600/x^{21/2}$ **3'.** (a) $x > 0, x \neq 1$ (b) $(e, e) =$ minimum point

405

4'. (a) $\frac{1}{2}\ln(e^2+1)\approx1.063$ (b) $e^2/[2(e^4-1)]\approx0.069$ **5'.** $2.2\cdot10^5$
6'. $\frac{3}{2}$

Chapter 8

Section 8.2
5. Several approximating points are $(\pm0.1,3.982)$, $(\pm1.0,2.434)$, $(\pm2.0,0.540)$, $(\pm3.0,0.043)$ **6.** sample points: (a) $(\pm1.0,6.28)$ (b) $(2.0,0.62)$, $(0,5.11)$, $(-1.0,3.21)$ (c) $(\pm1.0,-2.51)$ **7.** $x=\pm1$ gives points of inflection; the flexion is maximal at 0, minimal at $\pm\sqrt{3}$

Section 8.3
1. $y_0=\pm e^c$, Figure A-26 **3.** (a) $y=cx$ (b) $y=c/x$ (c) any differentiable function y satisfying $y^2+x^2=c$ (d) any differentiable function y satisfying $y^2-x^2=c$ (e) $y=-4/(x^4+c)$ **4.** (c) 4%

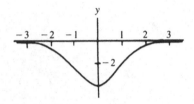

Fig. A-26

Section 8.5
1. (a) 64.4 years (b) 53.6 years **3.** 16.16 births per thousand

Section 8.7
3. $2/(x-3)-2/(x-2)$ **5.** $5/[4(x-3)]+3/[4(x+1)]$

Section 8.8
6. $d^2y/dt^2=(dy/dt)[c(y_E-2y)]=c^2y(y_E-y)(y_E-2y)$ **7.** Figure A-27
9. The numerically largest difference y_n-y is -1.5 which occurs when $n(=t)=6$ (also, $y_n-y=1.4$ at $n=18$).

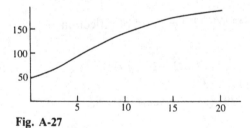

Fig. A-27

Section 8.9
9. (a) $y=c_1e^{-3x}+c_2e^{-4x}$ (b) $y=c_1+c_2e^{2t}+c_3e^{-3t}$ (c) $y=c_1e^{2x}+c_2e^{-2x}$ (d) $y=c_1+c_2e^{4x}$ (e) $y=c_1e^t+c_2e^{-t}+c_3e^{-2t}$ (f) $y=c_1e^{2t}+(c_2+c_3x)e^{-t}$
10. $y=1-e^{-3t}$ **12.** $y=(c_1+c_2x+c_3x^2)e^{-2x}$

Section 8.10
1. (a) If $y=F(t)$ is a solution of equation (26) and $k\neq0$, then for $y=k\cdot F(t)$, we have $A(d^2y/dt^2)+B(dy/dt)+Cy=kf(t)\neq f(t)$.

9. (a) $y = 2e^{-2x} + c_1 e^{-3x} + c_2 e^{-4x}$ (b) $y = x^2 - x + 2 + c_1 e^{-3x} + c_2 e^{-4x}$
(c) $y = 2e^{-2x} + x^2 - x + 2 + c_1 e^{-3x} + c_2 e^{-4x}$ **10.** $y = 5te^{-3t} + c_1 e^{-3t} + c_2 e^{-4t}$ **11.** $y = (c_1 - x/4)e^{-2x} + c_2 e^{2x}$ **12.** (b) $Y = ce^{k(1-l)t} + P/(1-l)$

Sample test

1. $1/(y^2 - 5y + 6) = 1/(y-3) - 1/(y-2)$, so that $dy/(y^2 - 5y + 6) = dx$
means $\ln[(y-3)/(y-2)] = x + c$ or $(y-3)/(y-2) = ke^x$ [note that
$y(0) = 4$ implies that, at least for values of x near 0, $y - 3 > 0$ and $y - 2 > 0$];
from $y(0) = 4$ it follows that $k = \frac{1}{2}$ and $y = (2e^x - 6)/(e^x - 2)$, $\lim_{x \to \infty} y =$
$\lim_{x \to \infty} (2 - 6/e^x)/(1 - 2/e^x) = 2$ **2.** $y = c_1 e^{3x} + c_2 e^{-3x}$ **3.** $y = c_1 + c_2 e^{9x}$ **4.** $y = -e^{2x} + c_1 e^{3x} + c_2 e^{-3x}$ **5.** $y = 9/x$ **6.** $y = (2x + c_1)e^{9x} + c_2$

Section 8.11

1. Carrying two-decimal-place accuracy, you should obtain $X_{10} = 100.00$,
$Y_{10} = 10.01$ **2.** (b) $X_{20} = 99.89$, $Y_{20} = 10.03$ (c) A large departure from an
equilibrium will cause instability in the other population as well.
4. (b) $z - x_{10} = 1.40 \cdot 10^{-1}$, $z - x_{100} = 1.62 \cdot 10^{-2}$ **6.** $x_1 = 1.100$, $y_1 = 0.100$;
$x_5 = 1.478$, $y_5 = 0.720$; $x_{10} = 1.666$, $y_{10} = 2.128$ **7.** $x_{10} = 1.100$, $y_{10} = 0.109$;
$x_{50} = 1.451$, $y_{50} = 0.783$; $x_{100} = 1.491$, $y_{100} = 2.273$ **8.** $x_1 = 0.100$, $y_1 = 0.800$;
$x_5 = 0.320$, $y_5 = 0.198$; $x_{10} = 0.332$, $y_{10} = -0.166$ **9.** $x_{10} = 0.091$, $y_{10} =$
0.809; $x_{50} = 0.293$, $y_{50} = 0.237$; $x_{100} = 0.312$, $y_{100} = -0.116$

Chapter 9

Section 9.2

2. (a) 8 (b) 24 (c) $\sqrt{6} - \sqrt{2}/3$ (d) $2\ln 3$ (e) $-\frac{1}{36}$ (f) $5(\sqrt{6} - \sqrt{2})$ (g) 261
(h) 52 (i) $\frac{1}{2}\ln 2$ (j) $\sqrt{6} - \sqrt{3}$ (k) $\frac{1}{12}$ **6.** 0

Section 9.5

7. (a) 130 lb (b) 800 lb-ft (c) 500 lb-ft

Section 9.6

1. 720, 2160 **2.** $96\frac{2}{3}$, 1710 **3.** $q = 20$, $p = 63$ **4.** $346\frac{2}{3}$, $573\frac{1}{3}$

Section 9.7

6. $23.89\frac{1}{3}\pi$ gm **7.** $2.7\pi \cdot 10^6$ cm

Section 9.10

5. 0 **6.** $4/(\ln 5)$ **7.** 17 lb-ft

Section 9.12

7. 272/15 oz **9.** $(\pi/6)(e^2 - e^{-2})$ **11.** $\frac{1}{2}(e^4 - e)$ oz-in.
13. $Gm[2\ln(8/3) - 5/4]$ dynes **15.** 200 oz **17.** 300π lb **19.** 23 ft/sec

21. 6/7 in.   **23.** (a) $\frac{2}{3}[(e^6-1)^{1/2}-(e^3-1)^{1/2}]$ (b) $\frac{1}{3}[\ln(3+e)-1]$ (c) 7/3
(d) $\frac{2}{3}(\sqrt{5}-2)$

Sample test
2. (a) $2x-3$ (b) $Gm(2\ln 3-1)$ **3.** 160 oz   **4.** 272π lb   **5.** 6.8 ft/sec
6. ≈ 3.7 in.

Section 9.13
3. $L_2=\frac{1}{32}\approx 0.03$, $L_4=\frac{98}{1024}\approx 0.10$; 0.17, 0.10; 1, $\frac{1}{2}$ **5.** (a) 1.61, 1.41, 1.31
(b) 0.61, 0.30, 0.15

Section 9.14
3. (a) 121.4062 (b) 48.4898 (c) 72.8227 (d) 54.5335   **4.** (a) 281.2500
(b) 6.5104 (c) 93.7500 (d) 0.4069   **5.** (a) 66.4062 (b) 6.5102 (c) 17.8227
(d) 0.4065   **9.** (a) 0.70833 (b) 0.69271 (c) 0.69702 (d) 0.69312
10. (a) 0.04167 (b) 0.00208 (c) 0.01042 (d) 0.00013   **11.** (a) 0.01519
(b) 0.00044 (c) 0.00388 (d) 0.00003   **13.** (a) 1432 (b) 143,109
14. (a) 28 (b) 272   **15.** 80   **17.** The results in the problem are correct to
within $2\cdot 10^{-5}$ and 10^{-6}.   **19.** $f'(x)=-xf(x)$, $f''(x)=(x^2-1)f(x)$,
$f'''(x)=(-x^3+3x)f(x)$, $f''''(x)=(x^4-6x+3)f(x)$; the error bounds
are $3.3\cdot 10^{-6}$, $3.3\cdot 10^{-8}$, $1.7\cdot 10^{-11}$ and $1.7\cdot 10^{-15}$ **20.** (d) With $y=y(1)=$
1.4106861, the errors of the Heun method are $y-y_{10}=-1.08\cdot 10^{-3}$ and
$y-y_{100}=-1.16\cdot 10^{-5}$; note the improvement, by a factor of 93, as
contrasted with 8.6 for the Euler method.

Chapter 10

Section 10.1
1. 2 sec, r cm, $-r$ cm, at point A, at the top and bottom   **4.** (a) $p=1$
(b) not periodic (c) $p=0.4$ (d) not periodic (e) $p=30$ (f) not periodic
(g) not periodic

Section 10.2
3. (a) $-90°$ (b) $135°$ (c) $-270°$ (d) $-1080°$ (e) $2.005°$ (f) $59.989°$
(g) $-180°$ (h) $2°$ (i) $572.958°$ (j) $57.296r$ **5.** (a) 8.378 (b) 3.491
(c) 16.755 (d) 20.944   **7.** (a) $\frac{1}{2}$ (b) $2/\pi$ (c) 4.584   **9.** (a) 206265
(b) $1.92\cdot 10^{13}$ (c) 3.27

Section 10.3
12. (c) $y=\sin 4\pi t$, $y=\cos 4\pi t$ **14.** (a) $\pi/2$ (b) $\pi/2$, $3\pi/2$ (c) $(\pi/28)$
$(1+8n)$ and $(\pi/28)(7+8n)$, for $n=0,1,2,3,4,5,6$ (d) $\pi/4$, $3\pi/4$, $5\pi/4$,
$7\pi/4$ (e) $\pi/8$, $7\pi/8$, $9\pi/8$, $15\pi/8$ (f) 1.22, 1.92   **15.** (a) $\pm\sqrt{3}/2$ (b) $\pm\frac{4}{5}$

(c) ± 1 (d) $\pm \frac{4}{5}$ (e) $\pm \frac{1}{2}$ (f) $\pm \frac{1}{2}$ (g) $\pm \sqrt{21}/5$ (h) $\pm \frac{3}{5}$ (i) 0 (j) $\pm \sqrt{3}/2$
(k) $\pm \frac{12}{13}$ (l) 0

Section 10.4

5. (a) $\pi/4 + n\pi$, where n is any integer (b) $n\pi$, where n is any integer
(c) $[(2n+1)/4]\pi$, where n is any integer (d) -3 **7.** 80 ft **11.** 98.5%
13. $\sqrt{37}$ **15.** $2\pi/3$ **17.** radius $= 7/\sqrt{3}$, area $= \frac{47}{8}\sqrt{3}$

Section 10.5

5. (a) $3\cos 3x$ (b) $\cos(x^2+1) - 2x^2 \sin(x^2+1)$ (c) $-1/\sin^2 x$ (d) $(x^2/2)$
$[1 - \sin(x^3/6)]$ (e) $-\sin x$ (f) $3\sin^2 x \cos x - 4\sin 4x$ (g) 0
(h) $[3\tan(3x+1)]/\cos(3x+1)$ (i) $1/[3\cos^2(x/3)] - 6x^2 \sin x^3 \cos x^3$
(j) $[15\sqrt{x}(x+3)\cos(5x^{3/2}) - 2\sin(5x^{3/2})]/[2(x+2)^2]$ (k) $(\cos x)/$
$[\cos^2(\sin x)]$ (l) $-(\cos x)/(\sin^2 x)$ (m) $-4/\sin^2 4x$ (n) $-[2x\cos(x^2)]/$
$[\sin^2(x^2)]$ (o) $-\sin x$ if $\cos x > 0$; $\sin x$ if $\cos x < 0$ **7.** $30/\pi + 15(4n+1)$,
where n is any integer

Section 10.6

3. (a) $2\tan 2x + c$ (b) $-\frac{1}{2}\cos(x^2-4) + c$ (c) $1/\cos u + c$ (d) $-(\cos 2t)/2 +$
$(\cos^3 2t)/6 + c$ (e) $(\sin^4 x)/4 + (\tan x^2)/2 + c$ (f) $-(\sin^4 x)/4 + c$
(g) $\ln(\sin v) + c$ if $\sin v > 0$; $\ln(-\sin v) + c$ if $\sin v < 0$ (h) $-(\cos 3x)/3 + c$
(i) $\sin u + c$ (j) $1/(3\cos 3x) + c$ **7.** minimum value of 0 at $x = n\pi$, n an
integer, no points of inflection, Figure A-28

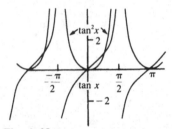

Fig. A-28

Section 10.7

2. (a) $\pi/2$ (b) 0 (c) $3\pi/4$ (d) $-\pi/2$ (e) $\pi/4$ (f) $\pi/2$ **3.** (a) 100° (b) 80°
(c) $-50°$ (d) 10° (e) $-40°$ (f) 10° **5.** (a) $4/(4-x^2)^{1/2}$ (b) $2x\arctan[(x$
$+1)/3] + 3x^2/[9 + (x+1)^2]$ (c) $-18(\arccos 2x)^2/(1-4x^2)^{1/2}$
(d) $-2/[(\arctan x)^2(1+x^2)]$ (e) $[(\arctan 5x)/(1-x^2)^{1/2} - (5\arcsin x)/$
$(1+25x^2)](\arctan 5x)^{-2}$ (f) $-2(1-4x^2)^{1/2}[6\arccos 2x + (1-4x^2)^{1/2}]$
7. (a) $\pi/2$ (b) $3\pi/8$ (c) $\pi/(4\sqrt{6})$ (d) $\pi/4$ **12.** (c) 0.76

Section 10.8

2. $3x/8 - [(\sin 2x)/4] + [(\sin 4x)/32] + c$ **3.** $-\cos x + \frac{2}{3}\cos^3 x -$
$\frac{1}{5}\cos^5 x + c$ **5.** (a) $\frac{1}{2}x + \frac{1}{12}\sin 6x + c$ (b) $\frac{1}{2}x(1-4x^2)^{1/2} + \frac{1}{4}\sin^{-1}x + c$
(c) $(\sqrt{2}/3)\tan^{-1}[(x^2-9)^{1/2}/3] + c$ (d) $27\{\sec^3[\tan^{-1}(x/3)] -$
$\sec[\tan^{-1}(x/3)]\} + c = (1/3)(9+x^2)^{3/2} - 9(9+x^2)^{1/2} + c$ **7.** (a) $(\pi/3)$
(b) $81\pi/16$ **9.** $(x-1)e^x + c$ **11.** 0.9460830726, 0.9460830704
13. 0.946083070, 0.946083071, 0.946083070

Section 10.9

5. (a) $x = \pi(t-1)$ (b) $g(t) = 2 + 2\sin[\pi(t-1)] - \cos[2\pi(t-1)]$
9. $1.5 - \sin x - 2\cos 3x$

Section 10.10

9. (a) $y = c_1\cos 4x + c_2\sin 4x$ (b) $y = e^{(5/2)u}[c_1\cos(u/2) + c_2\sin(u/2)]$
(c) $y = e^{-(1/2)t}[c_1\cos(t/\sqrt{2}) + c_2\sin(t/\sqrt{2})]$ (d) $y = e^t(c_1 + c_2\cos t + c_3\sin t)$
11. (a) $y = e^{-t}(1 + c_1\cos t + c_2\sin t)$ (b) $y = \frac{6}{25} + \frac{1}{5}t + e^{-2t}(c_1\cos t + c_2\sin t)$
(c) $y = 2 - 3t + 2t^2 + e^{-2t}(c_1\cos t + c_2\sin t)$ (d) $y = -t\cos t + c_1\cos t +$
$c_2\sin t$ (e) $y = \frac{1}{8}e^t\sin 2t + e^{-t}(c_1 + c_2 t)$ **13.** $y = 10e^{t/2}[c_1\cos(t/2) +$
$c_2\sin(t/2)]$, $x = e^{t/2}\{[(c_1 + c_2)/2]\cos(t/2) + [(c_2 - c_1)/2]\sin(t/2)\}$; no, the
populations would fluctuate wildly, and at least one of them would probably
disappear

Section 10.11

7. 5 is the length of the side opposite s; the other angles are s and $\pi - 2s$
9. (b) $1/\sqrt{2} + i/\sqrt{2}$ (c) $\cos(\pi/8) + i\sin(\pi/8) \cong 0.924 + 0.383i$
13. (a) $y = e^{-t}(c_1\cos 2t + c_2\sin 2t)$ (b) $y = c_1 e^t + c_2 e^{-2t}$ (c) $y = \frac{1}{10}\cos t$
$-\frac{1}{5}\sin t + c_1 e^{3t} + c_2 e^{-t}$ **15.** (a) $2/\cos^2 x$ (b) $-6\cos 3x \sin 3x$ (c) $[4(1 -$
$x^2)\arcsin x]^{-1/2}$ (d) $\tan 2x\sin^2 3x[9\cos 3x\tan 2x + (4\sin 3x)/\cos^2 2x]$
17. (a) $\frac{1}{3}\sin(3x + 2) + c$ (b) $(\tan^3 x)/3 + c$ (c) $\arctan x^2 + c$ (d) $\ln(\ln x) + c$
if $\ln x > 0$, $\ln(-\ln x) + c$ if $\ln x < 0$ (e) $\sin x - x\cos x + c$ (f) $x - 2$
$\arctan(x/2) + c$

Sample test

1. $10\sqrt{3}$ miles **2.** $\pi/6$ **3.** $n\pi$, $(\pi/3)(6n \pm 1)$, where n is any integer
4. $(\cos x)/(1 + \sin^2 x)$ **5.** $c_1\cos 3t + c_2\sin 3t$ **6.** $y = t/9 + \cos 3t - \sin 3t$
7. 1, 0

Appendix: Tables

Table A. Compound interest: $(1+r)^n$

AMOUNT OF ONE DOLLAR PRINCIPAL AT COMPOUND INTEREST AFTER n YEARS

n	2 %	2½%	3 %	3½%	4 %	4½%	5 %	6 %	7 %
1	1.0200	1.0250	1.0300	1.0350	1.0400	1.0450	1.0500	1.0600	1.0700
2	1.0404	1.0506	1.0609	1.0712	1.0816	1.0920	1.1025	1.1236	1.1449
3	1.0612	1.0769	1.0927	1.1087	1.1249	1.1412	1.1576	1.1910	1.2250
4	1.0824	1.1038	1.1255	1.1475	1.1699	1.1925	1.2155	1.2625	1.3108
5	1.1041	1.1314	1.1593	1.1877	1.2167	1.2462	1.2763	1.3382	1.4026
6	1.1262	1.1597	1.1941	1.2293	1.2653	1.3023	1.3401	1.4185	1.5007
7	1.1487	1.1887	1.2299	1.2723	1.3159	1.3609	1.4071	1.5036	1.6058
8	1.1717	1.2184	1.2668	1.3168	1.3686	1.4221	1.4775	1.5938	1.7182
9	1.1951	1.2489	1.3048	1.3629	1.4233	1.4861	1.5513	1.6895	1.8385
10	1.2190	1.2801	1.3439	1.4106	1.4802	1.5530	1.6289	1.7908	1.9672
11	1.2434	1.3121	1.3842	1.4600	1.5395	1.6229	1.7103	1.8983	2.1049
12	1.2682	1.3449	1.4258	1.5111	1.6010	1.6959	1.7959	2.0122	2.2522
13	1.2936	1.3785	1.4685	1.5640	1.6651	1.7722	1.8856	2.1329	2.4098
14	1.3195	1.4130	1.5126	1.6187	1.7317	1.8519	1.9799	2.2609	2.5785
15	1.3459	1.4483	1.5580	1.6753	1.8009	1.9353	2.0789	2.3966	2.7590
16	1.3728	1.4845	1.6047	1.7340	1.8730	2.0224	2.1829	2.5404	2.9522
17	1.4002	1.5216	1.6528	1.7947	1.9479	2.1134	2.2920	2.6928	3.1588
18	1.4282	1.5597	1.7024	1.8575	2.0258	2.2085	2.4066	2.8543	3.3799
19	1.4568	1.5987	1.7535	1.9225	2.1068	2.3079	2.5270	3.0256	3.6165
20	1.4859	1.6386	1.8061	1.9898	2.1911	2.4117	2.6533	3.2071	3.8697
21	1.5157	1.6796	1.8603	2.0594	2.2788	2.5202	2.7860	3.3996	4.1406
22	1.5460	1.7216	1.9161	2.1315	2.3699	2.6337	2.9253	3.6035	4.4304
23	1.5769	1.7646	1.9736	2.2061	2.4647	2.7522	3.0715	3.8197	4.7405
24	1.6084	1.8087	2.0328	2.2833	2.5633	2.8760	3.2251	4.0489	5.0724
25	1.6406	1.8539	2.0938	2.3632	2.6658	3.0054	3.3864	4.2919	5.4274
26	1.6734	1.9003	2.1566	2.4460	2.7725	3.1407	3.5557	4.5494	5.8074
27	1.7069	1.9478	2.2213	2.5316	2.8834	3.2820	3.7335	4.8223	6.2139
28	1.7410	1.9965	2.2879	2.6202	2.9987	3.4297	3.9201	5.1117	6.6488
29	1.7758	2.0464	2.3566	2.7119	3.1187	3.5840	4.1161	5.4184	7.1143
30	1.8114	2.0976	2.4273	2.8068	3.2434	3.7453	4.3219	5.7435	7.6123
31	1.8476	2.1500	2.5001	2.9050	3.3731	3.9139	4.5380	6.0881	8.1451
32	1.8845	2.2038	2.5751	3.0067	3.5081	4.0900	4.7649	6.4534	8.7153
33	1.9222	2.2589	2.6523	3.1119	3.6484	4.2740	5.0032	6.8406	9.3253
34	1.9607	2.3153	2.7319	3.2209	3.7943	4.4664	5.2533	7.2510	9.9781
35	1.9999	2.3732	2.8139	3.3336	3.9461	4.6673	5.5160	7.6861	10.6766
36	2.0399	2.4325	2.8983	3.4503	4.1039	4.8774	5.7918	8.1473	11.4239
37	2.0807	2.4933	2.9852	3.5710	4.2681	5.0969	6.0814	8.6361	12.2236
38	2.1223	2.5557	3.0748	3.6960	4.4388	5.3262	6.3855	9.1543	13.0793
39	2.1647	2.6196	3.1670	3.8254	4.6164	5.5659	6.7048	9.7035	13.9948
40	2.2080	2.6851	3.2620	3.9593	4.8010	5.8164	7.0400	10.2857	14.9745
41	2.2522	2.7522	3.3599	4.0978	4.9931	6.0781	7.3920	10.9029	16.0227
42	2.2972	2.8210	3.4607	4.2413	5.1928	6.3516	7.7616	11.5570	17.1443
43	2.3432	2.8915	3.5645	4.3897	5.4005	6.6374	8.1497	12.2505	18.3444
44	2.3901	2.9638	3.6715	4.5433	5.6165	6.9361	8.5572	12.9855	19.6285
45	2.4379	3.0379	3.7816	4.7024	5.8412	7.2482	8.9850	13.7646	21.0025
46	2.4866	3.1139	3.8950	4.8669	6.0748	7.5744	9.4343	14.5905	22.4726
47	2.5363	3.1917	4.0119	5.0373	6.3178	7.9153	9.9060	15.4659	24.0457
48	2.5871	3.2715	4.1323	5.2136	6.5705	8.2715	10.4013	16.3939	25.7289
49	2.6388	3.3533	4.2562	5.3961	6.8333	8.6437	10.9213	17.3775	27.5299
50	2.6916	3.4371	4.3839	5.5849	7.1067	9.0326	11.4674	18.4202	29.4570

Table B_1. Values of e^x and e^{-x}

x	e^x	x	e^{-x}
.05	1.051	.05	.951
.10	1.105	.10	.905
.15	1.162	.15	.861
.20	1.221	.20	.819
.25	1.284	.25	.779
.30	1.350	.30	.741
.35	1.419	.35	.705
.40	1.492	.40	.670
.45	1.568	.45	.638
.50	1.649	.50	.607
.6	1.822	.6	.549
.7	2.014	.7	.497
.8	2.226	.8	.449
.9	2.460	.9	.407
1.0	2.718	1.0	.368
1.1	3.004	1.1	.333
1.2	3.320	1.2	.301
1.3	3.669	1.3	.273
1.4	4.055	1.4	.247
1.5	4.482	1.5	.223
1.6	4.953	1.6	.202
1.7	5.474	1.7	.183
1.8	6.050	1.8	.165
1.9	6.686	1.9	.150
2.0	7.389	2.0	.135
2.1	8.166	2.1	.122
2.2	9.025	2.2	.111
2.3	9.974	2.3	.100
2.4	11.023	2.4	.091
2.5	12.182	2.5	.082
3.0	20.086	3.0	.050
3.5	33.115	3.5	.030
4.0	54.598	4.0	.018
4.5	90.017	4.5	.011
5.0	148.413	5.0	.0067
5.5	244.692	5.5	.0041
6.0	403.429	6.0	.0025
6.5	665.14	6.5	.0015
7.0	1096.6	7.0	.0009
7.5	1808.0	7.5	.0006
8.0	2981.0	8.0	.0003

Table B₂. Natural logarithms ($\ln x$)

x	0	1	2	3	4	5	6	7	8	9	1	2	3	4	5	6	7	8	9
1·0	0·0000	0100	0198	0296	0392	0488	0583	0677	0770	0862	10	19	29	38	48	57	67	77	86
1·1	0·0953	1044	1133	1222	1310	1398	1484	1570	1655	1740	9	17	26	35	44	52	61	70	79
1·2	0·1823	1906	1989	2070	2151	2231	2311	2390	2469	2546	8	16	24	32	40	48	56	64	72
1·3	0·2624	2700	2776	2852	2927	3001	3075	3148	3221	3293	7	15	22	30	37	45	52	60	67
1·4	0·3365	3436	3507	3577	3646	3716	3784	3853	3920	3988	7	14	21	28	35	42	48	55	62
1·5	0·4055	4121	4187	4253	4318	4383	4447	4511	4574	4637	6	13	19	26	32	39	45	52	58
1·6	0·4700	4762	4824	4886	4947	5008	5068	5128	5188	5247	6	12	18	24	30	36	43	49	55
1·7	0·5306	5365	5423	5481	5539	5596	5653	5710	5766	5822	6	11	17	23	29	34	40	46	52
1·8	0·5878	5933	5988	6043	6098	6152	6206	6259	6313	6366	5	11	16	22	27	33	38	43	49
1·9	0·6419	6471	6523	6575	6627	6678	6729	6780	6831	6881	5	10	15	21	26	31	36	41	46
2·0	0·6931	6981	7031	7080	7129	7178	7227	7275	7324	7372	5	10	15	20	24	29	34	39	44
2·1	0·7419	7467	7514	7561	7608	7655	7701	7747	7793	7839	5	9	14	19	23	28	33	37	42
2·2	0·7885	7930	7975	8020	8065	8109	8154	8198	8242	8286	4	9	13	18	22	27	31	36	40
2·3	0·8329	8372	8416	8459	8502	8544	8587	8629	8671	8713	4	9	13	17	21	26	30	34	38
2·4	0·8755	8796	8838	8879	8920	8961	9002	9042	9083	9123	4	8	12	16	20	25	29	33	37
2·5	0·9163	9203	9243	9282	9322	9361	9400	9439	9478	9517	4	8	12	16	20	24	28	31	35
2·6	0·9555	9594	9632	9670	9708	9746	9783	9821	9858	9895	4	8	11	15	19	23	26	30	34
2·7	0·9933	9969	0006	0043	0080	0116	0152	0188	0225	0260	4	7	11	15	18	22	26	29	33
2·8	1·0296	0332	0367	0403	0438	0473	0508	0543	0578	0613	4	7	11	14	18	21	25	28	32
2·9	1·0647	0682	0716	0750	0784	0818	0852	0886	0919	0953	3	7	10	14	17	20	24	27	31
3·0	1·0986	1019	1053	1086	1119	1151	1184	1217	1249	1282	3	7	10	13	16	20	23	26	30
3·1	1·1314	1346	1378	1410	1442	1474	1506	1537	1569	1600	3	6	10	13	16	19	22	25	29
3·2	1·1632	1663	1694	1725	1756	1787	1817	1848	1878	1909	3	6	9	12	15	18	22	25	28
3·3	1·1939	1969	2000	2030	2060	2090	2119	2149	2179	2208	3	6	9	12	15	18	21	24	27
3·4	1·2238	2267	2296	2326	2355	2384	2413	2442	2470	2499	3	6	9	12	15	17	20	23	26
3·5	1·2528	2556	2585	2613	2641	2669	2698	2726	2754	2782	3	6	8	11	14	17	20	23	25
3·6	1·2809	2837	2865	2892	2920	2947	2975	3002	3029	3056	3	5	8	11	14	16	19	22	25
3·7	1·3083	3110	3137	3164	3191	3218	3244	3271	3297	3324	3	5	8	11	13	16	19	21	24
3·8	1·3350	3376	3403	3429	3455	3481	3507	3533	3558	3584	3	5	8	10	13	16	18	21	23
3·9	1·3610	3635	3661	3686	3712	3737	3762	3788	3813	3838	3	5	8	10	13	15	18	20	23
4·0	1·3863	3888	3913	3938	3962	3987	4012	4036	4061	4085	2	5	7	10	12	15	17	20	22
4·1	1·4110	4134	4159	4183	4207	4231	4255	4279	4303	4327	2	5	7	10	12	14	17	19	22
4·2	1·4351	4375	4398	4422	4446	4469	4493	4516	4540	4563	2	5	7	9	12	14	16	19	21
4·3	1·4586	4609	4633	4656	4679	4702	4725	4748	4770	4793	2	5	7	9	12	14	16	18	21
4·4	1·4816	4839	4861	4884	4907	4929	4951	4974	4996	5019	2	4	7	9	11	13	16	18	20
4·5	1·5041	5063	5085	5107	5129	5151	5173	5195	5217	5239	2	4	7	9	11	13	15	18	20
4·6	1·5261	5282	5304	5326	5347	5369	5390	5412	5433	5454	2	4	6	9	11	13	15	17	19
4·7	1·5476	5497	5518	5539	5560	5581	5602	5623	5644	5665	2	4	6	8	11	13	15	17	19
4·8	1·5686	5707	5728	5748	5769	5790	5810	5831	5851	5872	2	4	6	8	10	12	14	17	19
4·9	1·5892	5913	5933	5953	5974	5994	6014	6034	6054	6074	2	4	6	8	10	12	14	16	18
5·0	1·6094	6114	6134	6154	6174	6194	6214	6233	6253	6273	2	4	6	8	10	12	14	16	18
5·1	1·6292	6312	6332	6351	6371	6390	6409	6429	6448	6467	2	4	6	8	10	12	14	16	17
5·2	1·6487	6506	6525	6544	6563	6582	6601	6620	6639	6658	2	4	6	8	10	11	13	15	17
5·3	1·6677	6696	6715	6734	6752	6771	6790	6808	6827	6845	2	4	6	7	9	11	13	15	17

For further values, e.g. $\ln 4560$, write $4560 = 4 \cdot 560 \times 10^3$, so that $\ln 4560 = \ln 4 \cdot 560 + \ln 10^3$ and use the table below.

x	1	2	3	4	5	6
$\ln 10^x$	2·3026	4·6052	6·9078	9·2103	11·5129	13·8155

413

Table B$_2$ (*continued*)

x	0	1	2	3	4	5	6	7	8	9	1	2	3	4	5	6	7	8	9
5·4	1·6864	6882	6901	6919	6938	6956	6974	6993	7011	7029	2	4	6	7	9	11	13	15	17
5·5	1·7047	7066	7084	7102	7120	7138	7156	7174	7192	7210	2	4	5	7	9	11	13	14	16
5·6	1·7228	7246	7263	7281	7299	7317	7334	7352	7370	7387	2	4	5	7	9	11	12	14	16
5·7	1·7405	7422	7440	7457	7475	7492	7509	7527	7544	7561	2	3	5	7	9	10	12	14	16
5·8	1·7579	7596	7613	7630	7647	7664	7681	7699	7716	7733	2	3	5	7	9	10	12	14	15
5·9	1·7750	7766	7783	7800	7817	7834	7851	7867	7884	7901	2	3	5	7	8	10	12	13	15
6·0	1·7918	7934	7951	7967	7984	8001	8017	8034	8050	8066	2	3	5	7	8	10	12	13	15
6·1	1·8083	8099	8116	8132	8148	8165	8181	8197	8213	8229	2	3	5	7	8	10	11	13	15
6·2	1·8245	8262	8278	8294	8310	8326	8342	8358	8374	8390	2	3	5	6	8	10	11	13	14
6·3	1·8405	8421	8437	8453	8469	8485	8500	8516	8532	8547	2	3	5	6	8	9	11	13	14
6·4	1·8563	8579	8594	8610	8625	8641	8656	8672	8687	8703	2	3	5	6	8	9	11	12	14
6·5	1·8718	8733	8749	8764	8779	8795	8810	8825	8840	8856	2	3	5	6	8	9	11	12	14
6·6	1·8871	8886	8901	8916	8931	8946	8961	8976	8991	9006	2	3	5	6	8	9	11	12	14
6·7	1·9021	9036	9051	9066	9081	9095	9110	9125	9140	9155	1	3	4	6	7	9	10	12	13
6·8	1·9169	9184	9199	9213	9228	9242	9257	9272	9286	9301	1	3	4	6	7	9	10	12	13
6·9	1·9315	9330	9344	9359	9373	9387	9402	9416	9430	9445	1	3	4	6	7	9	10	12	13
7·0	1·9459	9473	9488	9502	9516	9530	9544	9559	9573	9587	1	3	4	6	7	9	10	11	13
7·1	1·9601	9615	9629	9643	9657	9671	9685	9699	9713	9727	1	3	4	6	7	8	10	11	13
7·2	1·9741	9755	9769	9782	9796	9810	9824	9838	9851	9865	1	3	4	6	7	8	10	11	12
7·3	1·9879	9892	9906	9920	9933	9947	9961	9974	9988	0001	1	3	4	5	7	8	10	11	12
7·4	2·0015	0028	0042	0055	0069	0082	0096	0109	0122	0136	1	3	4	5	7	8	9	11	12
7·5	2·0149	0162	0176	0189	0202	0215	0229	0242	0255	0268	1	3	4	5	7	8	9	11	12
7·6	2·0281	0295	0308	0321	0334	0347	0360	0373	0386	0399	1	3	4	5	7	8	9	10	12
7·7	2·0412	0425	0438	0451	0464	0477	0490	0503	0516	0528	1	3	4	5	6	8	9	10	12
7·8	2·0541	0554	0567	0580	0592	0605	0618	0631	0643	0656	1	3	4	5	6	8	9	10	11
7·9	2·0669	0681	0694	0707	0719	0732	0744	0757	0769	0782	1	3	4	5	6	8	9	10	11
8·0	2·0794	0807	0819	0832	0844	0857	0869	0882	0894	0906	1	2	4	5	6	7	9	10	11
8·1	2·0919	0931	0943	0956	0968	0980	0992	1005	1017	1029	1	2	4	5	6	7	9	10	11
8·2	2·1041	1054	1066	1078	1090	1102	1114	1126	1138	1150	1	2	4	5	6	7	8	10	11
8·3	2·1163	1175	1187	1199	1211	1223	1235	1247	1258	1270	1	2	4	5	6	7	8	10	11
8·4	2·1282	1294	1306	1318	1330	1342	1353	1365	1377	1389	1	2	4	5	6	7	8	9	11
8·5	2·1401	1412	1424	1436	1448	1459	1471	1483	1494	1506	1	2	4	5	6	7	8	9	11
8·6	2·1518	1529	1541	1552	1564	1576	1587	1599	1610	1622	1	2	3	5	6	7	8	9	10
8·7	2·1633	1645	1656	1668	1679	1691	1702	1713	1725	1736	1	2	3	5	6	7	8	9	10
8·8	2·1748	1759	1770	1782	1793	1804	1815	1827	1838	1849	1	2	3	5	6	7	8	9	10
8·9	2·1861	1872	1883	1894	1905	1917	1928	1939	1950	1961	1	2	3	4	6	7	8	9	10
9·0	2·1972	1983	1994	2006	2017	2028	2039	2050	2061	2072	1	2	3	4	6	7	8	9	10
9·1	2·2083	2094	2105	2116	2127	2138	2148	2159	2170	2181	1	2	3	4	5	7	8	9	10
9·2	2·2192	2203	2214	2225	2235	2246	2257	2268	2279	2289	1	2	3	4	5	6	8	9	10
9·3	2·2300	2311	2322	2332	2343	2354	2364	2375	2386	2396	1	2	3	4	5	6	7	9	10
9·4	2·2407	2418	2428	2439	2450	2460	2471	2481	2492	2502	1	2	3	4	5	6	7	8	10
9·5	2·2513	2523	2534	2544	2555	2565	2576	2586	2597	2607	1	2	3	4	5	6	7	8	9
9·6	2·2618	2628	2638	2649	2659	2670	2680	2690	2701	2711	1	2	3	4	5	6	7	8	9
9·7	2·2721	2732	2742	2752	2762	2773	2783	2793	2803	2814	1	2	3	4	5	6	7	8	9
9·8	2·2824	2834	2844	2854	2865	2875	2885	2895	2905	2915	1	2	3	4	5	6	7	8	9
9·9	2·2925	2935	2946	2956	2966	2976	2986	2996	3006	3016	1	2	3	4	5	6	7	8	9

	0	1	2	3	4	5	6	7	8	9	1	2	3	4	5	6	7	8	9
1·0	·0000	0043	0086	0128	0170	0212	0253	0294	0334	0374	4	8	12	17	21	25	29	33	37
1·1	·0414	0453	0492	0531	0569	0607	0645	0682	0719	0755	4	8	11	15	19	23	27	30	34
1·2	·0792	0828	0864	0899	0934	0969	1004	1038	1072	1106	3	7	10	14	17	21	24	28	31
1·3	·1139	1173	1206	1239	1271	1303	1335	1367	1399	1430	3	6	10	13	16	19	23	26	29
1·4	·1461	1492	1523	1553	1584	1614	1644	1673	1703	1732	3	6	9	12	15	18	21	24	27
1·5	·1761	1790	1818	1847	1875	1903	1931	1959	1987	2014	3	6	8	11	14	17	20	22	25
1·6	·2041	2068	2095	2122	2148	2175	2201	2227	2253	2279	3	5	8	11	13	16	18	21	24
1·7	·2304	2330	2355	2380	2405	2430	2455	2480	2504	2529	2	5	7	10	12	15	17	20	22
1·8	·2553	2577	2601	2625	2648	2672	2695	2718	2742	2765	2	5	7	9	12	14	16	19	21
1·9	·2788	2810	2833	2856	2878	2900	2923	2945	2967	2989	2	4	7	9	11	13	16	18	20
2·0	·3010	3032	3054	3075	3096	3118	3139	3160	3181	3201	2	4	6	8	11	13	15	17	19
2·1	·3222	3243	3263	3284	3304	3324	3345	3365	3385	3404	2	4	6	8	10	12	14	16	18
2·2	·3424	3444	3464	3483	3502	3522	3541	3560	3579	3598	2	4	6	8	10	12	14	15	17
2·3	·3617	3636	3655	3674	3692	3711	3729	3747	3766	3784	2	4	6	7	9	11	13	15	17
2·4	·3802	3820	3838	3856	3874	3892	3909	3927	3945	3962	2	4	5	7	9	11	12	14	16
2·5	·3979	3997	4014	4031	4048	4065	4082	4099	4116	4133	2	3	5	7	9	10	12	14	15
2·6	·4150	4166	4183	4200	4216	4232	4249	4265	4281	4298	2	3	5	7	8	10	11	13	15
2·7	·4314	4330	4346	4362	4378	4393	4409	4425	4440	4456	2	3	5	6	8	9	11	13	14
2·8	·4472	4487	4502	4518	4533	4548	4564	4579	4594	4609	2	3	5	6	8	9	11	12	14
2·9	·4624	4639	4654	4669	4683	4698	4713	4728	4742	4757	1	3	4	6	7	9	10	12	13
3·0	·4771	4786	4800	4814	4829	4843	4857	4871	4886	4900	1	3	4	6	7	9	10	11	13
3·1	·4914	4928	4942	4955	4969	4983	4997	5011	5024	5038	1	3	4	6	7	8	10	11	12
3·2	·5051	5065	5079	5092	5105	5119	5132	5145	5159	5172	1	3	4	5	7	8	9	11	12
3·3	·5185	5198	5211	5224	5237	5250	5263	5276	5289	5302	1	3	4	5	6	8	9	10	12
3·4	·5315	5328	5340	5353	5366	5378	5391	5403	5416	5428	1	3	4	5	6	8	9	10	11
3·5	·5441	5453	5465	5478	5490	5502	5514	5527	5539	5551	1	2	4	5	6	7	9	10	11
3·6	·5563	5575	5587	5599	5611	5623	5635	5647	5658	5670	1	2	4	5	6	7	8	10	11
3·7	·5682	5694	5705	5717	5729	5740	5752	5763	5775	5786	1	2	3	5	6	7	8	9	10
3·8	·5798	5809	5821	5832	5843	5855	5866	5877	5888	5899	1	2	3	5	6	7	8	9	10
3·9	·5911	5922	5933	5944	5955	5966	5977	5988	5999	6010	1	2	3	4	6	7	8	9	10
4·0	·6021	6031	6042	6053	6064	6075	6085	6096	6107	6117	1	2	3	4	5	6	8	9	10
4·1	·6128	6138	6149	6160	6170	6180	6191	6201	6212	6222	1	2	3	4	5	6	7	8	9
4·2	·6232	6243	6253	6263	6274	6284	6294	6304	6314	6325	1	2	3	4	5	6	7	8	9
4·3	·6335	6345	6355	6365	6375	6385	6395	6405	6415	6425	1	2	3	4	5	6	7	8	9
4·4	·6435	6444	6454	6464	6474	6484	6493	6503	6513	6522	1	2	3	4	5	6	7	8	9
4·5	·6532	6542	6551	6561	6571	6580	6590	6599	6609	6618	1	2	3	4	5	6	7	8	9
4·6	·6628	6637	6646	6656	6665	6675	6684	6693	6702	6712	1	2	3	4	5	6	7	7	8
4·7	·6721	6730	6739	6749	6758	6767	6776	6785	6794	6803	1	2	3	4	5	5	6	7	8
4·8	·6812	6821	6830	6839	6848	6857	6866	6875	6884	6893	1	2	3	4	4	5	6	7	8
4·9	·6902	6911	6920	6928	6937	6946	6955	6964	6972	6981	1	2	3	4	4	5	6	7	8
5·0	·6990	6998	7007	7016	7024	7033	7042	7050	7059	7067	1	2	3	3	4	5	6	7	8
5·1	·7076	7084	7093	7101	7110	7118	7126	7135	7143	7152	1	2	3	3	4	5	6	7	8
5·2	·7160	7168	7177	7185	7193	7202	7210	7218	7226	7235	1	2	2	3	4	5	6	7	7
5·3	·7243	7251	7259	7267	7275	7284	7292	7300	7308	7316	1	2	2	3	4	5	6	7	7
	0	1	2	3	4	5	6	7	8	9	1	2	3	4	5	6	7	8	9

Table C (*continued*)

	0	1	2	3	4	5	6	7	8	9	1	2	3	4	5	6	7	8	9
5·4	·7324	7332	7340	7348	7356	7364	7372	7380	7388	7396	1	2	2	3	4	5	6	6	7
5·5	·7404	7412	7419	7427	7435	7443	7451	7459	7466	7474	1	2	2	3	4	5	5	6	7
5·6	·7482	7490	7497	7505	7513	7520	7528	7536	7543	7551	1	2	2	3	4	5	5	6	7
5·7	·7559	7566	7574	7582	7589	7597	7604	7612	7619	7627	1	2	2	3	4	5	5	6	7
5·8	·7634	7642	7649	7657	7664	7672	7679	7686	7694	7701	1	1	2	3	4	4	5	6	7
5·9	·7709	7716	7723	7731	7738	7745	7752	7760	7767	7774	1	1	2	3	4	4	5	6	7
6·0	·7782	7789	7796	7803	7810	7818	7825	7832	7839	7846	1	1	2	3	4	4	5	6	6
6·1	·7853	7860	7868	7875	7882	7889	7896	7903	7910	7917	1	1	2	3	4	4	5	6	6
6·2	·7924	7931	7938	7945	7952	7959	7966	7973	7980	7987	1	1	2	3	3	4	5	6	6
6·3	·7993	8000	8007	8014	8021	8028	8035	8041	8048	8055	1	1	2	3	3	4	5	5	6
6·4	·8062	8069	8075	8082	8089	8096	8102	8109	8116	8122	1	1	2	3	3	4	5	5	6
6·5	·8129	8136	8142	8149	8156	8162	8169	8176	8182	8189	1	1	2	3	3	4	5	5	6
6·6	·8195	8202	8209	8215	8222	8228	8235	8241	8248	8254	1	1	2	3	3	4	5	5	6
6·7	·8261	8267	8274	8280	8287	8293	8299	8306	8312	8319	1	1	2	3	3	4	5	5	6
6·8	·8325	8331	8338	8344	8351	8357	8363	8370	8376	8382	1	1	2	3	3	4	4	5	6
6·9	·8388	8395	8401	8407	8414	8420	8426	8432	8439	8445	1	1	2	3	3	4	4	5	6
7·0	·8451	8457	8463	8470	8476	8482	8488	8494	8500	8506	1	1	2	2	3	4	4	5	6
7·1	·8513	8519	8525	8531	8537	8543	8549	8555	8561	8567	1	1	2	2	3	4	4	5	5
7·2	·8573	8579	8585	8591	8597	8603	8609	8615	8621	8627	1	1	2	2	3	4	4	5	5
7·3	·8633	8639	8645	8651	8657	8663	8669	8675	8681	8686	1	1	2	2	3	4	4	5	5
7·4	·8692	8698	8704	8710	8716	8722	8727	8733	8739	8745	1	1	2	2	3	4	4	5	5
7·5	·8751	8756	8762	8768	8774	8779	8785	8791	8797	8802	1	1	2	2	3	3	4	5	5
7·6	·8808	8814	8820	8825	8831	8837	8842	8848	8854	8859	1	1	2	2	3	3	4	5	5
7·7	·8865	8871	8876	8882	8887	8893	8899	8904	8910	8915	1	1	2	2	3	3	4	4	5
7·8	·8921	8927	8932	8938	8943	8949	8954	8960	8965	8971	1	1	2	2	3	3	4	4	5
7·9	·8976	8982	8987	8993	8998	9004	9009	9015	9020	9025	1	1	2	2	3	3	4	4	5
8·0	·9031	9036	9042	9047	9053	9058	9063	9069	9074	9079	1	1	2	2	3	3	4	4	5
8·1	·9085	9090	9096	9101	9106	9112	9117	9122	9128	9133	1	1	2	2	3	3	4	4	5
8·2	·9138	9143	9149	9154	9159	9165	9170	9175	9180	9186	1	1	2	2	3	3	4	4	5
8·3	·9191	9196	9201	9206	9212	9217	9222	9227	9232	9238	1	1	2	2	3	3	4	4	5
8·4	·9243	9248	9253	9258	9263	9269	9274	9279	9284	9289	1	1	2	2	3	3	4	4	5
8·5	·9294	9299	9304	9309	9315	9320	9325	9330	9335	9340	1	1	2	2	3	3	4	4	5
8·6	·9345	9350	9355	9360	9365	9370	9375	9380	9385	9390	1	1	2	2	3	3	4	4	5
8·7	·9395	9400	9405	9410	9415	9420	9425	9430	9435	9440	0	1	1	2	2	3	3	4	4
8·8	·9445	9450	9455	9460	9465	9469	9474	9479	9484	9489	0	1	1	2	2	3	3	4	4
8·9	·9494	9499	9504	9509	9513	9518	9523	9528	9533	9538	0	1	1	2	2	3	3	4	4
9·0	·9542	9547	9552	9557	9562	9566	9571	9576	9581	9586	0	1	1	2	2	3	3	4	4
9·1	·9590	9595	9600	9605	9609	9614	9619	9624	9628	9633	0	1	1	2	2	3	3	4	4
9·2	·9638	9643	9647	9652	9657	9661	9666	9671	9675	9680	0	1	1	2	2	3	3	4	4
9·3	·9685	9689	9694	9699	9703	9708	9713	9717	9722	9727	0	1	1	2	2	3	3	4	4
9·4	·9731	9736	9741	9745	9750	9754	9759	9763	9768	9773	0	1	1	2	2	3	3	4	4
9·5	·9777	9782	9786	9791	9795	9800	9805	9809	9814	9818	0	1	1	2	2	3	3	4	4
9·6	·9823	9827	9832	9836	9841	9845	9850	9854	9859	9863	0	1	1	2	2	3	3	4	4
9·7	·9868	9872	9877	9881	9886	9890	9894	9899	9903	9908	0	1	1	2	2	3	3	4	4
9·8	·9912	9917	9921	9926	9930	9934	9939	9943	9948	9952	0	1	1	2	2	3	3	4	4
9·9	·9956	9961	9965	9969	9974	9978	9983	9987	9991	9996	0	1	1	2	2	3	3	3	4
	0	1	2	3	4	5	6	7	8	9	1	2	3	4	5	6	7	8	9

Table D. Trigonometric functions

SINES

	·0	·1	·2	·3	·4	·5	·6	·7	·8	·9	1·0		1	2	3	4	5	6	7	8	9
0°	·0000	0017	0035	0052	0070	0087	0105	0122	0140	0157	0175	89°	2	3	5	7	9	10	12	14	16
1	·0175	0192	0209	0227	0244	0262	0279	0297	0314	0332	0349	88	2	3	5	7	9	10	12	14	16
2	·0349	0366	0384	0401	0419	0436	0454	0471	0488	0506	0523	87	2	3	5	7	9	10	12	14	16
3	·0523	0541	0558	0576	0593	0610	0628	0645	0663	0680	0698	86	2	3	5	7	9	10	12	14	16
4	·0698	0715	0732	0750	0767	0785	0802	0819	0837	0854	0872	85	2	3	5	7	9	10	12	14	16
5	·0872	0889	0906	0924	0941	0958	0976	0993	1011	1028	1045	84	2	3	5	7	9	10	12	14	16
6	·1045	1063	1080	1097	1115	1132	1149	1167	1184	1201	1219	83	2	3	5	7	9	10	12	14	16
7	·1219	1236	1253	1271	1288	1305	1323	1340	1357	1374	1392	82	2	3	5	7	9	10	12	14	16
8	·1392	1409	1426	1444	1461	1478	1495	1513	1530	1547	1564	81	2	3	5	7	9	10	12	14	16
9	·1564	1582	1599	1616	1633	1650	1668	1685	1702	1719	1736	80	2	3	5	7	9	10	12	14	15
10	·1736	1754	1771	1788	1805	1822	1840	1857	1874	1891	1908	79	2	3	5	7	9	10	12	14	15
11	·1908	1925	1942	1959	1977	1994	2011	2028	2045	2062	2079	78	2	3	5	7	9	10	12	14	15
12	·2079	2096	2113	2130	2147	2164	2181	2198	2215	2233	2250	77	2	3	5	7	9	10	12	14	15
13	·2250	2267	2284	2300	2317	2334	2351	2368	2385	2402	2419	76	2	3	5	7	8	10	12	14	15
14	·2419	2436	2453	2470	2487	2504	2521	2538	2554	2571	2588	75	2	3	5	7	8	10	12	14	15
15	·2588	2605	2622	2639	2656	2672	2689	2706	2723	2740	2756	74	2	3	5	7	8	10	12	13	15
16	·2756	2773	2790	2807	2823	2840	2857	2874	2890	2907	2924	73	2	3	5	7	8	10	12	13	15
17	·2924	2940	2957	2974	2990	3007	3024	3040	3057	3074	3090	72	2	3	5	7	8	10	12	13	15
18	·3090	3107	3123	3140	3156	3173	3190	3206	3223	3239	3256	71	2	3	5	7	8	10	12	13	15
19	·3256	3272	3289	3305	3322	3338	3355	3371	3387	3404	3420	70	2	3	5	7	8	10	12	13	15
20	·3420	3437	3453	3469	3486	3502	3518	3535	3551	3567	3584	69	2	3	5	7	8	10	11	13	15
21	·3584	3600	3616	3633	3649	3665	3681	3697	3714	3730	3746	68	2	3	5	6	8	10	11	13	15
22	·3746	3762	3778	3795	3811	3827	3843	3859	3875	3891	3907	67	2	3	5	6	8	10	11	13	15
23	·3907	3923	3939	3955	3971	3987	4003	4019	4035	4051	4067	66	2	3	5	6	8	10	11	13	14
24	·4067	4083	4099	4115	4131	4147	4163	4179	4195	4210	4226	65	2	3	5	6	8	10	11	13	14
25	·4226	4242	4258	4274	4289	4305	4321	4337	4352	4368	4384	64	2	3	5	6	8	9	11	13	14
26	·4384	4399	4415	4431	4446	4462	4478	4493	4509	4524	4540	63	2	3	5	6	8	9	11	12	14
27	·4540	4555	4571	4586	4602	4617	4633	4648	4664	4679	4695	62	2	3	5	6	8	9	11	12	14
28	·4695	4710	4726	4741	4756	4772	4787	4802	4818	4833	4848	61	2	3	5	6	8	9	11	12	14
29	·4848	4863	4879	4894	4909	4924	4939	4955	4970	4985	5000	60	2	3	5	6	8	9	11	12	14
30	·5000	5015	5030	5045	5060	5075	5090	5105	5120	5135	5150	59	2	3	5	6	8	9	11	12	14
31	·5150	5165	5180	5195	5210	5225	5240	5255	5270	5284	5299	58	1	3	4	6	7	9	10	12	13
32	·5299	5314	5329	5344	5358	5373	5388	5402	5417	5432	5446	57	1	3	4	6	7	9	10	12	13
33	·5446	5461	5476	5490	5505	5519	5534	5548	5563	5577	5592	56	1	3	4	6	7	9	10	12	13
34	·5592	5606	5621	5635	5650	5664	5678	5693	5707	5721	5736	55	1	3	4	6	7	9	10	12	13
35	·5736	5750	5764	5779	5793	5807	5821	5835	5850	5864	5878	54	1	3	4	6	7	9	10	11	13
36	·5878	5892	5906	5920	5934	5948	5962	5976	5990	6004	6018	53	1	3	4	6	7	8	10	11	13
37	·6018	6032	6046	6060	6074	6088	6101	6115	6129	6143	6157	52	1	3	4	6	7	8	10	11	12
38	·6157	6170	6184	6198	6211	6225	6239	6252	6266	6280	6293	51	1	3	4	5	7	8	10	11	12
39	·6293	6307	6320	6334	6347	6361	6374	6388	6401	6414	6428	50	1	3	4	5	7	8	9	11	12
40	·6428	6441	6455	6468	6481	6494	6508	6521	6534	6547	6561	49	1	3	4	5	7	8	9	11	12
41	·6561	6574	6587	6600	6613	6626	6639	6652	6665	6678	6691	48	1	3	4	5	7	8	9	10	12
42	·6691	6704	6717	6730	6743	6756	6769	6782	6794	6807	6820	47	1	3	4	5	6	8	9	10	12
43	·6820	6833	6845	6858	6871	6884	6896	6909	6921	6934	6947	46	1	3	4	5	6	8	9	10	11
	1·0	·9	·8	·7	·6	·5	·4	·3	·2	·1	·0		1	2	3	4	5	6	7	8	9

COSINES

SINES

	·0	·1	·2	·3	·4	·5	·6	·7	·8	·9	1·0		1	2	3	4	5	6	7	8	9
44°	·6947	6959	6972	6984	6997	7009	7022	7034	7046	7059	7071	45°	1	2	4	5	6	7	9	10	11
45	·7071	7083	7096	7108	7120	7133	7145	7157	7169	7181	7193	44	1	2	4	5	6	7	9	10	11
46	·7193	7206	7218	7230	7242	7254	7266	7278	7290	7302	7314	43	1	2	4	5	6	7	8	10	11
47	·7314	7325	7337	7349	7361	7373	7385	7396	7408	7420	7431	42	1	2	4	5	6	7	8	9	11
48	·7431	7443	7455	7466	7478	7490	7501	7513	7524	7536	7547	41	1	2	3	5	6	7	8	9	10
49	·7547	7559	7570	7581	7593	7604	7615	7627	7638	7649	7660	40	1	2	3	5	6	7	8	9	10
50	·7660	7672	7683	7694	7705	7716	7727	7738	7749	7760	7771	39	1	2	3	4	6	7	8	9	10
51	·7771	7782	7793	7804	7815	7826	7837	7848	7859	7869	7880	38	1	2	3	4	5	7	8	9	10
52	·7880	7891	7902	7912	7923	7934	7944	7955	7965	7976	7986	37	1	2	3	4	5	6	7	8	10
53	·7986	7997	8007	8018	8028	8039	8049	8059	8070	8080	8090	36	1	2	3	4	5	6	7	8	9
54	·8090	8100	8111	8121	8131	8141	8151	8161	8171	8181	8192	35	1	2	3	4	5	6	7	8	9
55	·8192	8202	8211	8221	8231	8241	8251	8261	8271	8281	8290	34	1	2	3	4	5	6	7	8	9
56	·8290	8300	8310	8320	8329	8339	8348	8358	8368	8377	8387	33	1	2	3	4	5	6	7	8	9
57	·8387	8396	8406	8415	8425	8434	8443	8453	8462	8471	8480	32	1	2	3	4	5	6	7	8	8
58	·8480	8490	8499	8508	8517	8526	8536	8545	8554	8563	8572	31	1	2	3	4	5	5	6	7	8
59	·8572	8581	8590	8599	8607	8616	8625	8634	8643	8652	8660	30	1	2	3	4	4	5	6	7	8
60	·8660	8669	8678	8686	8695	8704	8712	8721	8729	8738	8746	29	1	2	3	3	4	5	6	7	8
61	·8746	8755	8763	8771	8780	8788	8796	8805	8813	8821	8829	28	1	2	2	3	4	5	6	7	7
62	·8829	8838	8846	8854	8862	8870	8878	8886	8894	8902	8910	27	1	2	2	3	4	5	6	6	7
63	·8910	8918	8926	8934	8942	8949	8957	8965	8973	8980	8988	26	1	2	2	3	4	5	5	6	7
64	·8988	8996	9003	9011	9018	9026	9033	9041	9048	9056	9063	25	1	2	2	3	4	5	5	6	7
65	·9063	9070	9078	9085	9092	9100	9107	9114	9121	9128	9135	24	1	1	2	3	4	4	5	6	7
66	·9135	9143	9150	9157	9164	9171	9178	9184	9191	9198	9205	23	1	1	2	3	3	4	5	6	6
67	·9205	9212	9219	9225	9232	9239	9245	9252	9259	9265	9272	22	1	1	2	3	3	4	5	5	6
68	·9272	9278	9285	9291	9298	9304	9311	9317	9323	9330	9336	21	1	1	2	3	3	4	4	5	6
69	·9336	9342	9348	9354	9361	9367	9373	9379	9385	9391	9397	20	1	1	2	2	3	4	4	5	6
70	·9397	9403	9409	9415	9421	9426	9432	9438	9444	9449	9455	19	1	1	2	2	3	3	4	5	5
71	·9455	9461	9466	9472	9478	9483	9489	9494	9500	9505	9511	18	1	1	2	2	3	3	4	4	5
72	·9511	9516	9521	9527	9532	9537	9542	9548	9553	9558	9563	17	1	1	2	2	3	3	4	4	5
73	·9563	9568	9573	9578	9583	9588	9593	9598	9603	9608	9613	16	0	1	1	2	2	3	3	4	4
74	·9613	9617	9622	9627	9632	9636	9641	9646	9650	9655	9659	15	0	1	1	2	2	3	3	4	4
75	·9659	9664	9668	9673	9677	9681	9686	9690	9694	9699	9703	14	0	1	1	2	2	3	3	3	4
76	·9703	9707	9711	9715	9720	9724	9728	9732	9736	9740	9744	13	0	1	1	2	2	2	3	3	4
77	·9744	9748	9751	9755	9759	9763	9767	9770	9774	9778	9781	12	0	1	1	2	2	2	3	3	3
78	·9781	9785	9789	9792	9796	9799	9803	9806	9810	9813	9816	11	0	1	1	1	2	2	2	3	3
79	·9816	9820	9823	9826	9829	9833	9836	9839	9842	9845	9848	10	0	1	1	1	2	2	2	3	3
80	·9848	9851	9854	9857	9860	9863	9866	9869	9871	9874	9877	9	0	1	1	1	1	2	2	2	3
81	·9877	9880	9882	9885	9888	9890	9893	9895	9898	9900	9903	8	0	1	1	1	1	2	2	2	2
82	·9903	9905	9907	9910	9912	9914	9917	9919	9921	9923	9925	7	0	0	1	1	1	1	2	2	2
83	·9925	9928	9930	9932	9934	9936	9938	9940	9942	9943	9945	6	0	0	1	1	1	1	1	2	2
84	·9945	9947	9949	9951	9952	9954	9956	9957	9959	9960	9962	5	0	0	1	1	1	1	1	1	2
85	·9962	9963	9965	9966	9968	9969	9971	9972	9973	9974	9976	4	0	0	0	1	1	1	1	1	1
86	·9976	9977	9978	9979	9980	9981	9982	9983	9984	9985	9986	3	0	0	0	0	1	1	1	1	1
87	·9986	9987	9988	9989	9990	9990	9991	9992	9993	9993	9994	2	0	0	0	0	0	0	1	1	1
88	·9994	9995	9995	9996	9996	9997	9997	9997	9998	9998	9998	1	0	0	0	0	0	0	0	0	0
89	·9998	9999	9999	9999	9999	0000	0000	0000	0000	0000	0000	0	0	0	0	0	0	0	0	0	0
	1·0	·9	·8	·7	·6	·5	·4	·3	·2	·1	·0		1	2	3	4	5	6	7	8	9

COSINES

Table D (*continued*)

TANGENTS

	·0	·1	·2	·3	·4	·5	·6	·7	·8	·9	1·0		1	2	3	4	5	6	7	8	9
0°	0·0000	0017	0035	0052	0070	0087	0105	0122	0140	0157	0175	89°	2	3	5	7	9	10	12	14	16
1	0·0175	0192	0209	0227	0244	0262	0279	0297	0314	0332	0349	88	2	3	5	7	9	10	12	14	16
2	0·0349	0367	0384	0402	0419	0437	0454	0472	0489	0507	0524	87	2	3	5	7	9	10	12	14	16
3	0·0524	0542	0559	0577	0594	0612	0629	0647	0664	0682	0699	86	2	4	5	7	9	11	12	14	16
4	0·0699	0717	0734	0752	0769	0787	0805	0822	0840	0857	0875	85	2	4	5	7	9	11	12	14	16
5	0·0875	0892	0910	0928	0945	0963	0981	0998	1016	1033	1051	84	2	4	5	7	9	11	12	14	16
6	0·1051	1069	1086	1104	1122	1139	1157	1175	1192	1210	1228	83	2	4	5	7	9	11	12	14	16
7	0·1228	1246	1263	1281	1299	1317	1334	1352	1370	1388	1405	82	2	4	5	7	9	11	12	14	16
8	0·1405	1423	1441	1459	1477	1495	1512	1530	1548	1566	1584	81	2	4	5	7	9	11	12	14	16
9	0·1584	1602	1620	1638	1655	1673	1691	1709	1727	1745	1763	80	2	4	5	7	9	11	13	14	16
10	0·1763	1781	1799	1817	1835	1853	1871	1890	1908	1926	1944	79	2	4	5	7	9	11	13	14	16
11	0·1944	1962	1980	1998	2016	2035	2053	2071	2089	2107	2126	78	2	4	5	7	9	11	13	15	16
12	0·2126	2144	2162	2180	2199	2217	2235	2254	2272	2290	2309	77	2	4	5	7	9	11	13	15	16
13	0·2309	2327	2345	2364	2382	2401	2419	2438	2456	2475	2493	76	2	4	6	7	9	11	13	15	17
14	0·2493	2512	2530	2549	2568	2586	2605	2623	2642	2661	2679	75	2	4	6	7	9	11	13	15	17
15	0·2679	2698	2717	2736	2754	2773	2792	2811	2830	2849	2867	74	2	4	6	8	9	11	13	15	17
16	0·2867	2886	2905	2924	2943	2962	2981	3000	3019	3038	3057	73	2	4	6	8	9	11	13	15	17
17	0·3057	3076	3096	3115	3134	3153	3172	3191	3211	3230	3249	72	2	4	6	8	10	12	13	15	17
18	0·3249	3269	3288	3307	3327	3346	3365	3385	3404	3424	3443	71	2	4	6	8	10	12	14	16	17
19	0·3443	3463	3482	3502	3522	3541	3561	3581	3600	3620	3640	70	2	4	6	8	10	12	14	16	18
20	0·3640	3659	3679	3699	3719	3739	3759	3779	3799	3819	3839	69	2	4	6	8	10	12	14	16	18
21	0·3839	3859	3879	3899	3919	3939	3959	3979	4000	4020	4040	68	2	4	6	8	10	12	14	16	18
22	0·4040	4061	4081	4101	4122	4142	4163	4183	4204	4224	4245	67	2	4	6	8	10	12	14	16	18
23	0·4245	4265	4286	4307	4327	4348	4369	4390	4411	4431	4452	66	2	4	6	8	10	12	15	17	19
24	0·4452	4473	4494	4515	4536	4557	4578	4599	4621	4642	4663	65	2	4	6	8	11	13	15	17	19
25	0·4663	4684	4706	4727	4748	4770	4791	4813	4834	4856	4877	64	2	4	6	9	11	13	15	17	19
26	0·4877	4899	4921	4942	4964	4986	5008	5029	5051	5073	5095	63	2	4	7	9	11	13	15	17	20
27	0·5095	5117	5139	5161	5184	5206	5228	5250	5272	5295	5317	62	2	4	7	9	11	13	16	18	20
28	0·5317	5340	5362	5384	5407	5430	5452	5475	5498	5520	5543	61	2	5	7	9	11	14	16	18	20
29	0·5543	5566	5589	5612	5635	5658	5681	5704	5727	5750	5774	60	2	5	7	9	12	14	16	18	21
30	0·5774	5797	5820	5844	5867	5890	5914	5938	5961	5985	6009	59	2	5	7	9	12	14	16	19	21
31	0·6009	6032	6056	6080	6104	6128	6152	6176	6200	6224	6249	58	2	5	7	10	12	14	17	19	22
32	0·6249	6273	6297	6322	6346	6371	6395	6420	6445	6469	6494	57	2	5	7	10	12	15	17	20	22
33	0·6494	6519	6544	6569	6594	6619	6644	6669	6694	6720	6745	56	3	5	8	10	13	15	18	20	23
34	0·6745	6771	6796	6822	6847	6873	6899	6924	6950	6976	7002	55	3	5	8	10	13	15	18	21	23
35	0·7002	7028	7054	7080	7107	7133	7159	7186	7212	7239	7265	54	3	5	8	11	13	16	18	21	24
36	0·7265	7292	7319	7346	7373	7400	7427	7454	7481	7508	7536	53	3	5	8	11	14	16	19	22	24
37	0·7536	7563	7590	7618	7646	7673	7701	7729	7757	7785	7813	52	3	6	8	11	14	17	19	22	25
38	0·7813	7841	7869	7898	7926	7954	7983	8012	8040	8069	8098	51	3	6	9	11	14	17	20	23	26
39	0·8098	8127	8156	8185	8214	8243	8273	8302	8332	8361	8391	50	3	6	9	12	15	18	21	23	26
40	0·8391	8421	8451	8481	8511	8541	8571	8601	8632	8662	8693	49	3	6	9	12	15	18	21	24	27
41	0·8693	8724	8754	8785	8816	8847	8878	8910	8941	8972	9004	48	3	6	9	12	16	19	22	25	28
42	0·9004	9036	9067	9099	9131	9163	9195	9228	9260	9293	9325	47	3	6	10	13	16	19	22	26	29
43	0·9325	9358	9391	9424	9457	9490	9523	9556	9590	9623	9657	46	3	7	10	13	17	20	23	27	30
	1·0	·9	·8	·7	·6	·5	·4	·3	·2	·1	·0		1	2	3	4	5	6	7	8	9

COTANGENTS

Table D (*continued*)

TANGENTS

	·0	·1	·2	·3	·4	·5	·6	·7	·8	·9	1·0		1	2	3	4	5	6	7	8	9
44°	0·9657	9691	9725	9759	9793	9827	9861	9896	9930	9965	0000	45°	3	7	10	14	17	21	24	27	31
45	1·0000	0035	0070	0105	0141	0176	0212	0247	0283	0319	0355	44	4	7	11	14	18	21	25	28	32
46	1·0355	0392	0428	0464	0501	0538	0575	0612	0649	0686	0724	43	4	7	11	15	18	22	26	29	33
47	1·0724	0761	0799	0837	0875	0913	0951	0990	1028	1067	1106	42	4	8	11	15	19	23	27	31	34
48	1·1106	1145	1184	1224	1263	1303	1343	1383	1423	1463	1504	41	4	8	12	16	20	24	28	32	36
49	1·1504	1544	1585	1626	1667	1708	1750	1792	1833	1875	1918	40	4	8	12	17	21	25	29	33	37
50	1·1918	1960	2002	2045	2088	2131	2174	2218	2261	2305	2349	39	4	9	13	17	22	26	30	35	39
51	1·2349	2393	2437	2482	2527	2572	2617	2662	2708	2753	2799	38	5	9	14	18	23	27	32	36	41
52	1·2799	2846	2892	2938	2985	3032	3079	3127	3175	3222	3270	37	5	9	14	19	24	28	33	38	42
53	1·3270	3319	3367	3416	3465	3514	3564	3613	3663	3713	3764	36	5	10	15	20	25	30	35	39	44
54	1·3764	3814	3865	3916	3968	4019	4071	4124	4176	4229	4281	35	5	10	16	21	26	31	36	41	47
55	1·4281	4335	4388	4442	4496	4550	4605	4659	4715	4770	4826	34	5	11	16	22	27	33	38	44	49
56	1·4826	4882	4938	4994	5051	5108	5166	5224	5282	5340	5399	33	6	11	17	23	29	34	40	46	52
57	1·5399	5458	5517	5577	5637	5697	5757	5818	5880	5941	6003	32	6	12	18	24	30	36	42	48	54
58	1·6003	6066	6128	6191	6255	6319	6383	6447	6512	6577	6643	31	6	13	19	26	32	38	45	51	58
59	1·6643	6709	6775	6842	6909	6977	7045	7113	7182	7251	7321	30	7	14	20	27	34	41	47	54	61
60	1·7321	7391	7461	7532	7603	7675	7747	7820	7893	7966	8040	29	7	14	22	29	36	43	50	58	65
61	1·8040	8115	8190	8265	8341	8418	8495	8572	8650	8728	8807	28	8	15	23	31	38	46	54	61	69
62	1·8807	8887	8967	9047	9128	9210	9292	9375	9458	9542	9626	27	8	16	25	33	41	49	57	66	74
63	1·9626	9711	9797	9883	9970	0057	0145	0233	0323	0413	0503	26	9	18	26	35	44	53	61	70	79
64	2·0503	0594	0686	0778	0872	0965	1060	1155	1251	1348	1445	25	9	19	28	38	47	57	66	75	85
65	2·1445	1543	1642	1742	1842	1943	2045	2148	2251	2355	2460	24	10	20	30	41	51	61	71	81	91
66	2·2460	2566	2673	2781	2889	2998	3109	3220	3332	3445	3559	23	11	22	33	44	55	66	77	88	99
67	2·3559	3673	3789	3906	4023	4142	4262	4383	4504	4627	4751	22									
68	2·4751	4876	5002	5129	5257	5386	5517	5649	5782	5916	6051	21									
69	2·6051	6187	6325	6464	6605	6746	6889	7034	7179	7326	7475	20									
70	2·7475	7625	7776	7929	8083	8239	8397	8556	8716	8878	9042	19									
71	2·9042	9208	9375	9544	9714	9887	0061	0237	0415	0595	0777	18									
72	3·0777	0961	1146	1334	1524	1716	1910	2106	2305	2506	2709	17									
73	3·2709	2914	3122	3332	3544	3759	3977	4197	4420	4646	4874	16									
74	3·4874	5105	5339	5576	5816	6059	6305	6554	6806	7062	7321	15									
75	3·7321	7583	7848	8118	8391	8667	8947	9232	9520	9812	0108	14									
76	4·0108	0408	0713	1022	1335	1653	1976	2303	2635	2972	3315	13									
77	4·3315	3662	4015	4373	4737	5107	5483	5864	6252	6646	7046	12									
78	4·7046	7453	7867	8288	8716	9152	9594	0045	0504	0970	1446	11									
79	5·1446	1929	2422	2924	3435	3955	4486	5026	5578	6140	6713	10									
80	5·671	5·730	5·789	5·850	5·912	5·976	6·041	6·107	6·174	6·243	6·314	9									
81	6·314	6·386	6·460	6·535	6·612	6·691	6·772	6·855	6·940	7·026	7·115	8									
82	7·115	7·207	7·300	7·396	7·495	7·596	7·700	7·806	7·916	8·028	8·144	7									
83	8·144	8·264	8·386	8·513	8·643	8·777	8·915	9·058	9·205	9·357	9·514	6									
84	9·514	9·677	9·845	10·019	10·199	10·385	10·579	10·780	10·988	11·205	11·430	5									
85	11·430	11·664	11·909	12·163	12·429	12·706	12·996	13·300	13·617	13·951	14·301	4									
86	14·301	14·669	15·056	15·464	15·895	16·350	16·832	17·343	17·886	18·464	19·081	3									
87	19·081	19·740	20·446	21·205	22·022	22·904	23·859	24·898	26·031	27·271	28·636	2									
88	28·636	30·145	31·821	33·694	35·801	38·188	40·917	44·066	47·740	52·081	57·290	1									
89	57·29	63·66	71·62	81·85	95·49	114·59	143·24	190·98	286·48	572·96		0									
	1·0	·9	·8	·7	·6	·5	·4	·3	·2	·1	·0		1	2	3	4	5	6	7	8	9

Use Linear Interpolation

COTANGENTS

Index

absolute value, 65
acceleration, 149
adjusted trapezoidal rule, 339
  error bound with, 339
angle measure, 346
antiderivatives, 194–205
  of constant times a function, 198
  of cos x, 363
  differ by a constant, 196
  of exponential functions, 249, 253
  involving trigonometric functions, 371–5
  of $1/x$, 266
  repeated, 202
  of sin x, 363
  of sum of functions, 198
  using the Chain Rule, 198, 222–4, 308–10
  of various trigonometric functions, 364–5
  of x^n, $n \neq -1$, 197
averages, 320–8

center of gravity, 326
Chain Rule, 159, 167
  proof of, 164
  special case of, 159
consumers' and producers' surplus, 315
cost
  average total or total unit, 121
  fixed, 120
  incremental, 121
  marginal, 121, 182
curve-fitting, 44–8, 53–6, 68, 270–4
curves,
  concave up (down), 152
  derived, 149

derivative
  of arccos x, 368
  of arcsin x, 367
  of arctan x, 369
  of a constant times a function, 96
  of cos x, 360
  of an exponential function, 244, 253, 268
  of a function, 93
  of a logarithmic function, 264, 269
  of a product, 179
  of a quotient, 180

  of sin x, 360
  of a sum, 96
  of tan x, 360
  of x^n, if n is an integer, 96
  of x^n, if n is a rational number, 173
differential equations
  approximate solutions (Euler method), 284, 287, 302
  approximate solutions (Heun method), 342
  with constant coefficients, 294, 299
  definition of, 283
  definition of a solution of, 283
  with trigonometric solutions, 380–6
  with variables separable, 286
differentials, 176
differentiation
  definition of, 93
  implicit, 172, 174, 185
  logarithmic, 265
  repeated, 148
displacement, 73
distance formula, 357
division, synthetic, 22

equations
  approximate solution of (bisection method), 143
  approximate solution of (Newton–Raphson method), 140, 142
  differential, *see* differential equations
  logistic, 290, 292
  solution of higher degree, 20
  solution of linear, 16, 139
  solution of quadratic, 18, 139
  solution of simultaneous linear, 16
error, relative or percentage, 246
exponents
  fractional, 9
  integral, 8

flexion, 148
force of attraction, 310
formulas
  different for different regions, 45

exponential, 46, 233, 271
mensuration, 30
polynomial, 44
power law, 47, 270
function
  complementary, 299
  continuous, 161
  cosine, 349
  decreasing (strictly), 112
  definition of, 56
  derived, 93
  differentiable, 125
  domain and range of a, 56
  extension of a, 57
  function of a, 60
  implicit form of a, 171
  increasing (strictly), 112
  inverse of a, 62, 257
  natural domain of a, 60
  periodic, 343
  restriction of a, 57
  sine, 349
  stationary, 113
  tangent, 354
functions
  exponential, 233
  logarithmic, 259
  periodic, 375–9
  products of, 60
  sums of, 60
Fundamental Theorem of Calculus
  applications of, 217–22, 310–17, 329–31
  statement of, 215

graphs
  with log–log scales, 269
  with semilog scales, 269
  with uniform scales, 34–41

increments, 114, 165
  approximate, 115, 116, 117
inflection, point of, 152
integral, 212
  indefinite, 224
interest, compound, 50, 80, 237–9

Index

interpolation
  graphical, 35
  linear, 41
intervals, notation for, 61
inverse trigonometric functions, 365–71

left-end-point rule, 335
  error bound with, 335
limit
  of a constant function, 82, 87
  of a function, 76, 85
  of $g(u) = u$, 82, 87
  of the product of two functions, 82, 87
  of the quotient of two functions, 82, 87
  of the quotient of two functions (supplement), 83
  of a sum (integral), 205–13
  of the sum of two functions, 82, 86
limits, guessing with a calculator, 101
linearization, 117
loads, 312
logarithms
  common, 259
  definition of, 27, 259
  laws of, 27, 259
  natural, 258

maxima and minima, 123
  absolute, 126

end-point, 126
  first test for, 127
  relative, 126
  second test for, 127
  third test for, 154
  turning-point, 126
Mean-Value Theorem, 110
moment of a force, 314

numbers
  critical, 126
  ordered pairs of, 16
  rational, 5
  real, 10

partial fractions, 291
particular integral, 299
polynomial
  continuity of a, 162
  definition of a, 20
  limit of a, 84
profit, 182
  maximization of, 183
progressions
  arithmetic, 24
  geometric, 25
projectiles thrown vertically, 202–4

quadrature, 334

radian measure, 346
radicals, 9
rate(s) of change
  average, 98, 166
  instantaneous, 98, 166
  related, 168, 174
  relative (or percentage), 246
revenue
  gross, 182
  marginal, 182

scale, economies (diseconomies) of, 122
slope
  average, 88
  at a point, 89
speed
  average, 72
  instantaneous, 75

tangent to a curve, 92
trapezoidal rule, 337
  error bound with, 338

value, incremental and marginal, 121
velocity
  average, 73
  instantaneous, 75, 77